The Chemistry and Biology of Volatiles

The Chemistry and Biology of Volatiles

Edited by

ANDREAS HERRMANN

Firmenich SA, Genève, Switzerland

WILEY

A John Wiley and Sons, Ltd, Publication

This edition first published 2010

Registered office
John Wiley & Sons Ltd, The Atrium, Southern Gate, Chichester, West Sussex, PO19 8SQ, United Kingdom

For details of our global editorial offices, for customer services and for information about how to apply for permission to reuse the copyright material in this book please see our website at www.wiley.com.

Library of Congress Cataloging-in-Publication Data

The chemistry and biology of volatiles / editor, Andreas Herrmann.
p. cm.
Includes bibliographical references and index.
ISBN 978-0-470-77778-7 (pbk.)
1. Volatile organic compounds.
QP550.I593 2010
612'.0157–dc22

2010013099

A catalogue record for this book is available from the British Library.

ISBN: 9780470777787

Set in 10/12pt, Times Roman by Thomson Digital, Noida, India

Contents

Foreword **xiii**
List of Contributors **xv**
Acknowledgements **xvii**
Abbreviations **xix**

1 Volatiles – An Interdisciplinary Approach **1**
Andreas Herrmann

1.1 Introduction 1
1.2 Geraniol – A Typical Example 2
1.3 Conclusion 8
References 8

2 Biosynthesis and Emission of Isoprene, Methylbutanol and Other Volatile Plant Isoprenoids **11**
Hartmut K. Lichtenthaler

2.1 Introduction 11
2.2 Plant Isoprenoids 12
2.3 Two IPP-Yielding Pathways in Plants 15
2.4 Prenyl Chain Formation and Elongation 16
2.5 Compartmentation of Plant Isoprenoid Biosynthesis 16
2.6 The Enzyme Steps of the Plastidic DOXP/MEP Pathway of IPP Formation 17
2.7 Cross-Talk Between the Two IPP Biosynthesis Pathways 19
2.8 Biosynthesis and Emission of Volatile Isoprene at High Irradiance 22
2.8.1 Regulation of Isoprene Emission 25
2.9 Inhibition of Isoprene Biosynthesis 26
2.9.1 Fosmidomycin and 5-Ketoclomazone 26
2.9.2 Diuron 27
2.10 Inhibition of Carotenoid and Chlorophyll Biosynthesis by Fosmidomycin and 5-Ketoclomazone 27
2.11 Biosynthesis and Emission of Methylbutenol at High Irradiance 28
2.12 Source of Pyruvate for Isoprene and Methylbutenol Biosynthesis 29
2.13 Branching Point of DOXP/MEP Pathway with Other Metabolic Chloroplast Pathways 30
2.14 Is There a Physiological Function of Isoprene and MBO Emission? 31
2.15 Biosynthesis and Emission of Monoterpenes, Sesquiterpenes and Diterpenes 33

2.15.1 Monoterpenes 35
2.15.2 Diterpenes 36
2.15.3 Sesquiterpenes 36
2.16 Some General Remarks on the Regulation of Terpene Biosynthesis in Plants 36
2.17 Volatile Terpenoids as Aroma Compounds of Wine 37
2.18 Function of Terpenes in Plant Defence 38
2.19 Conclusion 38
Acknowledgements 39
References 40

3 Analysis of the Plant Volatile Fraction **49**
Patrizia Rubiolo, Barbara Sgorbini, Erica Liberto, Chiara Cordero and Carlo Bicchi

3.1 Introduction 49
3.2 Sample Preparation 50
3.2.1 'Liquid' Phase Sampling 51
3.2.2 Headspace Sampling 51
3.2.3 Headspace–Solid Phase Microextraction 52
3.2.4 In-Tube Sorptive Extraction 54
3.2.5 Headspace Sorptive Extraction 55
3.2.6 Static and Trapped Headspace 56
3.2.7 Solid-Phase Aroma Concentrate Extraction 56
3.2.8 Headspace Liquid-Phase Microextraction 56
3.2.9 Large Surface Area High Concentration Capacity Headspace Sampling 59
3.3 Analysis 59
3.3.1 Fast-GC and Fast-GC-qMS EO Analysis 61
3.3.2 Qualitative Analysis 65
3.3.3 Quantitative Analysis 66
3.3.4 Enantioselective GC 70
3.3.5 Multidimensional GC Techniques 75
3.4 Further Developments 76
3.5 Conclusion 85
Acknowledgements 87
References 87

4 Plant Volatile Signalling: Multitrophic Interactions in the Headspace **95**
André Kessler and Kimberly Morrell

4.1 Introduction 95
4.2 The Specificity and Complexity of Herbivore-Induced VOC Production 97
4.2.1 Plant Endogenous Wound Signalling 99
4.2.2 Herbivore-Derived Elicitors of VOC Emission 102
4.3 Ecological Consequences of VOC Emission 104
4.3.1 Within-Plant Defence Signalling 104

4.3.2 Herbivore-Induced VOC Emission as Part of a Metabolic Reconfiguration of the Plant 105
4.3.3 Herbivores Use VOCs to Select Host Plants 107
4.3.4 VOCs as Indirect Defences Against Herbivores 108
4.3.5 VOCs in Plant–Plant Interactions 111
4.4 Conclusion 112
Acknowledgements 114
References 114

5 Pheromones in Chemical Communication **123**
Kenji Mori

5.1 Introduction 123
5.1.1 Definition of Pheromones 123
5.1.2 Classification of Pheromones 123
5.2 History of Pheromone Research 125
5.3 Research Techniques in Pheromone Science 127
5.3.1 The Collecting of Pheromones 127
5.3.2 Bioassay-Guided Purification 128
5.3.3 Structure Determination and Synthesis 128
5.3.4 Field Bioassay 129
5.3.5 Structure Elucidation of the Male-Produced Aggregation Pheromone of the Stink Bug *Eysarcoris lewisi* – A Case Study 129
5.4 Structural Diversity Among Pheromones 132
5.5 Complexity of Multicomponent Pheromones 137
5.6 Stereochemistry and Pheromone Activity 139
5.6.1 Only a Single Enantiomer is Bioactive and its Opposite Enantiomer Does Not Inhibit the Response to the Active Isomer 139
5.6.2 Only One Enantiomer is Bioactive, and its Opposite Enantiomer Inhibits the Response to the Pheromone 139
5.6.3 Only One Enantiomer is Bioactive, and its Diastereomer Inhibits the Response to the Pheromone 139
5.6.4 The Natural Pheromone is a Single Enantiomer, and its Opposite Enantiomer or Diastereomer is Also Active 140
5.6.5 The Natural Pheromone is a Mixture of Enantiomers or Diastereomers, and Both of the Enantiomers, or All of the Diastereomers are Separately Active 141
5.6.6 Different Enantiomers or Diastereomers are Employed by Different Species 141
5.6.7 Both Enantiomers are Necessary for Bioactivity 141
5.6.8 One Enantiomer is More Active Than the Other, but an Enantiomeric or Diastereomeric Mixture is More Active Than the Enantiomer Alone 141

5.6.9 One Enantiomer is Active on Males, While the Other is Active on Females 142
5.6.10 Only the *meso*-Isomer is Active 142
5.7 Pheromones With Kairomonal Activities 142
5.8 Mammalian Pheromones 143
5.9 Invention of Pheromone Mimics 145
5.10 Conclusion 147
Acknowledgements 147
References 147

6 Use of Volatiles in Pest Control **151**
J. Richard M. Thacker and Margaret R. Train

6.1 Introduction 151
6.2 Repellents (DEET, Neem, Essential Oils) 151
6.3 Volatile Synthetic Chemicals and Fumigants 154
6.4 Pheromones 158
6.5 Volatile Allelochemicals 165
6.6 Plant Volatiles and Behavioural Modification of Beneficial Insects 166
6.7 Concluding Comments 167
References 168

7 Challenges in the Synthesis of Natural and Non-Natural Volatiles **173**
Anthony A. Birkbeck

7.1 Introduction – The Art of Organic Synthesis 173
7.2 Overcoming Challenges in the Small-Scale Synthesis of Natural Volatile Compounds 174
7.2.1 D,L-Caryophyllene (1964) 174
7.2.2 β-Vetivone (1973) 175
7.3 Overcoming Challenges in the Large-Scale Synthesis of Nature Identical and Non-Natural Molecules 176
7.3.1 (*Z*)-3-Hexenol 176
7.3.2 Citral 177
7.3.3 (–)-Menthol 179
7.3.4 Habanolide 180
7.4 Remaining Challenges in the Large-Scale Synthesis of Natural and Non-Natural Volatiles 180
7.5 Design and Synthesis of Novel Odorants and Potential Industrial Routes to a Natural Product 182
7.5.1 Cassis (Blackcurrant) 182
7.5.2 Patchouli 184
7.5.3 Musk 187
7.5.4 Sandalwood 189
7.6 Other Challenges 193
7.7 Conclusion 193
Acknowledgements 194

Dedication 195
References 195

8 The Biosynthesis of Volatile Sulfur Flavour Compounds 203
Meriel G. Jones

8.1 Introduction: Flavours as Secondary Metabolites 203
8.2 Sulfur in Plant Biology 204
8.3 Sulfur Compounds as Flavour Volatiles 205
8.4 The Alk(en)yl Cysteine Sulfoxide Flavour Precursors 206
8.5 Biosynthesis of the Flavour Precursors of *Allium* 207
8.5.1 The Biosynthesis of *Allium* Flavour Precursors via γ-Glutamyl Peptides 208
8.5.2 The Biosynthesis of *Allium* Flavour Precursors via Cysteine Synthases 209
8.6 Formation of Volatiles from CSOs 210
8.6.1 *S*-Methyl-L-cysteine sulfoxide 210
8.6.2 Release of the *Allium* CSOs 211
8.7 The *Allium* Flavour Volatiles 212
8.8 The Enzyme Alliinase 213
8.9 The Enzyme Lachrymatory Factor Synthase 214
8.10 The Biological Roles of the Flavour Precursors 215
8.11 The Glucosinolate Flavour Precursors 216
8.12 GS and Their Biosynthetic Pathways 216
8.13 Release of Volatile GS Hydrolysis Products 218
8.14 The Biological Role of Glucosinolates 220
8.15 Application of Transgenic Technology to Applied Aspects of GS Biosynthesis 222
8.16 Volatile Sulfur Compounds from Other Plants 222
8.16.1 Complex Organic Sulfur Volatiles 222
8.16.2 Simple Sulfur Volatiles 223
8.16.3 Hydrogen Sulfide 223
8.16.4 Methanethiol 224
8.17 Conclusion 224
References 224

9 Thermal Generation of Aroma-Active Volatiles in Food 231
Christoph Cerny

9.1 Introduction 231
9.2 The Maillard Reaction 233
9.2.1 The Amadori Rearrangement 234
9.2.2 Deoxyosones 235
9.2.3 Retro-Aldolization 235
9.3 Formation of Aroma Compounds in the Later Stages of the Maillard Reaction 237
9.3.1 2-Furfurylthiol 237

9.3.2 4-Hydroxy-2,5-dimethyl-3(2*H*)-furanone 239
9.3.3 Alkyl and Alkenylpyrazines 239
9.3.4 2-Acetyl-1-pyrroline 241
9.4 The Strecker Degradation 241
9.5 Caramelization 244
9.6 Thiamin Degradation 246
9.7 Ferulic Acid Degradation 246
9.8 Fat Oxidation 247
9.9 Conclusion 250
References 250

10 Human Olfactory Perception **253**
Alan Gelperin

10.1 Introduction 253
10.2 Historical Perspective on Olfactory Perception 254
10.3 Human Olfactory Pathway 255
10.4 Functional Studies in Human Subjects 256
10.5 Functional Studies in Brain-Damaged Subjects 259
10.6 Single Odorants, Binary Mixtures and Complex Odour Objects 259
10.7 Olfactory Versus Trigeminal Odorant Identification 262
10.8 Orthonasal Versus Retronasal Odour Perception 263
10.9 Specific Anosmias 264
10.10 MHC-Correlated Odour Preferences in Human Subjects 265
10.11 Odour Deprivation and Odour Perception 266
10.12 Age-Related Decline in Olfactory Perception 267
10.13 New Neurons in Adult Brains 268
10.14 Epidemiological Studies of Human Olfaction 268
10.15 Active Sampling and Olfactory Perception 269
10.16 Human Olfactory Imagery 270
10.17 Top-Down Influences on Olfactory Perception 271
10.18 Reproductive State and Olfactory Sensitivity 272
10.19 Olfaction, Hunger and Satiety 273
10.20 Odour Perception Bias by Odour Names 274
10.21 Olfaction and Disease States 275
10.22 Prenatal and Postnatal Influences on Infant Odour/Flavour Preferences 276
10.23 Future Directions 277
Acknowledgements 277
References 278

11 Perfumery – The Wizardry of Volatile Molecules **291**
Christophe Laudamiel

11.1 The Big Picture 291
11.2 Wizardry No. 1: Full Holograms Create Real Emotions 292
11.3 Volatiles Need a Language Wizard 296

11.4 Wizardry No. 2: The Perfumer in the Jungle of Volatiles to Create Emotions 298
11.5 Wizardry No. 3: End Results Are Music to the Nose 303
References 304

12 Microencapsulation Techniques for Food Flavour **307**
Youngjae Byun, Young Teck Kim, Kashappa Goud H. Desai and Hyun Jin Park

12.1 Demands 307
12.2 Microencapsulation in the Food Industry 307
12.3 Techniques and Materials for Flavour Microencapsulation 308
12.3.1 Spray Drying 308
12.3.2 Extrusion 312
12.3.3 Cyclodextrin Inclusion Complexes 314
12.3.4 Helical Inclusion Complexes 316
12.3.5 Fluidized Bed Coating 318
12.3.6 Top Spray Fluidized Bed Coating 318
12.3.7 Bottom Spray System 318
12.3.8 Wurster System 320
12.3.9 Tangential Spray or Rotary Fluidized Bed Coating 320
12.3.10 Coacervation 320
12.3.11 Double or Multiple Emulsion with Freeze Drying 321
12.3.12 Co-Crystallization 322
12.3.13 Spray Chilling and Spray Cooling 322
12.3.14 Supercritical Fluids 323
12.3.15 Other Techniques 323
12.4 Conclusion and Future Trends 325
References 326

13 Profragrances and Properfumes **333**
Andreas Herrmann

13.1 Introduction 333
13.2 Release of Alcohols 335
13.2.1 Enzymatic Hydrolysis 335
13.2.2 Neighbouring-Group-Assisted, Non-Enzymatic Hydrolysis 340
13.3 Release of Carbonyl Derivatives 346
13.3.1 Oxidations 346
13.3.2 Reversible Systems 350
13.3.3 Retro 1,4-Additions 354
13.4 Profragrance and Properfume Strategies 356
13.4.1 Performance and Cost Efficiency 356
13.4.2 Stability 357
13.5 Conclusion 357
Acknowledgements 358
References 358

14 Reactions of Biogenic Volatile Organic Compounds in the Atmosphere **363**
Russell K. Monson

14.1 Introduction 363
14.2 The Relative Importance of Anthropogenic Versus Biogenic VOC Emissions to Atmospheric Chemistry 364
14.3 Overview of BVOC Oxidation 365
14.4 The Types of Emitted BVOCs and General Roles in Atmospheric Chemistry 370
14.5 Gas Phase Oxidation of BVOCs 372
14.6 Gas Phase Chemistry of BVOCs in Urban and Suburban Airsheds 374
14.7 Gas Phase Chemistry Within and Above Forests 375
14.8 BVOC Emissions and SOA Formation 377
14.9 Conclusion 381
References 381

Index **389**

Foreword

Volatile compounds play an important role in nature as messenger compounds to transmit selective information between species. The ubiquity of these compounds in our everyday environment has initiated a variety of research activities in the life sciences over recent decades. Both biologists and chemists became interested in exploring the role of bioactive volatile compounds in many different aspects. The evolution from molecular to supramolecular science has particularly influenced the research activities on the chemistry and biology of volatiles. The investigation of molecular properties beyond the single molecule required (and resulted in) numerous interdisciplinary efforts to answer important questions related to the role of these compounds in our direct environment.

Molecular recognition is one of the key aspects leading to the understanding of the biological processes involved in volatile signalling. In contrast to the investigation of host–guest interactions typically encountered in the area of pharmaceutical or biomedical research, which usually take place in aqueous solution, volatile compounds have to be diffused into the air and transported over large distances to reach their biological target. The specific feature of their volatility, as compared to other bioactive molecules, characterizes the behaviour of these molecules from their biogeneration, to their emission, analysis, release, transport, recognition and perception, up to their degradation in a specific environment.

The present book summarizes several aspects related to the chemistry and biology of volatile compounds in a structure-based approach and tries to give the reader an introduction to and general overview of the various research areas related to this particular class of molecules. It also provides perspectives along novel avenues of research and development. It should thus be of great interest to all those involved in the various facets of both basic and applied research on volatile compounds.

Jean-Marie Lehn
Strasbourg
November 2009

List of Contributors

Carlo Bicchi, Laboratory of Phytochemical Analysis, Dipartimento di Scienza e Tecnologia del Farmaco, Università degli Studi di Torino, Via Pietro Giuria 9, 10125 Torino, Italy

Anthony A. Birkbeck, Firmenich SA, Division Recherche et Développement, 1 Route des Jeunes, B. P. 239, 1211 Genève 8, Switzerland

Youngjae Byun, Department of Packaging Science, Clemson University, Clemson, SC 29634-0320, USA

Christoph Cerny, Firmenich SA, Division Recherche et Développement, 7 Rue de la Bergère, B. P. 148, 1217 Meyrin 2, Switzerland

Chiara Cordero, Laboratory of Phytochemical Analysis, Dipartimento di Scienza e Tecnologia del Farmaco, Università degli Studi di Torino, Via Pietro Giuria 9, 10125 Torino, Italy

Kashappa Goud H. Desai, Department of Food Technology, 306 School of Life Science and Biotechnology, Korea University, Seoul, 136-701, Korea

Alan Gelperin, Monell Chemical Senses Center, 3500 Market St., Philadelphia, PA 19104, USA, and Princeton Neuroscience Institute, Department of Molecular Biology, Princeton University, Princeton, NJ 08544, USA

Andreas Herrmann, Firmenich SA, Division Recherche et Développement, 1 Route des Jeunes, B. P. 239, 1211 Genève 8, Switzerland

Meriel G. Jones, The School of Biological Sciences, The University of Liverpool, Liverpool L69 7ZB, UK

André Kessler, Cornell University, Department of Ecology and Evolutionary Biology, E445 Corson Hall, Ithaca, NY 14853, USA

Young Teck Kim, Department of Packaging Science, Clemson University, Clemson, SC 29634-0320, USA

Christophe Laudamiel, DreamAir LLC, 210 Eleventh Ave, Suite 1002, New York, NY 10001, USA

Erica Liberto, Laboratory of Phytochemical Analysis, Dipartimento di Scienza e Tecnologia del Farmaco, Università degli Studi di Torino, Via Pietro Giuria 9, 10125 Torino, Italy

Hartmut K. Lichtenthaler, Botanisches Institut (Molecular Biology and Biochemistry of Plants), University of Karlsruhe, Kaiserstraße 12, 76133 Karlsruhe, Germany

Russell K. Monson, Department of Ecology and Evolutionary Biology, and Cooperative Institute for Research in Environmental Sciences, University of Colorado, Boulder, CO 80309, USA

Kenji Mori, The University of Tokyo, 1-20-6-1309, Mukogaoka, Bunkyo-ku, Tokyo 113-0023, Japan

Kimberly Morrell, Cornell University, Department of Ecology and Evolutionary Biology, E445 Corson Hall, Ithaca, NY 14853, USA

Hyun Jin Park, Department of Food Technology, 306 School of Life Science and Biotechnology, Korea University, Seoul, 136-701, Korea

Patrizia Rubiolo, Laboratory of Phytochemical Analysis, Dipartimento di Scienza e Tecnologia del Farmaco, Università degli Studi di Torino, Via Pietro Giuria 9, 10125 Torino, Italy

Barbara Sgorbini, Laboratory of Phytochemical Analysis, Dipartimento di Scienza e Tecnologia del Farmaco, Università degli Studi di Torino, Via Pietro Giuria 9, 10125 Torino, Italy

J. Richard M. Thacker, Biological Sciences, University of the West of Scotland, Paisley, PA1 2BE, UK

Margaret R. Train, Biological Sciences, University of the West of Scotland, Paisley, PA1 2BE, UK

Acknowledgements

First, I would like to thank Paul Deards from John Wiley & Sons, Ltd in Chichester for his invitation to edit this book; without his initiative, this project would not have been realized. Furthermore, I wish to thank Richard Davies, Gemma Valler, Rebecca Ralf and Mohan Tamilmaran, all from John Wiley & Sons, Ltd, for their continuous support during the entire publishing process.

Of course, particular thanks go to all of the authors for accepting the challenge to provide an overview of their research areas, for their enthusiasm and many stimulating discussions and exchanges during the editing process, as well as to Prof. Jean-Marie Lehn for agreeing to write a foreword to this book. As all the chapters were peer-reviewed, I would also like to thank the numerous referees for their suggestions and comments, which helped to improve the quality of the manuscripts.

Finally, I thank my superiors Dr. Daniel Benczédi, Dr. Maria-Inés Velazco and Dr. Antoine Gautier from Firmenich for supporting this project, and I thank my wife Anja for her patience. I hope this book provides an interesting and stimulating interdisciplinary approach to the chemistry and biology of volatile compounds, and that the readers forgive the errors that may have escaped the proofreading.

Andreas Herrmann
Genève
January 2010

Abbreviations

Ac	acetyl (in structural formula)
ACC	1-aminocyclopropane-1-carboxylic acid
ACS	American Chemical Society
ACSO	*S*-allyl-L-cysteine sulfoxide
AE	aroma extract
AEDA	aroma extract dilution analysis
AFNOR	Association Française de Normalisation
AMDIS	automatic mass spectral deconvolution
AMPI	acetylmethyl phosphinate
APC	anterior piriform cortex
approx.	approximatively
APS	adenosine-5′-phosphosulfate
aq.	aqueous
ASE	accelerated solvent extraction
ASES	aerosol solvent extraction system
ATP	adenosine-5′-triphosphate
BASF	Badische Anilin und Soda Fabrik
BINAP	2,2′-bis(diphenylphosphino)-1,1′-binaphthyl
BOLD	blood oxygenation level dependent
BOSS	beaver dam offspring study
Bu	butyl (in structural formula)
BVOC	biogenic volatile organic compound
ca.	*circa*
cAMP	cyclic adenosine monophosphate
CAN-BD	carbon dioxide assisted nebulization with a bubble dryer
CAR	carboxen
cat.	catalyst/catalytic
CBF	cerebral blood flow
CD	circular dichroism (spectroscopy)
CDs	cyclodextrins
CDP	cytidine-5′-diphosphate
CDP-ME	4-diphosphocytidyl-2-*C*-methyl-D-erythritol
CDP-ME2P	4-diphosphocytidyl-2-*C*-methyl-D-erythritol-2-phosphate
CDP-MEP	diphosphocytidyl-2-methyl-D-erythritol-2-phosphate
C-GC	conventional gas chromatography
CI	Criegee intermediate
CI-MS	chemical ionization–mass spectrometry
CITAC	Cooperation on International Traceability in Analytical Chemistry

CITES	Convention on International Trade in Endangered Species of Wild Fauna and Flora
CoA	coenzyme A
COBEL	children's olfactory behaviors in everyday life
conc.	concentrated
COSY	correlation spectroscopy
Cp	cyclopentadienyl
CPCSP	continuous powder coating spraying process
CRC	Chemical Rubber Company
CS	cysteine synthase
CSO	alk(en)ylcysteine sulfoxide
CTP	cytidine-5′-triphosphate
1D	one-dimensional
2D	two-dimensional
3D	three-dimensional
DADS	diallyl disulfide
DC	direct contact
DCMU	diuron
DEET	*N*,*N*-diethyl-2-toluamide
DELOS	depressurization of an expanded liquid organic solution
d.f.	film thickness
D-HS	dynamic headspace
DIBAL-H	diisobutylaluminium hydride
dil.	diluted
DMAPP	3,3-dimethylallyl diphosphate
DMSO	dimethylsulfoxide
DNA	deoxyribonucleic acid
DOX	1-deoxy-D-xylulose
DOXP	1-deoxy-D-xylulose 5-phosphate
DP	dual phase
DSC	differential scanning calorimetry
DTBP	di-*tert*-butyl peroxide
DVB	divinylbenzene
DXR	DOXP reductoisomerase
DXS	DOXP synthase
EAD	electroantennographic detection
EAG	electroantennogram
Ed.	editor/edition
EDGAR	emissions database for global atmospheric research
ee	enantiomeric excess
EEG	electroencephalogram
EHLS	epidemiology of hearing loss study
EO	essential oil
EPA	Environmental Protection Agency (USA)
er	enantiomeric ratio
ES-GC	enantioselective gas chromatography

ESP	epithiospecifier protein
ET	ethylene
Et	ethyl (in structural formula)
etc.	*et cetera*
EU	European Union
FACs	fatty acid–amino acid conjugates
FAO	Food and Agriculture Organisation (of the United Nations)
Fd	ferredoxin
FFNSC	flavour and fragrance natural and synthetic compounds
F-GC	fast gas chromatography
FID	flame ionization detector
fMRI	functional magnetic resonance imaging
FPP	farnesyl diphosphate
FQPA	Food Quality Protection Act (USA)
FSOT	fused silica open tubular
GA-3-P	glyceraldehyde-3-phosphate
GAS	gas (or supercritical fluids) anti-solvent
GC	gas chromatography
GC-FID	gas chromatography–flame ionization detection
GC-MS	gas chromatography–mass spectrometry
GC-qMS	gas chromatography–quadrupole mass spectrometry
GC-O	gas chromatography–olfactometry
GGPP	geranylgeranyl diphosphate
γGP	γ-glutamyl sulfoxide peptide derivative
GPP	geranyl diphosphate
GS	glucosinolate
GSH	reduced glutathione
GSSG	oxidised glutathione disulfide
HCC-HS	high concentration capacity headspace technique
HIPVs	herbivore-induced plant volatiles
HIV	human immunodeficiency virus
HLA	human leukocyte antigen
HMBC	heteronuclear multiple bond coherence
HMBPP	4-hydroxy-3-methyl-2-(*E*)-butenyl diphosphate
HMG	hydroxyl-methylglutaryl
HMPA	hexamethylphosphoramide
HMQC	heteronuclear multiple quantum coherence
HPC	hydroxypropyl cellulose
HPLC	high performance liquid chromatography
HPOD	hydroperoxyoctadienoate
HR	heart rate
HS	headspace
HS-LPME	headspace–liquid phase microextraction
HS-MS	headspace–mass spectrometry
HSQC	heteronuclear single quantum coherence
HSSE	headspace sorptive extraction

HS-SPDE	headspace–solid phase dynamic extraction
HS-SPME	headspace–solid phase microextarction
HS-SMSE	headspace–silicon membrane sorptive extraction
HS-STE	headspace–sorptive tape extraction
IAA	indole-3-acetic acid
IATA	International Air Transport Association
i.d.	inner diameter
IFF	International Flavors and Fragrances Inc.
IFRA	International Fragrance Association
INCAT	inside needle capillary adsorption trap
IOFI	International Organization of the Flavour Industry
IPP	isopentenyl diphosphate
IRMS	isotope ratio mass spectrometry
IspD	CDP-ME synthase
IspE	CDP-ME kinase
IspF	MEcPP synthase
IspG	HMBPP synthase
IspH	HMBPP reductase
ISTD	internal standard
IUPAC	International Union of Pure and Applied Chemistry
JA	jasmonic acid
LC	liquid chromatography
LDA	lithium diisopropylamine
LF	lachrymatory factor
LFS	lachrymatory factor synthase
LMCS	longitudinally modulated cryogenic system
LOD	limit of detection
LOQ	limit of quantification
LOX3	lipoxygenase 3 (gene)
MACR	methacrolein
MAE	microwave-assisted extraction
MA-HD	microwave-assisted hydrodistillation
MAM	methylthioalkylmalate
MBO	2-methylen-3-buten-2-ol
MCSO	*S*-methyl-L-cysteine sulfoxide
MD	multidimensional
ME	male equivalents
Me	methyl (in structural formula)
MEcPP	2-*C*-methyl-D-erythritol-2,4-cyclodiphosphate
MeJA	methyl jasmonate
MeSA	methyl salicylate
MEP	2-*C*-methylerythritol-4-phosphate
MESI	membrane extraction sorbent interface
MGL	methionine-γ-lyase
MHC	major histocompatibility complex
MHE	multiple headspace extraction

ML	maple lactone
MME	membrane microextraction
MS	mass spectrometry, mass spectrometer
Ms	mesyl (SO_2CH_3; in structural formula)
MVA	mevalonic acid/mevalonate
MVL	mevalonolactone
MVK	methyl vinyl ketone
MW	molecular weight
MYB	myeloblast
NADPH	nicotinamide adenine dinucleotide phosphate
NB	narrow bore (column)
NIST	National Institute of Standards and Technology
NMO	*N*-methylmorpholine-*N*-oxide
NMR	nuclear magnetic resonance
NOESY	nuclear Overhauser enhancement spectroscopy
NPQ	nonphotochemical quenching
NS	nosespace
Nu	nucleophile (in structural formula)
OAV	odour activity values
OB	olfactory bulb
OCO	oral cavity only (exposure)
OFC	orbitofrontal cortex
OP	other phytohormones
OPE	ozone production efficiency
OR	olfactive receptor
ORNs	olfactive receptor neurons
OV	Ohio Valley Speciality Chemical (brand of stationary phases)
OXP-01	2-decyl-1-oxaspiro[2.2]pentane
OXP-04	2-(4-hydroxybutyl)-1-oxaspiro[2.2]pentane
PAN	peroxyacetyl nitrate
PAPS	adenosine-3′-phosphate-5′-phosphosulfate
PCA	principal component analysis
PCSO	*S*-propyl-L-cysteine sulfoxide
pdf	portable document format
PeCSO	*trans*-*S*-1-propenyl-L-cysteine sulfoxide
PEG	poly(ethylene glycol)
PET	positron emission tomography
PBP1	pheromone binding protein 1
PD	Parkinson's disease
PDMS	poly(dimethylsiloxane)
PEP	phosphoenol pyruvate
PG	protecting group
PGA	phosphoglyceric acid
PGSS	particles from gas-saturated solutions
Ph	phenyl (in structural formula)
PLP	pyridoxal-5′-phosphate

PMHS	poly(methylhydrosiloxane)
pp.	pages
PPC	posterior piriform cortex
PTR-MS	proton transfer reaction–mass spectrometer
qMS	quadrupole mass spectrometry (detector)
quant.	quantitative
RA	retinoic acid
ref.	reference
RESS	rapid expansion of supercritical fluids
RNA	ribonucleic acid
RSD	relative standard deviation
r.t.	room temperature
RTL	retention time locking
RubisCO	ribulosebisphosphate carboxylase/oxygenase
SA	salicylic acid
SAA	supercritical assisted atomization
SAS	supercritical fluids (or gas) anti-solvent
SAT	serine acetyltransferase
SBSE	stir bar sorptive extraction
SC	skin conductance
SCC-GC	short capillary column gas chromatography
sc-CO_2	supercritical CO_2
SDE	simultaneous distillation–extraction
SDOIT	San Diego odor identification test
SEDS	solution enhanced dispersion by supercritical fluids
SFE	supercritical fluid extraction
SFEE	supercritical fluid extraction of emulsions
S-HS	static headspace
SIM	single ion monitoring
SIM-MS	single ion monitoring–mass spectrometry
SIM-qMS	single ion monitoring–quadrupole mass spectrometry
SiSTEx	solvent in silicone tube extraction
SMP	skimmed milk powder
SMSE	silicon membrane sorptive extraction
SOA	secondary organic aerosol
SPACE	solid phase aroma concentrate extraction
SPME	solid phase microextraction
SROs	stress-related odours
SSI	supercritical solvent impregnation
S&T-HS	static and trapped headspace
TAS	total analysis systems
TCD	thermal conductivity detector
TDS	thermodesorption system
tert	tertiary
THF	tetrahydrofuran
TIC	total ion current

TIC-MS	total ion current–mass spectrometry
TLC	thin-layer chromatography
TMS	trimethylsilyl (protecting group)
TMT	thiol methyltransferases
TOF	time of flight
TPLSM	two-photon laser scanning microscopy
TRGs	temperature-responsive gels
Ts	tosyl ($SO_2C_6H_4CH_3$; in structural formula)
p-TSA	*para*-toluenesulfonic acid
UFM-GC	ultra-fast module gas chromatography
UK	United Kingdom of Great Britain and Northern Ireland
UNEP	United Nations Environment Programme
USA	United States of America
UV	ultraviolet (spectroscopy)
UV/Vis	ultraviolet/visible (spectoscopy)
VOC	volatile organic compound
Vol.	volume
WOF	warmed-over flavour
WPC	whey protein (isolate) concentrate

Names of scientific journals are abbreviated according to the Chemical Abstracts Service Source Index

1

Volatiles – An Interdisciplinary Approach

Andreas Herrmann

Firmenich SA, Division Recherche et Développement, 1 Route des Jeunes, B. P. 239, 1211 Genève 8, Switzerland

1.1 Introduction

Volatiles, and in particular biogenic volatile organic compounds (VOCs), are everywhere. They directly and indirectly influence the lives of many plant and insect species, and even human beings in many ways. Transported by diffusion through the air, they perform numerous functions, for example as so-called 'semiochemicals', 'infochemicals' or 'pheromones' for the communication between insects and/or plants,[1,2] for (insect) mating[2–4] or even, as a consequence of their pleasant taste or smell to humans, as flavours and fragrances.[4,5] Without volatile compounds, life on earth as we know it would be impossible. The structural variety in these compounds, which are generally based on a hydrocarbon skeleton with oxygen, nitrogen and sulfur as the most common heteroatoms, is almost infinite and always perfectly adapted to the specific role these molecules play in nature.

Biogenic VOCs are usually highly selective for a given target. This selectivity is presumably the most important property of these different compounds which, of course, is defined by their molecular structure (and the spatial arrangement of the different functional groups from which they are composed) and usually results in a very low 'detection threshold' of a given compound to its target species.[6] This means that the receptor of the receiving species can selectively detect specific molecules at very low concentrations in the air (typically expressed in ng l^{-1} of air) which, in some cases, can be a few molecules.

The Chemistry and Biology of Volatiles Edited by Andreas Herrmann

In contrast to many other target-specific compounds found in nature, volatiles are characterized by (relatively) high vapour pressures, allowing their efficient evaporation from various surfaces. This enables their transport through the air and thus to reach their biological target. Nevertheless, the term 'volatile' is usually not well defined, and the vapour pressures of compounds considered to be volatile can vary over several orders of magnitude.[7] Some representative volatile compounds such as **1–18** are listed in Table 1.1. Their vapour pressures span nine orders of magnitude, ranging from the highly volatile methane thiol (**1**) to the relatively nonvolatile insect pheromone bombykol (**18**) from the silkworm moth *Bombyx mori*.

Furthermore, biogenic VOCs are generally characterized as being rather 'hydrophobic' which facilitates, among others, their efficient evaporation from water-based media into the air. The polarity of different compounds is usually expressed as the logarithm of their octanol/water partition coefficients (log $P_{o/w}$);[7,8] the corresponding values for compounds **1–18** are indicated in Table 1.1. Once again, one can see that these data vary considerably, ranging from values below 1 for relatively polar compounds **1** and **3** to highly apolar molecules such as **13** with a log $P_{o/w}$ above 9.

The numerous areas of research dealing with the investigation of volatile compounds are as varied as their structures and their physicochemical properties. Biologists and chemists have become interested in these compounds for various reasons. Because the same volatile compounds can have different functions, volatiles have been discussed separately by the specialists in different areas. Nevertheless, the same molecular structure is often of interest to a wide variety of quite different research topics such as the biosynthesis of the given volatile in plants, its analysis in compound mixtures by different techniques, its particular biological role as a signalling compound or pheromone, its use in pest control and, if associated with a pleasant taste or smell, as a flavour or fragrance, its chemical synthesis, the mechanism of its perception, its behaviour through encapsulation and processing, its controlled release, right up to its degradation in the environment (Scheme 1.1). In the following section the different interdisciplinary research aspects associated with a given compound are illustrated with the example of (*E*)-3,7-dimethyl-2,6-octadien-1-ol (geraniol, **11**) as a typical volatile molecule with an average vapour pressure and log $P_{o/w}$. At the same time, these aspects will allow the introduction of the various topics presented in the different chapters of this book and thus illustrate a typical lifecycle of volatile compounds from their biogeneration, via their release into the air, their role as semiochemicals, their specific recognition, through to their degradation in the atmosphere.

1.2 Geraniol – A Typical Example

Biogenic, bioactive volatiles, such as geraniol (**11**), are generated in plants from small precursor molecules in multistep enzymatic processes. As one of the main constituents of the essential oils of various rose species, monoterpene alcohol **11** is biosynthesized by condensation of isopentenyl diphosphate (IPP, **19**) and 3,3-dimethylallyl diphosphate (DMAPP, **20**) with the help of geranyl diphosphate (GPP) synthetase, followed by dephosphorylation of GPP (**21**),[9] as depicted in Scheme 1.2. GPP is of particular importance because it is the precursor of geraniol and of many different monoterpenes. The following chapter (Chapter 2) gives a general introduction to the various mechanisms involved in the

Table 1.1 *Vapour pressures and (logarithmic) octanol/water partition coefficients (log $P_{o/w}$) of a series of representative volatile compounds*

Name	Structure	Log $P_{o/w}$ [a]	Vapour pressure[a] [Pa]
Methanethiol (**1**)		0.78	2.01×10^{5}
Isoprene (**2**)	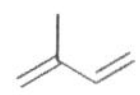	2.42	7.34×10^{4}
Methyl vinyl ketone (**3**)	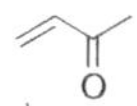	0.41	1.21×10^{4}
α-Pinene (**4**)		4.44	5.35×10^{2}
Frontalin (**5**)	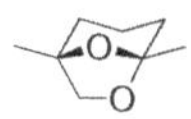	2.18	3.53×10^{2}
1,8-Cineol (**6**)	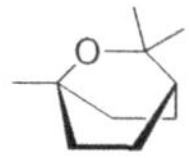	2.74	2.07×10^{2}
Limonene (**7**)	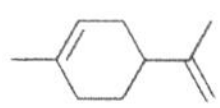	4.38	1.93×10^{2}
Diallyl disulfide (**8**)	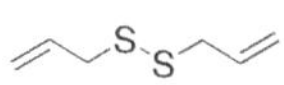	3.56	1.28×10^{2}
(*Z*)-3-Hexenol (**9**)	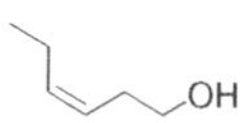	1.61	1.24×10^{2}
Citronellal (**10**)	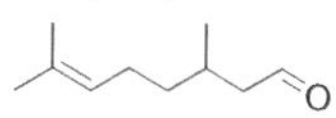	3.83	3.38×10^{1}
Geraniol (**11**)	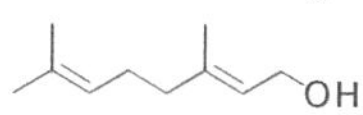	3.47	2.11×10^{0}
γ-Decalactone (**12**)	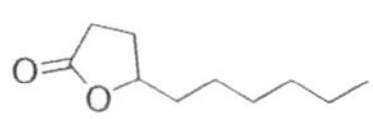	2.72	6.82×10^{-1}
Camphorene (**13**)	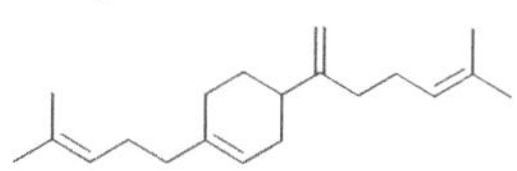	9.42	7.34×10^{-2}
Vanillin (**14**)	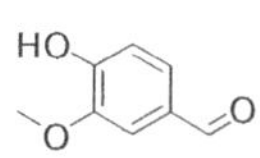	1.21	5.95×10^{-2}
Disparlure (**15**)	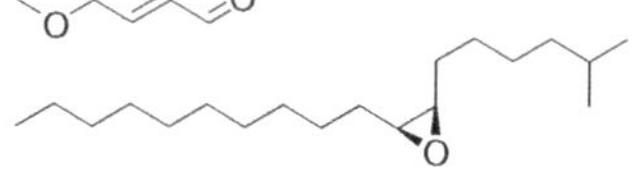	8.08	4.55×10^{-2}
Methyl jasmonate (**16**)	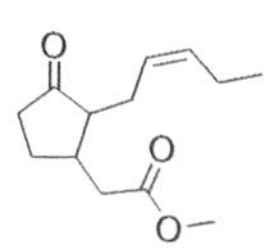	2.76	4.49×10^{-2}
5α-Androst-16-en-3-one (**17**)	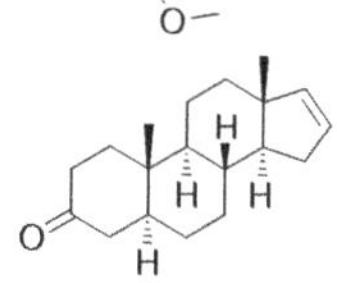	4.38	4.22×10^{-3}
Bombykol (**18**)	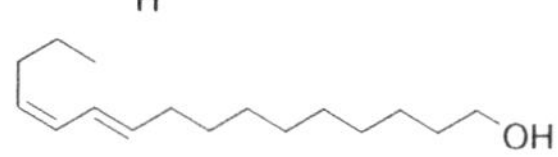	6.30	7.58×10^{-4}

[a]*Note:* Values calculated according to ref. 7.

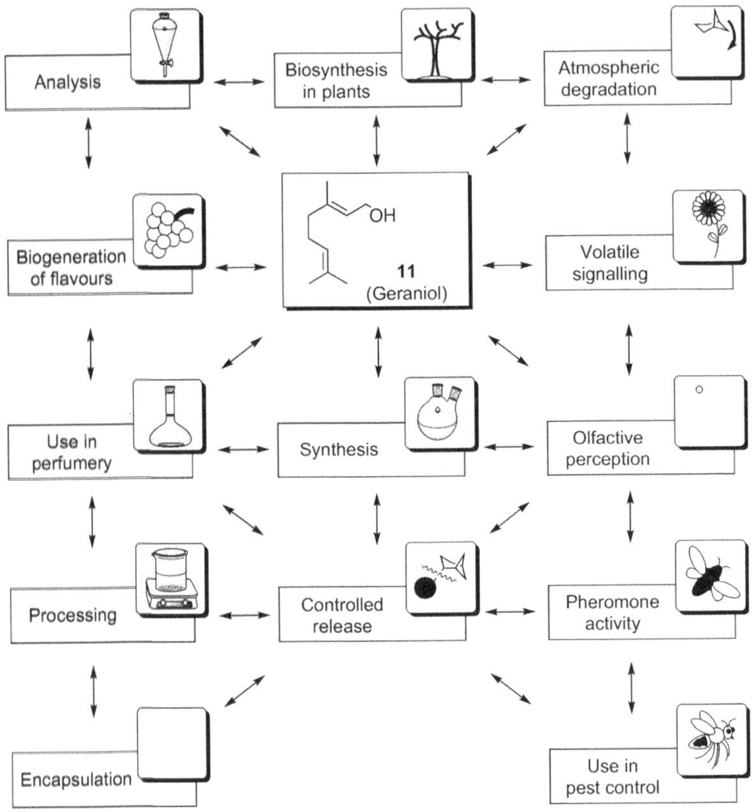

Scheme 1.1 Aspects of interdisciplinary research, using the example of geraniol

biosynthesis of plant isoprenoids and illustrate the structural variety of terpenes generated by plants.

The identification and quantification of the individual constituents isolated from plants is an important aspect in the understanding of the biochemical processes involved in their generation. Volatiles are analysed mainly by gas chromatography (GC), usually coupled with mass spectrometry (GC-MS), and some other more specific techniques. The volatility of such compounds allows specific sample preparation methods such as static and dynamic headspace analysis to be employed.[10] Chapter 3 gives an account of the numerous methods

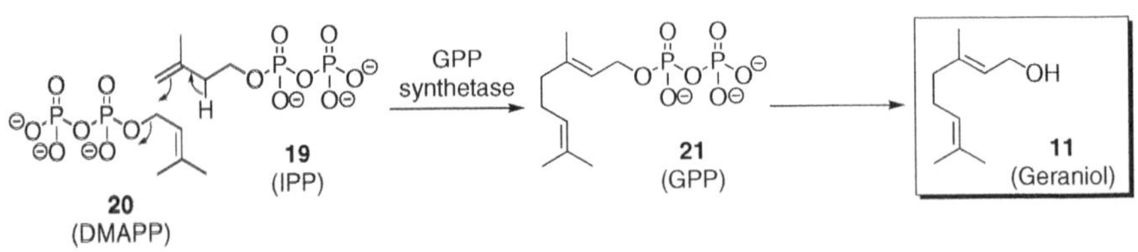

Scheme 1.2 Biosynthesis of geraniol

used for the analysis of volatile compounds emitted from plants by discussing the scopes and limitations of the different techniques.

Plants, insects and other animals use volatile compounds for their communication with the environment.[1] Besides its emission from flowers to attract pollinators, geraniol (**11**) is also a member of a class of so-called 'herbivore-induced plant volatiles' (HIPVs). Plants emit these compounds to defend themselves against herbivore attack by attracting natural enemies of the herbivores responsible for the plant damage. As an example, geraniol was found to attract wasps and flesh flies of the Braconidae and Sarcophagidae families, respectively.[11] The specific aspects and implications of volatile signalling for plants and insects are discussed in Chapter 4.

Geraniol (**11**) has also been identified in the secretions of the Nasonov gland of honeybees, where the compound, together with a series of other volatiles, serves as a pheromone to mark the entrance of the hive, for mating and orientation, as for example for swarm clustering or guidance to flowers.[12] Using insects and mammals as examples, Chapter 5 presents the classification, structural particularities and roles of pheromones in chemical communication.

Apart from acting as an attractant, geraniol was also found to repel certain insects, such as the malaria-transmitting mosquito *Anopheles gambiae*.[13] Essential oils (EOs) containing geraniol (and other insect-repellent compounds) have thus been used for protection against blood-feeding insects, whilst other volatiles have been identified as being useful for the protection of agricultural crops. The potential to selectively repel or kill certain insects is an important aspect of volatiles in the area of pest control, as documented in more detail in Chapter 6.

Although geraniol (**11**) is readily available from natural sources, several methods for its synthesis have been proposed, some of which are illustrated in Scheme 1.3. Typically the compound is prepared from other monoterpenes[14] such as myrcene (**22**), citral (**23**) or linalool (**24**) to usually afford mixtures of geraniol (**11**) and nerol (**25**). The selective preparation of specific isomers in high purities is therefore one of the major achievements of organic synthesis. Chapter 7 summarizes some of the challenges encountered in the synthesis of natural and non-natural fragrances. Besides allowing the preparation of natural compounds which are difficult to be accessed in large quantities, organic chemistry has

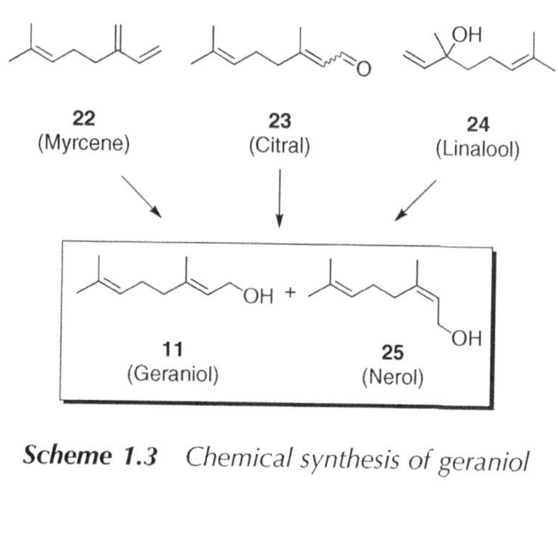

Scheme 1.3 *Chemical synthesis of geraniol*

delivered a multitude of new and entirely synthetic compounds, in particular for use in perfumery.

Certain volatile molecules are also part of our everyday life as flavours present in our daily nutrition. Monoterpenes, such as geraniol, contribute to the floral aroma of a series of grape varieties used in wine-making.[15] Many flavour volatiles, so-called 'secondary metabolites' in fruits or vegetables, are generated from fats or amino acid precursors during ripening. The mechanisms involved in the biogeneration and metabolism of these compounds are important for food preparation as well as in aspects of nutrition and health. Besides the terpenoid structures, sulfur- and nitrogen-containing compounds are particularly important flavour constituents. Whereas Chapter 8 discusses the biogeneration and the role of a variety of sulfur compounds encountered in onion and garlic, Chapter 9 focuses on the generation of flavours during food processing, as for example in the so-called 'Maillard reaction',[16] with particular focus on the thermal effects detected during cooking. Because not all volatiles have a pleasant taste or smell, the understanding of these processes is also important in allowing the efficient minimisation of undesired malodour formation under various conditions.

Once the volatiles have reached their target, they are recognized by specific receptors, which trigger an electric signal perceived by the brain. The mechanisms of olfactive perception are quite complicated in both insects and humans. The general importance of understanding the mechanisms of olfaction was underlined by the granting of the Nobel Prize in Physiology or Medicine to Axel and Buck in 2004 for their discoveries of odorant receptors and the organisation of the olfactory system,[17] which stimulated a general interest in exploring the chemistry and biology of volatile compounds in the life sciences.[18] Chapter 10 gives a general overview of the basic principles and mechanisms involved in (human) perception.

To humans geraniol (**11**) has a pleasant, sweet, floral smell and is therefore among the most frequently used perfumery compounds.[14,19] Perfumes are usually mixtures of many different volatile compounds created at the interface between art and science. In the general public, perfumery is most commonly associated with 'fine fragrances', but perfumes are also an important ingredient in body care and household products, such as shampoos, soaps, creams, deodorants, shower gels, surface cleaners, detergents, softeners and many others. Chapter 11 illustrates the particular artistic aspects of perfumery creation. For their creations, perfumers require a multitude of different molecules at their disposal. These compounds can be natural compounds extracted from natural raw materials, prepared selectively by biotechnical processes or by organic synthesis.

Many terpenes are sensitive to oxidation, either by oxygen in the air[20] or by a variety of bio-oxidation processes initiated by bacteria and fungi[21] (Scheme 1.4). To efficiently use

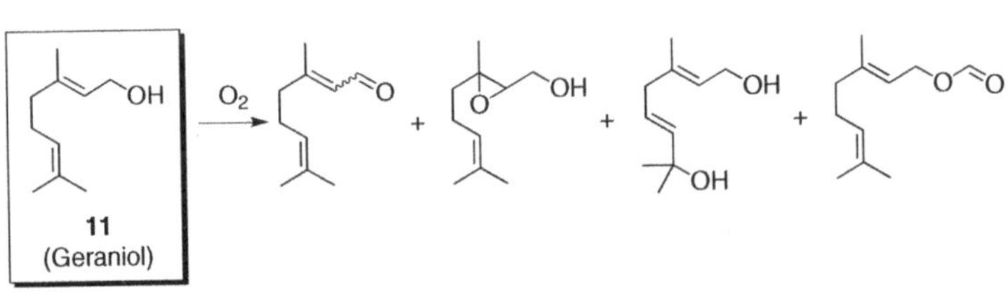

Scheme 1.4 *Compounds formed by air oxidation of geraniol*

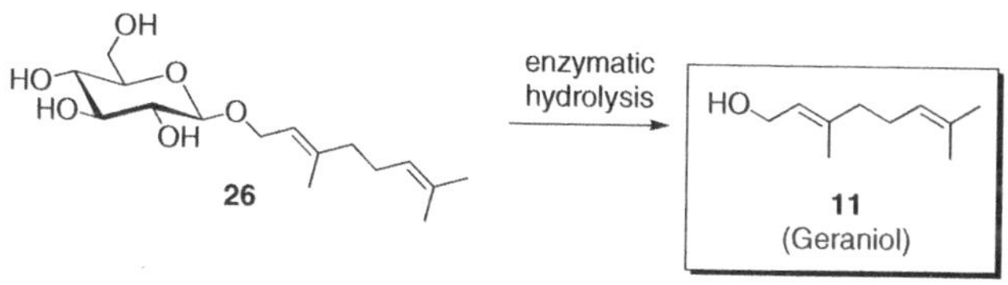

Scheme 1.5 *Release of geraniol by enzymatic cleavage of its glycoside conjugate*

susceptible terpenes such as geraniol (**11**) in commercial product formulations they have to be protected against (oxidative) degradation by using various encapsulation techniques.[22] Chapter 12 examines a series of flavour encapsulation processes which are most commonly used to increase the stability of food ingredients and to control their release in applications.

Besides the physical capturing of volatiles within capsules and other matrices, 'chemical' delivery systems have been developed to slowly release small quantities of volatiles to allow the duration of their perception to be increased. Nature stores and transports terpenes, such as geraniol (**11**) as hydrosoluble glycoside conjugates (e.g. **26**) before releasing them by enzymes into the environment,[23] which served as an inspiration to use fragrance conjugates as controlled release systems in practical applications (Scheme 1.5). Chapter 13 gives an account of the various techniques which have been developed for the release of fragrance molecules via covalent bond cleavage of different natural and non-natural fragrance precursors, so-called 'profragrances' or 'properfumes'. Of course, this concept can be generally applied to the controlled release of volatiles in other areas, as for example to attract or repel insects in pest control.

Another important area of interest concerns the (natural) biodegradation of volatiles in water or soil by the activity of various microorganisms. Nevertheless, as a consequence of their volatility, biogenic VOCs can also reach the atmosphere, where they are exposed to particular reaction conditions leading to their rapid degradation (Scheme 1.6).[24] The investigation of these processes becomes more and more important to understand the lifecycle of natural compounds and to estimate the impact of biogenic volatiles on our planet's climate. This book therefore concludes with a discussion of this important aspect with a summary of the gas phase chemistry of biogenic VOCs in the atmosphere (Chapter 14) as one aspect of their natural biodegradation.

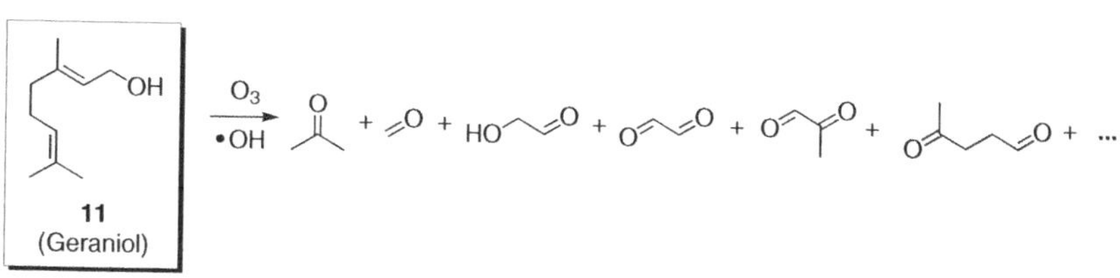

Scheme 1.6 *Degradation of geraniol in the atmosphere*

1.3 Conclusion

The different topics of research mentioned above for one representative molecule such as geraniol (**11**) indicate the broad variety of interest in volatile compounds. With the structures of the volatiles being the common link to all the different research areas presented in the following chapters, it is on purpose that the reader will find many common molecular structures illustrating the different chapters and thus referring to the different domains of research grouped together in this book. To date, these aspects have mainly been discussed separately by specialists from different areas, and several textbooks and reviews on the individual topics are available. However, the goal of this book is to provide an interdisciplinary approach on the various aspects of the chemistry and biology of volatile compounds to a reader generally interested in this area. The chapters constituting this book are not intended to give an extensive or comprehensive review of each of the specific topics, but rather represent a conceptual and quite general overview of a series of different research areas. They should give the reader the opportunity to discover the basic aspects of the different disciplines, to illustrate parallels between the different domains and, with the numerous references given in the different chapters, to invite further reading.

As depicted in Scheme 1.1, there is a certain overlap between the different research areas, which of course results from the discussion of similar aspects from different viewpoints in several of the chapters. At the same time, other important aspects of research on volatile compounds might only be briefly mentioned or even entirely omitted. However, to keep this book within a reasonable length, this seemed to be unavoidable.

The respective interest in volatile compounds from so many different angles is an excellent occasion for the exchange and common learning in chemistry and biology, and I hope that the present book will stimulate discussions and collaborations within a highly interdisciplinary research field.

References

1. For a selection of recent reviews, see for example: *(a)* N. Dudareva, E. Pichersky, J. Gershenzon, Biochemistry of plant volatiles, *Plant Physiol.*, **135**, 1893–1902 (2004); *(b)* J. K. Holopainen, Multiple functions of inducible plant volatiles, *Trends Plant Sci.*, **9**, 529–533 (2004); *(c)* N. Bakthavatsalam, P. L. Tandon, Interactions between plant chemicals and the entomophages, *Pestology*, **29**, 17–31 (2005); *(d)* Special section on plant volatiles, *Science*, **311**, 803–819 (2006); *(e)* M. D'Alessandro, T. C. J. Turlings, Advances and challenges in the identification of volatiles that mediate interactions among plants and arthropods, *Analyst*, **131**, 24–32 (2006); *(f)* L. G. W. Bergström, Chemical communication by behaviour-guiding olfactory signals, *Chem. Commun.*, 3959–3979 (2008); *(g)* S. B. Unsicker, G. Kunert, J. Gershenzon, Protective perfumes: the role of vegetative volatiles in plant defense against herbivores, *Curr. Opin. Plant Biol.*, **12**, 479–485 (2009).
2. For reviews, see for example: *The Chemistry of Pheromones and Other Semiochemicals*, S. Schulz (Ed.), Topics in Current Chemistry, Vols 239 and 240, Springer, Berlin, 2004 and 2005, and references cited therein.
3. For a selection of some recent reviews, see for example: *(a)* T. C. Baker, J. J. Heath, Pheromones: function and use in insect control, *Compr. Mol. Insect Sci.*, **6**, 407–459 (2005); *(b)* L. Stowers, T. F. Marton, What is a pheromone? Mammalian pheromones reconsidered, *Neuron*, **46**, 699–702 (2005); *(c)* A. Bigiani, C. Mucignat-Caretta, G. Montani, R. Tirindelli, Pheromone reception in mammals, *Rev. Physiol., Biochem., Pharmacol.*, **154**, 1–35 (2005); *(d)* P. A. Brennan, F. Zufall,

Pheromonal communication in vertebrates, *Nature*, **444**, 308–315 (2006); *(e)* K. D. Broad, E. B. Keverne, More to pheromones than meets the nose, *Nat. Neurosci.*, **11**, 128–129 (2008).

4. E. Breitmaier, *Terpenes – Flavors, Fragrances, Pharmaca, Pheromones*, Wiley-VCH, Weinheim, 2006.
5. See for example: *(a) Flavours and Fragrances – Chemistry, Bioprocessing and Sustainability*, R. G. Berger (Ed.), Springer Verlag, Berlin, 2007; *(b) Chemistry and Technology of Flavors and Fragrances*, D. J. Rowe (Ed.), Blackwell Publishing, Oxford, 2005.
6. *Standardized Human Olfactory Thresholds*, M. Devos, F. Patte, J. Rouault, P. Laffort, L. J. Van Gemert (Eds), Oxford University Press, Oxford, 1990.
7. Vapour pressures and log $P_{o/w}$ were calculated with the *EPI suite PBT calculator 1.0.0* based on the *EPIwin* program, US Environmental Protection Agency, Washington D. C., 2000.
8. *(a)* A. Leo, C. Hansch, D. Elkins, Partition coefficients and their use, *Chem. Rev.*, **71**, 525–616 (1971); *(b)* A. J. Leo, Calculating log P_{oct} from structures, *Chem. Rev.*, **93**, 1281–1306 (1993).
9. See for example: *(a)* T. Suga, T. Shishibori, The biosynthesis of geraniol and citronellol in *Pelargonium roseum* Bourbon, *Bull. Chem. Soc. Jpn.*, **46**, 3545–3548 (1973); *(b)* D. V. Banthorpe, D. R. S. Long, C. R. Pink, Biosynthesis of geraniol and related monoterpenes in *Pelargonium graveolens*, *Phytochemistry*, **22**, 2459–2463 (1983); *(c)* V. S. Dubey, An overview of geraniol biosynthesis in Palmarosa (*Cymbopogon martinii* Roxb. var. *motia*) and its biological and pharmacological properties, *Recent Progr. Med. Plants*, **11**, 75–97 (2006).
10. See for example: *(a)* K. Demeestere, J. Dewulf, B. De Witte, H. Van Langenhove, Sample preparation for the analysis of volatile organic compounds in air and water matrices, *J. Chromatogr. A*, **1153**, 130–144 (2007); *(b)* P. López, M. A. Huerga, R. Batlle, C. Nerin, Use of solid phase microextraction in diffusive sampling of the atmosphere generated by different essential oils, *Anal. Chim. Acta*, **559**, 97–104 (2006).
11. See for example: D. G. James, Further field evaluation of synthetic herbivore-induced plant volatiles as attractants for beneficial insects, *J. Chem. Ecol.*, **31**, 481–495 (2005).
12. See for example: *(a)* T. Schmitt, G. Herzner, B. Weckerle, P. Schreier, E. Strohm, Volatiles of foraging honeybees *Apis mellifera* (Hymenoptera: Apidae) and their potential role as semiochemicals, *Apidologie*, **38**, 164–170 (2007); *(b)* I. Lamprecht, E. Schmolz, B. Schricker, Pheromones in the life of insects, *Eur. Biophys. J.*, **37**, 1253–1260 (2008).
13. M. O. Omolo, D. Okinyo, I. O. Ndiege, W. Lwande, A. Hassanali, Repellency of essential oils of some Kenyan plants against *Anopheles gambiae*, *Phytochemistry*, **65**, 2797–2802 (2004).
14. H. Surburg, J. Panten, *Common Fragrance and Flavor Materials*, 5th edn, Wiley-VCH, Weinheim, 2006.
15. See for example: F. Luan, A. Mosandl, A. Münch, M. Wüst, Metabolism of geraniol in grape berry mesocarp of *Vitis vinifera* L. *cv.* Scheurebe: demonstration of stereoselective reduction, *E/Z*-isomerization, oxidation and glycosylation, *Phytochemistry*, **66**, 295–303 (2005).
16. See for example: *(a)* H. Nursten, *The Maillard Reaction: Chemistry, Biochemistry and Implications*, The Royal Society of Chemistry, Cambridge, 2005; *(b) The Maillard Reaction: Chemistry at the Interface of Nutrition, Aging, and Disease*, J. W. Baynes, V. M. Monnier, J. M. Ames, S. R. Thorpe (Eds), New York Academy of Sciences, New York, 2005; *(c) Process and Reaction Flavors: Recent Developments, ACS Symp. Ser.*, Vol. 905, D. K. Weerasinghe, M. K. Sucan (Eds), American Chemical Society, Washington D. C., 2005.
17. *(a)* R. Axel, Scents and sensibility: a molecular logic of olfactory perception (Nobel lecture) *Angew. Chem.*, **117**, 6264–6282 (2005); *Angew. Chem. Int. Ed.*, **44**, 6110–6127 (2005); *(b)* L. B. Buck, Unraveling the sense of smell (Nobel lecture) *Angew. Chem.*, **117**, 6283–6296 (2005); *Angew. Chem. Int. Ed.*, **44**, 6128–6140 (2005).
18. Special section on chemical sensing, *Nature*, **444**, 287–321 (2006).
19. S. Arctander, *Perfume and Flavor Chemicals*, published by the author, Montclair, 1969.
20. C. Bäcktorp, L. Hagvall, A. Börje, A.-T. Karlberg, P.-O. Norrby, G. Nyman, Mechanism of air oxidation of the fragrance terpene geraniol, *J. Chem. Theory Comput.*, **4**, 101–106 (2008).
21. J. L. Bicas, A. P. Dionísio, G. M. Pastore, Bio-oxidation of terpenes: an approach for the flavor industry, *Chem. Rev.*, **109**, 4518–4531 (2009).

22. *(a)* A. Madene, M. Jacquot, J. Scher, S. Desorby, Flavour encapsulation and controlled release – a review, *Int. J. Food Sci. Technol.*, **41**, 1–21 (2006); *(b)* M. A. Augustin, Y. Hemar, Nano- and micro-structured assemblies for encapsulation of food ingredients, *Chem. Soc. Rev.*, **38**, 902–912 (2009).
23. See for example: J.-E. Sarry, Z. Günata, Plant and microbial glycoside hydrolases: volatile release from glycosidic aroma precursors, *Food Chem.*, **87**, 509–521 (2004).
24. *(a)* F. M. N. Nunes, M. C. C. Veloso, P. A. de P. Pereira, J. B. de Andrade, Gas-phase ozonolysis of the monoterpenoids (*S*)-(+)-carvone, (*R*)-(–)-carvone, (–)-carveol, geraniol and citral, *Atmos. Environ.*, **39**, 7715–7730 (2005); *(b)* C. D. Forester, J. E. Ham, J. R. Wells, Geraniol (2,6-dimethyl-2,6-octadien-8-ol) reactions with ozone, and OH radical: rate constants and gas-phase products, *Atmos. Environ.*, **41**, 1188–1199 (2007).

2

Biosynthesis and Emission of Isoprene, Methylbutanol and Other Volatile Plant Isoprenoids

Hartmut K. Lichtenthaler

Botanisches Institut (Molecular Biology and Biochemistry of Plants), University of Karlsruhe, Kaiserstraße 12, 76133 Karlsruhe, Germany

2.1 Introduction

Various phytogenic volatile organic compounds (VOCs) emitted by plants are found in the atmosphere, such as alkanes, alkenes, alcohols (e.g. methanol),[1] aldehydes, ethers, esters, carboxylic acids[2] and various types of isoprenoids.[3,4] A major part of these phytogenic VOCs is of isoprenoid origin being emitted from herbaceous plants and to a very high extent from green leaves of a great number of trees.[4–6] The largest proportion of volatile isoprenoids emitted from green vegetation consists of the volatile hemiterpene isoprene (2-methyl-1,3-butadiene, **1**), several monoterpenes and, yet much less important, certain sesquiterpenes (i.e. C_{15} isoprenoid compounds). Plants without isoprene emission or without formation of other isoprenoid volatiles can accumulate particular diterpenes, isoprenoid C_{20} products, such as different ginkgolides in the green leaves of ginkgo trees; however, such diterpenes are mostly accumulated within the plants since they are not very or no longer volatile.

Today a lot of emphasis is placed on the emission of volatile phytogenic isoprene and monoterpenes, due to their high impact on atmospheric chemistry and ozone formation.[3,7,8] Isoprene contributes to photochemical smog and enhances the formation of ozone. In order to better predict and control this biogenic isoprenoid emission and ozone formation, it is

The Chemistry and Biology of Volatiles Edited by Andreas Herrmann

essential to understand the biological and physiological background of isoprene, monoterpene and sesquiterpene emission. Why and under which environmental conditions do plants emit these volatile isoprenoids? What types of change in the cell metabolism do and must occur in such plants to substantially reduce their amount of photosynthetically fixed carbon by the emission of isoprenoid VOCs, which essentially cuts down their productivity?

2.2 Plant Isoprenoids

Plants, animals and micro-organisms contain various primary isoprenoid compounds and partially also 'secondary' isoprenoid compounds (or natural products) that are composed of the C_5 units of 'active isoprene', which is known as isopentenyl diphosphate (IPP, **2**). These isoprenoid compounds are also termed terpenoids with hemiterpenes, monoterpenes, sesquiterpenes, diterpenes, triterpenes and tetraterpenes as C_5, C_{10}, C_{15}, C_{20}, C_{30} and C_{40} isoprenoids, respectively. Depending on the plant there exist: (*a*) hemiterpenes [C_5, e.g. isoprene (**1**)], (*b*) monoterpenes [C_{10}, e.g. geraniol (**3**), linalool (**4**), menthol (**5**), 1,4-cineol (**6**)], (*c*) sesquiterpenes [C_{15}, e.g. farnesol (**7**), bisabolol (**8**)], (*d*) diterpenes [C_{20}, e.g. phytol (**9**), camphorene (**10**), taxol (**11**), ginkgolides (e.g. ginkgolide A, **12**)], as shown in Figure 2.1, and (*e*) triterpenes [C_{30}, e.g. *Avena* saponins, oleanolic acid (**13**), cycloartenol (**14**)], (*f*) tetraterpenes [C_{40} compounds, such as the primary carotenoids β-carotene (**15**) and zeaxanthin (**16**), secondary carotenoids such as astaxanthin (**17**) and canthaxanthin (**18**) and the C_{20} apocarotenoid crocetin (**19**)], as shown in Figure 2.2. In several plant families very long-chain polyterpenes $(C_5H_8)_n$ are found (e.g. in rubber latex, gutta percha or in the chyle latex of several plant families, such as Euphorbiaceae or Asteraceae) whereby the number *n* for C_5 units ranges from about several thousands up to one million. Several plant compounds (i.e. the mixed prenyllipids), which are essential for the primary plant metabolism, obtain their lipophilic character by the possession of an isoprenoid side chain, such as the plastidic chlorophylls (phytyl chain), plastoquinone-9 (**20**, nona-prenyl side chain, Figure 2.2), phylloquinone K_1 (**21**, phytyl chain) or the mitochondrial ubiquinones Q-9 (**22**) and Q-10 (**23**) with nona-prenyl and deca-prenyl side chains, respectively.[9] In pure isoprenoid lipids (isoprene, phytol, menthol, cycloartenol, β-carotene) the complete carbon skeleton is composed of isoprenoid C_5 units, whereas in mixed prenyllipids (plastoquinone-9, phylloquinone K_1) the prenyl side chain is bound to a non-isoprenoid nucleus, such as a benzoquinone or naphthoquinone ring.

The 'biogenetic isoprene rule' (i.e. the composition of such terpenoid natural products from C_5 building blocks) was first detected in 1885 by Wallach,[10] and the head to tail addition of 'active C_5 units' was pointed out by Ruzicka *et al.*[11] This composition from a branched isoprenic C_5 unit is shown for several plant terpenoids in Figures 2.1 and 2.2. With respect to the biosynthesis of the isoprenoid chains, acetate[12] and acetyl-coenzyme A (acetyl-CoA)[13] were detected as biosynthetic precursors for cholesterol (triterpenoid derivatives) of mammals and fungi. Later mevalonic acid (**24**, MVA) was found as an intermediate[14] and IPP (**2**) as the active cellular biosynthetic C_5 unit (Figure 2.3).[15] Since ^{14}C-labelled acetate and acetyl-CoA were readily incorporated also into plant sterols and, in the case of the photosynthetic organism *Euglena*, also into carotenoids, it was generally accepted that all other isoprenoids of living cells and the large variety of different plant terpenoids including isoprene, monoterpernes, diterpenes, carotenoids

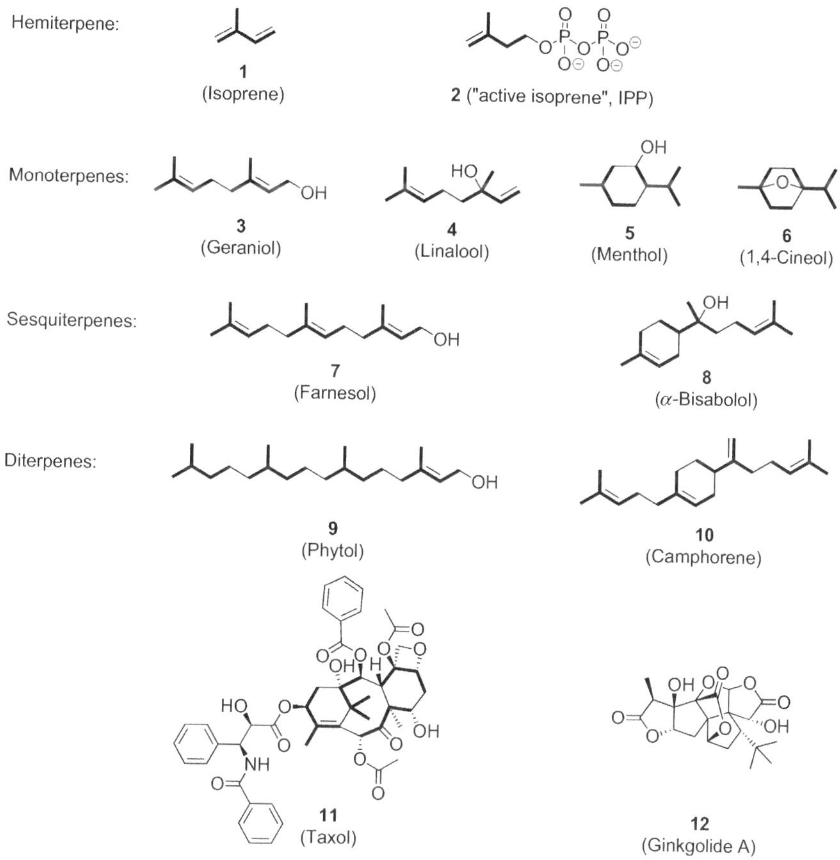

Figure 2.1 *Chemical structures of isoprene (**1**) and several other plant isoprenoids and terpenes (mono-, sesqui-, diterpenes) with indication of the isoprenoid C_5 units (printed in bold face) that make up the carbon skeleton of the final product*

and polyterpenes are synthesized via this isoprenoid pathway known as cytosolic acetate/mevalonate pathway.

For more than three decades it had been accepted that all plant terpenoids were synthesized from acetyl-CoA via this classical acetate/MVA pathway that proceeds in the cytosol of the plant cell catalysed by microsomes. This pathway is present in Plants, Fungi, Animals and Archaea, but generally not in Eubacteria.[16] An exception among the Eubacteria are only a few gram-positive cocci, such as *Streptococcus pneumonia*,[17] which obtained the MVA pathway apparently by lateral gene transfer. The acetate/MVA pathway starts from 3-acetyl-CoA, proceeds via hydroxyl-methylglutaryl-CoA (HMG-CoA) and mevalonate (MVA) to finally yield IPP.[18,19] This pathway can specifically be blocked by mevinolin (**25**) and other statins (e.g. cerivastatin, compactin, lovostatin), which inhibit the plant's microsomal enzyme HMG-CoA reductase, as first shown for mevinolin by Bach and Lichtenthaler.[20,21] Mevinolin (**25**) did, however, not block the bioynthesis and accumulation

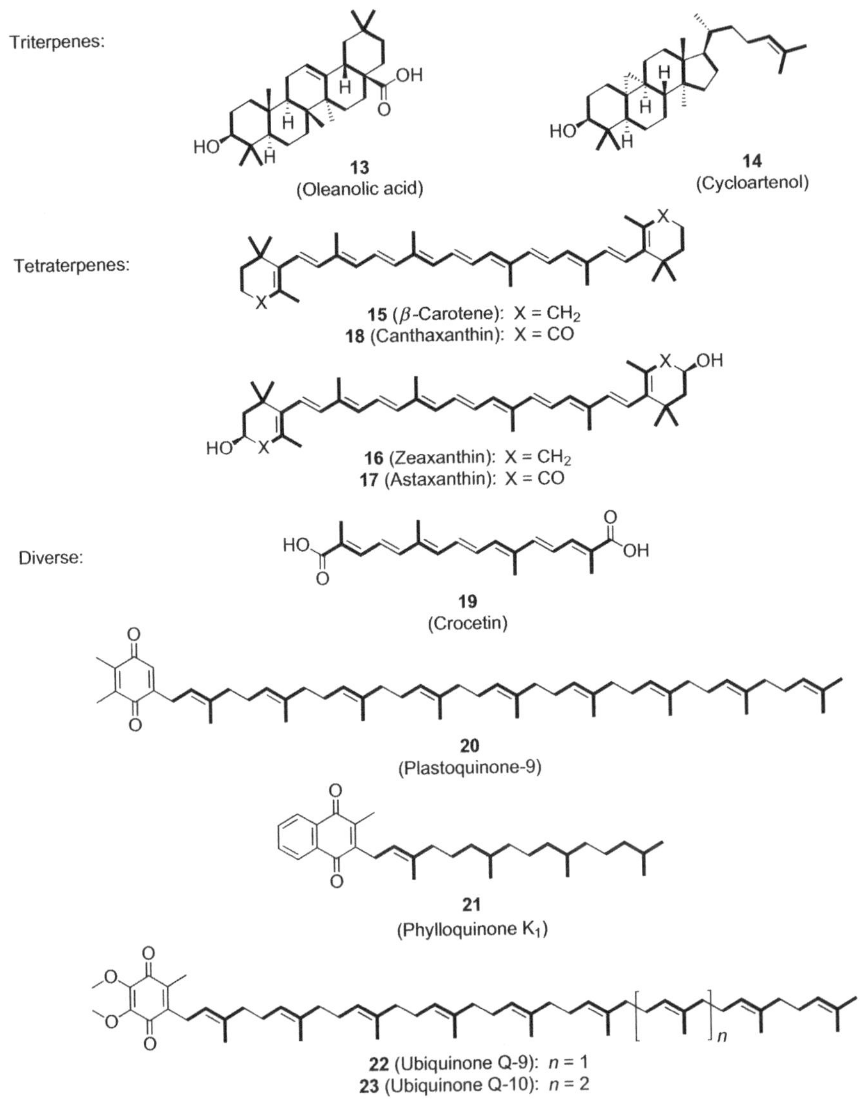

Figure 2.2 *Chemical structures of several plant isoprenoids and terpenes (tri- and tetraterpenes, some other structures) with indication of the isoprenoid* C_5 *units (printed in bold face) that make up the carbon skeleton of the final product*

of the isoprenoids in chloroplasts, such as carotenoids and chlorophylls (phytyl side chain).[22] This and other inconsistencies showing up in the labelling of plastidic isoprenoids (carotenoids, phytol chain of chlorophylls, prenyl side chain of plastoquinone-9), such as little or no labelling by ^{14}C-acetate or ^{14}C-mevalonate, finally led to the conclusion that chloroplasts might possess a separate biosynthetic pathway for IPP formation, as reviewed by Lichtenthaler *et al.*[19,23,24]

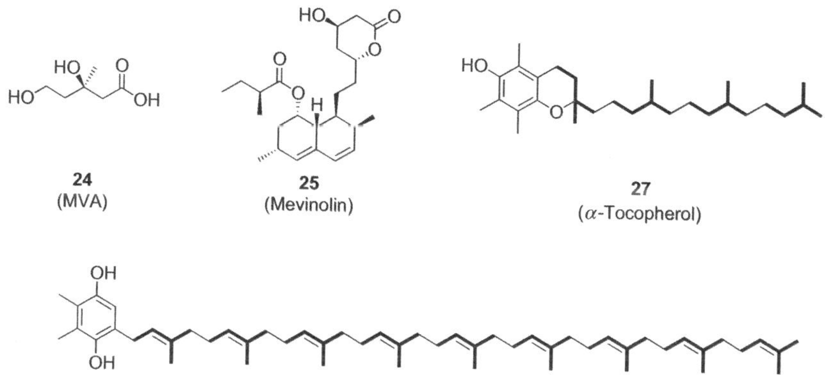

24 (MVA) **25** (Mevinolin) **27** (α-Tocopherol) **26** (Plastoquinol-9)

Figure 2.3 *Chemical structures of mevalonate (MVA,* ***24****), mevinolin (****25****), α-tocopherol (****27****) and plastoquinol-9 (****26****) with indication of the isoprenoid C_5 units (printed in bold face)*

2.3 Two IPP-Yielding Pathways in Plants

Today it is well known that plants possess two biochemically fully independent cellular pathways for the biosynthesis of isoprenoids:

- The *acetate/mevalonate* pathway (*acetate/MVA* pathway) in the cytosol that yields IPP for the biosynthesis of sterols, sesquiterpenes and ubiquinones.
- The *deoxyxylulose phosphate/methylerythritol phosphate* (*DOXP/MEP*) pathway in the chloroplasts that is responsible for the biosynthesis of carotenoids, chlorophylls (phytyl chain), plastoquinol-9 (**26**, nona-prenyl side chain), phylloquinone K_1 (**21**, phytyl side chain) and α-tocopherol (**27**, prenyl side chain) as well as isoprene, monoterpenes and diterpenes.

The DOXP/MEP pathway of chloroplasts was only detected in the early 1990s by the joint co-operation of the working groups of Lichtenthaler (Karlsruhe) and Rohmer (Mulhouse/Strasbourg) upon applying new labelling techniques.[23,25–27] When using (*a*) labelling with ^{13}C-glucose combined with high-resolution NMR spectroscopy and (*b*) deuterium-labelling complemented by mass spectroscopy, a new labelling pattern of plastidic isoprenoids was detected that did not at all correspond to a labelling via the cytosolic acetate/MVA pathway. This DOXP/MEP pathway of isoprenoid biosynthesis was shown to occur in all chloroplast-containing, photosynthetic oxygen-evolving organisms, such as green algae[28] and higher plants (first reported by Lichtenthaler *et al.*[23] and Zeidler *et al.*[29,30]) as well as in the free living prokaryotic cyanobacteria[25] which are supposed to have the same evolutionary ancestors as the endosymbiotic chloroplasts. Later, these findings were confirmed by several other research groups, as reviewed by Lichtenthaler[19,25,26] and Rohmer.[27] The only exception of a photosynthetic organism producing carotenoids from IPP synthesized via the acetate/mevalonate pathway is *Euglena* which, in the course of evolution, has lost its plastidic DOXP/MEP pathway of IPP biosynthesis.[16,25,31] Besides photosynthetic organisms, the DOXP/MEP pathway also occurs in the heterotrophic, mostly pathogenic eubacteria, such as *Escherichia coli, Haemophilus influenzae, Helicobacter pylori, Chlamydia pneumoniae, Mycobacterium tuberculosis* and *Vibrio cholerae*,[16,26] organisms which might have evolved from originally photosynthetic bacteria after loss of their photosynthetic competence.

2.4 Prenyl Chain Formation and Elongation

Both cellular isoprenoid pathways of photosynthetic organisms provide IPP which, via an isomerase, can be transferred to its isomer 3,3-dimethylallyl diphosphate (DMAPP, **28**, Scheme 2.1). By the action of the enzyme isoprene synthase, isoprene is directly formed from DMAPP. When the C_5 unit of IPP is added to the starter molecule DMAPP in a head to tail condensation, geranyl diphosphate (GPP, **29**), the C_{10} prenyl diphosphate, is formed. By head to tail condensation of one or two more IPP units to GPP, the C_{15} and C_{20} prenyl diphosphates are formed [i.e. farnesyl diphosphate (FPP, **30**) and geranylgeranyl diphosphate (GGPP, **31**); Scheme 2.1]. Sterols, being found in several cytosolic cell membranes, are synthesized via a tail to tail condensation of the C_{15} carbon skeletons of two FPP (dimerization), yielding first squalene (**32**) and then the triterpene cycloartenol (**14**, see Figure 2.2). The C_{30} isoprenoid structure is further modified, for example by the removal of three methyl groups, to form the final sterol ring structure. Carotenoids as tetraterpenoids are composed of eight isoprenic C_5 units. They are formed by a tail to tail condensation of two GGPP molecules followed by a successive desaturation of their C_{40} prenyl chain. This biogenetic relationship is summarized in Scheme 2.1. Higher prenyl chain homologues are synthesized via the addition of the C_5 carbon skeletons of several hundred or thousands of IPP molecules, a process which is the basis for the biosynthesis of polyterpenes, as found in the white latex of Euphorbiaceae and Asteraceae, for example in the latex of the natural rubber tree *Hevea brasiliensis* that primarily contains long-chain *cis*-1,4-polyisoprene molecules but also the plastidic isoprenoids plastoquinol-9 (**26**) and α-tocopherol (**27**).

2.5 Compartmentation of Plant Isoprenoid Biosynthesis

The compartmentation of isoprenoid biosynthesis in the plant cell is shown in Scheme 2.2. The *plastidic DOXP/MEP* pathway starts from glyceraldehyde-3-phosphate (GA-3-P, **33**, Figure 2.4) and pyruvate (**34**) with 1-deoxy-D-xylulose 5-phosphate (DOXP, **35**) as the first product that is transferred to DMAPP and IPP in six further enzymatic steps. By head to tail condensation the latter form GPP, to which more C_5 carbon skeletons of IPP are condensed to eventually form the isoprenoid chains of the different carotenoids and plastidic prenyl lipids, mono- and diterpenes, etc. In contrast, the *cytosolic acetate/mevalonate* pathway starts from 3-acetyl-CoA and provides IPP and DMAPP for cytosolic sterol biosynthesis, sesquiterpenes and polyterpernes as well as the prenyl side chains of ubiquinones Q-9 (**22**) and Q-10 (**23**). Both cellular isoprenoid pathways can specifically be inhibited: the acetate/mevalonate pathway by mevinolin (**25**) and other statins (target: HMG-CoA reductase),[20–22,32] whereas the DOXP/MEP pathway is specifically blocked by fosmidomycin (**36**; target: DOXP reductoisomerase, DXR)[26,30] and by 5-ketoclomazone (**37**; target: DOXP synthase, DXS),[26,33,34] as indicated in Scheme 2.2.

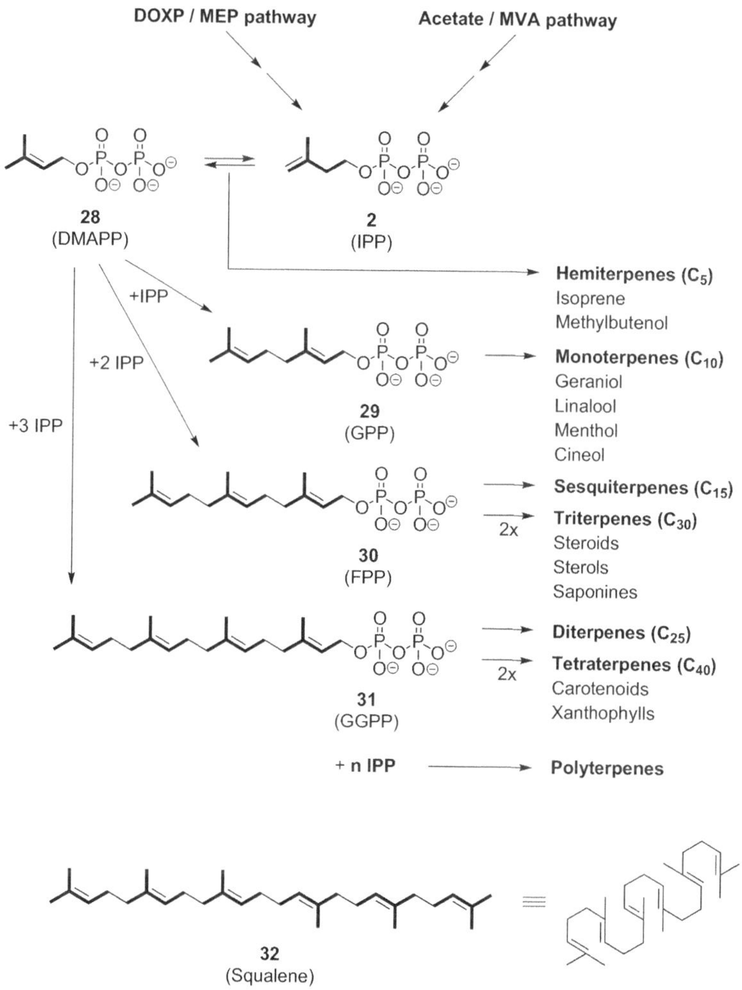

Scheme 2.1 *Biogenetic relationship of plant isoprenoids and terpenes being formed from the starter molecule dimethylallyl diphosphate (DMAPP) by head to tail condensation of one or more isopentenyl diphosphates (IPP). Depending on the type and cellular localization of the plant isoprenoid, the IPP molecules are derived either from the cytosolic acetate/mevalonate pathway or the plastidic DOXP/MEP pathway. Triterpenes and tetraterpenes are synthesized by dimerization (tail to tail condensation) of two farnesyl and geranylgeranyl diphosphates, respectively*[4,19]

2.6 The Enzyme Steps of the Plastidic DOXP/MEP Pathway of IPP Formation

The nonmevalonate DOXP/MEP pathway in plastids starting from GA-3-P (**33**) and pyruvate (**34**) requires seven enzymes, three ATP equivalents [adenosine-5′-triphosphate (ATP) or cytidine-5′-triphosphate (CTP)] and three nicotinamide adenine dinucleotide

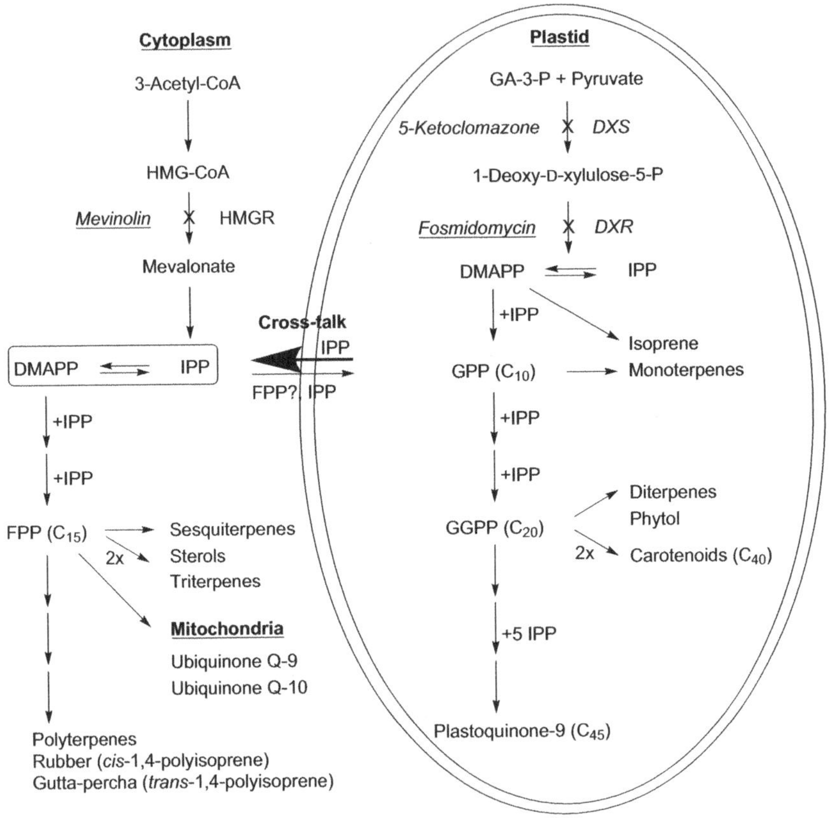

Scheme 2.2 *Compartmentation of the two isoprenoid biosynthesis pathways in the plant cell. (1) The chloroplastidic DOXP/MEP pathway for the biosynthesis of the active C_5 units (IPP) for chlorophylls (phytyl side chain), carotenoids, prenylquinones (isoprenoid side chains), mono- and diterpenes. (2) The cytosolic acetate/mevalonate pathway of IPP biosynthesis for the formation of sterols, sesquiterpenes, triterpenes, polyterpenes and the prenyl side chain of the mitochondrial ubiquinones Q-9 and Q-10. Specific inhibition is indicated for the acetate/mevalonate pathway by mevinolin (target: HMG-CoA reductase = HMGR), the DOXP/MEP pathway by 5-ketoclomazone (target: DOXP-synthase = DXS) and fosmidomycin (target: DOXP-reductase = DXR). Possible cross-talk between the two cellular biosynthetic isoprenoid pathways, which primarily consists of an export of active C_5 units from chloroplasts to the cytosol,[42] is accentuated*[23,25,30,33,34,42]

phosphates (NADPH), as shown in Scheme 2.3. The enzymes and their genes are well defined today. The DOXP synthase (DXS), a thiamine-dependent transketolase-type enzyme, is the *first enzyme* of this IPP-producing pathway. It condenses pyruvate with GA-3-P, yielding DOXP, the first C_5 chain of this pathway. 5-Ketoclomazone (**37**), an oxidation product of the herbicide clomazone, specifically inhibits the DOXP synthase.[33,34] *Enzyme 2* is the DOXP reductoisomerase (DXR) and reduces DOXP to 2-*C*-methylerythritol-4-phosphate (MEP, **38**).[33,35] This enzymic reaction comprises, besides the NADPH catalysed reduction step, an intramolecular rearrangement of the carbon

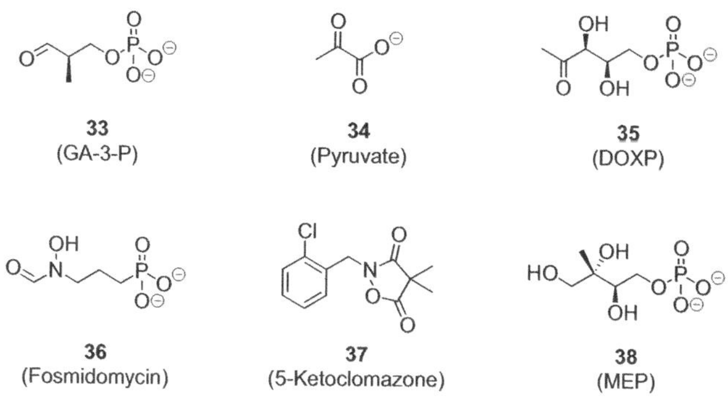

Figure 2.4 *Chemical structures of glyceraldehyde-3-phosphate (GA-3-P, **33**), pyruvate (**34**), deoxyxylulose phosphate (DOXP, **35**), 2-C-methylerythritol-4-phosphate (MEP, **38**) and enzyme inhibitors fosmidomycin (**36**) and 5-ketoclomazone (**37**)*

skeleton. The DOXP reductoisomerase is apparently the major regulatory step of the whole DOXP/MEP pathway. It is specifically blocked by fosmidomycin (**36**), as independently shown for plants[30,35] and for Eubacteria[36] that also possess the DOXP/MEP pathway. Fosmidomycin is a structural analogue of MEP, the intermediate in the enzymic reaction of DXR.

Enzyme 3, the CDP-ME synthase, catalyses the activation of MEP by CTP to form CDP-methyl-D-erythritol (CDP-ME, **39**, Figure 2.5).[37] *Enzyme 4*, CDP-ME kinase, phosphorylates CDP-ME to CDP-ME2P, which is transformed by *enzyme 5*, MEcPP synthase, to 2-*C*-methyl-D-erythritol-2,4-cyclo-diphosphate (MEcPP, **40**). *Enzyme 6*, HMBPP synthase, reduces MEcPP to 4-hydroxy-3-methyl-2-(*E*)-butenyl-diphosphate (HMBPP, **41**), as was independently verified by several groups.[38] *Enzyme 7*, the HMBPP reductase, was detected by several groups.[39] The HMBPP reductase (gene: *IspH*, formerly *lytB*) in one step yields two isoprenoid C_5 diphosphate products: IPP and its isomer DMAPP, usually at a ratio of 5:1 or 3:1. Depending on the cellular metabolic demand, IPP and DMAPP can be interconverted by a plastidic IPP isomerase.

All seven enzymes of the DOXP/MEP pathway have been isolated and their genes have been cloned in plants and eubacteria (as reviewed for the first five enzymes by Lichtenthaler[16,26] and for all seven enzymes by Rodríguez-Concepción and Boronat and by Eisenreich *et al.*[40]). These are *DXS*, *IspC* (*DXR*), *IspD* (*ygbP*), *IspE* (*ychB*), *IspF* (*ygbB*), *IspG* (*cpE*) and *IspH* (*lytB*), with the old names given in parentheses. The gene (*Idi*) for the plastidic IPP isomerase is also known. As in the case of many other chloroplast enzymes, the genes of the seven DOXP/MEP pathway enzymes are encoded by the nuclear genome, yet the final enzyme proteins operate in chloroplasts and also in non-green plastid forms. For this purpose the enzyme proteins of the DOXP/MEP pathway, being synthesized in the cytosol, possess a transit peptide sequence that directs them to their proper organelle, the chloroplast.

2.7 Cross-Talk Between the Two IPP Biosynthesis Pathways

There exists some cross-talk between the two cellular isoprenoid-forming pathways (see Scheme 2.2) that was first demonstrated by using cross-labelling of the plastidic isoprenoid

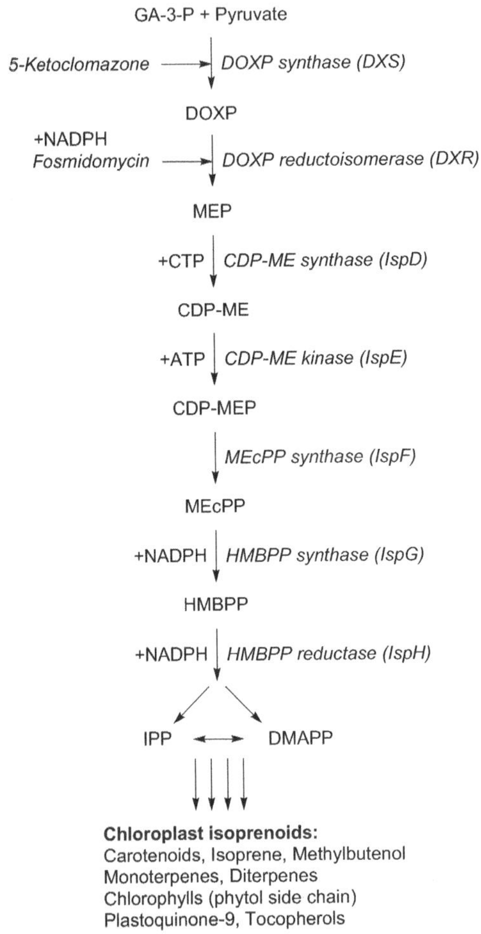

Scheme 2.3 *Biochemical sequence of the DOXP/MEP pathway of plastidic isoprenoid biosynthesis in plants yielding the active isoprenic C_5 units IPP and DMAPP used for biosynthesis of plastidic isoprenoids (chlorophylls, carotenoids) as well as isoprene, monoterpernes and diterpenes. The pathway, starting from pyruvate and GA-3-P, consists of seven enzymes and consumes three NADPH and three ATP equivalents to yield one active C_5 diphosphate (IPP or DMAPP). The first two enzymatic steps can specifically be inhibited by 5-ketoclomazone and fosmidomycin, respectively.[26,30,33] The genes of the seven enzymes have been cloned, their names are shown in italics in parentheses following the enzyme names. Via the enzyme IPP isomerase, whose gene (Idi) is also known, IPP and DMAPP can be converted into each other*

phytol (**9**, side chain of chlorophylls) and the cytosolic sterols from precursors of the plastidic DOXP/MEP pathway and the cytosolic mevalonate pathway in two algae and a higher plant (*Lemna gibba*) under photosynthetic conditions. When ^{14}C-labelled deoxyxylulose (^{14}C-DOX, ^{14}C-**42**, Figure 2.6) and tritium-labelled mevalonolactone (^{3}H-MVL, ^{3}H-**43**) were applied as tracers, the degree of radioactivity incorporation in phytol and sterols was determined. Phytol and sterols were labelled by their proper precursor

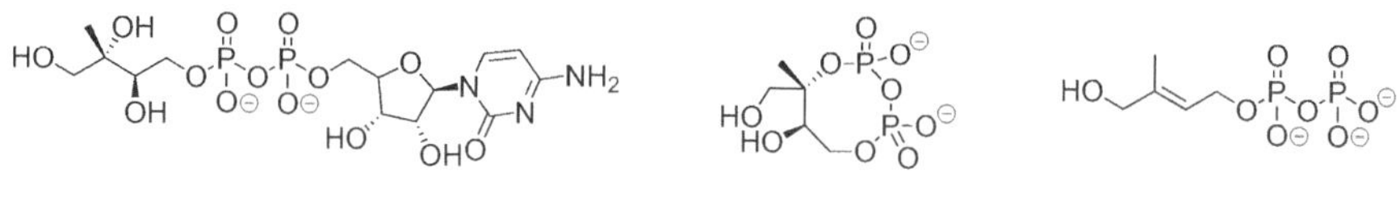

39
(CDP-ME)

40
(MEcPP)

41
(HMBPP)

Figure 2.5 *Chemical structures of CDP-methyl-D-erythritol (CDP-ME,* ***39****), 2-C-methyl-D-erythritol-2,4-cyclo-diphosphate (MEcPP,* ***40****) and 4-hydroxy-3-methyl-2-(E)-butenyl diphosphate (HMBPP,* ***41****)*

substances ^{14}C-DOX and ^{3}H-MVL.[41] The rather good labelling of the cytosolic sterols from ^{14}C-DOX indicated an export of isoprenoid units from the chloroplast to the cytosol. A labelling of phytol from ^{3}H-MVL occurred only at very low rates[42] indicating that under photosynthetic conditions the cross-talk worked primarily in a chloroplast to cytosol direction. Other approaches were performed by several authors using specific inhibitors of the two cellular IPP-producing pathways (e.g. mevinolin (**25**) on the one hand and fosmidomycin (**36**) or 5-ketoclomazone (**37**) on the other hand), as reviewed by Lichtenthaler.[43] However, an adequate compensation for the inhibited activity of one cellular isoprenoid pathway by the second isoprenoid pathway of the other cell compartment apparently did not and does not occur under physiological standard conditions of growth.

It is evident from the currently available data of several laboratories that, under photosynthetic conditions, there occurs a unidirectional, relatively high flow of plastidic C_5 isoprenoid units into cytosolic sterol and sesquiterpene or polyterpene biosynthesis, as shown and reviewed by Lichtenthaler[42] and Nagegowda *et al.*[44] This export of isoprenoid C_5 units into cytosolic isoprenoid biosynthesis is presumably mediated by the plastidial unidirectional proton symport system[45] that can transport IPP and geranyl diphosphate (GPP, **29**) in a plastid to cytosol direction. Thus, the biomembranes of the chloroplast envelope play an essential role in mediating the cross-talk between both IPP-producing pathways. In addition, it is well known that the chloroplast envelope biomembranes also possess many other biosynthetic capacities and mechanisms[46] to transport various compounds, for example triose phosphates and phosphates, pyruvate and organic acids (malate, oxalacetate, glutamate) via special transporter/antiporter systems (triosephosphate/phosphate translocator, dicarboxylate transporter, phosphoenolpyruvate/phosphate transporter) from the chloroplast to the cytosol and vice versa.

In illuminated green leaves chloroplasts yield the starter substances GA-3-P and pyruvate in the course of the photosynthetic CO_2 fixation process (Calvin cycle) and in the

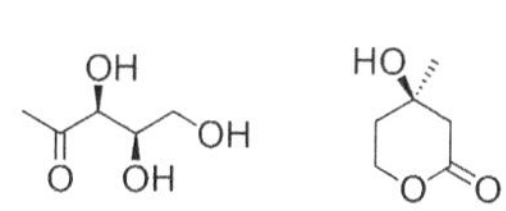

42
(DOX)

43
(MVL)

Figure 2.6 *Chemical structures of 1-deoxy-D-xylulose (DOX,* ***42****) and mevalonolactone (MVL,* ***43****)*

photosynthetic light reactions the three ATP and three NADPH required for plastidic IPP biosynthesis. Under light conditions green plant cells use this efficient IPP biosynthesis program of chloroplasts to also provide, via a fast export system, the C_5 building blocks (IPPs) required for the cytosolic sterol and sesquiterpene biosynthesis.[42,43] In contrast, in the dark and apparently also in most cases in non-green plant tissues, the cytosolic acetate/mevalonate pathway is fully responsible for supplying the IPP molecules for sterol and sesquiterpene biosynthesis. However, there are exceptions to this general rule. Thus, Dudareva *et al.*[47] unambiguously demonstrated that, in the chlorophyll-free epidermis of snapdragon petals, the DOXP/MEP pathway of the non-green plastids (leucoplasts) provided the IPP precursors not only for the synthesis of volatile plastidic monoterpenes, but also for the biosynthesis of cytosolic volatile sesquiterpenes. In addition, Hampel *et al.*[48] demonstrated in carrot leaves and carrot roots that monoterpenes are synthesized exclusively via the DOXP/MEP pathway, whereas the C_5 carbon skeletons of cytosolic sesquiterpenes are generated via the classical MVA pathway as well as via the plastidic DOXP/MEP pathway.

2.8 Biosynthesis and Emission of Volatile Isoprene at High Irradiance

The branched chain hydrocarbon isoprene (**1**) is emitted by a great number of higher plants and also by mosses, ferns and gymnosperms.[6] The isoprene emission by plants was independently discovered in the late 1950s and at the beginning of the 1960s by Sanadze and by Went.[49] In evolutionary terms, isoprene emission appears to be a very old trait of plants that may have been common to all photosynthetically active algae and the first land plants. During further evolution isoprene emission was apparently lost in many plants, but was retained in other plants as a major mechanism for survival during high-temperature, high-irradiance and other stress conditions. Its emission from plants amounts to hundreds of millions of metric tonnes to the global atmosphere; the estimates range from 180 to 450×10^{12} g carbon year^{-1} worldwide.[50] It is of great interest in this respect that more organic carbon is lost from plants as isoprene than as any other volatile plant molecule.[8] The light- and temperature-dependent isoprene emission by leaves preferentially occurs at high rates at temperatures above 28 °C and at high irradiance, such as full sun light, when the photosynthesis process with its light reactions and associated electron transport reactions is almost fully light-saturated.

Trees are responsible for the highest rates of isoprene emission of green leaves, for example *Acacia nigrescens* (African acacia), *Eucalyptus globulus* (blue gum), *Liquidambar styraciflua* (sweet gum), *Populus nigra* (black poplar) and other *Populus* species, *Quercus robur* (European oak, pedunculate oak), *Quercus coccinea* (scarlet oak), *Salix babylonica* (weeping willow) and other plants such as *Arundo donax* (giant reed), *Myrtus communis* (myrtle) and *Pueraria* spp. (kudzu), with emission rates of 30 to 300 μg isoprene g^{-1} dw h^{-1}.[4,51]

Isoprene is synthesized via the DOXP/MEP pathway and requires three ATP and three NADPH molecules (Scheme 2.4).[29,52] With its two conjugated double bonds it has a typical UV spectrum, and its time-dependent continuous emission at high light conditions can easily be measured in plant leaves via a photometric UV cuvette test system using leaf pieces (Figure 2.7).[53] A leaf section in the closed UV cuvette with its main leaf vein

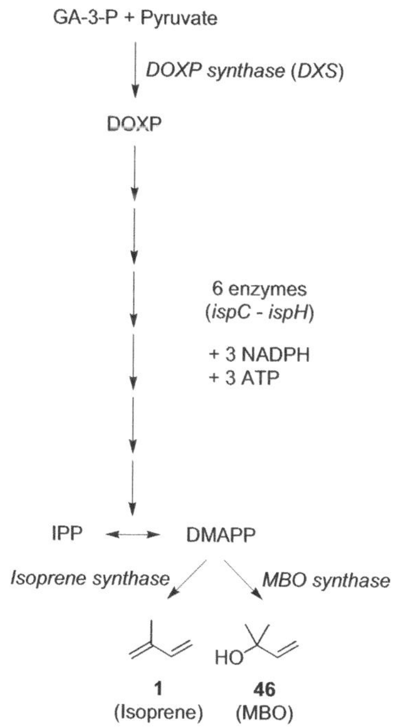

Scheme 2.4 *Biosynthesis of the volatile hemiterpene isoprene in chloroplasts from pyruvate and GA-3-P via the DOXP/MEP pathway. Seven enzymes (genes) are involved in the biosynthesis of the active isoprenoid C_5 unit IPP, and the cofactor requirements are three ATP and three NADPH. Isoprene (**1**) and MBO (**46**) are set free from the IPP isomer DMAPP through the action of plastidic hemiterpene synthases*

immersed in water is exposed from the side, perpendicular to the leaf surface, to high irradiance (above 2000 μmol m^{-2} s^{-1}) and a temperature above 30 °C (e.g. water bath). In a spectrophotometer one can register, via the measuring light beam that passes through the cuvette unhindered by the leaf, the continuous emission of isoprene after one, two or four hours. In this test system one can investigate the leaf capacity of different plants for isoprene emission. In addition, it can be checked if leaves of particular plants are able to produce isoprene or not. Another method to prove the isoprene emission of leaves, including various conifer needles, is the collection of the emitted isoprene gas in a closed system and its injection into a gas chromatography system combined with a mass spectrometer (GC-MS).[53]

There occurs a rapid path of photosynthetically fixed carbon into isoprenoid hydrocarbons, as shown by the rapid appearance of ^{14}C activity (after applying $^{14}CO_2$) in prenyl chains, such as carotenoids and the phytyl chain of chloroplasts, for example as shown for the green alga *Chlorella*.[54] This also applies to isoprene. Under high irradiance and at a temperature above 30 °C isoprene is instanteously formed de novo. In fact, the fast appearance of ^{13}C-labelling in isoprene from photosynthetically fixed $^{13}CO_2$ suggested that isoprene biosynthesis must be closely connected to intermediates of the photosynthetic

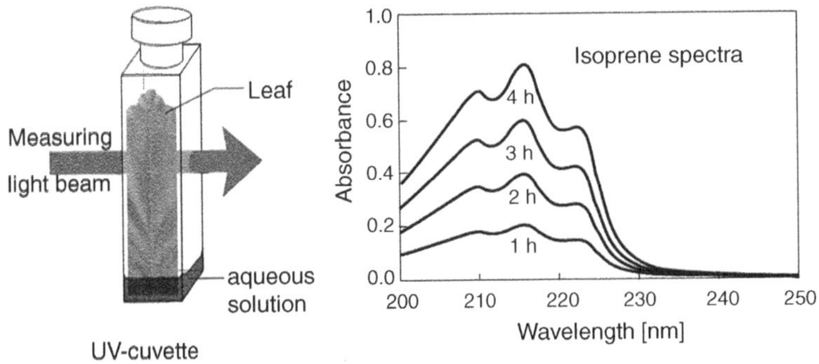

Figure 2.7 *Measurement of isoprene emission by leaves via the UV cuvette system. The isoprene emitted at high light (> 1500 µmol m^{-2} s^{-1}) and high temperature (> 30 °C) conditions is collected in a closed UV cuvette and its time-dependent emission measured in a spectrophotometer*[53,60]

CO_2 fixation pathway, known as the Calvin cycle.[55,56] In 1997 we demonstrated that the biosynthesis of isoprene proceeds via the plastidic DOXP/MEP pathway for isoprenoid biosynthesis from GA-3-P and pyruvate.[29,52] GA-3-P is an early product and an intermediate in the photosynthetic Calvin cycle. In several plants, for example spinach, pyruvate can directly be formed in chloroplasts from phosphoglyceric acid (PGA, **44**, Figure 2.8), another early product of the Calvin cycle.[57]

After detecting the plastidic DOXP/MEP pathway of IPP formation, we showed the specific incorporation of deuterium-labelled 1-deoxy-D-xylulose (^{2}H-DOX, ^{2}H-**42**) in the form of its xyluloside (methylglycoside) into isoprene and also into phytol, the prenyl side chain of chlorophylls (Scheme 2.5). This incorporation was verified via GC-MS and via high-resolution NMR spectroscopy.[29,52,53] DOX, the nonphosphorylated precursor substance, is rapidly phosphorylated in the plant cell by a cytosolic enzyme into DOXP (**31**), the actual intermediate precursor of the plastidic isoprenoid pathway.[58] DOXP can be transported from the cytosol into chloroplasts by the xylulose-5-phosphate translocator, a recently discovered plastidic transporter for pentose phosphates.[59]

Deuterium-labelled 1-deoxy-D-xylulose (^{2}H-DOX) was readily incorporated into isoprene by the green leaves of *Chelidonium majus* (greater celandine), *Eucalyptus globulus* (blue gum), *Platanus x acerifolia* (plane tree), *Populus nigra* (poplar), *Robinia pseudoacacia* (locust) and *Salix viminalis* (common willow), whereby a labelling degree of 30–80% was achieved.[60] This was proved using the GC-MS technique, whereby isoprene and other volatile

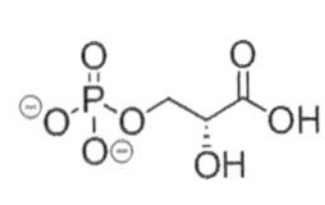

44a
(3-PGA)

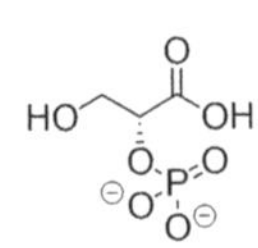

44b
(2-PGA)

Figure 2.8 *Chemical structures of 3- and 2-phosphoglyceric acid (PGA, **44a** and **44b**)*

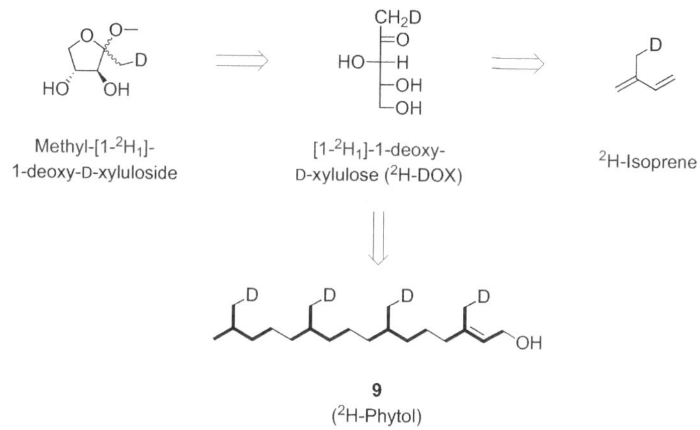

Scheme 2.5 *Specific incorporation of deuterium-labelled [1-^{2}H]-deoxyxylulose (D-DOX) into the volatile hemiterpene isoprene and phytol (side chain of chlorophylls). The deuterium label was found in the expected and specific methyl group of isoprene and in the four methyl groups of phytol, which proved DOX and DOXP to be a precursor and an intermediate, respectively, of the pastidic DOXP/MEP pathway[25,52] (D = deuterium label)*

isoprenoids emitted were collected by means of the solid phase microextraction (SPME) method with one drop of decane hanging in the closed experimental cuvette. In contrast, in blue gum leaves ^{2}H-DOX was not incorporated into the monoterpene 1,8-cineol which, after isoprene, was the next most frequent volatile isoprenoid.[60] This indicates that isoprene is spontaneously formed and emitted, whereas the emitted monoterpene cineol had been synthesized before and was released at high-light and high-temperature conditions apparently from a depot within the leaves.

The final step of the DOXP/MEP pathway yields both isoprenoid C_5 diphosphates, IPP and its isomer DMAPP. Isoprene is set free from DMAPP in a single enzymatic step via the plastidic isoprene synthase.[61] The latter exists in a thylakoid-bound form[62] and also as isoforms in the chloroplast stroma.[63] The enzyme isoprene synthase, a hemiterpene synthase, is related to monoterpene synthases found in other plants.[64] Its K_m is 10- to 100-fold higher for its DMAPP substrate than the related monoterpene synthases for their substrate geranyl diphosphate (GPP).[65] These findings indicate that isoprene biosynthesis apparently starts only when, under particular environmental conditions (e.g. high irradiance, heat stress), the concentration of DMAPP in the chloroplast rises to considerably higher levels. This assumption would also mean that isoprene biosynthesis starts when other processes that use either GA-3-P and pyruvate or IPP and DMAPP slow down, with the consequence that higher levels of IPP and DMAPP accumulate, finally starting the isoprene synthase.

2.8.1 Regulation of Isoprene Emission

The regulation of the light- and temperature-dependent isoprene emission apparently proceeds via the relative activity of the DOXP/MEP pathway and possibly via the concentration of DMAPP.[65,66] Moreover, in grey poplar leaves isoprene synthase and the second enzyme of the DOXP/MEP pathway, DXR, show distinct seasonal patterns and peak in the summer,[67] thus suggesting that the metabolic carbon flux through the DOXP/MEP pathway and isoprene emission are closely intercoordinated. Studies on the natural ^{13}C-carbon isotope composition of isoprene in several plants confirmed that isoprene is synthesized de novo from the currently formed primary photosynthates, yet a low percentage of carbon came from another carbon source,[68] possibly from cytosolic pyruvate imported to the chloroplast to be joined with GA-3-P in order to form DOXP, the first intermediate in the DOXP/MEP pathway (see Scheme 2.3). When photosynthetic carbon fixation was inhibited by CO_2-free air in these studies, the contribution of this alternative carbon source increased. Also, other leaf internal carbon pools (e.g. starch or xylem-fed labelled glucose) can be used as alternative carbon sources for isoprene emission, especially when, after abscisic acid application, the stomata close and CO_2 for photosynthetic carbon fixation is missing.[56] Xylem-transported glucose as an additional carbon source for leaf isoprene formation was also verified in pedunculate oak *Quercus robur*.[69] In this oak it was also demonstrated that isoprene synthase activity correlated well with the isoprene emission rates observed.[70] Recent research deals with various aspects of isoprene formation and the regulation of isoprene enzymes. Thus, the relationship of IPP isomerase activity (transforming IPP to DMAPP) with isoprene emission have been studied,[71] as well as the circadian rhythms of isoprene biosynthesis in grey poplar leaves[72] and the fact that transgenic, non-isoprene-emitting poplars 'don't like it hot'.[73] These transgenic non-isoprene-emitting poplars showed reduced rates of net assimilation and photosynthetic electron transport during heat stress, but not in the absence of stress, whereby the decrease in the efficiency of photochemistry was inversely correlated with the increase in heat dissipation of absorbed light energy, measured as nonphotochemical quenching (NPQ) of chlorophyll fluorescence. The downregulation of isoprene emission affected the thermotolerance of photosynthesis and induced an increased energy dissipation and, as expected, an increased formation of the xanthophyll cycle pigment zeaxanthin. In addition, studies were performed to link isoprene emission with plant thermotolerance, antioxidants and monoterpene emission.[74]

2.9 Inhibition of Isoprene Biosynthesis

2.9.1 Fosmidomycin and 5-Ketoclomazone

Biosynthesis and emission of isoprene in illuminated plant leaves is efficiently blocked by the herbicide fosmidomycin (**36**, see Figure 2.5) that specifically inhibits the second enzyme DXR of the plastidic DOXP/MEP pathway of IPP and isoprenoid formation, as verified also in the UV cuvette test system shown in Figure 2.9.[30] 5-Ketoclomazone (**37**), the specific inhibitor of the plastidic DOXP synthase, also efficiently blocks isoprene formation, as shown in the UV cuvette test system.[26,33,34,60] In these tests the leaf section is dipped either into the cuvette for 6–10 h at low light and room temperature conditions (below 24 °C) or overnight into the inhibitor solution and the controls into water. At temperatures above 30 °C and high irradiance (above 1200 $\mu mol\,m^{-2}\,s^{-1}$) a dose-dependent inhibition of isoprene emission is detected.

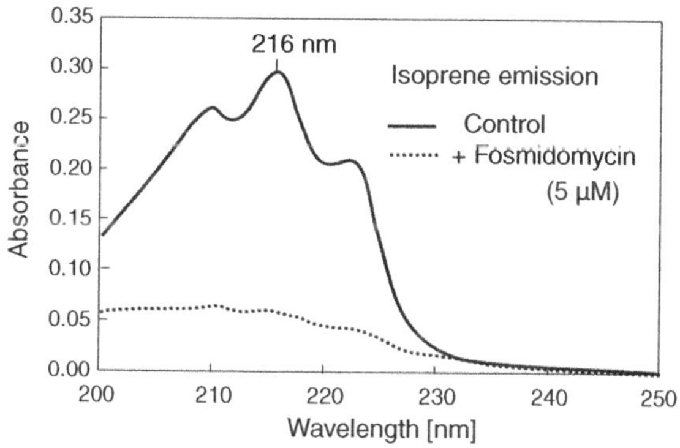

Figure 2.9 *Inhibition of light-induced isoprene emission in leaf pieces of Platanus x acerifolia by fosmidomycin (target: DOXP-reductase). The inhibitor (5 µM) was applied to the leaf in the dark in the cuvette test 12 h before starting the illumination. The amount of isoprene emission can be determined by measuring its UV spectrum in the closed cuvette*

2.9.2 Diuron

Using the herbicide diuron (DCMU, **45**, Figure 2.10), we demonstrated that the photosynthetic light reactions and associated electron transport reactions with their ATP and NADPH synthesis are as essential for isoprene biosynthesis and emission as the precursor substances GA-3-P and pyruvate. Diuron is known to specifically block the electron transport at the photosynthetic photosystem 2 by binding to the D1-protein instead of the endogenous electron acceptor plastoquinone-9,[75] an inhibition that blocks the formation of ATP and NADPH. The effect of diuron on isoprene emission was tested in the leaves of several plants, such as *Platanus x acerifolia*, *Chelidonium majus*, *Populus nigra* and *Populus alba x tremula*. After 24 h pre-incubation with diuron, by dipping the lower leaf part into a 10^{-5} M diuron solution, the isoprene emission was 70–80% inhibited. When the leaves were dipped into a 10^{-4} M diuron solution, the inhibition of the high-irradiance-induced isoprene emission was 100%.[60]

2.10 Inhibition of Carotenoid and Chlorophyll Biosynthesis by Fosmidomycin and 5-Ketoclomazone

The photosynthetic pigments, chlorophylls and carotenoids, are bound to thylakoids, the photosynthetic biomembranes.[76] In this context, it is of interest that, similar to isoprene

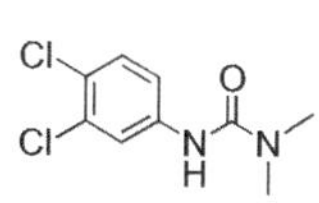

45
(Diuron)

Figure 2.10 *Chemical structure of the herbicide diuron (**45**)*

inhibition, the two specific inhibitors of the DOXP/MEP pathway of IPP biosynthesis, fosmidomycin (**36**) and 5-ketoclomazone (**37**), also affect and block the biosynthesis of carotenoids and chlorophylls (phytyl side chain) in chloroplasts in a dose-dependent manner.[30,33,34] This was investigated in etiolated barley seedlings which were dipped overnight with their cut ends into inhibitor solutions and the controls into water. During subsequent light exposure both inhibitors caused a dose-dependent, considerable reduction of the biosynthesis and accumulation of carotenoids and chlorophylls. In fact, the greening process of the leaves was retarded and an incomplete photosynthetic apparatus was obtained.[43] The latter is in the biosynthetic transition stage, as indicated by high values of the ratio of chlorophyll *a* to chlorophyll *b* (Chl *a*/*b* ratio) of 6.0 to 10.0, whereas normal values of green leaf tissue are in the range 2.7 to 3.3. The inhibition also causes a considerable increase in the weight ratios of total chlorophylls $(a+b)$ to total carotenoids $(x+c)$, with x corresponding to xanthophylls; and the values of $(a+b)/(x+c)$ range from 10.0 to 19.0. Green leaf tissue exhibits values of $(a+b)/(x+c)$ in the range 4.3 to 7.0, indicating an even stronger inhibition of carotenoid accumulation. For comparison, the normal pigment ratios of photosynthetic pigments in green leaf tissue,[42,77] and the changes in pigment ratios during the greening process are reported in the literature.[78]

2.11 Biosynthesis and Emission of Methylbutenol at High Irradiance

The needles of several pines of western North America (*Pinus ponderosa*, *P. contorta*, *P. sabiniana*) do not emit isoprene itself, but its partially oxidized form, the hemiterpene 2-methylen-3-buten-2-ol (MBO, **46**).[79] The MBO structure is shown in Scheme 2.4. Like isoprene, MBO is volatile and emitted in a light- and temperature-dependent manner at higher temperatures (heat stress) and high irradiance. Similar to isoprene, MBO can have a significant influence on the oxidative capacity of the atmosphere, for example through the consumption of hydroxyl radicals.

MBO is synthesized via the plastidic DOXP/MEP pathway, which was demonstrated by a high-rate incorporation of deuterium-labelled deoxy-D-xylulose (^{2}H-DOX) into MBO, as verified by mass spectrometry.[60,80] Incorporation of ^{2}H-DOX into MBO was studied in needles of *Pinus ponderosa* after 14 days preincubation with ^{2}H-DOX. The degree of MBO-labelling was at least 50% and was determined via collection of MBO with the SPME drop method followed by GC-MS. In addition, the formation of MBO is inhibited by fosmidomycin, which specifically blocks the DOXP reductase, the second enzyme in the plastidic DOXP/MEP pathway,[30] and also via 5-ketoclomazone, which specifically blocks the DOXP synthase.[26] Moreover, when offering ^{13}C-labelled MVA to ponderosa pine needles we did not find any incorporation of the ^{13}C-label into MBO, as checked using mass spectrometry. Like isoprene, the hemiterpene MBO is not formed from IPP but from its isomer DMAPP; this occurs in one step by the action of the enzyme MBO synthase. The two hemiterpene synthases, isoprene synthase and MBO synthase, use the same substrate DMAPP (Scheme 2.4), but the chemical mechanism for cleavage of the C_5 carbon structure from the diphosphate of DMAPP is different,[42] resulting in different endproducts.

Various conditions for MBO emission have been studied in detail. At temperatures above 25 °C the MBO emission and photosynthetic rates of ponderosa pine needles increased with light intensity and neither process showed any light saturation, even at an irradiance of 2000 $\mu mol\ m^{-2}\ s^{-1}$, which is close to full sunlight.[81] Water stress (closure of stomata) strongly reduced the photosynthetic rates but had no effect on MBO emission.

2.12 Source of Pyruvate for Isoprene and Methylbutenol Biosynthesis

Isoprene and methylbutenol are both synthesized from DMAPP produced via the DOXP/MEP pathway in chloroplasts. As volatile compounds they are not stored in the leaf. The rather large amounts of isoprene and MBO emitted by herbaceous plants as well as broadleaf trees and pine forests at higher irradiances and elevated summer temperatures derive from spontaneous de novo biosynthesis under photosynthetic conditions starting from GA-3-P, an intermediate of the Calvin cycle, and pyruvate (Scheme 2.6). Pyruvate in chloroplasts can derive from three sources. It can be formed in the chloroplast, at least in some plants such as spinach, directly from photosynthetically fixed carbon, that is from the first photosynthetic product 3-phosphoglyceric acid (3-PGA, **44a**) via 2-phosphoglyceric acid (2-PGA, **44b**) and phosphoenol pyruvate (PEP, **47**, Figures 2.8, 2.11).[57] In addition, pyruvate can be imported to chloroplasts from the cytosol where it is formed in the glycolytic pathway. As long as photosynthetic CO_2 fixation works well, both volatile hemiterpenes are formed from newly synthesized GA-3-P (Calvin cycle) and pyruvate that derives from its successive oxidation and transformation. However, pyruvate is also formed as a byproduct of ribulosebisphosphate carboxylase/oxygenase (RubisCO) activity.[82] Theoretically there is the possibility that, at higher temperatures and a certain shortage of CO_2, the enzyme RubisCO may yield a higher amount of pyruvate. This means that a higher yield of pyruvate by an enhanced RubisCO activity could stimulate the DOXP/MEP pathway as well as the biosynthesis and emission of isoprene and MBO. In fact, when the endogenous level of pyruvate in chloroplasts is increased, as shown below in the experiment with the inhibitor acetylmethyl phosphinate (AMPI, **48**, Figure 2.11),[60] this resulted in enhanced isoprene emission. However, the hypothesis that RubisCO at high-irradiance conditions, higher temperatures and lower CO_2 levels produces more pyruvate has yet to be analytically investigated and proved.

However, the particular environmental conditions of isoprene and MBO emission lead sooner or later to a certain water stress and, as a consequence, to partial or full closure of the leaf stomata. Since, under water stress and stomata closure, the CO_2 assimilation via the Calvin cycle is successively reduced, the major part of the starter molecules of the DOXP/MEP pathway (i.e. GA-3-P and pyruvate) must then come from sources other than the process of photosynthetic carbon fixation and reduction. In contrast to photosynthetic CO_2 assimilation, the two photosynthetic light reactions and associated electron transport reactions that provide ATP and NADPH for the DOXP/MEP pathway still take place even at water stress and stomata closure. At such conditions the main carbon sources for isoprene and MBO biosynthesis are: (*a*) GA-3-P from the breakdown of starch and (*b*) pyruvate that is formed by a breakdown of GA-3-P. Most probably pyruvate is formed in the cytosol by glycolytic breakdown of GA-3-P and is imported by chloroplasts. The question has yet to be determined as to whether chloroplasts of isoprene-emitting leaves and ponderosa pine needles possess the competence to form pyruvate directly from GA-3-P, as may be possible in some plants.

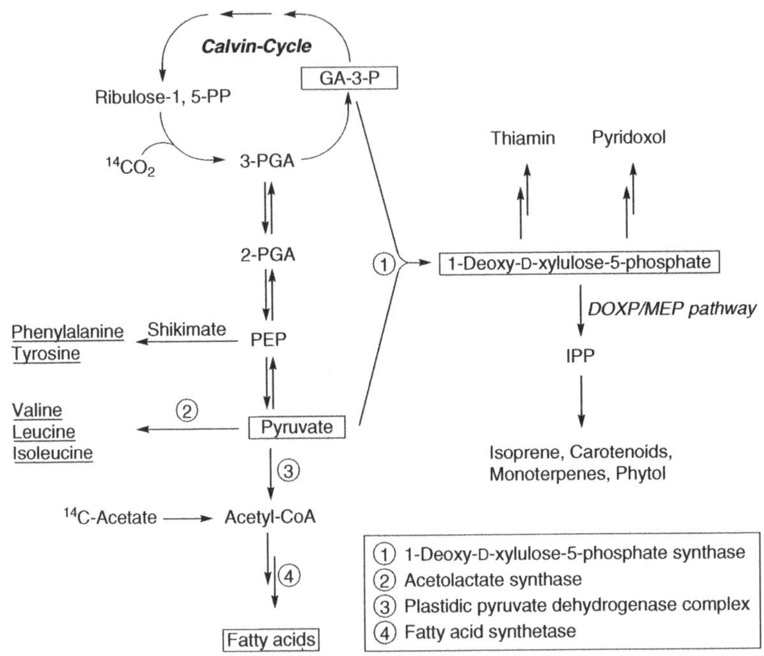

Scheme 2.6 *Biosynthetic pathways in chloroplasts with branching points for photosynthetic intermediates and products. The flow is indicated for carbon metabolites from the photosynthetic CO_2 reduction cycle (Calvin cycle) into different end-products, such as IPP and isoprenoids, or into fatty acids and amino acids, as well as from DOXP into thiamine and pyridoxal. The central role of GA-3-P and pyruvate in the DOXP/MEP pathway for biosynthesis of IPP and plastidic isoprenoids is emphasized. In some plants pyruvate can directly be made in chloroplasts from photosynthetically formed 3-PGA; in others it comes from the cytosol and is transported into the chloroplast*[25]

2.13 Branching Point of DOXP/MEP Pathway with Other Metabolic Chloroplast Pathways

Besides the photosynthetic carbon reduction cycle (Calvin cycle, Scheme 2.6), the performance of the photosynthetic light reactions (ATP, NADPH) and the DOXP/MEP pathway for isoprenoid biosynthesis, chloroplasts possess several other biosynthetic activities, such

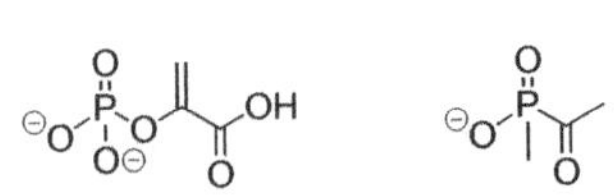

47
(PEP)

48
(AMPI)

Figure 2.11 *Chemical structures of phosphoenol pyruvate (PEP,* ***47****) and acetylmethyl phosphinate (AMPI,* ***48****)*

as the biosynthesis of chlorophyll (porphyrin ring), de novo fatty acid biosynthesis[83] and also the biosynthesis of particular amino acids, such as aromatic amino acids and branched-chain amino acids. Some of these pathways compete for the same substrates and are dependent on the carbon metabolite flow from intermediates or products of the Calvin cycle (Scheme 2.6).

In view of the operation of the plastidic DOXP/MEP pathway, in which GA-3-P and pyruvate are direct substrates of DOXP synthase, the early observations are understood very well today, confirming that ^{14}C-labelled CO_2, GA-3-P and pyruvate are better precursors of chloroplast isoprenoids than ^{14}C-MVA or ^{14}C-acetate.[84] The two latter compounds are no substrates of the DOXP/MEP pathway, and within the chloroplasts ^{14}C-acetate is rapidly incorporated into fatty acids via the plastidic de novo fatty acid synthetase.[83] In contrast, photosynthetically fixed $^{14}CO_2$ is rapidly transformed in the Calvin cycle to 3-PGA and reduced to GA-3-P.[85] For the reasons mentioned above, the ^{14}C-label of $^{14}CO_2$ rapidly appears via GA-3-P and pyruvate in β-carotene and various other plastidic isoprenoids, as described by various authors.[54] Also, the fast ^{13}C-labelling of isoprene from photosynthetically fixed $^{13}CO_2$ is easily understood in view of the DOXP/MEP pathway that is fed by products of the Calvin cycle.[55,56]

Two other biosynthetic pathways in chloroplasts compete for distant products of the carbon reduction cycle. Thus, PEP (**47**) is a substrate of the shikimic acid pathway that yields the aromatic amino acids phenylalanine and tyrosine. Pyruvate, in turn, is also a substrate of the biosynthesis of branched-chain amino acids, such as isoleucine, leucine and valine, a pathway that also proceeds in chloroplasts. Moreover, when the plastidic pyruvate dehydrogenase complex yielding acetate is blocked by the specific inhibitor AMPI (**48**),[86] as shown in Scheme 2.7, biosynthesis and emission of the isoprene is enhanced by 30% at 5×10^{-4} M AMPI and by 60% at 10^{-4} M AMPI applied to leaf pieces.[60] These results indicate that there exist several branching points in the metabolite flow of Calvin cycle products to pyruvate and different other chloroplast products. This requires a very special fine regulation of chloroplast metabolism concerning the use of the photosynthetic products. This fine regulation must respond in the daily changing climatic conditions (temperature, irradiance) to the availability of CO_2 (e.g. open and closed stomata). In any case, enhanced or reduced isoprene emission can be a very effective valve for a fine-tuning of carbon flow and chloroplast metabolism.[42] In addition, there also exist particular regulations of the activity of the DOXP/MEP pathway and its enzymes, as recently reviewed.[87]

2.14 Is There a Physiological Function of Isoprene and MBO Emission?

The biosynthesis and emission of both volatile plant hemiterpenes depend on the chloroplastidic production of DMAPP. In fact, plant species with the highest potential for isoprene and MBO production also possess an elevated light-dependent production of DMAPP.[66] The physiological meaning of the emission of isoprene and MBO is not yet fully clear. Isoprene provides the leaves with a certain thermotolerance to heat damage.[88] In addition, by functioning as a potential scavenger of radicals both volatile hemiterpenes can protect thylakoid lipids and various chloroplast constituents from ozone and other reactive oxygen species,[68,89–91] thereby preventing photo-oxidation of the photosynthetic apparatus at high irradiance conditions. When in *Phragmites australis* leaves isoprene formation was

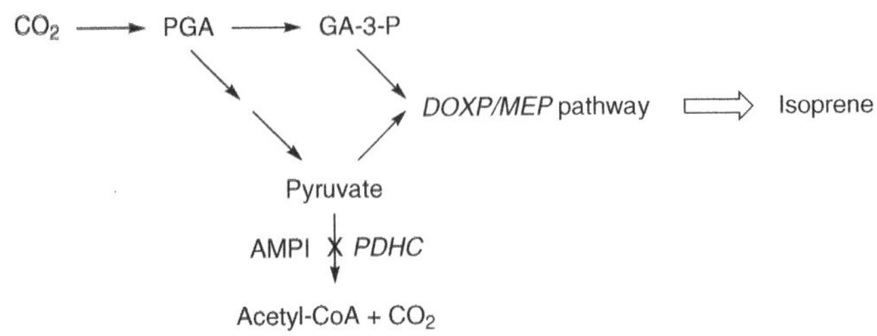

Scheme 2.7 *Enhanced flow of pyruvate into isoprene when the plastidic pyruvate dehydrogenase is inhibited by AMPI,*[60,86] *a specific inhibitor of the plastidic pyruvate dehydrogenase*

blocked by the inhibitor fosmidomycin, leaves became sensitive to ozone, as seen in the decline of photosynthesis, stomatal conductance and chlorophyll fluorescence parameters and the start of membrane lipid peroxidation.[90] These results clearly indicate that isoprene biosynthesis and emission exert a protective effect on the photosynthetic apparatus and its biomembranes.

Concerning the mechanism how isoprene exerts this protective effect, one also has to consider that the enhanced de novo biosynthesis of isoprene and MBO requires a continuous supply of ATP and NADPH being formed in the photosynthetic light reactions (see Scheme 2.4), a process which keeps the two photosynthetic photosystems 'busy' and intact by avoiding overreduction and photo-oxidative damage under excess light conditions. Thus, isoprene and MBO emissions may be a 'safety valve', similar to the process of photorespiration, to protect the photosynthetic pigment apparatus with its photosystems and light-harvesting pigment proteins[92] from photo-oxidation. The only disadvantage is that isoprene and MBO emissions are a waste of the previously photosynthetically fixed reduced carbon. Although the ATP and NADPH consumption through isoprene and MBO biosynthesis should not be overestimated, it essentially contributes, besides other mechanisms, to the stability of the photosynthetic pigment apparatus under high irradiance conditions.

Recently Rosenstiel *et al.*[66] proposed a special hypothesis why plants emit isoprene: the isoprene synthase converting DMAPP to isoprene and pyrophosphate would prevent DMAPP to rise to such high levels that would unnecessarily sequester phosphate. However, this concept is not convincing and should be ignored. The starting point of this hypothesis is the assumption that plants would accumulate more and more DMAPP at high-light and high-temperature conditions. However, this hypothesis fully neglects the fact that, from DMAPP and its isomer IPP, various other isoprenoid compounds can be and are produced de novo under high light- and heat-stress conditions, such as β-carotene (**15**) and the xanthophyll zeaxanthin (**16**), which accumulate at high rates, as has been proved for several plants.[42] Zeaxanthin is known to play an essential role in the heat dissipation of absorbed light energy by disconnecting the photosynthetic light-harvesting pigment proteins from the photosynthetic reaction centres. Moreover, various plants that have no competence for isoprene emission in their leaves accumulate nonvolatile diterpenes under such conditions, for example different ginkgolides in the green leaves of ginkgo trees or various other diterpenes. Also, the high export rates of IPP from chloroplasts to the cytosol for sterol formation[41] need to be considered. In any case, the available data show that, depending on the type of plant under high incident photon flux and elevated temperatures, rather high amounts of photosynthetically fixed carbon are channelled within the chloroplasts into the biosynthesis of either volatile isoprene and MBO or carotenoids, volatile monoterpenes as well as sterols, volatile sesquiterpenes or other isoprenoid compounds. All those biosynthetic programmes consume the ATP and NADPH produced during the course of the photosynthetic light reactions, which keeps those light reactions going. Thus, the emission and/or accumulation of volatile and nonvolatile isoprenoids preserve the two photosynthetic photosystems PS1 and PS2 under light- and heat-stress conditions when (e.g. at closed stomata) the photosynthetic CO_2 assimilation and the performance of the Calvin cycle are blocked.

This is an extremely essential physiological function of the emission of isoprene and other volatile isoprenoids and also of the accumulation of nonvolatile isoprenoid compounds in plants with no competence for isoprene emission. Also, several other metabolic processes occur, particularly in plants without isoprene emission, such as photorespiration, efficient quenching of excess light (heat emission of absorbed light energy) or the process of photoinhibition, all of which have in common that they more or less protect the photosynthetic photosystems from destruction and photo-oxidation. Thus, as soon as the stress conditions are over, the leaves can immediately restart photosynthesis and photosynthetic carbon fixation because the two photosystems PS1 and PS2 still exist. This immediate switch from the carbon-wasting emission of isoprene (and other isoprenoids) to a functional, photosynthetic carbon gain is an essential advantage for the leaves of green plants and thus has been preserved and even enhanced during evolution. Other plants that lost their isoprene synthase and isoprene emission capacity during evolution had to develop other mechanisms to protect their two photosynthetic photosystems.

2.15 Biosynthesis and Emission of Monoterpenes, Sesquiterpenes and Diterpenes

There are various plant families that produce particularly volatile monoterpenes and sesquiterpenes as part of their essential oils ('etherische Öle') in specialized cells, such as the glandular cells or trichomes, secretory cells in flower petals, inflorescences, fruits, orange peels, rhizomes or seeds (e.g. Lamiaceae, Lauraceae, Myrtaceae, Pinaceae, Rosaceae, Rutaceae, Umbelliferae). There are schizogenic and lysogenic compartments for the deposition of the essential oils. Many of these volatile isoprenoids are monoterpenes, such as the *acyclic monoterpenes* geraniol (**3**), linalool (**4**), citronellol (**49**), nerol, (**50**), citral (**51**) and myrcene (**52**, Figures 2.1, 2.12). *Monocyclic monoterpenes* are menthol (**5**), cineol (**6**), limonene (**53**), α- and β-phellandrene (**54**, **55**), α-terpineol (8-terpineol, **56**), menthone (**57**), pulegone (**58**), *p*-cymene (**59**), thymol (**60**) and carvacrol (**61**). There also exist *bicyclic monoterpenes*, such as sabinene (**62**), thujol (**63**), carene (**64**), 1,8-cineol (eucalyptol, **65**), α- and β-pinene (**66**, **67**), borneol (**68**) and camphor (**69**). *Sesquiterpenes* in essential oils are farnesol (**7**), bisabolol (**8**), nerolidol (**70**), zingiberene (**71**) and caryophyllene (**72**, see Figures 2.1, 2.13). Rosemary leaves primarily contain 1,8-cineol and borneol and, to a lesser degree, camphor and limonene. Lavender flowers possess linalool as a major volatile ingredient, together with cineol and geraniol. Peppermint leaves exhibit menthol as a major substance, together with menthone, cineol, limonene and α-pinene. The

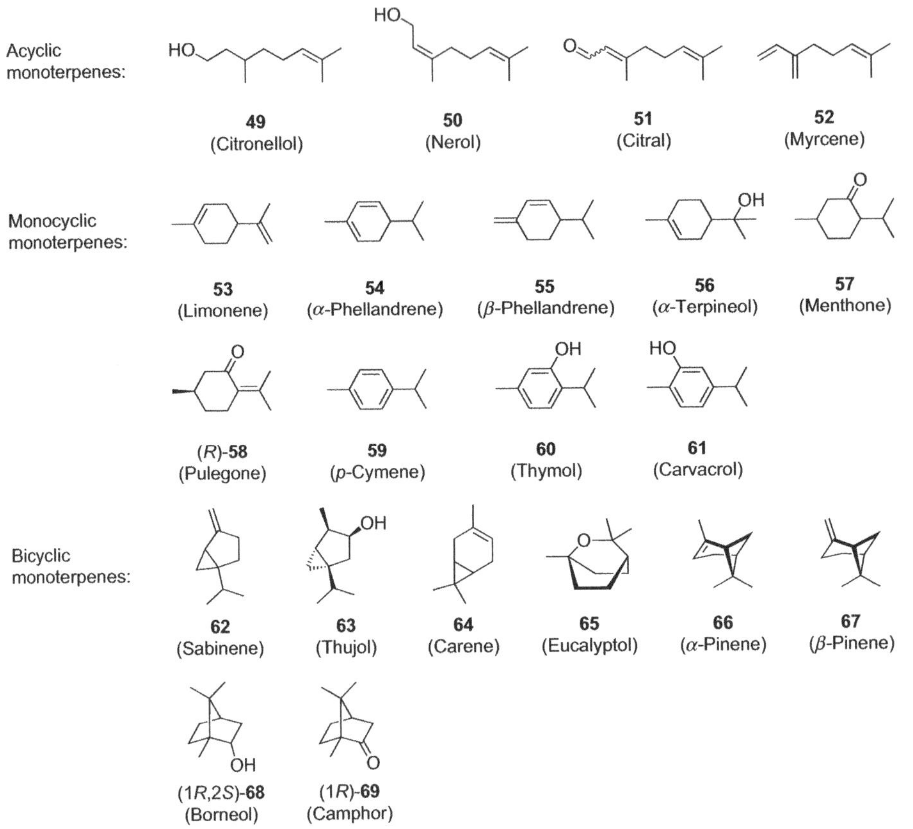

Figure 2.12 *Chemical structures of a series of acyclic, monocyclic and bicyclic monoterpenes*

scales and peels of citrus fruits contain high levels of limonene and, in lower amounts, geraniol, nerol, linalool and citral. The essential oil of rose flowers consists of citronellol (40–50%) and geraniol (20%). Eucalyptus leaves possess high amounts of the monoterpene 1,8-cineol. Only a few volatile diterpenes exist in plants. These are acyclic C_{20} isoprenoids (e.g. phytol derivatives), monocyclic (e.g. camphorene, **10**) or dicyclic isoprenoids (e.g. geranyllinalool, **73**). Most diterpene derivatives are, however, nonvolatile diterpenic carbonic acids as components of bitter substances ('Bitterstoffe') and resins. Monoterpenes, sesquiterpenes and diterpenes are formed and accumulated in particular cell structures of inflorescences, leaves and seeds or in special compartments between the cell walls of neighbouring cells. Since such volatile isoprenoids are formed only in members of certain plant families and in specialized cells, their contribution to the total amount of volatile organic compounds emitted into the air by plants is much smaller than the amount of isoprene, which is formed in many plants from different systematic positions and not only in specialized leaf cells but in the cells of all green leaves. An exception to this rule, that primarily isoprene is the main source of organic plant volatiles, may be oak forests in Mediterranean areas and eucalyptus forests in Australia, where monoterpene and other

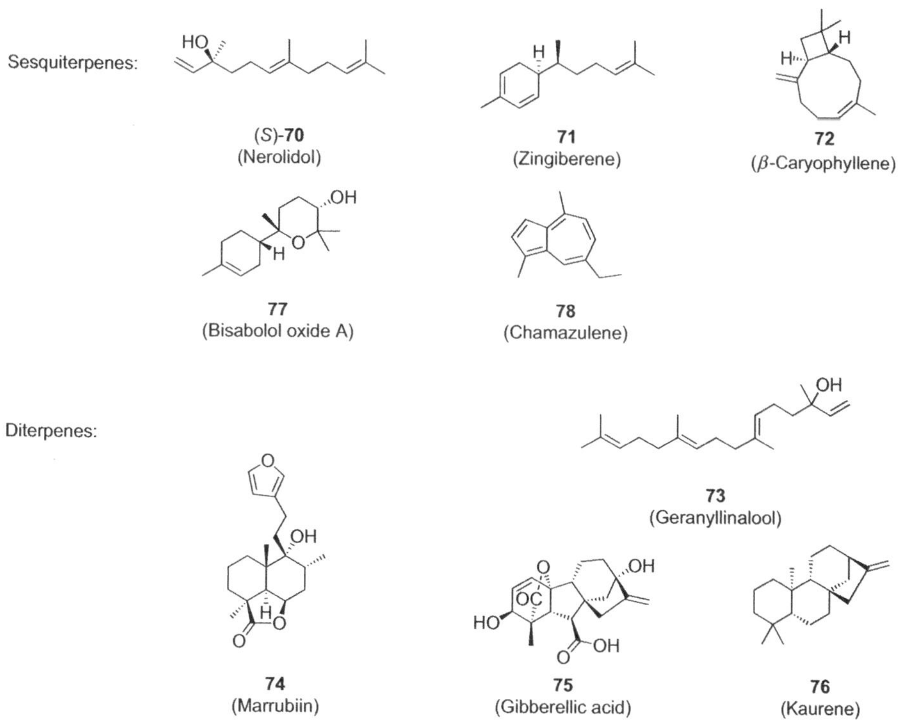

Figure 2.13 *Chemical structures of selected sesqui- and diterpenes*

isoprenoids, besides isoprene, may essentially contribute to the emitted phytogenic organic compounds in the atmosphere.

2.15.1 Monoterpenes

The biosynthesis of monoterpenes also proceeds in plastids via the DOXP/MEP pathway of IPP. After detection of the DOXP/MEP pathway, this was verified by different working groups for various monoterpenes, such as menthone, pulegone, geraniol and thymol.[47,93] It was shown that limonene synthase, the enzyme responsible for the synthesis of the monoterpene limonene (**53**), is localized in peppermint in the leucoplasts of the oil gland secretory cells.[94] Additional enzymes of monoterpene biosynthesis have been cloned and characterized in peppermint, such as menthone reductase[95] as well as geranyl diphosphate synthase, limonene-6-hydroxylase, isopiperitenol dehydrogenase and pulegone reductase.[96] Geranyl diphosphate synthase functions as a heterodimer.[97] This has also been documented in snapdragon (*Antirrhinum majus*) and fairyfan (*Clarkia breweri*) flowers.[98] Other publications describe seasonal changes of monoterpene emission and monoterpene synthase activities.[99] A recent investigation concentrated on the emission of biogenic volatile organic compounds in local urban vegetation which contribute to urban ozone and aerosol formation. The monoterpene emissions of 11 ornamental tree species, three conifers and nine angiosperms were dominated by α- and β-pinene, myrcene, α- and β-phellandrene, carene, limonene and eucalyptol.[100]

2.15.2 Diterpenes

The biosynthesis of diterpenes proceeds within the plastids from IPP produced via the DOXP/MEP pathway. This was first shown for phytol (**8**, side chain of chlorophylls, Scheme 2.5).[23,25] Later it was proved for the diterpenes marrubiin (**74**, Figure 2.13)[101] and taxol (**11**)[102] as well as ginkgolide A (**12**), gibberellins (e.g. gibberellic acid, **75**), kaurene (**76**) and other compounds.[103]

2.15.3 Sesquiterpenes

These are produced in the cytosol and derive from farnesyl diphosphate (FPP, **26**) which is usually made via the acetate/MVA pathway (Scheme 2.1). However, under photosynthetic conditions, the IPP molecules produced in chloroplasts via the DOXP/MEP pathway are exported to the cytosol. They make an essential contribution not only to the biosynthesis of sterols,[41,42] but also to that of sesquiterpenes (e.g. in snapdragon flowers).[44,47] Of the sesquiterpenes, bisabolol oxide A (**77**, Figure 2.13) and chamazulene (**78**) from chamomile, 50% of the isoprenic units came from the acetate/MVA pathway, whereas the other 50% were derived from the DOXP/MEP pathway of IPP formation.[104] Other sesquiterpenes made from IPP of the DOXP/MEP pathway are found in eucalyptus leaves, clematis and in barley roots. Recent research has been concentrating on the characterization and cloning of particular enzymes involved in sesquiterpene biosynthesis, such as the farnesyl diphosphate synthase in the big sagebrush *Artemisia tridentata*.[105]

2.16 Some General Remarks on the Regulation of Terpene Biosynthesis in Plants

Concerning the regulation of monoterpenes, sequiterpenes and diterpenes, it has to be taken into consideration that their biosynthesis is differently regulated from that of the spontaneously formed and immediately emitted highly volatile hemiterpenes isoprene and MBO. The C_{10}, C_{15} and C_{20} prenyl derivatives are usually synthesized and accumulated during or after full development of the proper plant tissues where they are formed more or less continuously. Their synthesis can also be induced as a defensive response. Their biosynthesis is not as specifically bound to high temperatures and high light conditions as that of isoprene and MBO. Thus, at de novo biosynthesis conditions of isoprene in blue gum (*Eucalytus globulus*), as proved by specific incorporation of deuterium ^{2}H-DOX into isoprene, there was no incorporation into the monoterpene cineol.[60] Moreover, the biosynthesis of monoterpenes and diterpene resin acids are induced in the stems of lodgepole pine (*Pinus contorta* var. *latifolia*) saplings when wounded and inoculated with the blue-stain fungus *Ceratocystis clavigera* or treated with carbohydrate elicitors.[106] In addition, the biochemical regulation of monoterpene, sequiterpene and diterpene formation, accumulation and release from conifer needles of Norway spruce (*Picea abies*) can be induced as a defensive response against potential herbivores and pathogens via methyl jasmonate (**79**, Figure 2.14).[107] These are just a few examples for the independent regulation of terpene biosynthesis in comparison to that of isoprene and MBO.

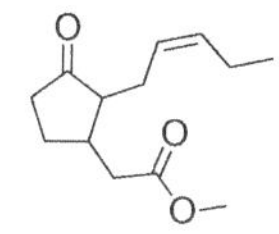

79
(Methyl jasmonate)

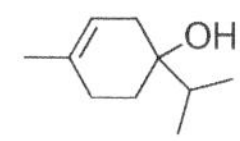

80
(Terpinen-4-ol)

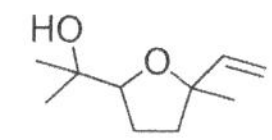

81
(Linalool oxide)

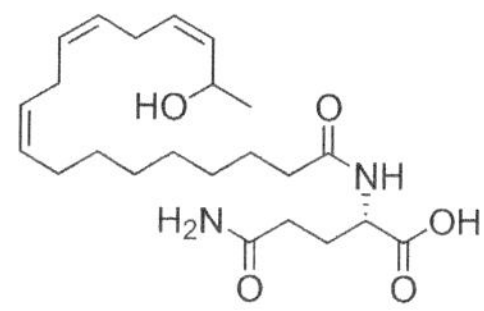

82
(Volicitin)

*Figure 2.14 Chemical structures of methyl jasmonate (**79**), terpinen-4-ol (**80**), linalool oxide (**81**) and volicitin (**82**)*

2.17 Volatile Terpenoids as Aroma Compounds of Wine

Various kinds of terpenoid compounds have been found in the different grape varieties and in wine, where they are essential components of the wine aroma and wine taste. Most of these compounds are volatile monoterpene alcohols, such as geraniol, linalool, nerol and α-terpineol. However, in certain wines the volatile sesquiterpene alcohol farnesol is also an important constituent of the wine aroma. Although not all terpene volatiles of grapes have been particularly checked, it is clear from specific labelling studies and the localization of special terpene synthase enzymes in plastids that the volatile monoterpenes of ripening grapes are synthesized via the plastidic DOXP/MEP pathway of IPP formation, whereas the sesquiterpenes are formed from IPP molecules deriving from the cytosolic acetate/MVA pathway. Yet under photosynthetic conditions IPP molecules, exported from chloroplasts, can also contribute to the biosynthesis of grape sesquiterpenes.

Essential for the particular aroma of a special grape variety and its wine is apparently the relative level of geraniol with respect to the other terpenes. The grape terpenoids of *Vitis vinifera* are usually synthesized rather late in grape berry development and particularly accumulate during ripening of the grapes. Thus, the very characteristic and rather strong aroma of Muscat wines is due to the three monoterpene alcohols geraniol, linalool and nerol. An investigation of the Albariño wines, found in north-west Spain, showed that the terpenes (here linalool and terpineol) are mostly present in the free alcohol form and very little in a bound or ester form.[108] In a broad investigation of Chardonnay wines from all regions of California, one found particularly linalool and α-terpineol, however in minor amounts also terpinen-4-ol (**80**, Figure 2.14), geraniol, nerol and linalool oxide (**81**).[109] In the present research of wine breeding institutions terpenes are screened as markers for potential new grape wine varieties. Other studies deal with the overexpression of particular enzymes of the isoprenoid and terpene biosynthesis machinery to increase the terpene level of grapes. Since the level of volatile terpenes decreases during wine ageing together with the decline of aroma and taste, recent research also concerns the search for inhibitors of the disappearance of volatile aroma in wine.[110]

2.18 Function of Terpenes in Plant Defence

Volatile plant terpenes play an essential ecological role and have a variety of functions in mediating antagonistic and beneficial interactions among organisms.[111] The emission of volatile terpenes from flowers to attract insects for pollination is widely known. Volatile terpenes can also be released at low rates from leaves and other vegetative plant tissues such as roots. However, the level and profile of released volatile terpenes from leaves and other plant tissues is markedly different when plant leaves are damaged, for example by insect feeding. Then newly synthesized volatile terpenes can become part of a defensive mechanism, for example either by direct toxic action against the plant pathogen and the feeding insect or by very special interactions. Thus, in the past two decades it has been shown that volatile plant terpenes mediate very specific interactions between plants and herbivores and other organisms, such as arthropods, nematodes and also other plants. Such interactions have been detected and documented in recent years by various groups;[112–114] and certainly many more interactions will be detected in future.

When plants are damaged (e.g. by herbivorus insects) many more terpene volatiles and also non-isoprenoid volatiles are synthesized de novo and set free. Such volatile terpenes attract both parasitic and predatory insects that are natural enemies of herbivores, as described in the update reviews by Paré and Tumlinson[112] and Arimura *et al.*[113] This is then of great advantage for the attacked plant which via the volatile terpene signal 'asks' for help and can indirectly defend itself. In addition, the emitted volatile plant isoprenoids can also induce defensive responses in neighbouring plants. Such specific interactions and communications are often based on volatile mono- and sesquiterpenes. It is of interest in this respect that, in all plants investigated so far, even in phylogenetically distant plants, there exist notable similarities in the structure of volatile terpene and nonterpenoid volatiles that are released from insect-damaged leaves and from leaves distal to the site of the damage.[111,112] The production of such volatile terpenes is generally induced by an interaction of special elicitors such as volicitin (**82**) in the oral secretion of insect herbivores with the damaged leaf tissue. In these elicitor-induced responses for synthesis and release of volatile defensive terpenes various special compounds, such as jasmonates and oxylipins, seem to be involved. More recently, the drimane sesquiterpenes occurring in plants, fungi and some marine organisms have found special attention in defensive responses.[115] So far, very little is known on the production of volatile terpenes in plant roots as indirect defensive compounds against nematodes and other plant herbivores. However, one can expect a function similar to that in leaf tissues.

2.19 Conclusion

Plants possess two systems for the biosynthesis of IPP and isoprenoid formation; and the DOXP/MEP pathway in the chloroplasts operates independently of the acetate/MVA pathway in the cytosol. The DOXP/MEP pathway is required to supply the IPP C_5 units necessary for the synthesis of functional compounds in the photosynthetic apparatus, such as carotenoids, prenyl side chains of chlorophylls, plastoquinone-9, α-tocopherol and phylloquinone K_1. This plastidic isoprenoid pathway is also involved in the chloroplast adaptation response to high or low light conditions by providing the isoprenoid C_5 building blocks to form sun- and shade-type chloroplasts that are characterized by particular chlorophyll *a*/*b* ratios as well as carotenoid and prenylquinone composition.[42] At high photon flux densities the DOXP/MEP pathway also participates in the relatively rapid de novo biosynthesis of high additional amounts of the xanthophyll cycle carotenoids (zeaxanthin, antheraxanthin) that protect the photosynthetic apparatus from photoinhibition as well as from photo-oxidation. In sun-exposed leaves the DOXP/MEP pathway also supplies the active C_5 IPP units for the continuous accumulation of plastoquinone-9 and α-tocopherol in the osmiophilic plastoglobuli.[42] At high irradiances the chloroplast DOXP/MEP pathway operates efficiently and uses photosynthetically formed ATP, NADPH and newly fixed carbon to provide a major part of the C_5 IPP units necessary for the biosynthesis of the cytosolic sterols. In addition, starting with intermediates of the Calvin cycle, the chloroplastidic isoprenoid pathway serves the biosynthesis and emission or accumulation of volatile isoprenoids, such as isoprene, methylbutenol, monoterpenes, diterpenes (see Scheme 2.2), and particular sesquiterpenes, all of which can be regarded as direct primary photosynthetic products and secondary plant products or natural products. The data reviewed above indicate that, depending on the light and temperature conditions, enormous amounts of freshly fixed photosynthetic carbon flow into various volatile and nonvolatile isoprenoid compounds. Thus, chloroplast isoprenoid biosynthesis via the IPP-forming DOXP/MEP pathway appears to be a ‘metabolic valve’ for regulating the photosynthetic carbon flow as well as fine-tuning for chloroplast and cell metabolism. This chloroplast isoprenoid pathway consumes large amounts of photosynthetically formed ATP and NADPH and may also serve as a ‘safety valve’ in order to avoid overreduction and photoinhibition of the photosynthetic apparatus as well as photo-oxidation of the photosynthetic pigment systems PS1 and PS2. The essential significance of terpene synthesis in plants is underlined by the fact that terpenoid compounds (prenyl pigments, prenylquinones) not only have specific functions in primary cell metabolism, but that the synthesis and emission of various mono- and sesquiterpenes in plants serve as chemical defensive mechanisms against herbivores, pathogens and competitors. Currently various genetic and metabolic engineering studies are being performed in many laboratories on the biosynthetic isoprenoid pathways of plants, using genetically transformed bacterial and plant test systems. This research is essential in order to better understand the biological function of these plant terpenoids and to modulate the production of carotenoids, vitamin E and particular mono-, sesqui- and diterpenes of commercial and pharmaceutical interest.

Acknowledgements

I am grateful to several of my former PhD students and group members who carried out essential parts of the research on plant isoprenoid biosynthesis (T. J. Bach, C. Müller, J. Schwender, J. Zeidler). I also wish to thank Ms Gabrielle Johnson for English language assistance.

References

1. (*a*) A. M. Nonomura, A. A. Benson, The path of carbon in photosynthesis: improved crop yields with methanol, *Proc. Natl. Acad. Sci. USA*, **89**, 9794–9798 (1992); (*b*) R. Fall, A. A. Benson, Leaf methanol – the simplest natural product from plants, *Trends Plant Sci.*, **1**, 296–301 (1996); (*c*) A. Folkers, K. Hüve, C. Amman, T. Dindorf, J. Kesselmeier, E. Kleist, U. Kuhn, R. Uerlings, J. Wildt, Methanol emissions from deciduous tree species: dependence on temperature and light intensity, *Plant Biol.*, **10**, 76–85 (2008).
2. J. Peñuelas, J. Llusià, Plant VOC emissions: making use of the unavoidable, *Trends Ecol. Evol.*, **19**, 402–404 (2004).
3. F. Loreto, J. Kesselmeier, J.-P. Schnitzler, Volatile organic compounds in the biosphere–atmosphere system: a preface, *Plant Biol.*, **10**, 2–7 (2008).
4. H. K. Lichtenthaler, J. Zeidler, Isoprene and terpene biosynthesis, in *Trace Gas Exchange in Forest Ecosystems*, R. Gasche, H. Papen, H. Rennenberg (Eds), Kluwer Academic Publishers, Dordrecht, 2002, pp. 79–99.
5. (*a*) G. Helas, J. Slanina, R. Steinbrecher, *Biogenic Volatile Organic Compounds in the Atmosphere*, SPB Academic Publishing, Amsterdam, The Netherlands, 1997; (*b*) T. D. Sharkey, E. A. Holland, H. Mooney, *Trace Gas Emissions by Plants*, Academic Press, San Diego, 1991; (*c*) G. A. Sanadze, Biogenic isoprene (a review), *Russ. J. Plant Physiol.*, **51**, 729–741 (2004).
6. (*a*) J. Kesselmeier, M. Staudt, Biogenic volatile organic compounds (VOC): an overview on emission, physiology and ecology, *J. Atmos. Chem.*, **33**, 23–88 (1999); (*b*) T. D. Sharkey, S. Yeh, Isoprene emission from plants, *Annu. Rev. Plant Physiol. Plant Mol. Biol.*, **52**, 407–436 (2001).
7. M. Trainer, E. P. Williams, D. D. Parrish, M. P. Buhr, E. J. Allwine, H. H. Westberg, F. C. Fehsenfeld, S. Liu, Models and observations of the impact of natural hydrocabons on rural ozone, *Nature*, **329**, 705–707 (1987).
8. M. Lerdau, A. Günther, R. Monson, Plant production, emission of volatile organic compounds, *BioScience*, **47**, 373–383 (1997).
9. (*a*) T. W. Goodwin, The prenyllipids of the membranes of higher plants, in *Lipids and Lipid Polymers in Higher Plants*, M. Tevini, H. K. Lichtenthaler (Eds), Springer-Verlag, Berlin, 1977, pp. 29–44; (*b*) H. K. Lichtenthaler, Regulation of prenylquinone biosynthesis in higher plants, in *Lipids and Lipid Polymers in Higher Plants*, M. Tevini, H. K. Lichtenthaler (Eds), Springer-Verlag, Berlin, 1977, pp. 231–258.
10. O. Wallach, Zur Kenntnis der Terpene und der ätherischen Öle, *Liebigs Ann. Chem.*, **227**, 277–302 (1885).
11. (*a*) L. Ruzicka, Die Architektur der Polyterpene, *Angew. Chem.*, **51**, 5–11 (1938); (*b*) L. Ruzicka, A. Eschenmoser, H. Heusser, The isoprene rule and the biogenesis of isoprenoid compounds, *Experientia*, **9**, 357–396 (1953).
12. H. N. Little, K. Bloch, Studies on the utilization of acetic acid for the biological synthesis of cholesterol, *J. Biol. Chem.*, **138**, 33–46 (1950).
13. F. Lynen, E. Reichart, L. Rueff, Zum biologischen Abbau der Essigsäure VI, 'aktivierte Essigsäure', ihre Isolierung aus Hefe und ihre chemische Natur, *Liebigs Ann. Chem.*, **574**, 1–32, (1951).
14. D. E. Wolf, C. H. Hoffmann, P. E. Aldrich, H. R. Skeggs, L. D. Wright, K. Folkers, β-Hydroxy-β-methyl-δ-dihydroxy-β-methylvaleric acid (divalonic acid), a new biological factor, *J. Am. Chem. Soc.*, **78**, 4499 (1956).
15. S. Chaykin, J. Law, A. H. Philipps, K. Bloch, Phosphorylated intermediates in the synthesis of squalene, *Proc. Natl. Acad. Sci. USA*, **44**, 998–1004 (1958).
16. H. K. Lichtenthaler, Evolution of carotenoid and sterol biosynthesis in photosynthetic and non-photosynthetic organisms, in *Proceedings of the 16th Plant Lipid Symposium*, P. Biacs, P. Gerely (Eds), Mete Publisher, Budapest, 2004, pp. 11–24, www.mete.mtesz.hu/pls/proceedings/index.htm.

17. E. I. Wilding, J. R. Brown, A. P. Bryant, A. F. Chalker, D. J. Holmes, K. A. Ingraham, S. Iordanescu, C. Y. So, M. Rosenberg, M. N. Gwynn, Identification, evolution, and essentiality of the mevalonate pathway for isopentenyl diphosphate biosynthesis in gram-positive cocci, *J. Bacteriol.*, **182**, 4319–4327 (2000).
18. T. J. Bach, A. Boronat, N. J. Campos, A. Ferrer, K.-U. Vollack, Mevalonate biosynthesis in plants, *Crit. Rev. Biochem. Mol. Biol.*, **34**, 107–122 (1999).
19. H. K. Lichtenthaler, The plants' 1-deoxy-D-xylulose-5-phospate pathway for biosynthesis of isoprenoids, *Fett/Lipid*, **100**, 128–138 (1998).
20. T. J. Bach, H. K. Lichtenthaler, Mevinolin a highly specific inhibitor of microsomal 3-hydroxy-methyl-glutaryl-coenzyme A reductase of radish plants, *Z. Naturforsch.*, **37c**, 46–50 (1982).
21. T. J. Bach, H. K. Lichtenthaler, Mechanisms of inhibition by mevinolin (MK 803) of microsome-bound radish and of partially purified yeast HMG-CoA reductase, (EC. 1.1.1.34), *Z. Naturforsch.*, **37c**, 212–219 (1983).
22. T. J. Bach, H. K. Lichtenthaler, Inhibition by mevinolin of plant growth, sterol formation and pigment accumulation, *Physiol. Plant.*, **59**, 50–60 (1983).
23. H. K. Lichtenthaler, J. Schwender, A. Disch, M. Rohmer, Biosynthesis of isoprenoids in higher plant chloroplasts proceeds via a mevalonate independent pathway, *FEBS Lett.*, **400**, 271–274 (1997).
24. H. K. Lichtenthaler, M. Rohmer, J. Schwender, Two independent biochemical pathways for isopentenyl diphosphate (IPP) and isoprenoid biosynthesis in higher plants, *Physiol. Plant.*, **101**, 643–652 (1997).
25. H. K. Lichtenthaler, The 1-deoxy-D-xylulose-5-phosphate pathway of isoprenoid biosynthesis in plants, *Annu. Rev. Plant Physiol. Plant Mol. Biol.*, **50**, 47–65 (1999).
26. H. K. Lichtenthaler, The non-mevalonate isoprenoid biosynthesis: enzymes, genes and inhibitors, *Biochem. Soc. Trans.*, **28**, 787–792 (2000).
27. M. Rohmer, The discovery of a mevalonate-independent pathway for isoprenoid biosynthesis in bacteria, algae and higher plants, *Nat. Prod. Rep.*, **16**, 565–574 (1999).
28. (*a*) H. K. Lichtenthaler, J. Schwender, M. Seemann, M. Rohmer, Carotenoid biosynthesis in green algae proceeds via a novel biosynthetic pathway, in *Photosynthesis: from Light to Biosphere*, P. Mathis (Ed.), Kluwer Academic Publishers, Amsterdam, 1995, pp. 115–118; (*b*) J. Schwender, H. K. Lichtenthaler, M. Seemann, M. Rohmer, Biosynthesis of isoprenoid chains of chlorophylls and plastoquinone in *Scenedesmus* by a novel pathway, in *Photosynthesis: from Light to Biosphere*, P. Mathis (Ed.), Kluwer Academic Publishers, Amsterdam, 1995, pp. 1001–1004; (*c*) J. Schwender, M. Seeman, H. K. Lichtenthaler, M. Rohmer, Biosynthesis of isoprenoids (carotenoids, sterols, prenyl side chains of chlorophyll and plastoquinone) via a novel pyruvate/glycero-aldehyde-3-phosphate non-mevalonate pathway in the green alga *Scenedesmus obliguus*, *Biochem. J.*, **316**, 73–80 (1996).
29. J. G. Zeidler, H. K. Lichtenthaler, H. U. May, F. W. Lichtenthaler, Is isoprene emitted by plants synthesized via the novel isopentenylpyrophosphate pathway? *Z. Naturforsch.*, **52c**, 15–23 (1997).
30. J. G. Zeidler, J. Schwender, C. Müller, J. Wiesner, C. Weidemeyer, E. Beck, H. Jomaa, H. K. Lichtenthaler, Inhibition of the non-mevalonate 1-deoxy-D-xylulose-5-phosphate pathway of plant isoprenoid biosynthesis by fosmidomycin, *Z. Naturforsch.*, **53c**, 980–986 (1998).
31. A. Disch, J. Schwender, C. Müller, H. K. Lichtenthaler, M. Rohmer, Distribution of the mevalonate and glyceraldehyde phosphate/pyruvate pathways for isoprenoid biosynthesis in unicellular algae and the cyanobacterium *Synechocystis* PCC 6714, *Biochem. J.*, **333**, 381–388 (1998).
32. S. Schindler, T. J. Bach, H. K. Lichtenthaler, Differential inhibition by mevinolin of prenyllipid accumulation in radish seedlings, *Z. Naturforsch.*, **40c**, 208–214 (1985).
33. C. Müller, J. Schwender, J. Zeidler, H. K. Lichtenthaler, Properties and inhibition of the first two enzymes of the non-mevalonate pathway of isoprenoid biosynthesis, *Biochem. Soc. Trans.*, **28**, 794–795 (2000).

34. (*a*) J. Zeidler, J. Schwender, C. Müller, H. K. Lichtenthaler, The isoprenoid biosynthesis of plants as test-system for drugs against malaria and pathogenic bacteria, *Biochem. Soc. Trans.*, **28**, 798–800 (2000); (*b*) H. K. Lichtenthaler, J. Zeidler, J. Schwender, C. Müller, The non-mevalonate isoprenoid biosynthesis of plants as a test-system for new herbicides and drugs against pathogenic bacteria and the malaria parasite, *Z. Naturforsch.*, **55c**, 305–313 (2000).
35. J. Schwender, C. Müller, J. Zeidler, H. K. Lichtenthaler, Cloning and heterologous expression of a cDNA encoding 1-deoxy-D-xylulose-5-phosphate reductoisomerase of *Arabidopsis thaliana*, *FEBS Lett.*, **455**, 140–144 (1999).
36. T. Kuzuyama, T. Shimizu, H. Seto, Fosmidomycin, a specific inhibitor of 1-deoxy-D-xylulose 5-phosphate reductoisomerase in the nonmevalonate pathway for terpenoid biosynthesis, *Tetrahedron Lett.*, **39**, 7913–7916 (1998).
37. (*a*) S. Herz, J. Wungsintaweekul, C. A. Schuhr, S. Hecht, H. Lüttgen, S. Sagner, M. A. Fellermeier, W. Eisenreich, M. H. Zenk, A. Bacher, F. Rohdich, Biosynthesis of terpenoids: YgbB protein converts 4-diphosphocytidyl-2*C*-methyl-D-erythritol 2-phosphate to 2*C*-methyl-D-erythritol 2,4-cyclodiphosphate, *Proc. Natl. Acad. Sci. USA*, **97**, 2486–2490 (2000); (*b*) F. Rohdich, J. Wungsintaweekul, M. Fellermeier, S. Sagner, S. Herz, K. Kis, W. Eisenreich, A. Bacher, M. H. Zenk, Cytidine 5′-triphosphate-dependent biosynthesis of isoprenoids: YgbP protein of *Escherichia coli* catalyzes the formation of 4-diphosphocytidyl-2-*C*-methylerythritol, *Proc. Natl. Acad. Sci. USA*, **96**, 11758–11763 (1999).
38. (*a*) B. Altincicek, A.-K. Kollas, S. Sanderbrand, J. Wiesner, M. Hintz, E. Beck, H. Jomaa, *GcpE* is involved in the 2*C*-methyl-D-erythritol 4-phosphate pathway of isoprenoid biosynthesis in *Escherichia coli*, *J. Bacteriol.*, **183**, 2411–2416 (2001); (*b*) N. Campos, M. Rodríguez-Concepción, M. Seemann, M. Rohmer, A. Boronat, Identification of *gcpE* as a novel gene of the 2*C*-methyl-D-erythritol 4-phosphate pathway for isoprenoid biosynthesis in *Escherichia coli*, *FEBS Lett.*, **488**, 170–173 (2001); (*c*) M. Seemann, N. Campos, M. Rodríguez-Concepción, J.-F. Hoeffler, C. Grosdemange-Billiard, A. Boronat, M. Rohmer, Isoprenoid biosynthesis via the methylerythritol phosphate pathway: accumulation of 2-*C*-methyl-D-erythritol 2,4-cyclodiphosphate in a *gcpE* deficient mutant of *Escherichia coli*, *Tetrahedron Lett.*, **43**, 775–778 (2002); (*d*) J. Querol, N. Campos, S. Imperial, A. Boronat, M. Rodríguez-Concepción, Functional analysis of the *Arabidopsis thaliana* GCPE protein involved in plastid isoprenoid biosynthesis, *FEBS Lett.*, **514**, 343–346 (2002).
39. (*a*) P. Adam, S. Hecht, W. Eisenreich, J. Kaiser, T. Gräwert, D. Arigoni, A. Bacher, F. Rohdich, Biosynthesis of terpenes: studies on 1-hydroxy-2-methyl-2-(*E*)-butenyl 4-diphosphate reductase, *Proc. Natl. Acad. Sci. USA*, **99**, 12108–12113 (2002); (*b*) B. Altincicek, E. C. Duin, A. Reichenberg, R. Hedderich, A.-K. Kollas, M. Hintz, S. Wagner, J. Wiesner, E. Beck, H. Jomaa, LytB protein catalyzes the terminal step of the 2*C*-methyl-D-erythritol-4-phosphate pathway of isoprenoid biosynthesis, *FEBS Lett.*, **532**, 437–440 (2002); (*c*) F. Rohdich, S. Hecht, K. Gärtner, P. Adam, C. Krieger, S. Amslinger, D. Arigoni, A. Bacher, W. Eisenreich, Studies on the nonmevalonate terpene biosynthetic pathway: metabolic role of IspH (LytB) protein, *Proc. Natl. Acad. Sci. USA*, **99**, 1158–1163 (2002).
40. (*a*) M. Rodríguez-Concepción, A. Boronat, Elucidation of the methylerythritol phosphate pathway for isoprenoid biosynthesis in bacteria and plastids. A metabolic milestone achieved through genomics, *Plant Physiol.*, **130**, 1079–1089 (2002); (*b*) W. Eisenreich, A. Bacher, D. Arigoni, F. Rohdich, Biosynthesis of isoprenoids via the non-mevalonate pathway, *Cell Mol. Life Sci.*, **61**, 1401–1426 (2004).
41. J. Schwender, H. K. Gemünden, H. K. Lichtenthaler, Chlorophyta exclusively use the 1-deoxyxylulose 5-phosphate/2-*C*-methylerythritol 4-phosphate pathway for the biosynthesis of isoprenoids, *Planta*, **212**, 416–423 (2001).
42. H. K. Lichtenthaler, Biosynthesis, accumulation and emission of carotenoids, α-tocopherol, plastoquinone and isoprene in leaves under high photosynthetic irradiance, *Photosynth. Res.*, **92**, 163–181 (2007).
43. For a review, see: H. K. Lichtenthaler, The DOXP/MEP pathway of chloroplast isoprenoid biosynthesis, in *The Chloroplast: Basics and Applications*, Chapter 7, C. A. Rebeiz, C. Benning,

H. J. Bohnert, H. Daniell, K. J. Hoober, H. K. Lichtenthaler, A. R. Portis, B. C. Tripathy (Eds), Springer, Dordrecht, 2010, pp. 95–118.

44. D. A. Nagegowda, D. Rhodes, N. Dudareva, The role of the methyl-erythritol-phosphate pathway in rhythmic emission of volatiles, in *The Chloroplast: Basics and Applications*, Chapter 10, C. A. Rebeiz, C. Benning, H. J. Bohnert, H. Daniell, K. J. Hoober, H. K. Lichtenthaler, A. R. Portis, B. C. Tripathy (Eds), Springer, Dordrecht, 2010, pp. 139–153.
45. J. A. Bick, B. M. Lange, Metabolic cross talk between cytosolic and plastidial pathways of isoprenoid biosynthesis: unidirectional transport of intermediates across the chloroplast envelope membrane, *Arch. Biochem. Biophys.*, **415**, 146–154 (2003).
46. (*a*) U. Heber, H. W. Heldt, The chloroplast envelope: structure, function and role in leaf metabolism, *Annu. Rev. Plant Physiol.*, **322**, 139–168 (1981); (*b*) U. Flügge, H. W. Heldt, Metabolite translocators of the chloroplast envelope, *Annu. Rev. Plant Physiol. Plant Mol. Biol.*, **42**, 129–144 (1991); (*c*) K. Fischer, B. Kammerer, M. Gutensohn, B. Arbinger, A. Weber, R. E. Häusler, U.-I. Flügge, A new class of plastidic phosphate translocators: a putative link between primary and secondary metabolism by the phosphoenolpyruvate/phosphate antiporter, *Plant Cell*, **9**, 453–462 (1997); (*d*) J. Joyard, E. Tessier, C. Miège, D. Berny-Seigneurin, E. Maréchal, M. A. Block, A.-J. Dorne, N. Rolland, G. Ajlani, R. Douce, The biochemical machinery of plastid envelope membranes, *Plant Physiol.*, **118**, 715–723 (1998).
47. N. Dudareva, S. Andersson, I. Orlova, N. Gatto, M. Reichelt, D. Rhodes, W. Boland, J. Gershenzon, The nonmevalonate pathway supports both monoterpene and sesquiterpene formation in snapdragon flowers, *Proc. Natl. Acad. Sci. USA*, **102**, 933–938 (2005).
48. D. Hampel, A. Mosandl, M. Wüst, Biosynthesis of mono- and sesquiterpenes in carrot roots and leaves (*Daucus carota* L.): metabolic cross talk of cytosolic mevalonate and plastidial methylerythritol phosphate pathways, *Phytochemistry*, **66**, 305–311 (2005).
49. (*a*) G. A. Sanadze, Light and gaseous organic metabolites from plants, *Rep. Akad. Nauk. GSSR*, **22**, 449–454 (1959); (*b*) G. A. Sanadze, G. M. Dolidze, Mass-spectrometric identification of C_5H_8 (isoprene)-type compound from gaseous excrete from plant leaves, *Rep. Akad. Nauk. GSSR*, **27**, 747–750 (1961); (*c*) F. W. Went, Organic material in the atmosphere and its possible relation to petroleum formation, *Proc. Natl. Acad. Sci. USA*, **46**, 212–221 (1960); (*d*) R. A. Rasmussen, F. W. Went, Volatile organic material of plant origin in the atmosphere, *Proc. Natl. Acad. Sci. USA*, **53**, 215–220 (1965).
50. See for example: (*a*) R. H. Rasmussen, M. A. K. Khalil, Isoprene over the Amazon Basin, *J. Geoph. Res.*, **93**, 1417–1421 (1998); (*b*) T. D. Sharkey, Isoprene emission by plants and animals, *Endeavour*, **20**, 74–78 (1996).
51. C. N. Hewitt, H. Stewart, R. A. Street, P. A. Scholefield, *Isoprene, Monoterpene-Emitting Species Survey, 1997*, http://www.es.lancs.ac.uk/cnhgroup/download.html.
52. J. Schwender, J. Zeidler, R. Gröner, C. Müller, M. Focke, S. Braun, F. W. Lichtenthaler, H. K. Lichtenthaler, Incorporation of 1-deoxy-D-xylulose into isoprene and phytol by higher plants and algae, *FEBS Lett.*, **414**, 129–134 (1997).
53. J. Zeidler, H. K. Lichtenthaler, Two simple methods for measuring isoprene emission of leaves by UV-spectroscopy and GC-MS, *Z. Naturforsch.*, **53c**, 1087–1089 (1998).
54. H. K. Grumbach, H. K. Lichtenthaler, K. H. Erismann, Incorporation of $^{14}CO_2$ in photosynthetic pigments of *Chlorella pyrenoidosa*, *Planta*, **140**, 37–43 (1978).
55. C. F. Delwiche, T. D. Sharkey, Rapid appearance of ^{13}C in biogenic isoprene when $^{13}CO_2$ is fed to intact leaves, *Plant, Cell Environ.*, **16**, 587–591 (1993).
56. J.-P. Schnitzler, M. Graus, J. Kreuzwieser, U. Heizmann, H. Rennenberg, A. Wisthaler, A. Hansel, Contribution of different carbon sources to isoprene biosynthesis in poplar leaves, *Plant Physiol.*, **135**, 152–160 (2004).
57. (*a*) D. Schulze-Siebert, D. Heinecke, H. Scharf, G. Schulze, Pyruvate-derived amino acids in spinach chloroplasts, *Plant Physiol.*, **76**, 465–471 (1984); (*b*) D. Schulze-Siebert, G. Schulze, β-Carotene synthesis in isolated chloroplasts, *Plant Physiol.*, **84**, 1233–1237 (1987).
58. A. Hemmerlin, D. Tritsch, M. Hartmann, K. Pacaud, J.-F. Hoeffler, A. van Dorsselaer, M. Rohmer, T. J. Bach, A cytosolic Arabidopsis D-xylulose kinase catalyzes the phosphorylation

of 1-deoxy-D-xylulose into a precursor of the plastidial isoprenoid pathway, *Plant Physiol.*, **142**, 441–457 (2006).
59. U. I. Flügge, W. Gao, Transport of isoprenoid intermediates across chloroplast envelope membranes, *Plant Biol.*, **7**, 91–97 (2005).
60. J. Zeidler, Precursors and inhibitors of isoprene synthesis in plants (in German), *Karlsruhe Contr. Plant Physiol.*, **38**, 1–157 (2001).
61. G. M. Silver, R. Fall, Characterization of aspen isoprene synthase, an enzyme responsible for leaf isoprene emission to the atmosphere, *J. Biol. Chem.*, **270**, 13010–13016 (1995).
62. M. C. Wildermuth, R. Fall, Light-dependent isoprene emission (Characterization of a thylakoid-bound isoprene synthase in *Salix discolor* chloroplasts), *Plant Physiol.*, **112**, 171–182 (1996).
63. M. C. Wildermuth, R. Fall, Biochemical characterization of stromal and thylakoid-bound isoforms of isoprene synthase in willow leaves, *Plant Physiol.*, **116**, 1111–1123 (1998).
64. T. D. Sharkey, S. Yeh, A. E. Wiberley, T. G. Falbel, D. Gong, D. E. Fernandez, Evolution of the isoprene biosynthetic pathway in kudzu, *Plant Physiol.*, **137**, 700–712 (2005).
65. M. Wolfertz, T. D. Sharkey, W. Boland, F. Kuhnemann, Rapid regulation of the methylerythritol 4-phosphate pathway during isoprene synthesis, *Plant Physiol.*, **135**, 1939–1945 (2004).
66. (*a*) T. N. Rosenstiel, A. J. Fisher, R. Fall, R. K. Monson, Differential accumulation of dimethylallyl diphosphate in leaves and needles of isoprene- and methylbutenol-emitting and nonemitting species, *Plant Physiol.*, **129**, 1276–1284 (2002); (*b*) T. N. Rosenstiel, A. L. Ebbets, W. C. Khatri, R. Fall, R. K. Monson, Induction of poplar leaf nitrate reductase: a test of extrachloroplastidic control of isoprene emission rate, *Plant Biol.*, **6**, 12–21 (2004).
67. S. Mayrhofer, M. Teuber, I. Zimmer, S. Louis, R. J. Fischbach, J.-P. Schnitzler, Diurnal and seasonal variation of isoprene biosynthesis-related genes in grey poplar leaves, *Plant Physiol.*, **139**, 474–484 (2005).
68. H. P. Affek, D. Yakir, Natural abundance carbon isotope composition of isoprene reflects incomplete coupling between isoprene synthesis and photosynthetic carbon flow, *Plant Physiol.*, **131**, 1727–1736 (2003).
69. J. Kreuzwieser, M. Graus, A. Wisthaler, A. Hansel, H. Rennenberg, J.-P. Schnitzler, Xylem-transported glucose as an additional carbon source for leaf isoprene formation in *Quercus robu*, *New Phytol.*, **156**, 171–178 (2002).
70. A. Lehning, I. Zimmer, R. Steinbrecher, N. Brüggemann, J.-P. Schnizler, Isoprene synthase activity and its relation to isoprene emission in *Quercus robur* L. leaves, *Plant, Cell Environ.*, **22**, 495–504 (2002).
71. N. Brüggemann, J.-P. Schnitzler, Relationship of isopentenyl diphosphate (IDP) isomerise activity to isoprene emission of oak leaves, *Tree Physiol.*, **20**, 799–805 (2002).
72. M. Loivamäki, S. Louis, G. Cinege, I. Zimmer, R. J. Fischbach, J.-P. Schnitzler, Circadian rhythms of isoprene biosynthesis in grey poplar leaves, *Plant Physiol.*, **143**, 540–551 (2007).
73. K. Behnke, B. Ehlting, M. Teuber, M. Bauerfeind, S. Louis, R. Hänsch, A. Polle, J. Bohlmann, J.-P. Schnitzler, Transgenic, non-isoprene emitting poplars don't like it hot, *Plant J.*, **51**, 485–499 (2007).
74. J. Peñuelas, J. Llusià, D. Asensio, S. Munné-Bosch, Linking isoprene with plant thermotolerance, antioxidants and monoterpene emissions, *Plant, Cell Environ.*, **28**, 278–286 (2005).
75. H. K. Lichtenthaler, M. Lang, M. Sowinska, P. Summ, F. Heiseland, A. Miehé, Uptake of the herbicide diuron (DCMU) as visualized by the fluorescence imaging technique, *Bot. Acta*, **110**, 158–163 (1997).
76. H. K. Lichtenthaler, R. B. Park, Chemical composition of chloroplast lamellae from spinach, *Nature*, **198**, 1070–1072 (1963).
77. (*a*) H. K. Lichtenthaler, F. Babani, Light adaption and senescence of the photosynthetic apparatus: changes in pigment composition, chlorophyll fluorescence parameters and photosynthetic activity during light adaptation and senescence of leaves, in *Chlorophyll Fluorescence: a Signature of Photosynthesis*, G. Papageorgiou and Govindjee (Eds), Springer, Dordrecht, 2004, Chapter 30, pp. 713–736; (*b*) G. Sarijeva, M. Knapp, H. K. Lichtenthaler, Differences in photosynthetic activity, chlorophyll and carotenoid levels, and in chlorophyll fluorescence

parameters in green sun and shade leaves of *Ginkgo* and *Fagus*, *J. Plant Physiol.*, **164**, 950–955 (2007).

78. (*a*) F. Babani, H. K. Lichtenthaler, Light-induced and age-dependent development of chloroplasts in etiolated barley leaves as visualized by determination of photosynthetic pigments, CO_2 assimilation rates and different kinds of chlorophyll fluorescence ratios, *J. Plant Physiol.*, **148**, 555–566 (1996); (*b*) C. Schindler, P. Reith, H. K. Lichtenthaler, Differential levels of carotenoids and decrease of zeaxanthin cycle performance during leaf development in a green and an aurea variety of tobacco, *J. Plant Physiol.*, **143**, 500–507 (1994).
79. (*a*) P. Harley, V. Fridd-Stroud, J. Greenberg, A. Guenther, P. Vasconcellos, Emission of 2-methyl-3-buten-2-ol by pines: a potential large source of reactive carbon to the atmosphere, *J. Geophys. Res. D*, **103**, 25479–25486 (1998); (*b*) G. W. Schade, A. H. Goldstein, D. W. Gray, M. T. Lerdau, Canopy and leaf level 2-methyl-3-buten-2-ol fluxes from a ponderosa pine plantation, *Atmos. Environ.*, **34**, 3535–3544 (2000).
80. J. Zeidler, H. K. Lichtenthaler, Biosynthesis of 2-methyl-3-buten-2-ol emitted from needles of *Pinus ponderosa* via the non-mevalonate DOXP/MEP pathway of isoprenoid formation, *Planta*, **213**, 323–326 (2001).
81. D. W. Gray, M. T. Lerdau, A. H. Goldstein, Influence of temperature history, water stress, and needle age on methylbutenol emissions, *Ecology*, **84**, 765–776 (2002).
82. T. J. Andrews, H. J. Kane, Pyruvate is a by-product of catalysis by ribulosebisphosphate carboxylase/oxigenase, *J. Biol. Chem.*, **266**, 9447–9452 (1991).
83. (*a*) P. K. Stumpf, Fatty acid biosynthesis in higher plants, in *Fatty Acid Metabolism and its Regulation*, S. Numa (Ed.), Elsevier Science Publishers BV, Amsterdam, 1984, pp. 155–179; (*b*) A. Golz, M. Focke, H. K. Lichtenthaler, Inhibitors of de novo fatty acid biosynthesis in higher plants, *J. Plant Physiol.*, **143**, 426–433 (1994).
84. (*a*) G. D. Braithwaite, T. W. Goodwin, Studies on carotenogenesis. 27. Incorporation of [2-^{14}C]acetate, D,L-[2-^{14}C]mevalonate and $^{14}CO_2$ into carrot-root preparations, *Biochem. J.*, **76**, 194–197 (1960); (*b*) T. W. Goodwin, Studies on carotenogenesis. 25. The incorporation of $^{14}CO_2$, [2-^{14}C]acetate and D,L-[2-^{14}C]mevalonic acid into β-carotene by illuminated etiolated maize seedlings, *Biochem. J.*, **70**, 612–617 (1958).
85. M. Calvin, J. A. Bassham, *The Photosynthesis of Carbon Compounds*, W. A. Benjamin Co., New York, 1962.
86. A. Golz, H. K. Lichtenthaler, Inhibition of the plastidic pyruvate dehydrogenase complex in isolated plastids of oat, *Z. Naturforsch.*, **49c**, 421–426 (1994).
87. A. Boronat, The methylerythritol 4-phosphate pathway: regulatory role in plastid isoprenoid biosynthesis, in *The Chloroplast: Basics and Applications*, Chapter 8, C. A. Rebeiz, C. Benning, H. J. Bohnert, H. Daniell, K. J. Hoober, H. K. Lichtenthaler, A. R. Portis, B. C. Tripathy (Eds), Springer, Dordrecht, 2010, pp. 119–126.
88. T. D. Sharkey, E. L. Singsaas, Why plants emit isoprene, *Nature*, **374**, 769 (1995).
89. H. P. Affek, D. Yakir, Protection by isoprene against singlet oxygen in leaves, *Plant Physiol.*, **129**, 269–277 (2002).
90. F. Loreto, V. Velikova, Isoprene produced by leaves protects the photosynthetic apparatus against ozone damage, quenches ozone products, and reduces lipid peroxidation of cellular membranes, *Plant Physiol.*, **127**, 1781–1787 (2001).
91. F. Loreto, M. Mannozzi, C. Maris, P. Nascetti, F. Ferranti, S. Pasqualini, Ozone quenching properties of isoprene and its antioxidant role in leaves, *Plant Physiol.*, **126**, 993–1000 (2001).
92. (*a*) J. P. Thornber, Chlorophyll-proteins: light-harvesting and reaction center components of plants, *Annu. Rev. Plant Physiol.*, **26**, 127–158 (1975); (*b*) H. K. Lichtenthaler, U. Prenzel, G. Kuhn, Carotenoid composition of chlorophyll-carotenoid-proteins from radish chloroplasts, *Z. Naturforsch.*, **37c**, 10–12 (1982); (*c*) H. K. Lichtenthaler, G. Kuhn, U. Prenzel, C. Buschmann, D. Meier, Adaptation of chloroplast-ultrastructure and of chlorophyll-protein levels to high-light and low-light growth conditions, *Z. Naturforsch.*, **37c**, 464–475 (1982).

93. (*a*) W. Eisenreich, S. Sagner, M. H. Zenk, A. Bacher, Monoterpene essential oils are not of mevalonate origin, *Tetrahedron Lett.*, **39**, 3889–3892 (1997); (*b*) S. Sagner, C. Latzel, W. Eisenreich, A. Bacher, M. H. Zenk, Differential incorporation of 1-deoxy-D-xylulose into monoterpenes and carotenoids in higher plants, *Chem. Commun.*, 221–222 (1998).
94. G. Turner, J. Gershenzon, E. E. Nielson, J. E. Froelich, R. Croteau, Limonene synthase, the enzyme responsible for monoterpene biosynthesis in peppermint, is localized to leucoplasts of oil gland secretory cells, *Plant Physiol.*, **120**, 879–886 (1999).
95. E. M. Davis, K. L. Ringer, M. E. McConkey, R. Croteau, Monoterpene metabolism. Cloning, expression and characterization of menthone reductases from peppermint, *Plant Physiol.*, **137**, 873–881 (2005).
96. G. W. Turner, R. Croteau, Organisation of monoterpene biosynthesis in Mentha. Immunocytochemical localizations of geranyl diphosphate synthase, limonene-6-hydroxylase, isopiperitenol dehydrogenase, and pulegone reductase, *Plant Physiol.*, **136**, 4215–4227 (2004).
97. C. C. Burke, M. R. Wildung, R. Croteau, Geranyl diphosphate synthase: cloning, expression, and characterization of this prenyltransferase as a heterodimer, *Proc. Natl. Acad. Sci. USA*, **96**, 13062–13067 (1999).
98. D. Tholl, C. M. Kish, I. Orlova, D. Sherman, J. Gershenzon, E. Pichersky, N. Dudareva, Formation of monoterpenes in *Antirrhinum majus* and *Clarkia breweri* flowers involves heterodimeric geranyl diphosphates synthases, *Plant Cell*, **16**, 977–992 (2004).
99. (*a*) F. Loreto, J. Robert, R. J. Fischbach, J.-P. Schnitzler, P. Ciccioli, E. Brancaleoni, C. Calfapietra, G. Seufert, Monoterpene emission and monoterpene synthase activities in the Mediterranean evergreen oak *Quercus ilex* L. grown at elevated CO_2 concentrations, *Global Change Biol.*, **7**, 709–717 (2001); (*b*) R. J. Fischbach, M. Staudt, I. Zimmer, S. Rambal, J.-P. Schnitzler, Seasonal pattern of monoterpene synthase activities in leaves of the evergreen tree *Quercus ilex*, *Physiol. Plant.*, **114**, 354–360 (2002).
100. S. M. Noe, J. Peñuelas, Ü. Niinemets, Monoterpene emissions from ornamental trees in urban areas: a case study of Barcelona, Spain, *Plant Biol.*, **10**, 163–169 (2008).
101. W. Knöss, B. Reuter, J. Zapp, Biosynthesis of the labdane diterpene marrubiin in *Marubium vulgare* via a non-mevlonate pathway, *Biochem. J.*, **326**, 449–454 (1997).
102. W. Eisenreich, B. Menhard, P. J. Hylands, M. H. Zenk, A. Bacher, Studies on the biosynthesis of taxol: the taxane skeleton is not of mevalonoid origin, *Proc. Natl. Acad. Sci. USA*, **93**, 6431–6436 (1996).
103. K. Nabeta, T. Ishikawa, H. Okuyama, Sesqui- and di-terpene biosynthesis from ^{13}C labelled acetate and mevalonate in cultured cells of *Heteroscyphus planus*, *J. Chem. Soc., Perkin Trans. 1*, 3111–3115 (1995).
104. K.-P. Adam, J. Zapp, Biosynthesis of the isoprene units of chamomile sesquiterpenes, *Phytochemistry*, **48**, 953–959 (1998).
105. A. Hemmerlin, S. B. Rivera, H. K. Erickson, C. D. Poulter, Enzymes encoded by the farnesyl diphosphate synthase gene family in the big sagebrush *Artemisia tridentate* ssp. *spiciformis*, *J. Biol. Chem.*, **278**, 32132–32140 (2003).
106. R. Croteau, S. Gurkewitz, M. A. Johnson, H. J. Fisk, Biochemistry of oleoresinosis: monoterpene and diterpene biosynthesis in lodgepole pine saplings infected with *Ceratocystis clavigera* or treated with carbohydrate elicitors, *Plant Physiol.*, **85**, 1123–1128 (1987).
107. (*a*) D. Martin, D. Tholl, J. Gershenzon, J. Bohlmann, Methyl jasmonate induces traumatic resin ducts, terpenoid resin biosynthesis, and terpenoid accumulation in developing xylem of Norway spruce stems, *Plant Physiol.*, **129**, 1003–1018 (2002); (*b*) D. M. Martin, J. Gershenzon, J. Bohlmann, Induction of volatile terpene biosynthesis and diurnal emission by methyl jasmonate in foliage of Norway spruce, *Plant Physiol.*, **132**, 1568–1599 (2003).
108. M. Vilanova, C. Sieiro, Determination of free and bound terpene compounds in Albariño wine, *J. Food Compos. Anal.*, **19**, 694–697 (2006).
109. S. P. Arrhenius, L. P. McCloskey, M. Sylvan, Chemical markers of *Vitis vinifera* var. Chardonnay regional wines, *J. Agric. Food Chem.*, **44**, 1085–1090 (1996).
110. S. Sergianitis, I. G. Roussis, Protection of volatile esters and terpenes during storage of a white wine and a model wine medium by a mixture of *N*-acetyl-cysteine and caffeic acid, *Eur. Food Res. Technol.*, **227**, 643–647 (2007).

111. J. Gershenzon, N. Dudareva, The function of natural terpene products in the natural word, *Nat. Chem. Biol.*, **3**, 408–414 (2007).
112. P. W. Paré, J. H. Tumlinson, Plant volatiles as a defense against insect herbivores, *Plant Physiol.*, **121**, 325–332 (1999).
113. G. Arimura, C. Kost, W. Boland, Herbivore-induced, indirect plant defense, *Biochim. Biophys. Acta*, **1734**, 91–111 (2005).
114. J. Degenhardt, Ecological roles of vegetative terpene volatiles, in *Induced Plant Resistance to Herbivory*, A. Schaller (Ed.), Springer Science, Dordrecht, 2009, pp. 433–442.
115. B. J. M. Jansen, A. de Groot, Occurrence, biological activity and synthesis of drimane sesquiterpenoids, *Nat. Prod. Rep.*, **21**, 449–477 (2004).

3

Analysis of the Plant Volatile Fraction

*Patrizia Rubiolo, Barbara Sgorbini, Erica Liberto, Chiara Cordero and Carlo Bicchi**

Laboratory of Phytochemical Analysis, Dipartimento di Scienza e Tecnologia del Farmaco, Università degli Studi di Torino, Via Pietro Giuria 9, 10125 Torino, Italy

3.1 Introduction

The volatile fraction of a matrix of vegetable origin is, in general, defined as a mixture of volatiles that can be sampled because of their ability to vaporize spontaneously and/or under suitable conditions or by employing appropriate techniques. The term 'volatile fraction' is therefore an umbrella term that includes a range of approaches and/or techniques, which produce samples that are representative of the volatiles characterizing a vegetable matrix, for example headspace, essential oils, flavours, fragrances, aromas and extracts prepared through specific techniques. The compositions of a volatile fraction may thus vary depending on the approach adopted; because they are based on completely different principles, the resulting samples cannot be compared directly. These considerations are not obvious, since the literature still includes articles that (erroneously) compare essential oil and headspace compositions directly, or even worse that do not even distinguish between the two, although their definitions are absolutely clear. In agreement with AFNOR (Association Française de Normalisation) 1998[1] and European Pharmacopoeia 2008[2] an 'essential oil' (EO) is the product obtained by hydro-, steam- or dry-distillation or by a suitable mechanical process without heating (for citrus fruits) of a plant or of some parts of it, while 'headspace sampling' is a solventfree technique aimed at sampling the gaseous or vapour phase in equilibrium (or not) with a solid or liquid matrix in order to characterize its composition.[3]

The volatile fraction emitted from a plant is now recognized to be an important biosensor, diagnostic not only of its metabolism and changes, but also because it can, in some cases, be

*Note: corresponding author.

The Chemistry and Biology of Volatiles Edited by Andreas Herrmann

related to the compositions of other fractions, including nonvolatile ones. The most important task in its analysis is therefore to detect, identify and, when necessary, quantify the volatile component(s) (in general secondary metabolites) that are marker(s) of the investigated biological phenomenon. This requires analytical methods and technologies to be adopted that are sensitive enough (*sensu lato*) to detect variations in the composition of the emitted volatile fraction, and to enable the dynamics of the reaction(s) of a vegetable organism to be monitored when its metabolism is altered. This requirement can only be satisfied by applying an analytical set-up consisting of sampling techniques that maximize the recovery over time of biologically interesting metabolites and analytical techniques that minimize the analysis time.

The approach to the analysis of the volatile fraction has changed radically over the past 15–20 years, mainly thanks to the dramatic increase in the number of chemical analyses a laboratory must run. This need has strongly influenced the development of both the main steps of an analytical procedure, that is sample preparation and analysis, and, in consequence, has definitively induced the inclusion of data elaboration as a third active step, able to produce a 'higher' level of information from the samples investigated. This has resulted in a trend for both routine and research laboratories (in particular for those where large numbers of homogeneous samples must be analysed) whose aim has been to develop, wherever possible, automatic systems in which the above three steps are merged into a single one (i.e. the so-called 'total analysis system', TAS) to reduce the workload to a minimum.

This chapter is a critical overview of the approaches and techniques used to study the composition of the volatile fraction of a vegetable matrix on the basis of the most recent achievements in sample preparation, analysis and data elaboration.

3.2 Sample Preparation

From the beginning of the 1990s, sample preparation of the volatile fraction has evolved considerably, and highly innovative and effective techniques now flank those previously adopted. The number of conventional techniques available to sample the volatile fraction in or by a 'liquid' phase is quite large. Examples for distillates include: vacuum-, dry-, steam- or hydro-distillation for essential oils,[4] simultaneous distillation–extraction (SDE)[5] and solvent extraction offline combined with normal pressure or vacuum distillation. Examples for extracts include: highly effective extraction techniques such as ultrasound or microwave-assisted extraction–hydrodistillation (MAE, MA-HD)[6] and selective and/or pressurized (or accelerated) solvent extraction (ASE) or supercritical fluid extraction (SFE).[7]

However, the most important improvements in the sampling of the volatile fraction have been achieved in the 'vapour' phase, better known as headspace (HS) sampling. Traditionally, HS sampling operates either in static (S-HS) or dynamic mode (D-HS): the principles of these two approaches were defined so clearly that, paradoxically, they slowed down the development of other HS techniques. The renewed interest in HS sampling by the end of the 1980s coincided to a great extent with the development of solvent-free sample preparation techniques, that are techniques in which the analyte(s) is (are) isolated from a matrix, minimizing the use of a liquid solvent.[8] This approach perfectly meets the definition of HS sampling (see above)[3] and it lays at the origin of the introduction of high concentration capacity headspace techniques (HCC-HS).[9,10] HCC-HS techniques act as a 'bridge' between S-HS and D-HS techniques and are based on either the static or the dynamic accumulation of volatiles on polymers operating in sorption and/or adsorption modes, or, less frequently, on solvents. They were immediately and quite unexpectedly successful, mainly because they are as simple, fast, easy to automate and as reliable as S-HS but, at the same time, their analyte concentration capability is often comparable to that offered by D-HS.

The development of HCC-HS techniques was also favoured by the ever increasing knowledge of sorption, mainly due to Sandra and coworkers and Cramer and coworkers. Sorption is partition based on an analyte's dissolution in a liquid retaining polymer [in particular poly(dimethylsiloxane) (PDMS)] and together with adsorption, it is the main phenomenon involved in the recovery of an analyte from a liquid or vapour sample by a polymeric phase.[11]

3.2.1 'Liquid' Phase Sampling

The so-called 'liquid phase sample preparation techniques' are all now long established, and their use is reported in official methods, making their detailed description redundant. A technique whose performance has been underestimated is SFE probably because of the cost of the system compared to that of conventional extraction techniques, but it shows several characteristics that can be very useful in the recovery of the volatile fraction. SFE is an environmentally friendly technique that not only enables selective extractions to be carried out by varying the polarity of CO_2, either chemically through a polar modifier, or physically through its density (tuned by temperature and pressure), and concentrated samples to be obtained without residual polluting solvents; it is also possible to fractionate the resulting samples by differential depressurization or by online combination with SPE affording the isolation of fractions homogeneous for polarity and volatility.[7,12]

3.2.2 Headspace Sampling

The HCC-HS approach has stimulated the introduction of several new techniques to sample the volatile fraction in the vapour phase. This approach has also favoured, whenever possible, the adoption of headspace sampling to replace distillation and extraction techniques, the latter being more time-consuming and almost impossible to combine online to the analytical step in total analysis systems; HCC-HS also avoids pollution problems related to the hydrocarbons or chlorinated solvents used in solvent extraction.

Several HCC-HS-based techniques are now available to sample the volatile fraction of vegetable matrices, in addition to conventional S-HS and D-HS samplings: HS-solid-phase microextraction (HS-SPME),[13] in-tube sorptive extraction [inside needle capillary adsorption trap (INCAT), headspace solid-phase dynamic extraction (HS-SPDE)],[14–16] headspace sorptive extraction (HSSE),[17,18] static and trapped headspace (S&T-HS),[19–21] solid-phase aroma concentrate extraction (SPACE),[22] headspace liquid-phase microextraction (HS-LPME)[23,24] and large surface area HCC-HS sampling [membrane extraction sorbent interface (MESI), membrane microextraction (MME), headspace-sorptive tape extraction (HS-STE)].[25–27] HCC-HS, HS-SPME and D-HS techniques in the sample preparation of the plant volatile fraction, their advantages and limits, were recently reviewed by the authors' group.[9,10,28] The principles underlying the above mentioned techniques are briefly described below.

3.2.3 Headspace–Solid Phase Microextraction

The first and most popular HCC-HS technique to appear was HS-SPME presented by Zhang and Pawliszyn in 1993 as an extension of SPME,[13] in turn introduced by Arthur and Pawliszyn in 1990;[29] SPME and HS-SPME were also the first routine techniques to operate in the microextraction mode. They are based on the accumulation of the investigated analytes onto a film of a sorbent and/or an adsorbent coated onto a mobile fused silica fibre, which is part of a special stainless steel needle assembled onto a customized device; in HS-SPME, the accumulated analytes are recovered by thermal desorption directly into the body of a conventional split/splitless gas chromatography (GC) injector. Figure 3.1 shows a diagram of an SPME holder and a fused silica fibre. Zhang and Pawliszyn also advanced a theory for SPME applied to HS sampling[13,30] in which they showed that analyte recovery from headspace by a fibre depends on two closely related but distinct equilibria: the first is the matrix/headspace equilibrium responsible for the headspace composition (measured by its distribution coefficient, K_2), the second is the headspace/polymeric fibre coating equilibrium (measured by its distribution coefficient, K_1). The HS-SPME approach (and the underlying theory) has been, and still is, the basis for the development of new HCC-HS sampling techniques. Pawliszyn and coworkers have reviewed the theory, technology, evolution and applications of SPME together with some additional specific topics.[8,31] Figure 3.2 shows the fast gas chromatography–mass spectrometry (GC-MS) profile of the headspace of a sample of juniper berries sampled by SPME, together with the structures of compounds **1–23** identified for the different peaks. An extensive review on applications of HS-SPME in the plant field was recently published by Belliardo and coworkers.[28]

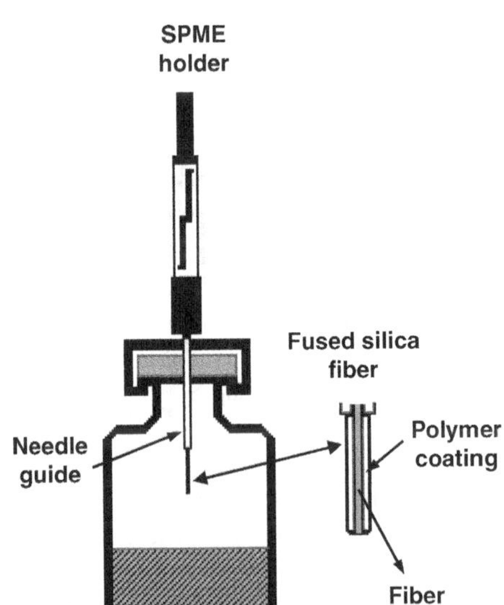

Figure 3.1 *Diagram of a SPME holder for HS sampling and a fused silica fibre*

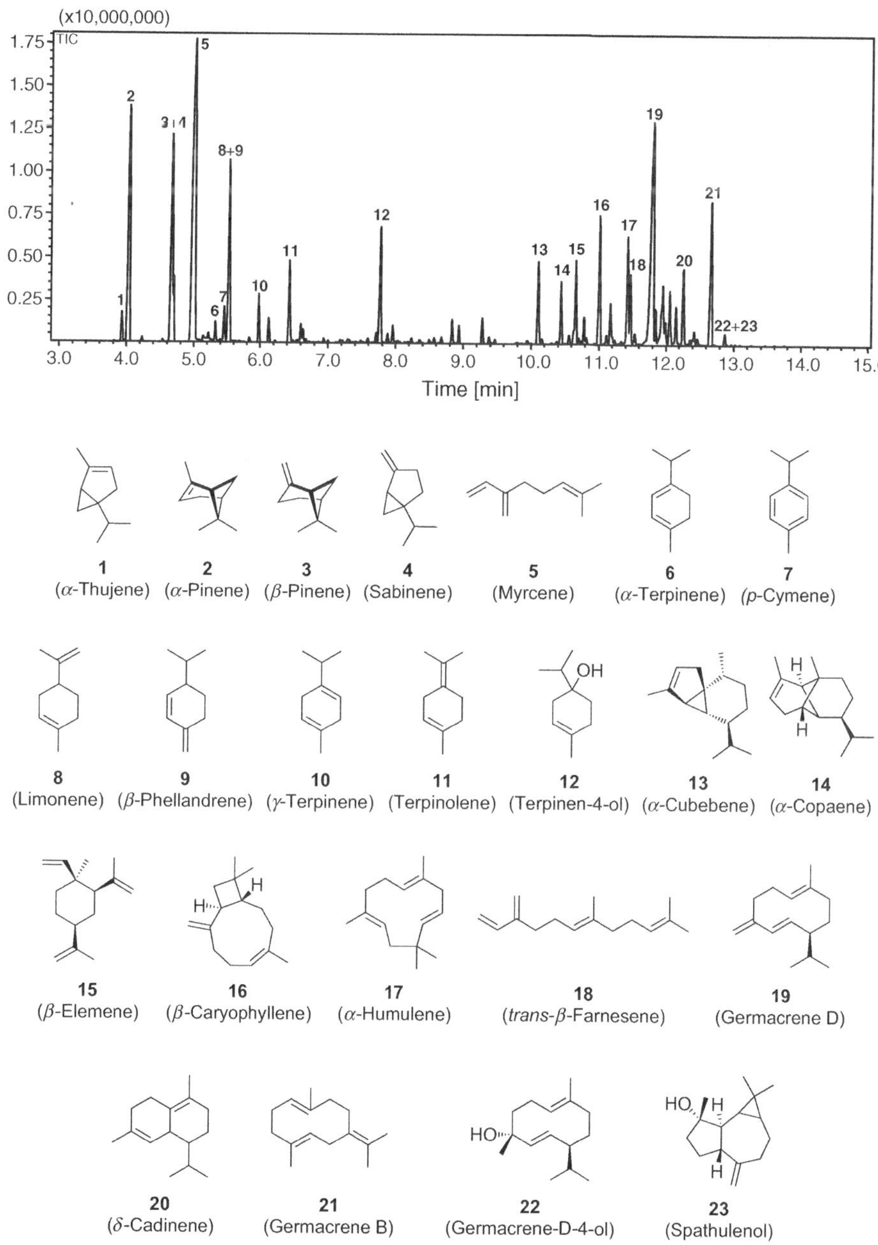

Figure 3.2 *HS-SPME-F-GC-MS profile of juniper berries EO (Juniperus communis L.) and structures of the identified compounds **1–23**. HS-SPME sampling: fibre: 2 cm DVB/CAR/PDMS; incubation temperature 50 °C, equilibration time: 5 min, sampling time: 10 min. GC-MS system: Shimadzu QP2010 GC-MS system provided with Shimadzu GC-MS Solution 2.51 software (Shimadzu, Milan, Italy). Columns: SE52 10 m × 0.10 mm × 0.10 μm (MEGA, Legnano, Italy). Temperature rates: from 40 °C (1.46 min) to 250 °C (1.46 min) at 10.25 °C min^{-1}; injection mode: split; split ratio: 1 : 50; temperatures: injector: 230 °C, transfer line: 250 °C; ion source: 200 °C; carrier gas: He, flow rate: 0.5 ml min^{-1}. MS ionization mode: electron impact (EI) at 70 eV, data rate: 20 Hz, mass range: 35–350 m/z*

3.2.4 In-Tube Sorptive Extraction

In-tube sorptive extraction is an approach based on D-HS, which was developed to overcome the relatively limited concentration capability of HS-SPME.

The inside needle capillary adsorption trap (INCAT), introduced in 1997 by McComb *et al.*,[14] is a pre-concentration device consisting of a hollow needle, with either a short length of GC capillary column placed inside it, or an internal coating of carbon. Sampling is achieved by passing the gaseous or liquid sample through the device, either actively with a syringe, or passively via diffusion. The sampled analytes are recovered by direct thermal desorption into the GC injector.

Solid-phase dynamic extraction (SPDE), also known as 'the magic needle', was introduced by Lipinsky[32] in 2001 and applied to HS sampling by Musshoff *et al.*[15] SPDE is a technique in which the analytes from the vapour phase (HS-SPDE) or from the liquid phase (IS-SPDE) are concentrated on a thick film (50 μm, about 4.5 μl) of a polymer coated onto the inside wall of the stainless steel needle (5.5 or 7.5 cm long) of a gas-tight syringe (2.5 ml) by pulling in and pushing out a fixed volume of HS to be sampled, through the gas-tight syringe, for an appropriate number of times within a fixed time (Figure 3.3). The trapped analytes are then thermally desorbed, transferred online by a fixed volume of carrier gas into the GC injector body, and analysed by GC or GC-MS. Several polymeric coatings are available: PDMS, PDMS/activated charcoal, PDMS/OV 225, PDMS/phenyl-methylpolysiloxane, poly(ethylene glycol) (PEG) and PDMS/7% phenyl- or 7% cyanopropyl (OV 1701). Bicchi *et al.*[16] applied HS-SPDE to the analysis of aromatic plants and food matrices, in particular rosemary, banana, green and roasted coffee and red and white wine, all with a PDMS and 10% of activated carbon-coated needle.

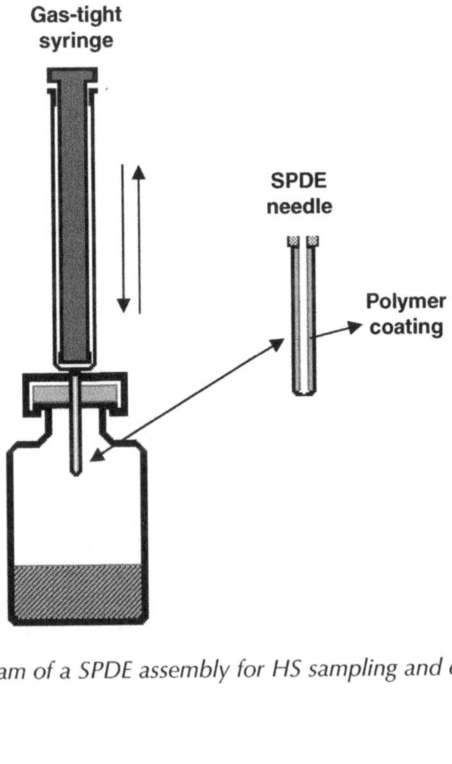

Figure 3.3 Diagram of a SPDE assembly for HS sampling and of a 'magic needle'

3.2.5 Headspace Sorptive Extraction

Headspace sorptive extraction (HSSE) is an extension of stir bar sorptive extraction (SBSE)[33] introduced in 2000 by Bicchi *et al.*[17,34] and Tienpont *et al.*[18] to increase the concentration capability of SPME. In HSSE, the analytes are accumulated onto a thick film of PDMS coating a glass-coated magnetic stir bar (better known by the name Twister), whose PDMS volume ranges from 25 to 250 μl depending on its size. HS components are recovered statically by suspending the PDMS stir bar in the vapour phase, in equilibrium or not with the matrix, for a fixed time. After sampling, the stir bar is placed in a glass tube and transferred to a thermo-desorption system where the analytes are thermally recovered and analysed by GC or GC-MS. More recently, Bicchi *et al.*[35] introduced dual-phase twisters (DP twisters) mainly to overcome the chief limit of PDMS (i.e. an apolar polymer) in the recovery of both polar analytes from complex or multi-ingredient matrices and very volatile components (C_1–C_4 analytes). DP twisters consist of a short PDMS tube (2 cm), closed at both ends with two magnetic stoppers, thus creating an inner cavity that is packed with different adsorbents (Carbopack B, Tenax GC, a bisphenol-PDMS copolymer and Carbopack coated with 5% of Carbowax). Figure 3.4 gives schemes of a HSSE sampling system and a conventional PDMS twister (*a*) and a DP twister (*b*). The concentration capability of DP twisters is therefore the result of the sorption of the analyte(s) onto PDMS from the liquid or vapour phase, followed by analyte diffusion through the PDMS layer and adsorption onto the inner phase. HSSE has been successfully applied in HS sampling of vegetable matrices in the food, cosmetic and pharmaceutical fields.

The concentration capability of PDMS can also be improved by modifying its polarity with the help of a solvent impregnating the polymer. This approach was first developed for samplings in the liquid phase by Jánská *et al.*[36] [solvent in silicone tube extraction (SiSTEx)] and Sandra and coworkers[37] [silicon membrane sorptive extraction (SMSE)]; more recently, Sgorbini *et al.*[38] applied it to samplings in the vapour phase. The sampling device consists of PDMS tubing closed at both ends by glass stoppers, the resulting inner cavity being filled with an organic solvent. In liquid phase sampling, analytes are concentrated in the inner solvent where they diffuse from the sample through the PDMS, which acts as a selective membrane. In headspace sampling, the analytes accumulate in the PDMS, whose polarity is modified by the inner solvent diffusing through it, and are recovered by thermal desorption and analysed by GC or GC-MS. Figure 3.5 shows a diagram of a HS-SMSE system and a sampling device. HS-SMSE was successfully applied

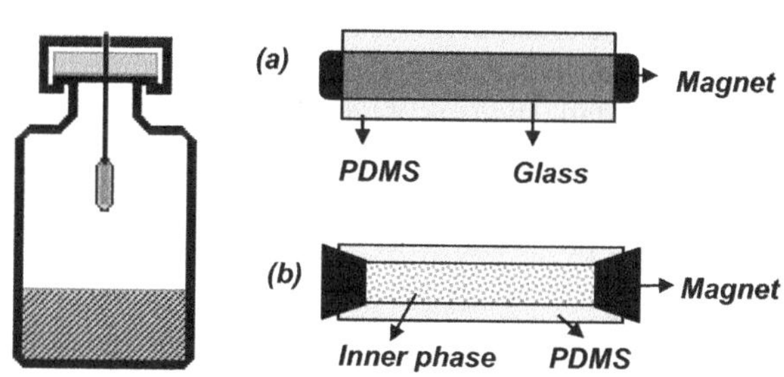

Figure 3.4 *Diagram of a HSSE sampling system and (a) a conventional PDMS twister and (b) a DP twister*

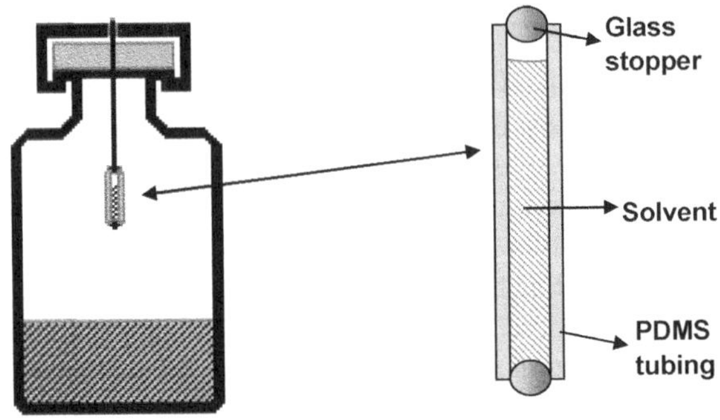

Figure 3.5 *Diagram of a HS-SMSE system and a sampling device*

to headspace sampling of: (*a*) sage and thyme with ethyl acetate and cyclohexane as PDMS modifiers, (*b*) coffee with PDMS/ethyl acetate and (*c*) fatty acids with PDMS/ethyl acetate at pH 8. Figure 3.6 reports the GC-FID profiles of the headspace of a sample of dried sage leaves containing compounds **2**, **12–17** and **24–30** sampled by an empty PDMS tubing (*a*) and by SMSE using cyclohexane (*b*) and ethyl acetate (*c*) as solvents.

3.2.6 Static and Trapped Headspace

This technique was mainly developed to overcome the limit with the injection of large gas volumes in classical GC injectors with S-HS and the inaccuracy in the measurement of the total gas volume with purge and trap-headspace sampling.[19–21] Static and trapped headspace (S&T-HS) try to combine advantages of both techniques, since after equilibration, the vapour phase is evacuated from the HS chamber by the pressure of a piston and trapped by an adsorbent in a cartridge through which the gas flows. The trap is then transferred to a thermal desorber from where the recovered analytes are online injected in a GC unit. The sensitivity mainly depends on the sampled gas volume which depends on the cell volume. This technique was shown to be very effective for quantitation both by combining the determination of the analyte air to liquid partition coefficients to multiple headspace extraction (MHE)[19] or by the standard addition procedure.[20]

3.2.7 Solid-Phase Aroma Concentrate Extraction

Solid-phase aroma concentrate extraction (SPACE), introduced by Ishikawa *et al.* in 2004,[22] bases analyte recovery on a stainless steel rod coated with a mixture of adsorbents, mainly graphite carbon. Sampling in static mode is obtained by suspending the SPACE rod in the matrix headspace for a fixed time; the sampled analytes are then recovered by thermal desorption online to the GC or GC-MS system. This technique has been shown to be successful with roasted coffee beans and other vegetable matrices.[39]

3.2.8 Headspace Liquid-Phase Microextraction

Liquid-phase microextraction (LPME) was introduced about 15 years ago by Jeannot and Cantwell[40] for sampling in liquid phase; its use was extended to S-HS sampling by Theis *et al.*[23] and Tankeviciute *et al.*[24] in 2001. In HS-LPME the analytes are accumulated into a

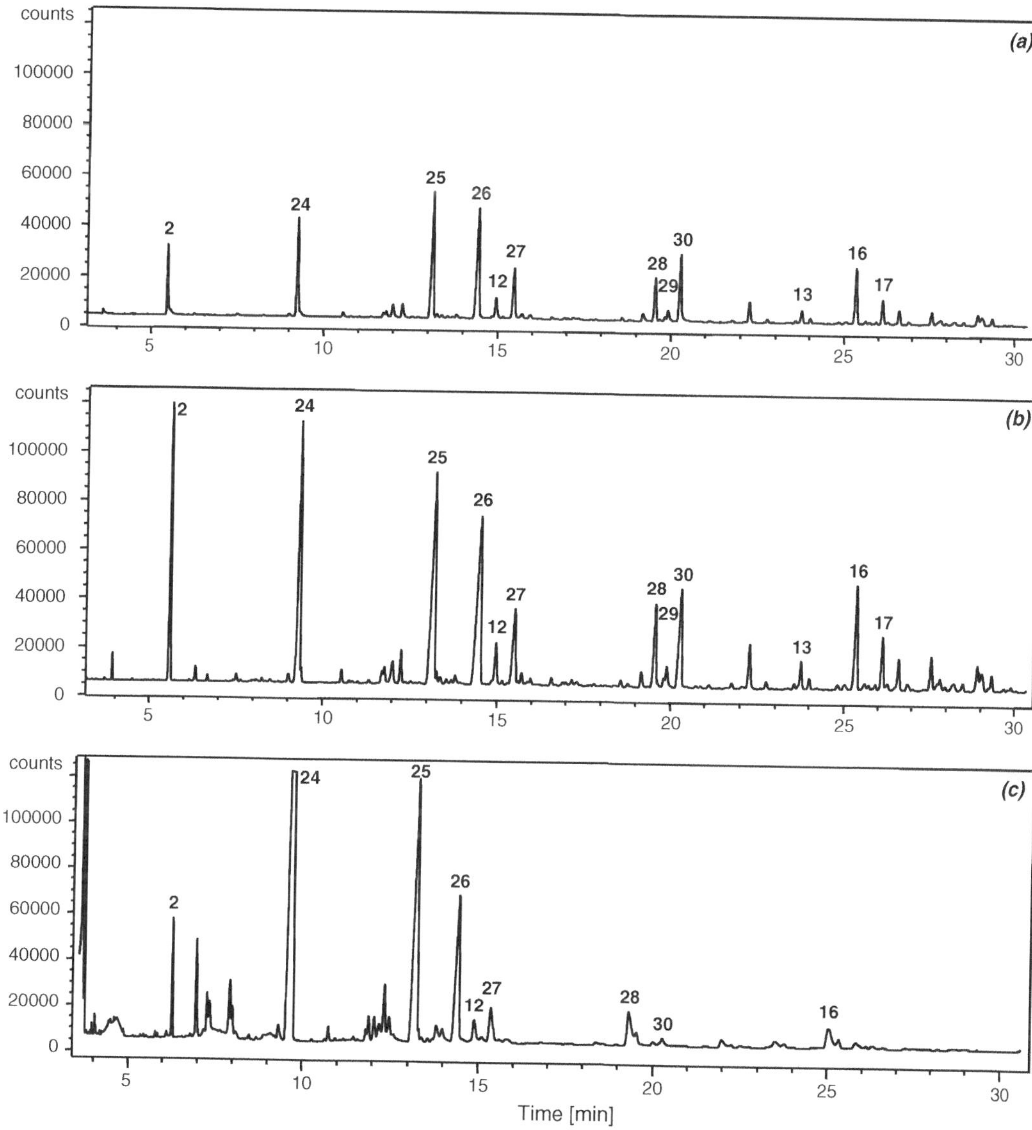

Figure 3.6 *HS-SMSE-GC-FID profile of sage dried leaves (Salvia lavandulifolia Vahl.) and structures of the identified compounds* **2**, **12–17** *(see Figure 3.2) and* **24–30** *sampled by: (a) empty PDMS tubing, (b) PDMS tubing filled with cyclohexane, (c) PDMS tubing filled with ethyl acetate. HS-SMSE sampling: PDMS tubing: 4 cm long, 1.5 i.d., 0.5 mm thick. Sample amount: 20 mg. Vial volume: 20 ml. Sampling temperature: 50 °C. Sampling time: 20 min. Solvent volume: 120 μl (20 μl for cyclohexane). GC-FID system: HP 5890 series II. Column: FSOT OV-1 (MEGA, Legnano, Italy) d.f. 0.25 μm, i.d. 0.25 mm, 125 m. Carrier gas: He; flow rate: 1.0 ml min^{-1}. Oven: temperature programme: from 50 °C (1 min) to 180 °C at 3 °C min^{-1} then to 270 °C (5 min) at 20 °C min^{-1}. TDS conditions: desorption: from 40 °C to 250 °C (5 min) at 60 °C min^{-1}; flow mode: splitless; transfer line: 250 °C. Injection system: Gerstel CIS-4 PTV injector, cryogenic fluid: CO_2; split mode, split ratio 1 : 10, injection temperature: from –50 °C to 280 °C (5 min) at 600 °C min^{-1}*

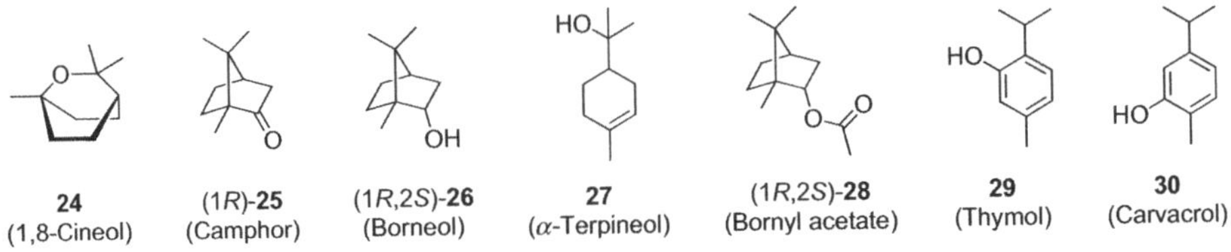

Figure 3.6 *(continued)*

drop generated at the tip of a microsyringe needle, which enables the drop to be retracted back into the needle after sampling, and then to be directly injected into the GC system. Figure 3.7 shows a diagram of a HS-LPME sampling system.

Since its introduction HS-LPME appeared to be very attractive, because it is cheap, easy to apply, to automate and to combine online and offline with other extraction and distillation sampling techniques and can reliably be used for quantitative analysis. But, its most important advantage is flexibility related to the wide range of nonpolar, polar and water miscible solvents that can be used. The main limitation on the solvent is that its vapour pressure must be low enough to avoid evaporation during sampling but, at the same time, it must be compatible with GC analysis, and produce stable drops to avoid its falling from the needle.

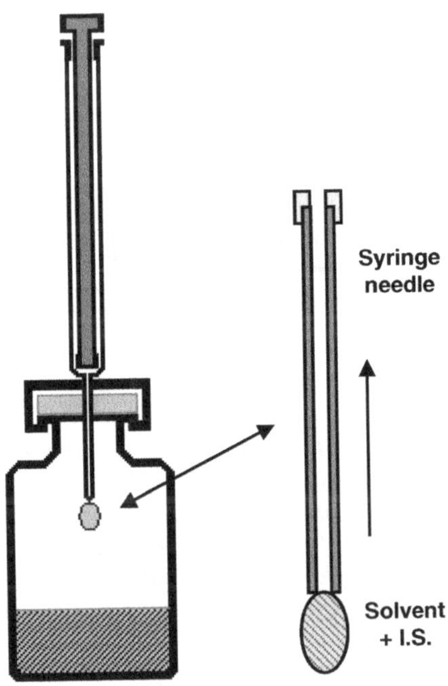

Figure 3.7 *Diagram of a HS-LPME sampling system*

HS-LPME in S-HS mode has been widely used to analyse the volatile fraction of plant matrices, or those of products derived from their transformation: its applications have recently been reviewed by Bicchi *et al.*[10] The most used solvents for plant matrices are 1-octanol and hydrocarbons with different volatility

3.2.9 Large Surface Area High Concentration Capacity Headspace Sampling

One of the chief parameters conditioning analyte recovery in HCC-HS techniques is the area of sorbent (adsorbent) exposed to the HS.[26,27] A number of techniques focussing on this parameter have been proposed. The first described technique was the membrane extraction sorbent interface method (MESI) proposed by Yang *et al.*[41] in 1994 for in-solution sampling and applied to HS sampling in 2000 by Segal *et al.*[25] MESI is a technique whereby a PDMS membrane acts as a selective 'filter' and the analytes are dynamically accumulated on a conventional polymeric trap.[42] The technique comprises four steps: (*a*) extraction of the analytes from the vapour phase of a matrix flowing on a thin film PDMS membrane, (*b*) diffusion through the membrane, (*c*) stripping from the other side of the membrane into a flowing gas and (*d*) concentration by a sorption (adsorption) trap. The sampled analytes are then recovered by thermal desorption of the trap and online analysed by GC or GC-MS. Wang *et al.*[43] applied MESI to monitoring the volatiles emitted by *Eucalyptus dunnii* into indoor air and developed a portable system for on-site monitoring of biogenic emissions from its leaves, consisting of a MESI device (PDMS membrane, Tenax microtrap) coupled online with a GC.[44]

The large surface of PDMS membranes has also been exploited directly to achieve high yields in recovering analytes from both liquid and vapour phases. Bruheim *et al.*[26] showed that a higher extraction efficiency in a shorter time and with better sensitivity could be achieved with a PDMS thin sheet than with a thick-film PDMS-coated SPME fibre because of the larger surface area/extraction phase volume ratio, for both in-solution and S-HS sampling.

More recently, Sisalli *et al.*[45] introduced sorptive tape extraction (STE) to study sebum composition and hence to determine the effect of a matifying product on skin shininess through *in vivo* sampling at the human skin surface. Sebum was sampled before and after cosmetic treatment by means of a thin flexible PDMS tape directly in contact with the skin surface for a fixed time. Analysis was by GC or GC-MS after analyte recovery by either thermal or solvent desorption. The increased recovery and sampling speed obtained with the wide exchange surface offered by PDMS tapes was applied by Bicchi *et al.* to analyse plant volatile fractions, in particular to monitor chemical messages emitted from plants (or animals) when a variation in their metabolism occurred as a consequence of stress.[27] Figure 3.8 is a scheme of an HS- and a direct contact (DC)-STE sampling system and a PDMS tape. PDMS tapes were used for sampling aromatic plants and fruits both in static HS mode and in combination with DC surface sampling.

3.3 Analysis

As for all fields, the analysis of the volatile fraction also consists of a number complementary steps: separation, identification, quantitative determination of components (when

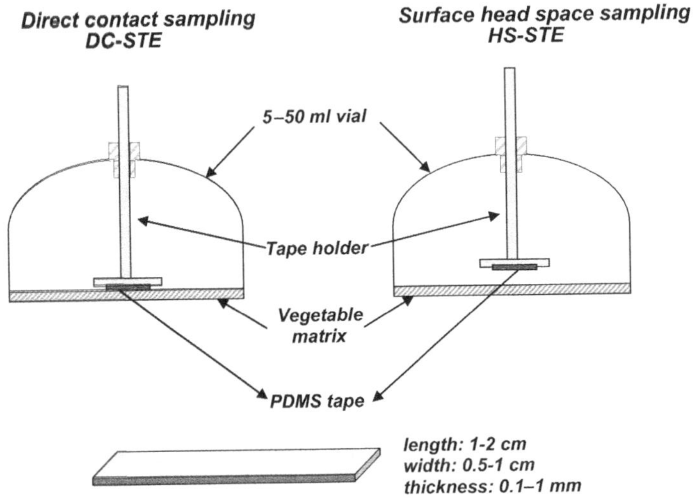

Figure 3.8 *Diagram of an HS- and a direct contact (DC)-STE sampling system and a PDMS tape*

required) and further data processing to increase the amount of information provided by the analysis. The volatile fraction of many plants is complex and is characterized by groups of volatile and medium to low polarity components with similar structures and physicochemical characteristics (e.g. monoterpenoids): these characteristics make capillary GC the technique of choice for volatile fraction analysis. In spite of the high intrinsic GC separation efficiency, the homogeneous composition and complexity of the volatile fraction of many plants requires two different-polarity stationary phases to concur so as to separate it exhaustively and reliably. The most widely used apolar stationary phases in routine analysis are those based on methyl polysiloxanes (SE30, OV-1, OV 101, DB-1, HP-1, PS 347.5, etc.) and methyl-phenyl-polysiloxanes (SE-52, SE-54, DB-5, HP-5, PS-086, etc.) while the most popular polar phases are those with PEG (PEG-20M, CW-20M, DB-Wax, etc.). Recently, the medium-polarity cyanopropyl-phenyl polysiloxanes (i.e. OV-17, OV-1701, DB 1701 and similar) have also gained ground because their separation capability lies between those of the apolar and polar stationary phases.

Identification is usually done by GC-MS, that is by combining chromatographic (Kováts indices, linear retention indices, relative retention time, locked retention times) and mass spectrometric data; the introduction of highly reliable, sensitive and relatively cheap benchtop mass spectrometric detectors is gradually phasing out the use of universal detectors such as a flame ionization detector (FID) or a thermal conductivity detector (TCD), because of the fundamental contribution that the mass spectrum gives to reliable component identification.[46] Fourier transform infrared spectroscopy (FT-IR) has also been proposed as a detector for GC (GC-FT-IR)[47] but, although it is closely complementary to MS in component identification, its use has quite unexpectedly been almost abandoned after the encouraging success at its introduction in the mid-1980s.

A satisfactory survey on the state of the art of analysis of the volatile fraction cannot leave out some basic considerations on modern approaches to analysis. An updated view of the analysis of the volatile fraction must be based on the extended concept of dimension that, as described below, cannot be limited to separation but must involve all steps of the analytical process, that is sample preparation, separation, detection (identification and/or quantitation) and data elaboration, provided that they can directly or indirectly be connected online. Moreover, the application of modern chemiometric and statistical methods and software evolution can provide information in greater depth, thus giving a decisive contribution not only to the interpretation of the meaning of the volatile fraction or biological phenomenon involved, but also to defining the analysis strategy. For instance, thanks to the discrimination capability of chemometric methods, sampling can be connected to MS detection directly, thus avoiding the separation step that is indispensable in a conventional analysis (e.g. HS-MS without GC separation, or direct flow injection-MS without previous LC separation). The next sections of this chapter discuss the most recent achievements for routine work, in fast-GC (F-GC) combined with FID and MS, reliable automatic identification approaches, multidimensional GC, enantioselective-GC (ES-GC) combined with FID and/or MS (ES-GC-MS), comprehensive two dimensional gas chromatography (GC × GC) and some of the new approaches recently introduced. Last but not least, special attention must be paid to quantitative analysis, both because its interest is continually increasing and because its present status is not entirely clear.

3.3.1 Fast-GC and Fast-GC-qMS EO Analysis

The demand for analysis control is continually increasing, even with the volatile fraction. One way to meet this demand is to adopt fast chromatographic techniques, which not only enable sample throughput and laboratory productivity to be increased, but also reduce analysis costs. The aim of fast chromatographic techniques [and in particular that of fast-GC (F-GC)] is to reduce analysis time to a minimum, while keeping the separation and producing reliable qualitative and quantitative results. Although investigated since the early 1960s, only in the past decade high-speed GC has been used for routine analysis,[48] in particular the analysis of essential oils. The speed of a GC analysis is defined by average peak width, analysis time and temperature rate.[49,50] Today, it is generally accepted that a GC analysis is 'fast' when it runs in less than 10 min, with column inner diameters (i.d.) between 0.25 and 0.1 mm, length 5 to 15 m, temperature programmes 20–60 °C min^{-1} and peak widths in the 0.5 to 5.0 s range. Further, a GC analysis is considered 'ultra fast' (known as ultra fast module GC, UFM-GC) for analyses of 1 min or less carried out with short (2–10 m) narrow bore (NB) columns (0.1–0.05 mm i.d.), temperature rates above 1 °C s^{-1}, and peak widths of 0.05 to 0.5 s.[51] Temperature rates above 100 °C min^{-1} cannot be achieved with a conventional GC (C-GC) oven, but can be achieved only by resistively heated columns.[51]

Two approaches are now available to speed up the analysis of a plant volatile fraction. The first and most popular one is based on short columns with narrow i.d. (0.1 mm or less) F-GC, it was first applied to essential oil analysis by Proot and Sandra.[52] The second approach adopts short capillary columns with conventional i.d. (SCC-GC).[53] In both cases, speeding up the GC separation is achieved with shorter than usual columns that, however, must maintain a suitable efficiency to achieve the required separation. The other parameters affecting the speed of a GC analysis are stationary phase, film thickness, temperature programme and linear carrier gas velocity. The various theoretical and practical aspects involved with F-GC have been reviewed by Cramers *et al.*[48] and David *et al.*[54]

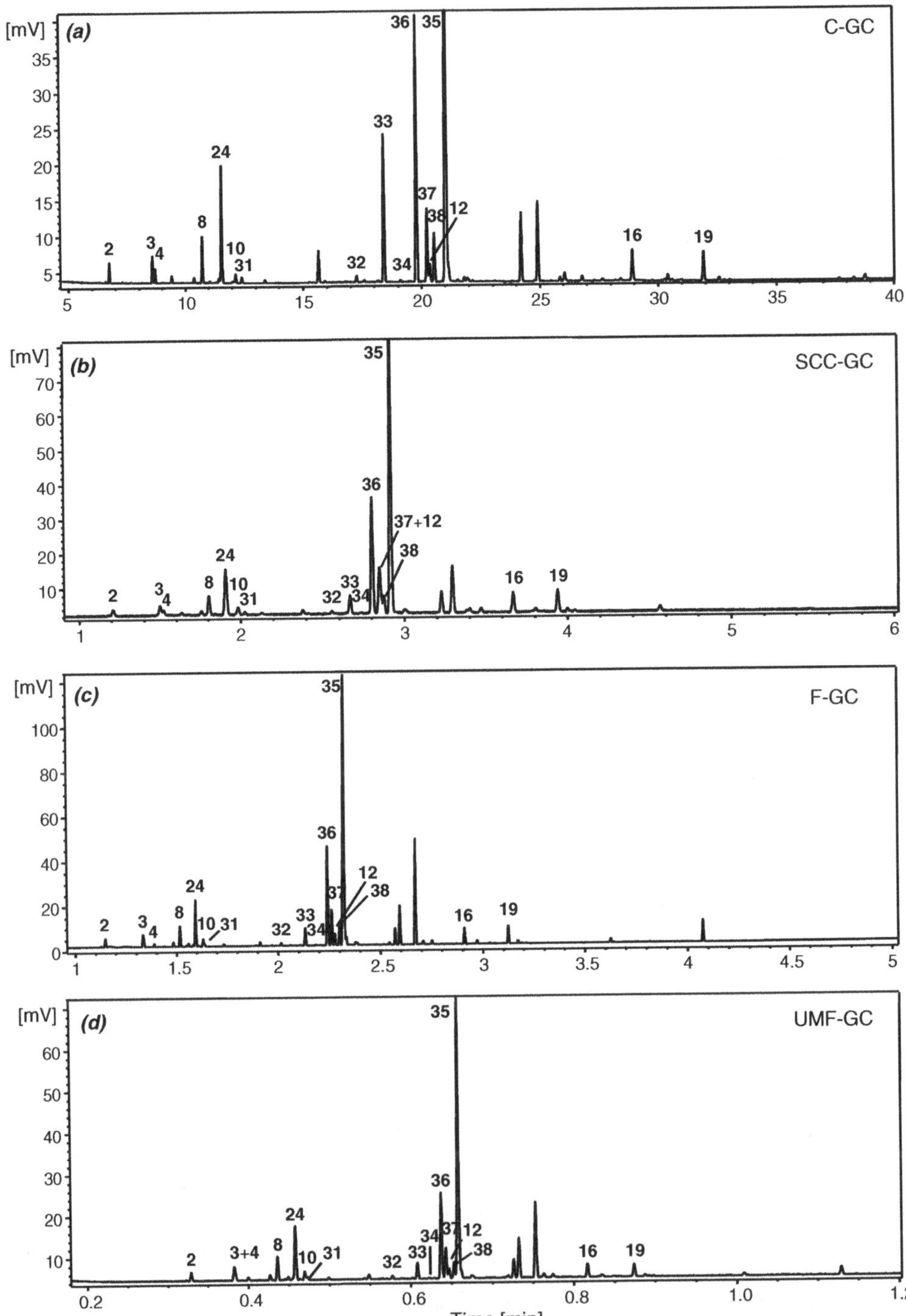
[mV]
(a)
C-GC
36 35
33
24
37
38 12
8
16
19
2
3 4
10
31
32
34
5 10 15 20 25 30 35 40
(b)
SCC-GC
35
36
37+12
38
24
16
19
8 10
33
2
3 4
31
32 34
1 2 3 4 5 6
(c)
F-GC
35
36
12
24
37 38
8
16
19
2
3 4
10 31
32
33 34
1 1.5 2 2.5 3 3.5 4 4.5 5
(d)
UMF-GC
35
36
24
34 37 12
3+4
8
10 31
32
33
38
16
19
2
0.2 0.4 0.6 0.8 1.0 1.2
Time [min]

The relationship between the main parameters is illustrated by the following simplified equation, which expresses retention time as a function of the main parameters influencing the speed of a GC analysis:

$$t_r = \frac{L}{u}(k+1)$$

where L is the column length, u the linear carrier gas velocity, and k the retention factor. This equation shows that the speed of a GC analysis with a conventional GC instrument can be increased by applying a higher linear velocity u, and lower k values, as well as decreasing column length. Columns (5–10 m) with narrower i.d. (0.1 mm) can be used to compensate for the loss of efficiency due to the increased u, while higher temperature programmes (20–60 °C min^{-1}) and thinner film stationary phases are to reduce analyte k values. Under these conditions, F-GC can reduce analysis time by a factor of between four and eight compared to a conventional GC analysis, depending on the complexity of the matrix investigated, and can also successfully and effectively be combined with a conventional quadrupole MS detector (qMS).[48,50,54]

The SCC-GC approach is based on the consideration that (*a*) the efficiency of a capillary column is frequently much higher than necessary for a given separation and (*b*) a *rational* reduction of efficiency (i.e. column length) can produce a considerable saving of analysis time. This approach can successfully be applied to routine quali-quantitative analysis of medium-complexity volatile fractions (up to about 30–40 components). In this case, too, effective separation can be achieved with an analysis time reduced by a factor between four and eight by adopting a column of 5 m (instead of 30 m) in combination with a suitable temperature programme, although the efficiency (i.e. the number of theoretical plates) of the chromatographic system is reduced by a similar factor. If the efficiency of the chromatographic system is insufficient for a given separation, it is possible to overcome this lack by a careful selectivity tuning by adopting a stationary phase more suitable for that specific separation.[53] Figure 3.9 reports the GC-FID profiles of a peppermint essential oil analysed by (*a*) conventional GC, (*b*) SCC-GC, (*c*) F-GC and (*d*) UFM-GC and shows that the analysis time can be reduced from 30 min for conventional GC to 60 s for

Figure 3.9 *Peppermint (Mentha x piperita L.) EO GC-FID chromatograms obtained by: (a) Conventional GC column: OV-1701 25 m × 0.25 mm i.d. × 0.3 μm d.f. (MEGA, Legnano, Italy); carrier gas: H_2, constant flow: 1.5 ml min^{-1}; injection mode: split, split ratio: 1/30; temperature programme: from 50 °C (1 min) to 250 °C (1 min) at 3 °C min^{-1}; (b) Short column C-GC column: OV-1701 5 m × 0.25 mm i.d. × 0.3 μm d.f. (MEGA, Legnano, Italy); carrier gas: H_2, constant flow: 1.5 ml min^{-1}; injection mode: split, split ratio: 1/50; tempertaure programme: from 50 °C (1 min) to 220 °C (1 min) at 30 °C min^{-1}; (c) F-GC-GC Column: narrow bore OV-1701 10 m × 0.10 mm i.d. × 0.10 μm d.f. (MEGA, Legnano, Italy); carrier gas: H_2, constant flow: 0.5 ml min^{-1}; injection mode: split, split ratio: 1/300; temperature programme: from 50 °C (0.3 min) to 250 °C (1 min) at 40 °C min^{-1}; (d) UFM-GC-Column: OV-1701 5 m × 0.10 mm i.d. × 0.10 μm d.f. (MEGA, Legnano, Italy); carrier gas: H_2, constant flow: 0.5 ml min^{-1}; injection mode: split, split ratio: 1/300; temperature programme: from 50 °C (0.1 min) to 250 °C (1 min) at 150 °C min^{-1}. All analyses were carried out on Trace GC 2000 series (Thermo, Milan, Italy) by applying an injection temperature of 230 °C and a FID temperature of 250 °C*

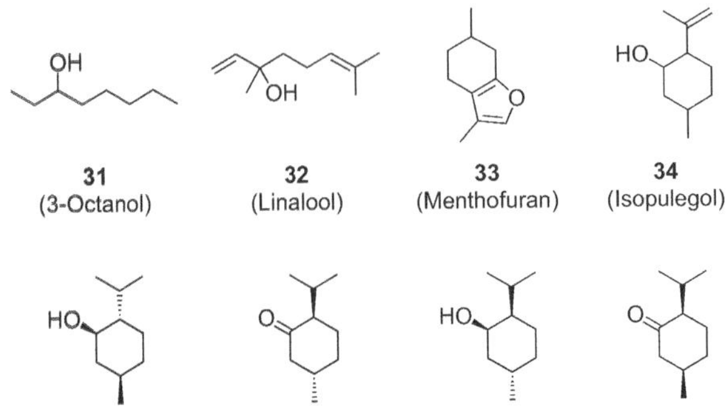

31 (3-Octanol) **32** (Linalool) **33** (Menthofuran) **34** (Isopulegol)

(–)-**35** (Menthol) (+)-**36** (Menthone) (–)-**37** (Neomenthol) (+)-**38** (Isomenthone)

Figure 3.10 *Structures of compounds found in a typical commercial peppermint EO sample*

UFM-GC while maintaining good separation of the 17 different compounds indicated (**2–4**, **8**, **10**, **12**, **16**, **19**, **24**, **31–38**, Figure 3.10).

A fundamental contribution to the use of F-GC in routine analysis was given by Klee and Blumberg,[55] who introduced the concept of method translation, from which it is possible to derive F-GC analysis conditions automatically from those of the corresponding conventional GC method; dedicated software is available on the internet.[56] This approach is based on the classification of parameters influencing the speed of a GC analysis into translatable and nontranslatable. Translatable parameters are column length, i.d. and film thickness (d.f.), carrier gas, flow rate, proportional changes in heating rates, duration of temperature plateau and detector working at reduced pressure (MS). Nontranslatable parameters are stationary phase, phase ratio and initial and plateau temperatures. The concept of method translation is based on the void time (t_M), which can be taken as the fundamental time unit in chromatography and can be used to express time-related components in all chromatographic metrics. In a normalized temperature programme, the duration of each temperature plateau and the heating rates are expressed in the units t_M, measured at the same temperature. Two methods are mutually translatable if they have identical nontranslatable parameters and the same normalized temperature programme.

The routine use of F-GC has also been made possible by the improvement of instrumentation, in particular by the introduction into routine, about 15 years ago, of electronic pressure control of the mobile phase, highly effective oven temperature regulation and controls, direct heating of columns for UFM-GC, high-speed detectors such as high-frequency FID, high-speed qMS and time of flight (TOF) mass spectrometers.

F-GC analysis has also benefited from its perfect compatibility with mass spectrometers having quadrupole analysers, that is the most popular MS detectors used in routine analysis due to their reliability and acceptable cost. The state of the art of the F-GC-qMS combination was exhaustively and critically reviewed by Maštovská and Lehotay.[57] Rubiolo *et al.*[58] recently investigated in depth the compatibility of F-GC-qMS in essential oil analysis, in a study dealing with the separation, identification and quantitation of ten components characteristic of peppermint essential oils. Their results showed that, operating at a suitable scan speed in the total ion current (TIC) mode (from 999 to 11 000 amu s^{-1}) or at suitable dwell time in the single ion monitoring (SIM) mode (10 to 50 ms), the results obtained with F-GC-qMS with a 10 m, 0.1 mm i.d. column combined with temperature programmes from 20 to 60 °C min^{-1} are fully comparable, and in some cases better, than those obtained by conventional GC qMS, while the analysis time is reduced by a factor of about ten (from about 35 to 3–4 min). Scan speed and dwell time play a fundamental role because:

- They influence the separation performance of the F-GC-qMS system at higher temperature programmes. This is, because they contribute to the correct definition of the peak shape, through the number of points per peak that, in turn, defines peak width and, as a consequence, GC separation.
- They condition the correct peak area determination in both TIC and SIM-qMS modes at different F-GC speeds.
- Their value depends on the required speed of the GC separation.

A further reduction of analysis time can be obtained by applying a temperature programme above that indicated by the translation software, although to the detriment of the separation power of the chromatographic system and with possible peak overlapping. In target analysis, the adoption of a qMS detector can overcome this limit by operating in SIM with suitable diagnostic ions (see below).

3.3.2 Qualitative Analysis

Identification of a component after GC separation is most often achieved by MS, and the GC contribution is often overlooked, not least because identification through a mass spectrum is highly reliable and specific. However, in the analysis of a plant volatile fraction, MS may fail because the volatile fraction often consists of complex groups of mono- and sesquiterpenoids with very similar molecular weights, structures, polarities and volatilities that, in consequence, give very similar mass spectra. In these cases, GC data play a fundamental role in correct identification. Retention indices (*I*s) are the most reliable and effective tool to define the GC behaviour of an analyte; they were first introduced by Kováts[59] for isothermal analysis and then by van den Dool and Kratz[60] for temperature programmed analysis (in which case they are better known as linear retention indices, I^Ts). The I^T specifically contributes to analyte identification because it 'measures' its physicochemical interaction with the adopted chromatographic system (or better with a given stationary phase). Analyte identification by I^T and mass spectrum together in GC-MS is therefore highly reliable because the two methods are highly orthogonal, being based on completely different principles. Most GC-MS software still does not include I^T as an identification criterion, and only a few packages include I^Ts in their library as 'inactive' data, appearing in the legend of each identification record. This means that I^T is only useful for further or additional offline confirmation. On the contrary, the 'interactive' use of I^Ts and mass spectral matching can be very useful, since it offers the possibility of operating simultaneously and synergically with two independent tools to identify a compound. The use of I^T in the analysis of essential oils has recently been reviewed by d'Acampora Zellner *et al.*[61] Only a few commercially available mass spectral library software packages include retention index information to facilitate peak assignment.[62–65] The flavour and fragrance natural and synthetic compounds (FFNSC) MS Library[63] system developed by Costa *et al.*[66] is equipped with an interactive tool to calculate I^Ts automatically, and it incorporates I^Ts as an active part of the matching criteria in characterizing and identifying flavour and fragrance components, thus dramatically increasing the reliability of the identification. A second software package actively using I^Ts is automatic mass spectral deconvolution (AMDIS),[64] which is often used in combination with NIST Mass Spectral Libraries[65] (both developed by the National Institute of Standards and Technology, USA). This programme also enables mass spectral libraries containing retention index information to be used through a retention index matching window on raw data automatically converted into the AMDIS format.

A further method for compound identification in programmed temperature analysis using only retention behaviour without index calculation is retention time locking (RTL).[67] The RTL approach is based on analyte retention behaviour for programmed temperature analysis and on a careful tuning of the carrier gas inlet pressure to provide identical retention time for the same compound in any system equipped with the same 'nominal' column. This method operates only in constant pressure mode and affords only small deviations in column size and stationary phase loading parameters. A commercially available RTL-based software package (Flavfid) is available; it can be combined with an additional mass spectral library when operating in GC-MS based on retention time-locked GC-FID and GC-MS data in the identification process.[68]

3.3.3 Quantitative Analysis

The demand for the quantitative analysis of volatile fraction components is growing continually, mainly due to the increasing need for controls to verify the quality, safety and biological activity of vegetable matrices. It is not a simple matter to deal exhaustively with quantitation in the analysis of a volatile fraction, not least because it is often approached ambiguously. As mentioned above, a quantitative analysis must involve both steps of the whole analytical procedure: sample preparation and analysis itself. For vegetable matrices, whatever method is used to sample the volatile fraction (steam distillation, hydrodistillation, mechanical procedures or HCC-HS techniques) some basic precautions must be adopted, in addition to the usual ones, to achieve reliable quantitative results: (*a*) the natural variability of a plant matrix must be reduced to that of the plant itself by rigorous standardization, during both collection and sample treatments, (*b*) a suitable number of samples must be analysed to obtain a representative composition of the volatile fraction of the species investigated (i.e. the average composition from at least three samples from *distinct* populations).

This section only discusses the analysis step, since an exhaustive discussion of sample preparation would require a dedicated chapter. Quantitative analysis of essential oils has recently been reviewed by Bicchi *et al.*[69] The quantitative composition of most essential oils or headspaces is very often reported in the literature as relative percentage abundances although, unfortunately, these data only give an approximate indication of the ratio between components, and *only* for the sample under investigation.

A single absolute approach to quantitation cannot be proposed because of the different physical states and chemical characteristics of the matrices investigated and because of the wide variety of data required. A rapid survey of the relative literature showed that the most widely used approaches with EOs are: (*a*) relative percentage abundance, (*b*) internal standard normalized percentage abundance, (*c*) 'absolute' or true quantitation of one or more components (target analysis) and (*d*) quantitation by a validated method.

As already mentioned, relative percentage abundance is the most commonly used approach, although it is very often used incorrectly, in particular when compositional data are used to compare samples from the same species. The only correct use for relative percentage abundance results is to evaluate relative component ratios within the same sample. When a group of samples is compared, raw data must first be normalized versus an internal standard (or at least versus an external standard if an automatic injector is available) and the percentage abundance *must* be calculated versus the sum of the areas of a fixed number of selected components taken as markers, usually common to all investigated samples. When normalized percentage abundance is applied, the nature of the detectors has also to be considered. The most critical detector characteristics are linearity and analyte response factors, which must be as close as possible to one. FID is the most popular GC detector, being universal, highly sensitive and robust, but it is well known that its response factors with some compounds can be quite far from one, depending on the compound's structure, for example for some esters it can be as high as 1.6, compared to *n*-nonane, taken as the internal standard.[70] The use of MS as a detector for quantitation is continually increasing, because of its simultaneous contribution to component identification, although it cannot be used for normalized percentage abundance, because the analyte structure conditions, not only the formation of ions but also their abundances, are mass-sensitive. In quantitation, MS can correctly be used only in SIM mode, which also increases the reliability of identification, since it combines diagnostic ions with chromatographic retention data (i.e. linear retention indices, locked retention times, relative retention times). The recent introduction of GC-MS systems operating alternately in SIM and scan modes on the same peak, made possible by the high scan rate and short interscan delay of the latest generation quadrupole systems, opens promising perspectives to achieve simultaneous identification and quantitation of an analyte.

In any case a normalized percentage abundance (i.e. a semi-quantitative comparison of GC profiles) is not sufficient to solve all quantitation problems concerning the volatile fraction, in particular with essential oils. An essential oil, or a complex matrix containing it, such as cosmetics, food or pharmaceuticals, must very frequently be characterized through the concentration (ppm or mg l^{-1}) or, less frequently, through the absolute amount of one or more of its quality markers or components restricted by legislation.

This approach involves the true quantitation of a marker, in general by official and/ or literature methods, as for instance is the case of the determination of α- and β-thujone (**39** and **40**, Figure 3.11) in bitter liqueurs,[71] or suspected allergens in cosmetics.[72] The marker concentration or absolute amount in a real-world sample is determined from its chromatographic area normalized versus an internal (or external) standard and calculated

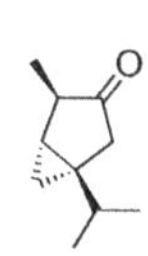
39
(α-Thujone)

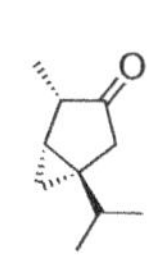
40
(β-Thujone)

Figure 3.11 *Structures of* α*- (**39**) and* β*-thujone (**40**)*

via a calibration curve constructed from amounts of pure marker standard in the operative concentration range. Since pure standards are not always available commercially, or are difficult and time-consuming to isolate, an accepted compromise is to adopt compounds belonging to the same class (hydrocarbons, aldehydes, alcohols, esters, etc.) and with structures as similar as possible to the analyte under investigation.[70] Cicchetti *et al.*[73] recently compared the methods usually applied in flavour, fragrance and essential oil quantitative analysis. The best results were obtained by true internal standardization in MS and FID, although it is a time-consuming procedure for mixtures of many components. They also showed the importance of analyte response factors for a reliable quantitation and proposed an approach to build up a response factor database. In addition, for GC-MS analysis, specific isotope-labelled internal standards (in general ^{2}H- or ^{13}C-labelled compounds) can also be used: although it is expensive, a labelled standard assures a response factor equal to or very close to that of the corresponding marker, and it gives the same fragmentation pattern, but with a known increase in molecular weight, easily detectable by MS. The main practical advantage of using labelled standards is the possibility of obtaining a recovery of the target analyte equal (or at least very similar) to that of the labelled standard, thus avoiding its determination.

It must be stressed that true quantitation of all components in an EO is, in general, unrealistic for several reasons, including: (*a*) the unacceptably long time needed to develop the method, in particular for complex essential oils, (*b*) the commercial unavailability of the standards of most components, (*c*) the very high EO amounts necessary to construct reliable calibration curves.

Validated quantitative methods for biologically active EO components are increasingly required, in particular for those used in the pharmaceutical industry. The development of a method meeting all the conditions and parameters required for a 'pharmaceutical' validation is time-consuming (up to three months) and expensive, because of the large number of experiments and data to be collected.

Full validation of a quantitation method for all EO components is even more unrealistic and must therefore be limited to those applications where it is expressly required and, above all, to a small number of selected representative markers. The validation of a method must follow dedicated guidelines established by the international regulatory bodies and committees (Eurachem, CITAC, IUPAC, etc.);[74] its performance is evaluated through parameters such as selectivity, specificity, linearity over the working range, repeatability, precision, intermediate precision, reproducibility, limit of detection (LOD), limit of quantitation (LOQ), accuracy and assessment of uncertainty. A reliable analytical procedure must be developed before submitting it to validation; and the choice of the parameters to investigate must therefore be rational and depends on the nature of the matrix. Repeatability, intermediate precision and linearity in the working range are indispensable for all quantitation methods. LOD and LOQ are mainly important for trace analysis; and reproducibility is necessary when different laboratories are involved in control.

Table 3.1 shows the quantitative results concerning a group of ten markers (**3**, **8**, **10**, **24**, **31–36**, Figures 3.2, 3.10) of peppermint EO, six of them indicated by the European Pharmacopoeia and included here to show the effectiveness of F-GC-qMS in true quantitation. A calibration curve was constructed for each marker, using six different calibration levels in the range 2–100 mg l^{-1} (i.e. 2–100 ppm); three more calibration levels had to be included for menthol (**35**) and menthone (**36**) because of their EO abundances.

Table 3.1 *Quantitative results calculated by SIM (target ion given in bold) of ten markers in a commercial peppermint EO sample obtained by conventional GC (C-GC) and fast GC (F-GC) expressed as g/100 g and relative standard deviation (RSD) calculated versus C-CC*

		Conventional GC 3 °C min^{-1}			Fast GC 20 °C min^{-1}				Fast GC 60 °C min^{-1}			
Compound	SIM-MS ions	g/100 g[a]	LOD [mg l^{-1}]	LQD [mg l^{-1}]	g/100 g[b]	RSD[c]	LOD [mg l^{-1}]	LQD [mg l^{-1}]	g/100 g[b]	RSD[c]	LOD [mg l^{-1}]	LQD [mg l^{-1}]
β-Pinene (**3**)	69, 91, **93**	0.89	1.23	2.55	0.90	1.01	0.48	1.74	1.32	27.60[d]	0.09	0.60
Limonene (**8**)	**68**, 93, 136	1.32	1.79	3.70	1.33	0.46	0.95	2.53	1.31	0.34	0.23	0.87
γ-Terpinene (**10**)	91, 93, **121**	0.30	1.23	2.55	0.30	1.51	0.62	1.74	0.30	1.51	0.16	0.60
1,8-Cineol (**24**)	43, **81**, 108	5.31	3.08	6.38	5.22	1.20	1.71	4.36	4.94	5.06	0.47	1.50
3-Octanol (**31**)	55, **59**, 83	0.28	3.57	7.40	0.28	1.34	1.19	5.06	0.29	3.17	0.31	1.75
Linalool (**32**)	71, **93**, 121	0.28	3.76	7.78	0.27	1.64	1.10	5.32	0.28	0.54	0.23	1.84
Menthofuran (**33**)	79, **108**, 150	2.21	1.23	2.55	2.21	0.10	0.48	1.74	2.17	1.17	0.09	0.60
Isopulegol (**34**)	71, **81**, 121	0.12	12.69	26.29	0.12	0.63	4.62	17.97	0.13	7.25	1.05	6.20
Menthone (**35**)	69, **112**, 139	18.00	3.88	8.04	17.91	0.35	1.43	5.50	17.87	0.50	0.23	1.90
Menthol (**36**)	71, **81**, 95	43.14	4.56	9.44	43.54	0.65	1.67	6.46	42.87	0.44	0.31	2.23

Note: [a]Dwell time: 50 ms, temperature programme: 3 °C min^{-1}.
[b]Dwell time: 10 ms, temperature rates: 20 and 60 °C min^{-1}.
[c]Relative to C-GC taken as a reference.
[d]Coelution with sabinene (**4**).

Tridecane (ISTD1) and octadecane (ISTD2) were used as internal standards. The analyses were carried out by GC-FID and GC-MS in SIM and scan modes.

3.3.4 Enantioselective GC

The possibility of a routine chiral recognition of volatile fraction components is one of the most important advances in this field over recent decades. Enantiomer separation and the determination of the enantiomeric excess (*ee*) or ratio (*er*) are important not only because different enantiomers can have different biological activities (e.g. different odours, specificity with insects, etc.), but also because, through the definition of the biosynthetic pathway of their marker compounds, they can help with plant classification, confirm geographical origins, and/or contribute to defining the technological treatments a vegetable matrix has undergone, and/or its authenticity or possible frauds. Chiral recognition of underivatized analytes in GC analysis (ES-GC) has mainly been made possible in routine work thanks to the introduction of differently derivatized cyclodextrins as stationary phases. Cyclodextrins were first introduced by Sibilska and Koscielski in 1983,[75] who separated pinane (**41**) and carane (**42**) derivatives with a column packed with α-cyclodextrin (**43**) in dimethylformamide (Figure 3.12). Juvancz *et al.* in 1987[76] and Schurig and Nowotny in 1988[77] almost simultaneously used β-cyclodextrin (**44**) for capillary GC. In 1989, Nowotny *et al.*[78] first proposed diluting cyclodextrin derivatives in moderately polar polysiloxane (OV-1701) to improve their chromatographic properties and extend their range of operating temperatures. Since then, several hundreds of articles have been published dealing with the theory of ES-

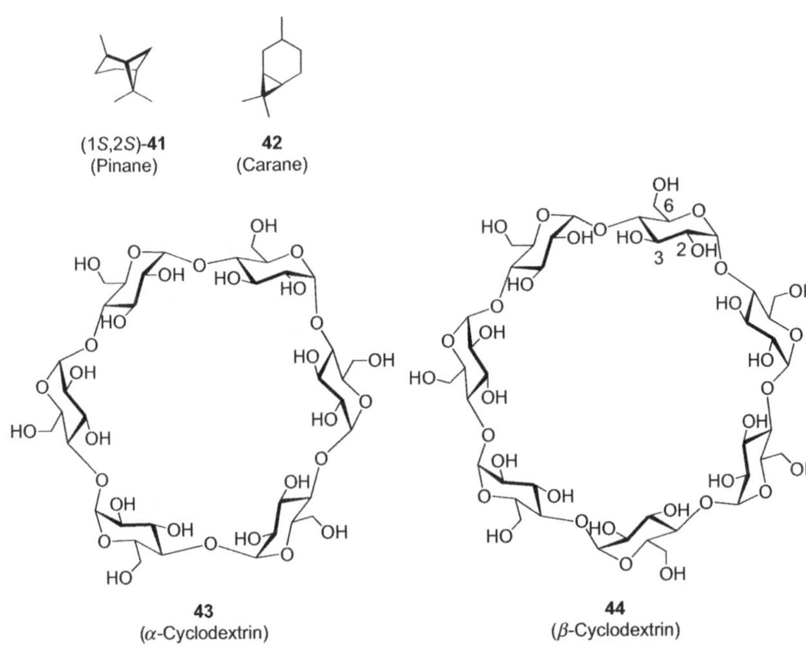

Figure 3.12 *Structures of pinane (**41**) and carane (**42**), and of α- (**43**) and β-cyclodextrin (**44**)*

GC chiral recognition with cyclodextrins, synthesis of new cyclodextrin derivatives, their enantioselectivity and applications; many of these concern the flavour and fragrance field.[79]

Chiral separation with cyclodextrins is based on the rather small difference in the energy of the host–guest interaction that each enantiomer of a racemate establishes with the cyclodextrin selector.[80] The intrinsic mechanism of chiral recognition in GC, which is not yet fully understood, is also one of the reasons for the lack of a universal cyclodextrin derivative (column) that can separate most of the significant racemates in this field. A laboratory must therefore be provided with at least two columns coated with different cyclodextrin derivatives if it is to be able to separate at least 80% of the commoner racemates in the flavour and fragrance field. The most effective cyclodextrin derivatives are at the present thought to be those consisting of β-cyclodextrins substituted at C(6) (i.e. the narrow side of the cyclodextrin) with a bulky group (*tert*-butyldimethylsilyl- or *tert*-hexyldimethylsilyl-)[81] and with alkylated and acylated groups at C(2) and C(3) (mainly methyl, ethyl, acetyl) on its wider side. The effectiveness of cyclodextrin derivatives as chiral selector for ES-GC is clearly shown by the separation of the chiral test components (**45–51**) developed in the authors' laboratory[82] analysed on 2,3-diethyl-6-*tert*-butyl-dimethyl-silyl-β-cyclodextrin (2,3DE6TBDMS-β-CD) diluted in PS-086, as reported in Figure 3.13.

Chiral recognition of marker compounds in complex real-world samples, such as those in the flavour and fragrance field, often requires a two-dimensional approach, because ES-GC may double the number of peaks of optically active analytes, thus making some parts of the chromatogram even more complex and, as a consequence, increasing the probability of incorrect *ee* and/or *er* determination. Two complementary but distinct strategies can therefore be adopted, beside the classical but time-consuming fractionation and isolation of the marker compounds:

- The first and best known strategy is based on a second dimension *in separation*. Chiral recognition is generally carried out either by conventional heart-cut GC-GC,[83] or by comprehensive GC (see next paragraph) when very complex samples or/and a very large number of components must be investigated simultaneously.[84]
- The second approach is based on a second dimension *in identification* using MS as a detector.

In this latter case, the enantiomer is located and identified by MS detection used interactively (or not) with I^Ts. Single- or multiple-ion monitoring MS (SIM-MS), carried out after careful selection of suitable diagnostic ions of the optically active marker(s) of the sample under investigation, can then be applied to overcome any peak jam and overlap, due to enantiomer peak doubling. This makes it possible to 'clean' the part of the chromatogram where the compounds of interest elute, making correct *ee* and/or *er* determination possible. Mass spectrometry is well known to be unable to discriminate between optical isomers, not being a selective chiral probe in this sense, and therefore giving indistinguishable spectra. As a consequence, MS cannot be used alone to determine which enantiomer is present in a sample, nor to establish the predominant one nor measure its *ee* and/or *er*. This makes it indispensable to combine it with chromatographic data (I^Ts) obtained with a column coated with a chiral selector suitable to separate the two enantiomers. In ES-GC-MS, the unequivocal identification of a given enantiomer in a complex mixture can therefore be achieved by using mass spectra (or diagnostic ion monitoring) to locate the enantiomers in

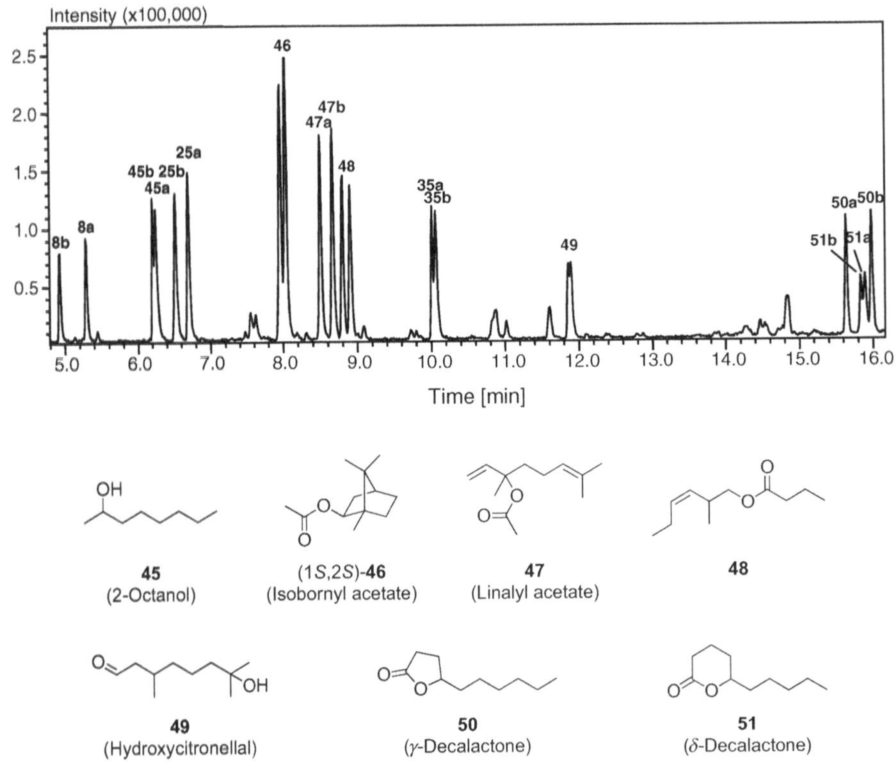

Figure 3.13 *ES-GC-MS chiral test profile and structures of compounds* ***45–51****. Peak identification: (a) the (R) enantiomer, (b) the (S) enantiomer. Analysis conditions: GC-MS system: Shimadzu QP2010 GC-MS system provided with Shimadzu GCMS Solution 2.51 software (Shimadzu, Milan, Italy). Column: 30% 2,3-di-O-ethyl-6-O-tBDMS-β-cyclodextrin in PS086, 5 m × 0.10 mm i.d. × 0.10 μm d.f. (MEGA, Legnano, Italy); injection mode: split; split ratio: 1 : 50; temperatures: injector: 220 °C, transfer line: 230 °C; ion source: 200 °C; carrier gas: He, flow rate: 1.0 ml min^{-1}. Temperature programme: from 50 to 220 °C (2 min) at 5 °C min^{-1}. MS ionization mode: electron impact (EI) at 70 eV, scan rate: 666 amu s^{-1}, mass range: 35–350 m/z*

the chromatogram and I^Ts to identify them. When a new compound is identified, the synthesis of at least one of the enantiomers is necessary to assign the correct elution order of the enantiomers in the chromatogram. Using this approach, Liberto *et al.*[85] described the construction of a MS library specific for the identification of enantiomeric compounds in the flavour and fragrance field using I^Ts in interactive combination with MS spectra. Figure 3.14 reports the enantioselective GC-MS profiles of: (*a*) linalool (**32**) in an adulterated bergamot EO (*Citrus bergamia* Risso et Poiteau) and (*b*) its racemate standard (- - -) analysed on a 2,3DE6TBDMS-β-cyclodextrin column.

MS as a second dimension in identification has also been successfully exploited to speed up ES-GC analysis.[86] A limitation of ES-GC in routine quality control is the long analysis time, since separation depends on the small difference in the energy of association between

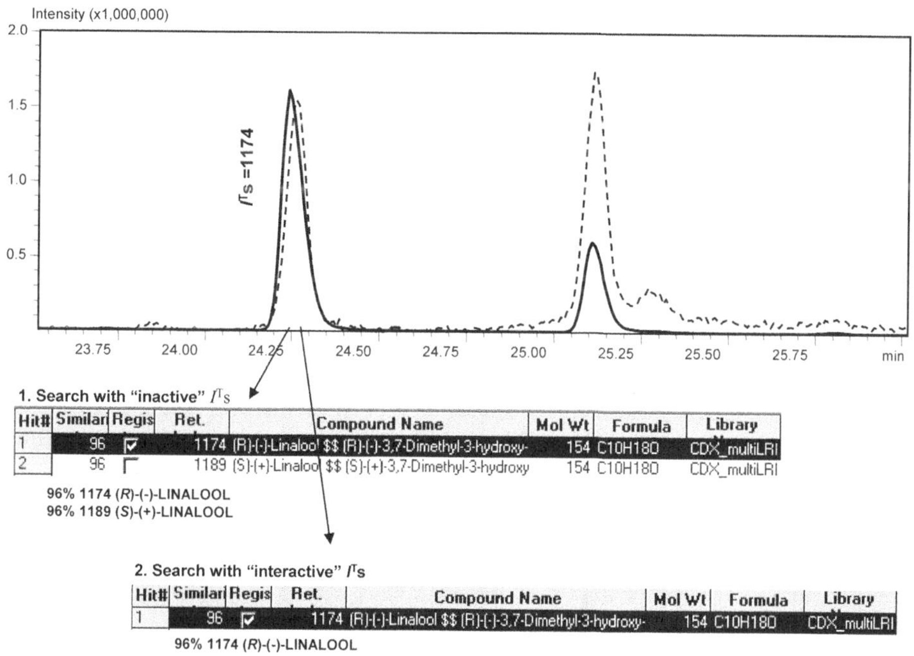

Figure 3.14 *Chiral recognition by ES-GC-MS of linalool (**32**) in adulterated bergamot (Citrus bergamia Risso et Poiteau) essential oil (——) and of its racemate standard (···). Library search with 'inactive' (1) and 'interative' (2) I^T values. Analysis conditions: ES-GC-MS system: Shimadzu QP2010 GC-MS system provided with Shimadzu GCMS Solution 2.51 software (Shimadzu, Milan, Italy). Column: 30% 2,3DE6TBDMS-β-CD in PS086 (25 m × 0.25 mm i. d. × 0.25 μm d.f.; MEGA, Legnano, Italy). Temperature programme: from 50 to 220 °C (2 min) at 2 °C min^{-1}. Injection mode: split; split ratio: 1 : 50. Temperatures: injector: 220 °C, transfer line: 230 °C; ion source: 200 °C; carrier gas: He, flow rate: 1.0 ml min^{-1}. MS ionization mode: electron impact (EI) at 70 eV, scan rate: 666 amu s^{-1}, mass range: 35–350 m/z*

each enantiomer and the cyclodextrin chiral selectors. Separation is thus strongly conditioned by temperature and requires very high chromatographic efficiency. As for F-GC, one of the commonest approaches to shortening analysis time is to adopt short conventional i.d. NB columns, which not only increase analysis speed and analyte detectability due to peak sharpening,[87] but also reduce enantiomer elution temperature, resulting in a gain of enantioselectivity that compensates (in full or in part) for the loss of efficiency (N). Fast-ES-GC can thus usually be achieved by acting on flow rate and short NB columns in combination with rather low temperature rates, which can only be increased at the expense of the extra base-line enantiomer resolution. This approach can therefore successfully be applied using: (*a*) short conventional and NB columns, (*b*) mass spectrometry as a detector to operate in both TIC and extract ion modes, to avoid problems due to coelution with other components in the *ee* and *er* determination, to locate the enantiomers in the chromatograms and I^Ts to identify them; (*c*) temperature rates up to 10 °C min^{-1} and (*d*) fixing a resolution limit of 1.5 to enable correct *ee* and *er* determination. The combination of these conditions decreased analysis time by as much as 85% compared to conventional i.d. columns, while keeping a resolution suitable for correct *ee* or *er* determination; thus, the number of samples analysable in a routine laboratory can be increased very markedly. The validity of this approach was illustrated through a series of experiments carried out on a series of pesticide and racemate standards and real-world samples in the flavour and fragrance field analysed with a set of columns consisting of a conventional i.d. column (25 m length, 0.25 mm i.d., 0.15 µm and 0.25 µm d.f.) coated with 30% 2,3DE6TBDMS-β-cyclodextrin in PS-086. These were taken as a reference, for comparison with those of a conventional i.d. short column (5 m length, 0.25 mm i.d., 0.15 µm d.f.) and different length NB columns (1, 2, 5 and 10 m long, 0.10 mm i.d., 0.10 µm d.f.).[87] The NB columns offering the most effective compromise between separation efficiency and analysis time were shown to be the 5 and 2 m columns, in combination with mass spectrometry as a detector, they were applied to lavender and bergamot essential oil analyses and enabled a threefold or greater reduction of analysis time, while separation of chiral markers remained unaltered. Figure 3.15 reports 2 m NB column ES-GC-MS profiles of bergamot (*Citrus bergamia* Risso et Poiteau) EO (top) and an extract ion profile of limonene (bottom).

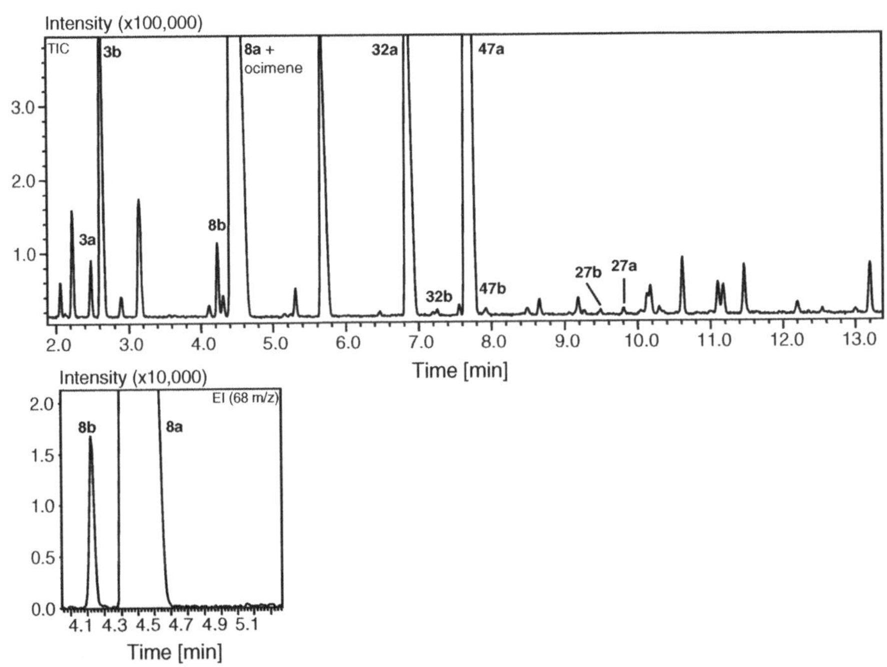

Figure 3.15 *ES-GC-MS profile of bergamot (Citrus bergamia Risso et Poiteau) EO (top) and extract ion profile of limonene (**8**, bottom). Peak identification: (a) the (R) enantiomer, (b) the (S) enantiomer. Analysis conditions: ES-GC-MS system: Shimadzu QP2010 GC-MS system provided with Shimadzu GCMS Solution 2.51 software (Shimadzu, Milan, Italy). Column: 30% 2,3DE6TBDMS-β-CD in PS086 (25 m × 0.25 mm i.d. × 0.25 µm d.f.; MEGA, Legnano, Italy). Temperature programme: from 50 to 220 °C (2 min) at 2 °C min^{-1}. Injection mode: split; split ratio: 1 : 50. Temperatures: injector: 220 °C, transfer line: 230 °C; ion source: 200 °C; carrier gas: He, flow rate: 1.0 ml min^{-1}. MS ionization mode: electron impact (EI) at 70 eV, scan rate: 666 amu s^{-1}, mass range: 35–350 m/z*

However, ES-GC alone is not always sufficient to detect the authenticity of a vegetable matrix, in particular with racemates of natural origin or when racemization is a consequence of processing and/or storage of the original product, or, more probably, when blending with a synthetic enantiomer has occurred. The most effective approach currently available to prove a vegetable sample authenticity is to combine ES-GC (or even better ES-GC-GC) with isotope ratio mass spectrometry (IRMS), since enantiomers from the same natural source can be expected to have the same ratio between stable isotopes (δ-^{13}C), even with partially racemized chiral molecules since, in a single organism, racemic compounds are generally formed through the same biochemical pathways. In an online coupled GC-IRMS system, analytes eluting from the GC column are combusted to carbon dioxide in a specially designed oven and directly introduced into an isotope ratio mass spectrometer, where the 44 ($^{12}C^{16}O_2$), 45 ($^{13}C^{16}O_2$, $^{12}C^{16}O^{17}O_2$) and 46 ($^{12}C^{16}O^{18}O$) mass ions are simultaneously measured with high precision ($\leq$0.3‰) in the nanomolar range, and the peak ratio is determined by the area ratio of two isotope peaks and compared to a standard value. A detailed discussion of this technique is outside the scope of this chapter.[88]

3.3.5 Multidimensional GC Techniques

The most exhaustive definition of multidimensional (MD) separation was given by Giddings in 1987,[89] who defined MD as ‘an orthogonal two column separation, with complete transfer of solute from the separation system 1 (column 1) to the separation system 2 (column 2), such that the separation performance from each system (column) is preserved’. This definition also stresses the concept of orthogonality of the two dimensions, that is the need for the phenomena underlying the separation achieved by the stationary phases coating the two columns to be different in nature.

Until a decade ago, multidimensional techniques were usually adopted to analyse those components of complex samples not separable by a mono-dimensional chromatographic system: the first column (first dimension) was used as a sort of ‘sophisticated clean-up’, producing a preliminary separation of the sample investigated (i.e. isolation of one or of very few components or fractions) which were then transferred online by dedicated time-programmable interfaces[90] to the second column (second dimension) coated with a different stationary phase, for further separation. With this approach, known as heart-cut GC-GC, only critical pairs or selected groups of compounds or markers could be analysed on the second dimension. In the flavour and fragrance field, heart-cut GC-GC is still widely used in chiral recognition:[83,91] the first column, coated with a conventional phase, locates the peak (s) of the racemate(s) to be transferred; and the second column, coated with a cyclodextrin stationary phase, separates the enantiomers.

However, heart-cut GC-GC does not fully meet Giddings’ definition, whereby *each peak* eluting from the first dimension has to be online and automatically reinjected into the second dimension. In 1991, Liu and Phillips[92] introduced GC × GC, with the aim of also fulfilling this condition. In GC × GC, each peak eluting from the first dimension is ‘cut’ into thin slices, each of which takes a fixed time (4–8 s), determined by cryogenic focusing achieved by a peak modulator. Each slice is then injected online into the second column, where it is analysed in a time equal to that of the successive modulation. A conventional GC × GC system therefore generally consists of a 0.25 mm i.d. column producing peaks at least 6–8 s wide, combined with a short NB column (less than 1.5–2.0 m) to analyse each peak slice, within the time of one modulation. A number of modulators have been proposed, including a dual-stage heated modulator, a sweeper modulator, a longitudinally modulated cryogenic system (LMCS), a dual cryogenic jet modulator and a loop modulator, the latter two now being those most widely used in routine analyses.[93] Whichever device is adopted, reliable results can only be obtained when at least the 99% of the original peak area is modulated roughly corresponding to four modulations per peak.[94] The orthogonality of the stationary phases of the column set is fundamental, because it is responsible for the optimization of the separation producing a spreading of the peaks over the chromatographic plane; the percentage separation space used is one of the parameters employed to evaluate the effectiveness of a GC × GC separation.[95]

Comprehensive GC × GC, in particular when combined with MS, is the most powerful separation technique now available. Since the F-GC separation in the second dimension produces peak widths never exceeding 200–300 ms, it requires high-speed detectors such as high-frequency FID, high-speed qMS and TOF MS. The resulting peak widths are at the limit of the performance of high-speed qMS, and thus it is often necessary to reduce the mass range so as to obtain a sufficient number of spectra for correct definition of the peak shape.

This technique was immediately successful for the analysis of ultra-high complex mixtures, in particular in the mineral oil industry, where it has proved capable of separating and detecting more than 15 000 components in a single run. Comparable success has been achieved in the flavour and fragrance fields, where very often samples with between 300 and 1000 components (e.g. coffee aroma) must be analysed; many of these components are highly significant in defining the quality or the sensory properties of the original vegetable matrix. In chiral recognition, GC × GC has also been successfully used when a high number of chiral compounds in a complex mixture must be separated but, in this case, the column 'geometry' must be inverted, because of the high efficiency required by the columns coated with cyclodextrins to obtain reliable separations (see above). Thus, the first column, coated with the enantioselective cyclodextrin stationary phase, and the second column, with a conventional stationary phase, together provide a complementary analyte separation based on volatility.[84]

The separation power of comprehensive GC × GC is evident when complex volatile fractions are analysed. Figure 3.16 reports the GC × GC contour plot of the volatile fraction of a standard roasted sample of hazelnut (*Corylus avellana* L.) from Piedmont (Italy) with more than 300 separated components (for further detail see below).[96]

3.4 Further Developments

The strategies used to analyse the volatile fraction have been very much conditioned by the evolution that has taken place in analysis techniques over the past 15–20 years. The online combination and integration, in a single step, of sample preparation, analysis and data elaboration, resulting in the TAS, has dramatically changed the approach to analysis, particularly in routine work. The elaboration of analytical data by chemometric and statistical methods has made a large contribution to increasing the amount of information obtainable. Appropriate data mining in combination with statistical elaboration can provide not only reliable sample discrimination and characterization, but also the 'identification' of

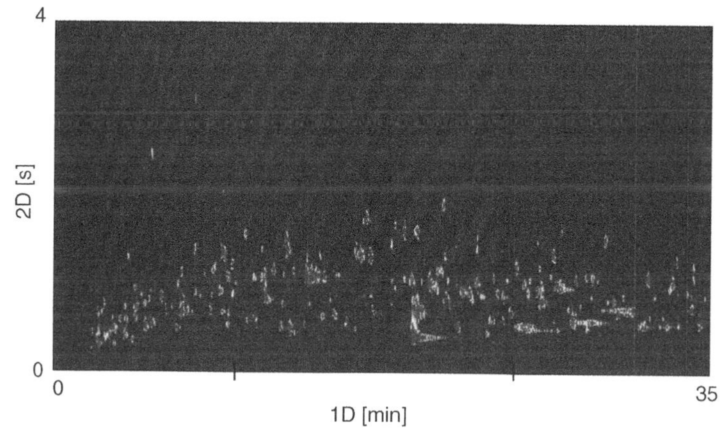

Figure 3.16 *2D patterns of the volatile fraction of a standard roasted hazelnut (Corylus avellana L.) sample from Piedmont obtained with a 1D CW20M column (100% polyethylene glycol; 30.0 m × 250 μm i.d., 0.25 μm d.f.) coupled with a 2D OV1701 (1.0 m × 100 μm i.d., 0.10 μm d.f.), Temperature programme from 50 °C (1 min) to 260 °C (5 min) at 3 °C min^{-1}; modulation period: 4 s. Reprinted with permission from [96]. Copyright 2008 American Chemical Society*

compounds, peaks or even diagnostic fragments, within a complex analytical profile through which it is possible to differentiate between samples within a group.

The *in toto* chromatographic profile (or MS profiles, see below), defined by the abundances (better if normalized percentage data) of a selected number of markers characteristic of the vegetable matrix under investigation, can therefore be considered a further distinctive parameter of a sample to be processed statistically, in particular for samples characterized by several uncorrelated components. The most widely used approach is multivariate analysis, in particular principal component analysis (PCA), a method that can explain the differences within a set of samples characterized by a suitable number of components (variables) through the linear combination of those explaining most of the variability.[97] This method not only provides classification and differentiation among groups of samples within a set, but can also find those variables (components or MS fragments) that have the highest statistical significance in sample discrimination. Figure 3.17 reports the PCA discrimination for quality control of 42 samples of *Artemisia umbelliformis* belonging to two different chemotypes (CH-1, CH-2) resulting from the GC-FID analysis of their headspace sampled by SPME using normalized percentage abundances of 28 components present in all the samples after their standardization and/or normalization to a single internal standard. The two chemotypes can be distinguished unequivocally with a total variability explained by the two principal components of about 74%.[71]

These methods have not only increased the importance of the role played by data processing in the entire analytical process, but have also influenced analysis strategy and stimulated the introduction of new approaches. A usual online method to analyse the volatile fraction of a vegetable matrix combines headspace sampling for sample preparation, GC for separation and MS to identify and quantify as many components as possible (or at least the characterizing markers). The possibility of using data elaboration as an active analysis tool has made it possible to avoid one of the above steps (i.e. GC separation or MS component

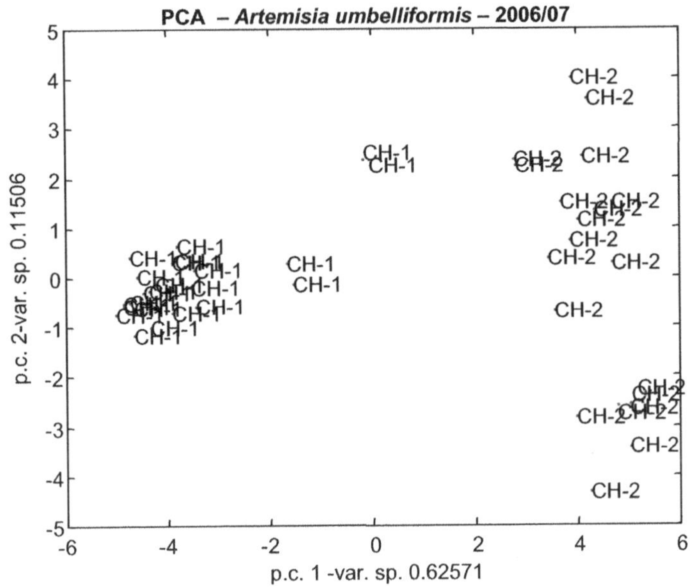

Figure 3.17 *PCA scores plot of the HS-SPME-GC-FID profiles of 42 samples of Artemisia umbelliformis Lam. belonging to two different chemotypes (CH-1, CH-2)*

identification) depending on the desired results. A typical example has been the development of methods in which the GC separation step is omitted, that is in which the headspace of the vegetable matrix is directly transferred into an MS system that provides an MS profile, comprising a significant fingerprint of the sample, and data processing acts as a 'tool' for sample discrimination. Specific instrumentation provided with dedicated statistic software has therefore been developed.[98] Figure 3.18 reports (*a*) the TIC-MS and (*b*) the MS fragmentation profiles of the headspace sampled by SPME of a set of wine samples from the same grapes (chardonnay) but submitted to different fermentation processes and their discrimination by PCA elaboration of their MS fragmentation profiles. This method can successfully be applied to the classification of samples or groups of samples within a set, in particular to discriminate them by origin, chemotype and/or the technological treatments to which they have been submitted; it can also be used in quantitative analysis, provided that characteristic markers can be quantified in the SIM-MS mode through specific diagnostic ions selected to avoid interference with those of other components.

This approach can also be used for *in vivo* analysis of volatile flavour release[99] and to correlate a biological characteristic with the chemical(s) responsible for it with a one-step analysis, as in the case, for instance, of the MS analysis of the nosespace (NS).[100] It is well known that volatiles can reach the olfactory epithelium either through the nostrils from the front via the orthonasal pathway or through the oral cavity and pharynx, via the retronasal pathway. The nosespace comprises the odorous compounds that are released in the mouth and can be perceived during eating and drinking, rather than the total content of volatiles in food products; its composition differs from the source owing to the complex dynamic processes occurring in the mouth.

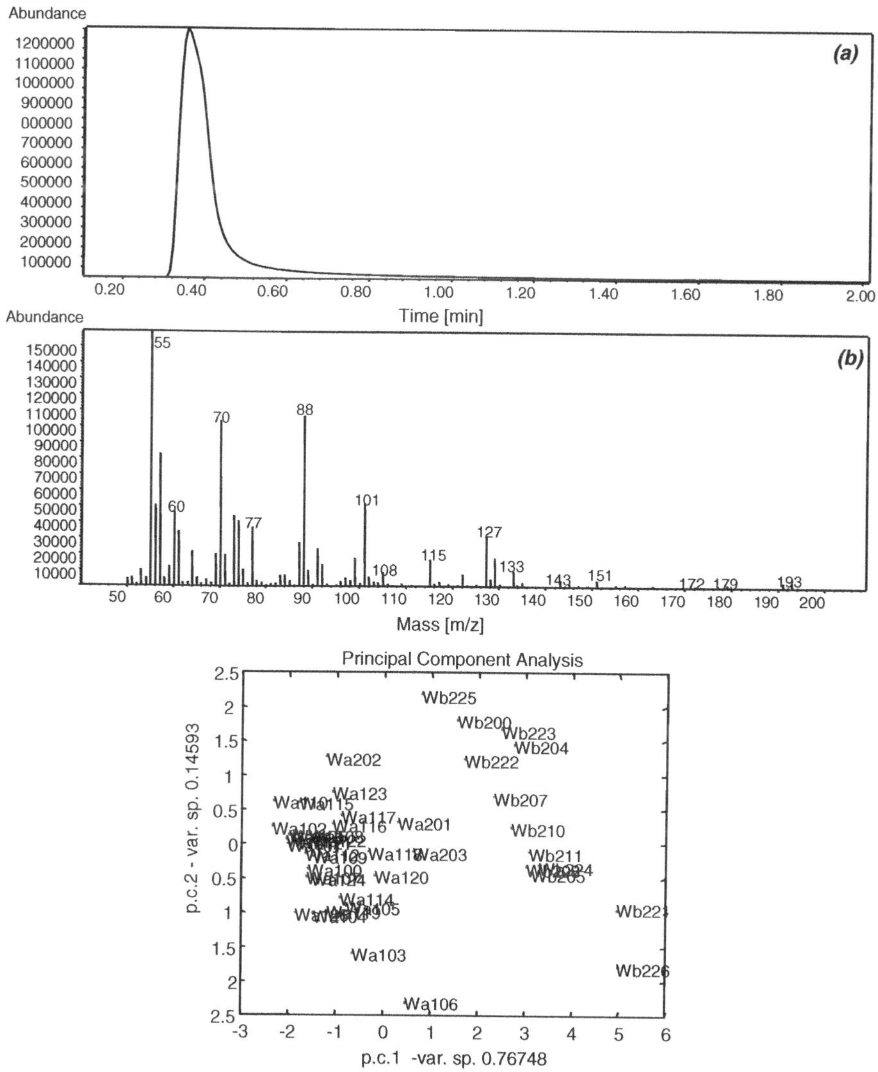

Figure 3.18 (a) HS-SPME-TIC-MS and (b) the MS fragmentation profiles of a set of wine samples and their discrimination by PCA elaboration of their MS fragment ions m/z. HS-SPME sampling: Fiber: 2 cm DVB/CAR/PDMS. Sample preparation: 5 ml of wine + 1.5 g of NaCl; vial volume: 20 ml; temperature: 50 °C; sampling time: 20 min. Desorption time: 4 min in the GC injector. GC-MS system: Agilent GC 6890 – MSD 5973N. Column: empty capillary deactivated tube: 10 m × 0.1 mm i.d. (MEGA, Legnano, Italy). Carrier gas: He; flow rate: 0.5 ml min^{-1}. Oven: 150 °C. Injection: split mode, split ratio 1 : 50, temp.: 250 °C. MS ionization mode: electron impact (EI) at 70 eV, mass range: 50–200 m/z; ion source temperature: 230 °C

A clear example of this approach is the *in vivo* breath by breath analysis of volatiles released in the mouth while eating ripe and unripe banana, studied by Mayra *et al.*[101] to determine the dynamic aspects of the in-mouth aroma, that is its evolution over time during

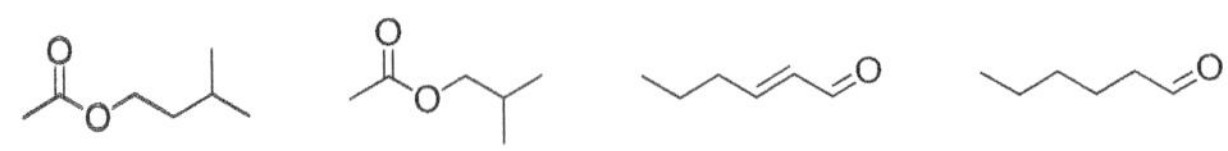

52 53 54 55

Figure 3.19 Structures of compounds **52–55** analysed in ripe and unripe bananas

mastication, and how it differs from an HS aroma. The experiments were carried out by introducing a small fraction of the air exhaled through the nose directly into a proton transfer reaction mass spectrometer (PTR-MS) via a nosepiece and monitoring online the time-intensity profiles of a series of selected markers, in this case isopentyl and isobutyl acetate (**52**, **53**) for ripe banana and (*E*)-2-hexenal and hexanal (**54**, **55**) for unripe banana (Figure 3.19). Figure 3.20 reports the experimental set-up for breath by breath nosespace analysis while eating banana. PTR-MS is a soft, sensitive and efficient chemical ionization MS (CI-MS) technique using protonated water as the reactant gas, thus avoiding the interference of humidity with ionization, since a fixed volume of gas is continuously introduced into the chemical ionization cell that contains buffer gas and a controlled ion density of H_3O^+. Volatiles with proton affinities above that of water (166.5 kcal mol^{-1}) are ionized by proton transfer from H_3O^+ and analysed by MS.[102] Figure 3.21 reports the breath by breath, nosespace PTR-MS profiles of six selected masses while eating unripe (left traces) and ripe (right traces) banana. These results provide a clear example of how conventional studies to define the markers which characterize the sensory properties of the matrix under investigation can complement new approaches to evaluate a biological property effectively.

The performance of data processing becomes increasingly important as the complexity of the volatile fraction under investigation increases. In particular its role is crucial for analytical profiles resulting from comprehensive two-dimensional (2D) gas chromatography (GC × GC). The high practical peak capacity and the ability to apply two different

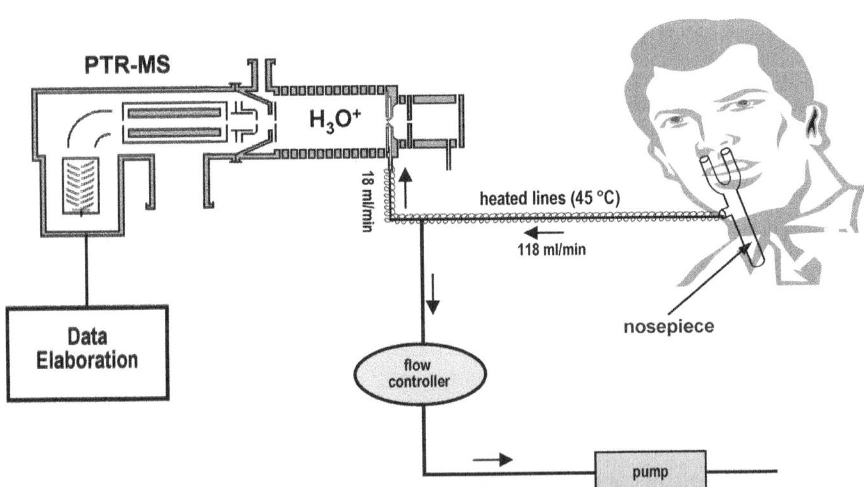

Figure 3.20 Experimental set-up for breath by breath NS analysis while eating banana

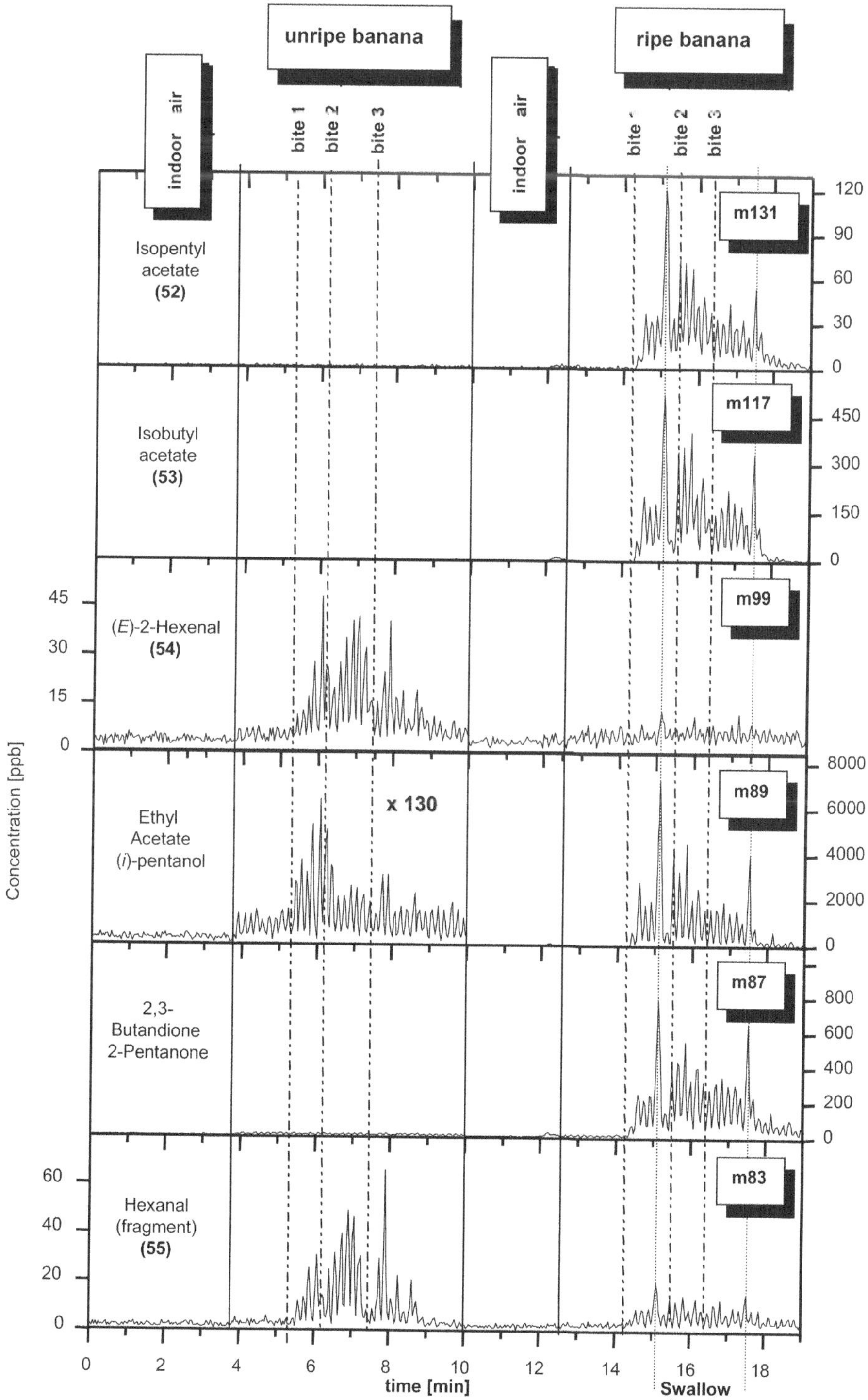

Figure 3.21 Breath by breath, NS-PTR-MS profiles of six selected masses while eating unripe (left traces) and ripe (right traces) banana. Reprinted with permission from [101]. Copyright Elsevier

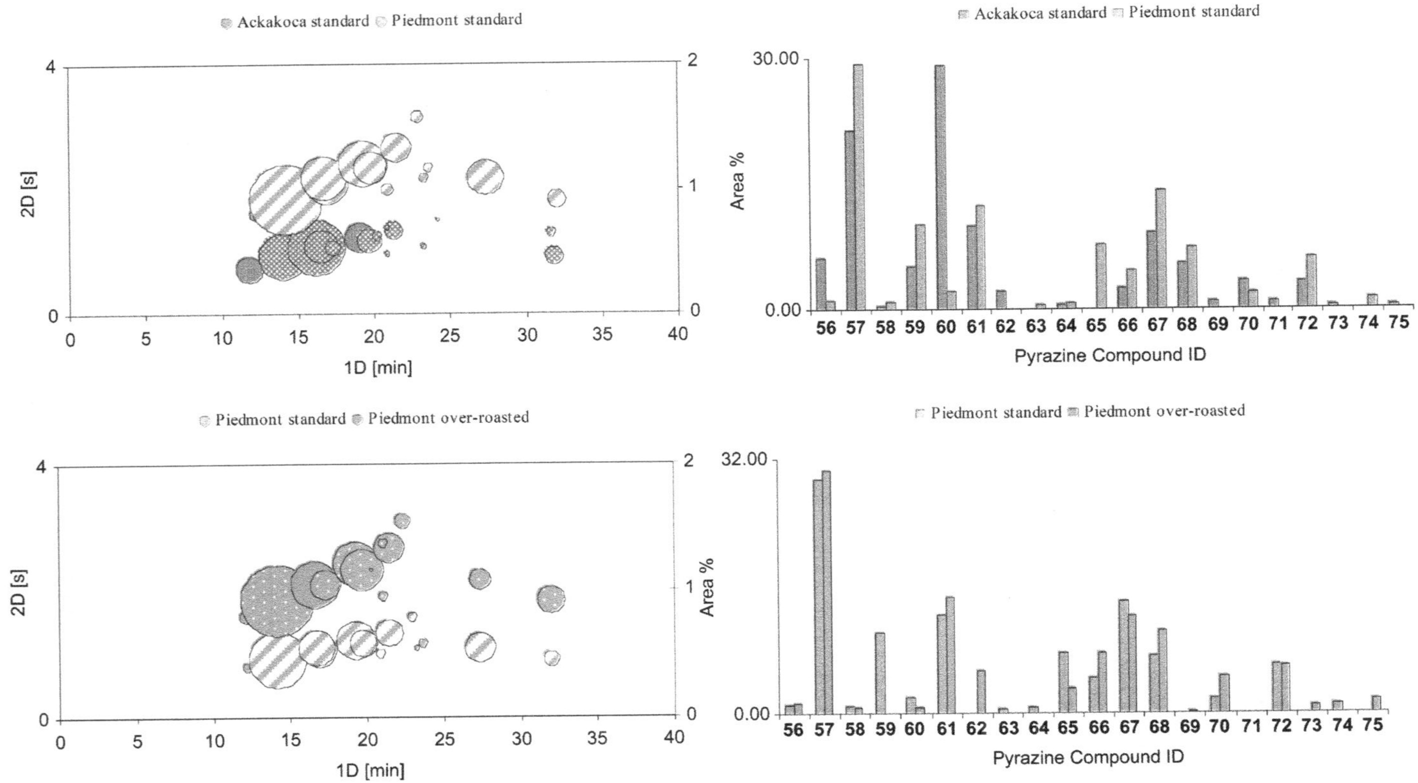

Figure 3.22 *Pyrazine 2D patterns of roasted Ackakoca (i.e. Turkish) and Piedmont hazelnut samples submitted to a standard and an over-roasted thermal treatment analysed with Column Set 3. Histograms report the area (%) of each congener while bubble plot graphs describe the components' location over the 2D plane. Reprinted with permission from [96]. Copyright 2008 American Chemical Society*

separation principles, one for each chromatographic dimension, enables GC × GC to give rationalized spatial domains for chemically correlated groups of compounds, and this provides specific separation patterns useful for component identification and characterizing a 2D fingerprint of each sample, thus providing more information than is offered by a one-dimensional (1D) profile.[103] Two specific GC × GC approaches, *group-type* and *fingerprint-type* analyses, have successfully been adopted in the authors' laboratory to classify and characterize samples;[96] these approaches were originally developed for the petrochemical field but can be successfully extended to the volatile fraction of vegetable matrices. The *group-type* approach can be adopted when identification of all individual components of a complex fraction is neither necessary nor possible, and the interest is focused on classes of compounds diagnostic for specific matrices. The 2D fingerprint analysis (i.e. the description of the specific 2D profile produced by all the separated and detected components of a sample) is useful as a primary tool to compare samples versus a reference model and to correlate them without the need to identify all their components. GC × GC coupled with MS detection (fast-scanning qMS or TOF-MS) can give unique opportunities for data processing and may be applied in chemical speciation and in the differentiation and correlation of fractions in complex samples of vegetable origin, based on the analyte distribution of the sample components over the chromatographic plane (fingerprint), through the different information it can provide, that is 1D and 2D retention times, detector response and the MS spectrum.

An interesting example is given by the characterization of roasted hazelnuts (*Corylus avellana* L.) from different geographical origins and submitted to different thermal treatments.

In this case, the group type approach enables the specific patterns of pyrazine series (i.e. one of the groups of compounds characterizing the roasting process) to be isolated; and the resulting quali-quantitative profiles enable the study of the effect of the roasting treatments on samples of different origin. Figure 3.22 shows the effectiveness of HS-SPME-GC × GC-qMS in reporting the comparison of bubble profiles of the homologous series of pyrazines **56–75** (Figure 3.23) from a Piedmont (Italy) roasted hazelnut and a Turkish sample.

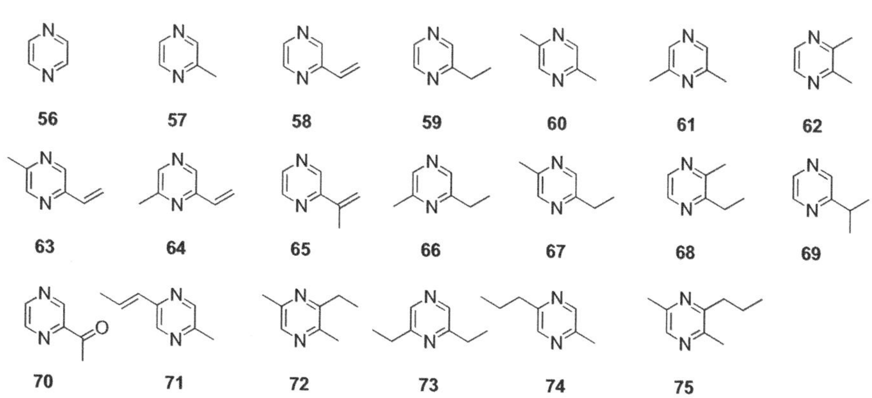

Figure 3.23 *Structures of pyrazines* ***56–75*** *of roasted Ackakoca and Piedmont hazelnut samples*

Further, the fingerprint-type approach enables two plots to be compared without requiring component identification and can successfully be used to discriminate GC × GC plots on the basis of peak superposition, in terms of differences in intensity, or by comparison with a reference template. Figure 3.24 reports the 2D fingerprint (i.e. the differential image) produced by comparing two Piedmont hazelnut samples submitted to two different thermal processes, that is *standard analysed* (i.e. standard Piedmont hazelnut) and *over-roasted reference* (i.e. over-roasted Piedmont hazelnut). In the enlarged area of the 2D plot in the fuzzy difference visualization, brighter spots correspond to analytes that were present in larger amounts in the over-roasted Piedmont hazelnut sample. Dot-plot circles indicate pyrazine identification.

Also, fingerprint comparisons can be made by adopting a suitable template from a reference sample which includes a selection or all the separated peaks. The resulting

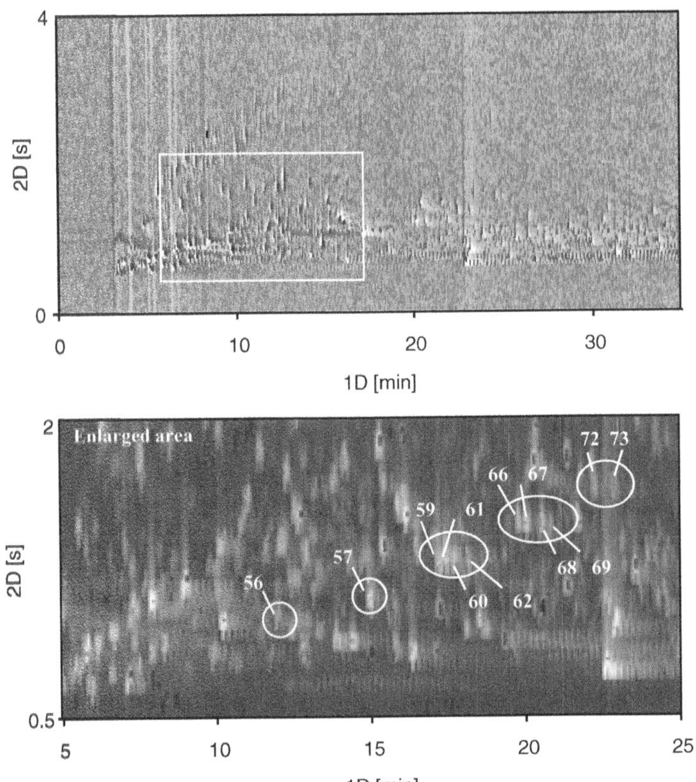

Figure 3.24 *Resulting 2D fingerprint, i.e. the difference or differential image, produced by comparing two Piedmontese hazelnut samples submitted to two different thermal processes, that is standard analysed (i.e. standard Piedmont hazelnut) and over-roasted reference (i.e. over-roasted Piedmont hazelnut). In the enlarged area of the 2D plot in the fuzzy difference visualization, brighter spots correspond to those analytes that were present in larger amount in the over-roasted Piedmont hazelnut sample. Reprinted with permission from [96]. Copyright 2008 American Chemical Society*

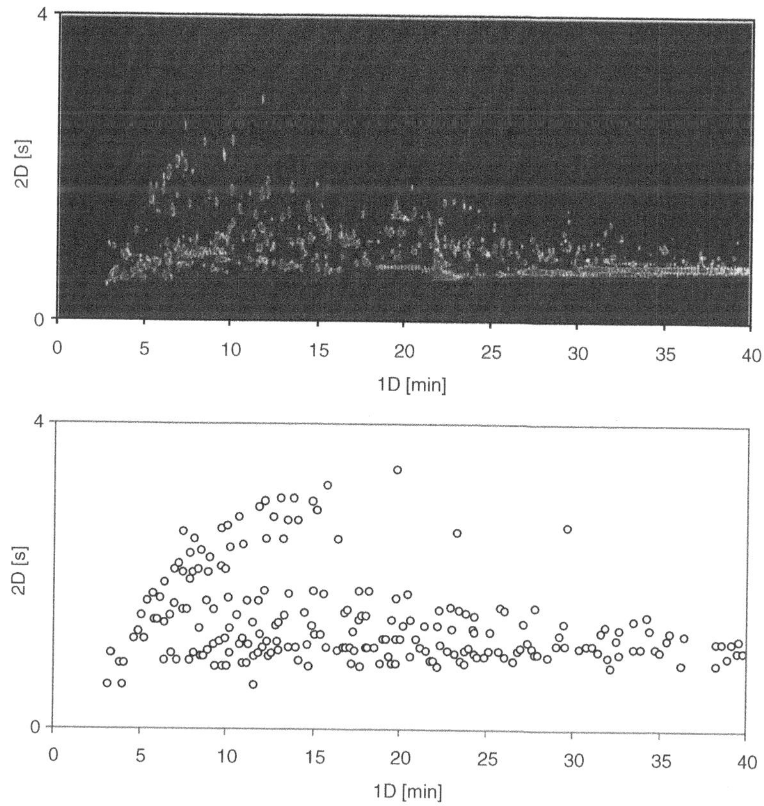

Figure 3.25 2D plot and graphical representation of the 231 template peaks chosen from a standard roasted Roman hazelnut (i.e. arbitrarily considered as reference). Reprinted with permission from [96]. Copyright 2008 American Chemical Society

reference template can also be used as a term of comparison between samples. Figure 3.25 reports the 2D plot and graphical representation of the 231 template peaks chosen from a standard roasted Roman hazelnut (i.e. arbitrarily considered as reference) and Table 3.2 shows the results of template matching with the above reference template and a group of samples (standard and over-roasted Piedmont and Turkish hazelnuts) giving the number of peaks matched, normalized number of peaks matched and the matching maximum and minimum absolute distances.[96]

3.5 Conclusion

The ever-increasing need to understand the biological and technological significance of the volatile fraction of a vegetable matrix has resulted in a growing demand for exhaustive knowledge of its composition. As a consequence, volatile-fraction analysis has evolved dramatically over the past two decades but, despite the widespread view that everything has already been done in this field, several topics still remain to be investigated in depth. In the authors' opinion, research efforts must be concentrated on developing and

Table 3.2 *Template matching with the standard roasted Roman hazelnut (i.e. arbitrarily considered as reference) and a group of samples (i.e. standard and over-roasted Piedmont and Turkish hazelnuts) giving the number of peaks matched, normalized number of peaks matched, and the matching maximum and minimum absolute 2D peak distances. Reprinted with permission from [96]. Copyright 2008 American Chemical Society*

Template Roman standard hazelnut	Total template peaks:	231
Sample 1 Piedmont standard hazelnut	Number of peaks matched:	165
	Number of peaks matched (normalized):	0.71
	Matching maximum distance:	12.60
	Matching minimum distance:	0.62
	Matching average distance:	2.37
Sample 2 Piedmont over-roasted hazelnut	Number of peaks matched:	176
	Number of peaks matched (normalized):	0.76
	Matching maximum distance:	11.11
	Matching minimum distance:	0.40
	Matching average distance:	1.64
Sample 3 Turkish standard hazelnut	Number of peaks matched:	157
	Number of peaks matched (normalized):	0.68
	Matching maximum distance:	11.23
	Matching minimum distance:	0.53
	Matching average distance:	4.89
Sample 4 Turkish over-roasted hazelnut	Number of peaks matched:	187
	Number of peaks matched (normalized):	0.81
	Matching maximum distance:	12.58
	Matching minimum distance:	0.23
	Matching average distance:	3.30

adopting total analysis systems, whenever possible, to enable a reliable and standardized production of data. This goal will only be achieved by acting on all steps of the analytical procedure:

- Sample preparation: highly effective, fast and easy to automate solvent less techniques must be developed that are easy to combine online with separation techniques.
- Analysis: high-speed, high-efficiency GC separation methods (F-GC, GC-GC, GC × GC) are required that provide highly reliable qualitative and quantitative data and at the same time meaningful profiles. This will involve the increased merging of chromatographic and MS data with their processing, and in particular will need highly effective deconvolution software. Special attention will also have to be paid to chromatographic data, whose importance has drastically grown in particular for further data elaboration (see next point) and whose contribution to component identification and sample profile definition is often underestimated.
- Data elaboration: ever increasing use of information technology and data mining will be necessary to achieve the so-called 'higher level of information' as a contribution to understanding the biological meaning of the volatile fraction. Highly effective sample

preparation and analysis techniques should provide as many data as possible, and data processing should enable those data concerning the biological property or properties of interest to be isolated, so as to obtain suitable metabolite profiling through which to understand the biological phenomenon investigated.

Acknowledgements

The authors are indebted to the project entitled: 'Sviluppo di metodologie innovative per l'analisi di prodotti agroalimentari' (FIRB Cod.: RBIP06SXMR_002) of the Ministero dell'Istruzione, dell'Università e della Ricerca (MIUR), Italy.

References

1. AFNOR NF T 75-006 (Feb 1998).
2. European Pharmacopoeia, 6th Edn., European Directorate for the Quality of Medicines, 2008.
3. B. Kolb, L. Ettre, *Static Headspace Gas Chromatography. Theory and Practice*, Wiley-VCH, New York, 1997.
4. *(a)* J. F. Clevenger, Apparatus for the determination of volatile oil, *J. Am. Pharm. Assoc.*, **17**, 346–349 (1928); *(b)* J. F. Clevenger, Report on (the analysis of) fluidextract of ginger, *J. Assoc. Off. Agric. Chem.*, **11**, 332–335 (1928).
5. *(a)* S. Likens, G. Nickerson, Detection of certain hop oil constituents in brewing products, *Proc. Amer. Soc. Brev. Chem.*, **5**, 13–16 (1964); *(b)* M. Godefroot, P. Sandra, M. Verzele, New method for quantitative essential oil analysis, *J. Chromatogr.*, **203**, 325–335 (1981); *(c)* M. Godefroot, M. Stechele, P. Sandra, M. Verzele, A new method for the quantitative analysis of organochlorine pesticides and polychlorinated biphenyls, *J. High Resolut. Chromatogr. Chromatogr. Commun.*, **5**, 75–79 (1982); *(d)* A. Chaintreau, Simultaneous distillation–extraction: from birth to maturity – review, *Flavour Fragr. J.*, **16**, 136–148 (2001).
6. *(a)* M. A. Ferhat, B. Y. Meklati, J. Smadja, F. Chemat, An improved microwave Clevenger apparatus for distillation of essential oils from orange peel, *J. Chromatogr. A*, **1112**, 121–126 (2006); *(b)* M. E. Lucchesi, F. Chemat, J. Smadja, An original solvent free microwave extraction of essential oils from spices, *Flavour Fragr. J.*, **19**, 134–138 (2004).
7. C. Bicchi, A. Binello, P. Rubiolo, Supercritical CO_2 in combination with silica gel to fractionate essential oils, *Phytochem. Anal.*, **10**, 7–21 (1999), and references cited therein.
8. J. Pawliszyn, *Solid Phase Microextraction – Theory and Practice*, 1st Ed., Wiley-VCH, New York, 1997.
9. C. Bicchi, C. Cordero, P. Rubiolo, High concentration capability headspace sampling techniques in the analysis of flavours and fragrances, *J. Chromatogr. Sci.*, **42**, 402–409 (2004).
10. C. Bicchi, C. Cordero, E. Liberto, B. Sgorbini, P. Rubiolo, Headspace sampling of the volatile fraction of vegetable matrices, *J. Chromatogr. A*, **1184**, 220–233 (2008).
11. *(a)* E. Baltussen, P. Sandra, F. David, H. G. Janssen, C. Cramers, Equilibrium sorptive enrichment on poly(dimethylsiloxane) particles for trace analysis of volatile compounds in gaseous samples, *Anal. Chem.*, **71** 5193–5198 (1999); *(b)* E. Baltussen, *New Concepts in Sorption Based Sample Preparation for Chromatography*, PhD thesis, Technische Universiteit Eindhoven, Eindhoven (The Netherlands), 2000, pp. 10–11; *(c)* E. Baltussen, C. Cramers, P. Sandra, Sorptive sample preparation – a review, *Anal. Bioanal. Chem.*, **373**, 3–22 (2002); *(d)* E. Baltussen, F. David, P. Sandra, H. G. Janssen, C. Cramers, Sorption tubes packed with polydimethylsiloxane: a new and promising technique for the preconcentration of volatiles and semi-volatiles from air and gaseous samples, *J. High Resolut. Chromatogr.*, **21**, 332–340 (1998).

12. C. Bicchi, A. Binello, P. Rubiolo, Determination of phenolic diterpene antioxidants in rosemary (*Rosmarinus officinalis* L.) with different methods of extraction and analysis, *Phytochem. Anal.*, **11**, 236–242 (2000).
13. Z. Zhang, J. Pawliszyn, Headspace solid-phase microextraction, *Anal. Chem.*, **65**, 1843–1852. (1993).
14. M. E. McComb, R. D. Oleschuk, E. Giller, H. D. Gesser, Microextraction of volatile organic compounds using the inside needle capillary adsorption trap (INCAT) device, *Talanta*, **44**, 2137–2143 (1997).
15. F. Musshoff, D. W. Lachenmeier, L. Kroener, B. Madea, Automated headspace solid-phase dynamic extraction for the determination of amphetamines and synthetic designer drugs in hair samples, *J. Chromatogr. A*, **958**, 231–238 (2002).
16. C. Bicchi, C. Cordero, E. Liberto, P. Rubiolo, B. Sgorbini, Automated headspace solid-phase dynamic extraction (HS-SPDE) to analyse the volatile fraction of food matrices, *J. Chromatogr. A*, **1024**, 217–226 (2004).
17. C. Bicchi, C. Cordero, C. Iori, P. Rubiolo, P. Sandra, Headspace sorptive extraction (HSSE) in the headspace analysis of aromatic and medicinal plants, *J. High Resolut. Chromatogr.*, **23**, 539–546 (2000).
18. B. Tienpont, F. David, C. Bicchi, P. Sandra, High capacity headspace sorptive extraction, *J. Microcol. Sep.*, **12**, 577–584 (2000).
19. A. Chaintreau, A. Grade, R. Muñoz-Box, Determination of partition coefficients and quantitation of headspace volatile compounds, *Anal. Chem.*, **34**, 3300–3304 (1995).
20. A. Ott, M. Baumgartner, A. Chaintreau, Aroma comparisons of traditional and mild yogurt: headspace-GC quantitation of volatiles and origin of α-diketones, *J. Agric. Food Chem.*, **47**, 2379–2385 (1999).
21. A. Chaintreau, Sample preparation, headspace techniques, in *Encyclopedia of Analytical Chemistry*, R. A. Meyers (Ed.), John Wiley & Sons, Chichester, 2000, pp. 4229–4246.
22. M. Ishikawa, O. Ito, S. Ishizaki, Y. Kurobayashi, A. Fujita, Solid-phase aroma concentrate extraction (SPACE): a new headspace technique for more sensitive analysis of volatiles, *Flavour Fragr. J.*, **19**, 183–187 (2004).
23. A. L. Theis, A. J. Waldack, S. M. Hansen, M. A. Jeannot, Headspace solvent microextraction, *Anal. Chem.*, **73**, 5651–5654 (2001).
24. A. Tankeviciute, R. Kazlauskas, V. Vickackaite, Headspace extraction of alcohols into a single drop, *Analyst*, **126**, 1674–1677 (2001).
25. A. Segal, T. Gorecki, P. Mussche, J. Lips, J. Pawliszyn, Development of membrane extraction with a sorbent interface–micro gas chromatography system for field analysis, *J. Chromatogr. A*, **873**, 13–27 (2000).
26. I. Bruheim, X. Liu, J. Pawliszyn, Thin-film microextraction, *Anal. Chem.*, **75**, 1002–1010 (2003).
27. C. Bicchi, C. Cordero, E. Liberto, B. Sgorbini, P. Sandra, P. Rubiolo, Sorptive tape extraction (STE) in the analysis of the volatile fraction emitted from biological solid matrices, *J. Chromatogr. A*, **1148**, 137–144 (2007).
28. F. Belliardo, C. Bicchi, C. Cordero, E. Liberto, P. Rubiolo, B. Sgorbini, Headspace-solid phase microextraction (HS-SPME) in the analysis of the volatile fraction of aromatic and medicinal plants, *J. Chromatogr. Sci.*, **44**, 416–429 (2006).
29. C. L. Arthur, J. Pawliszyn, Solid phase microextraction with thermal desorption using fused silica optical fibers, *Anal. Chem.*, **62**, 2145–2148 (1990).
30. T. Górecki, J. Pawliszyn, Sample introduction approaches for solid phase microextraction/rapid GC, *Anal. Chem.*, **67**, 3265–3274 (1995).
31. *(a)* H. Lord, J. Pawliszyn, Evolution of solid-phase microextraction technology, *J. Chromatogr. A.*, **885**, 153–193 (2000); *(b)* J. Pawliszyn, *Sampling and Sample Preparation for Field and Laboratory*, Elsevier, Amsterdam, 2002, pp. 389–477.
32. J. Lipinski, Automated solid phase dynamic extraction: extraction of organics using a wall coated syringe needle, *Fresenius' J. Anal. Chem.*, **369**, 57–62 (2001).

33. E. Baltussen, P. Sandra, F. David, C. Cramers, Stir bar sorptive extraction (SBSE), a novel extraction technique for aqueous samples: theory and principles, *J. Microcolumn Sep.*, **11**, 737–747 (1999).
34. C. Bicchi, E. Liberto, C. Cordero, B. Sgorbini, P. Rubiolo, Stir bar sorptive extraction (SBSE) and headspace sorptive extraction (HSSE): an overview, *LC-GC North America*, **27**, 376–391 (2009).
35. *(a)* C. Bicchi, C. Cordero, E. Liberto, P. Rubiolo, B. Sgorbini, F. David, P. Sandra, Dual-phase twisters: a new approach to headspace sorptive extraction and stir bar sorptive extraction, *J. Chromatogr. A*, **1094**, 9–16 (2005); *(b)* C. Bicchi, C. Cordero, E. Liberto, B. Sgorbini, F. David, P. Sandra, P. Rubiolo, Influence of polydimethylsiloxane outer coating and packing material on analyte recovery in dual-phase headspace sorptive extraction, *J. Chromatogr. A*, **1164**, 33–39 (2007).
36. M. Jánská, S. J. Lehotay, K. Maštovská, J. Hajšlová, T. Alon, A. Amirav, A simple and inexpensive 'solvent in silicone tube extraction' approach and its evaluation in the gas chromatographic analysis of pesticides in fruits and vegetables, *J. Sep. Sci.*, **29**, 66–80 (2006).
37. (*a*) E. Van Hoeck, E. Dumont, P. Sandra, in *HPLC 2007*, Book of Abstracts, Ghent (Belgium), June 17–21, 2007, p. 837; (*b*) E. Van Hoeck, F. David, P. Sandra, in *Proceedings of 32nd International Symposium on Capillary Chromatography*, T. Sandra, P. Sandra (Eds), Riva del Garda (Italy), 2008, p. 195 and p. 196.
38. B. Sgorbini, D. Budziak, C. Cordero, E. Liberto, P. Rubiolo, P. Sandra, C. Bicchi, 32nd International Symposium on Capillary Chromatography, May 26–30, 2008, Riva del Garda (Italy), Poster C21, p. 167.
39. S. Ishizaki, O. Itoh, A. Fujita, Y. Kurobayashi, M. Ishikawa, A new headspace technique for more sensitive analysis of volatiles, in *Flavour Research at the Dawn of the Twenty-first Century*, Proceedings of the 10th Weurman Flavor Research Symposium, J.-L. Le Quéré, P. X. Etiévant (Eds), Tec & Doc, Paris, 2003, pp. 634–637.
40. *(a)* M. A. Jeannot, F. F. Cantwell, Solvent microextraction into a single drop, *Anal. Chem.*, **68**, 2236–2240 (1996); *(b)* M. A. Jeannot, F. F. Cantwell, Mass transfer characteristics of solvent extraction into a single drop at the tip of a syringe needle, *Anal. Chem.*, **69**, 235–239 (1997).
41. M. J. Yang, S. Harms, Y. Z. Luo, J. Pawliszyn, Membrane extraction with a sorbent interface for capillary gas chromatography, *Anal. Chem.*, **66**, 1339–1346 (1994).
42. M. Mulder, *Basic Principles of Membrane Technology*, Kluwer, Dordrecht, 1991.
43. L. Wang, H. Lord, R. Morehead, F. Dorman, J. Pawliszyn, Sampling and monitoring of biogenic emissions by eucalyptus leaves using membrane extraction with sorbent interface (MESI) *J. Agric. Food Chem.*, **50**, 6281–6286 (2002).
44. X. Liu, R. Pawliszyn, L. Wang, J. Pawliszyn, On-site monitoring of biogenic emissions from *Eucalyptus dunnii* leaves using membrane extraction with sorbent interface combined with a portable gas chromatograph system, *Analyst*, **129**, 55–62 (2004).
45. S. Sisalli, A. Adao, M. Lebel, I. Le Fur, P. Sandra, Sorptive tape extraction – a novel sampling method for the *in vivo* study of skin, *LC-GC Europe*, **19**, 33–39 (2006).
46. International Organization of the Flavour Industry (IOFI), Statement on the identification in nature of flavouring substances, made by the working group on methods of analysis of the International Organization of the Flavour Industry (IOFI), *Flavour Fragr. J.*, **21**, 185 (2006).
47. W. Herres, *HRGC-FTIR: Capillary Gas Ghromatography–Fourier Transform Infrared Spectroscopy. Theory and Applications*, Huethig Verlag, Heidelberg, 1987.
48. C. A. Cramers, H. G. Janssen, M. M. Van Deursen, P. A. Leclercq, High-speed gas chromatography: an overview of various concepts, *J. Chromatogr. A*, **856**, 315–329 (1999).
49. L. M. Blumberg, M. S. Klee,Theory and practice of fast capillary GC. Efficiency and speed of analysis, in *Proceedings of 20th International Symposium of Capillary Chromatography*, P. Sandra, A. J. Rackstaw (Eds), Riva del Garda (Italy), 1998, PL9, and references cited therein.
50. P. Magni, R. Facchetti, D. Cavagnino, S. Trestianu, Ultra fast gas chromatography with conventional instruments using direct resistively heated capillary columns in *Proceedings of 25th International Symposium of Capillary Chromatography*, P. Sandra (Ed.), Riva del Garda (Italy), 2002, KLN05, and references cited therein.

51. C. Bicchi, C. Brunelli, C. Cordero, P. Rubiolo, M. Galli, A. Sironi, Direct resistively heated column – gas chromatography (ultrafast module-GC) for high speed analysis of essential oils of differing complexities, *J. Chromatogr. A*, **1024**, 195–207 (2004), and references cited therein.
52. M. Proot, P. Sandra, High speed capillary GC on 10 m × 100 μm i.d. FSOT columns, *J. High Resolut. Chromatogr.*, **9**, 618–623 (1986).
53. C. Bicchi, C. Brunelli, M. Galli, A. Sironi, Conventional inner diameter short capillary columns: an approach to speeding up gas chromatographic analysis of medium complexity samples, *J. Chromatogr. A*, **931**, 129–140 (2001), and references cited therein.
54. F. David, D. R. Gere, F. Scanlan, P. Sandra, Instrumentation and applications of fast high-resolution capillary gas chromatography, *J. Chromatogr. A*, **842**, 309–319 (1999).
55. M. Klee, L. M. Blumberg, Theoretical and practical aspects of fast gas chromatography and method translation, *J. Chromatogr. Sci.*, **40**, 234–247 (2002), and references cited therein.
56. http://www.chem.agilent.com.
57. K. Maštovská, S. J. Lehotay, Practical approaches to fast gas chromatography – mass spectrometry, *J. Chromatogr. A*, **1000**, 153–180 (2003), and reference cited therein.
58. P. Rubiolo, E. Liberto, B. Sgorbini, R. Russo, J. L. Veuthey, C. Bicchi, Fast-GC-conventional quadrupole mass spectrometry in essential oil analysis, *J. Separ. Sci.*, **31**, 1074–1084 (2008).
59. E. Kováts, Gaschromatographische Charakterisierung organischer Verbindungen. I. Retentions Indices aliphatischer Halogenide, Alkohole, Aldehyde und Ketone, *Helv. Chim. Acta*, **41**, 1915–1932 (1958).
60. *(a)* H. Van den Dool, P. D. Kratz, A generalization of the retention index system including linear temperature programmed gas–liquid partition chromatography, *J. Chromatogr.*, **11**, 463–471 (1963); *(b)* H. Van den Dool, *Standardisation of Gas Chromatographic Analysis of Essential Oils*, PhD thesis, Groningen (The Netherlands), 1974.
61. B. d'Acampora Zellner, C. Bicchi, P. Dugo, P. Rubiolo, G. Dugo, L. Mondello, Linear retention indices in gas chromatographic analysis: a review, *Flavour Fragr. J.*, **23**, 297–314 (2008).
62. *(a)* L. Mondello, P. Dugo, A. Basile, G. Dugo, K. Bartle, Interactive use of linear retention indices, on polar and apolar columns, with a MS-library for reliable identification of complex mixtures, *J. Microcolumn Sep.*, **7**, 581–591 (1995); *(b)* R. Shellie, P. Marriott, G. Zappia, L. Mondello, G. Dugo, Interactive use of linear retention indices on polar and apolar columns with an MS-library for reliable characterization of Australian tea tree and other Melaleuca sp. oils, *J. Essent. Oil Res.*, **15**, 305–312 (2003); *(c)* R. P. Adams, *Identification of Essential Oil Components by Gas Chromatography/Mass Spectrometry*, Allured Publications, Carol Stream, 2007; *(d)* A. Van Asten, The importance of GC and GC-MS in perfume analysis, *Trends Anal. Chem.*, **21**, 698–708 (2002).
63. FFNSC MS Library ver. 1.3, Chromaleont, Messina (Italy).
64. AMDIS http://chemdata.nist.gov/mass-spc/amdis/, Version 2.65, (2007).
65. NIST 05 – NIST/EPA/NIH Mass Spectral Library, National Institute of Standards and Technology, Gaithersburg, MD (USA).
66. R. Costa, M. De Fina, M. R. Valentino, P. Dugo, L. Mondello, Reliable identification of terpenoids and related compounds by using linear retention indices interactively with mass spectrometry search, *Nat. Prod. Commun.*, **2**, 413–418 (2007).
67. L. M. Blumberg, M. Klee, Method translation and retention time locking in partition GC, *Anal. Chem.*, **70**, 3828–3839 (1998).
68. Flavors RTL Databases for GC-FID and GC-MS, Agilent Technologies, Santa Clara (USA), 2004.
69. C. Bicchi, E. Liberto, M. Matteodo, B. Sgorbini, L. Mondello, B. d'Acampora Zellner, R. Costa, P. Rubiolo, Quantitative analysis of essential oils: a complex task, *Flavour Fragr. J.*, **23**, 382–391 (2008).
70. R. Costa, B. d'Acampora Zellner, M. L. Crupi, M. R. De Fina, M. R. Valentino, P. Dugo, G. Dugo, L. Mondello, GC MS, GC-O and enantio-GC investigation of the essential oil of *Tarchonanthus camphoratus* L., *Flavour Fragr. J.*, **23**, 40–48 (2008).

71. P. Rubiolo, M. Matteodo, C. Bicchi, G. Appendino, G. Gnavi, C. M. Bertea, M. E. Maffei, Chemical and biomolecular characterization of *Artemisia umbelliformis*, an important ingredient of the alpine liqueur Genepi, *J. Agric. Food Chem.*, **57**, 3436–3443 (2009).
72. *(a)* A. Chaintreau, D. Joulain, C. Marin, C.-O. Schmidt, M. Vey, GC-MS quantitation of fragrance compounds suspected to cause skin reactions, *J. Agric. Food Chem.*, **51**, 6398–6403 (2003); *(b)* C. Cordero, C. Bicchi, D. Joulain, P. Rubiolo, Identification, quantitation and method validation for the analysis of suspected allergens in fragrances by comprehensive two-dimensional gas chromatography coupled with quadrupole mass spectrometry and with flame ionization detection, *J. Chromatogr. A*, **1150**, 37–49 (2007).
73. E. Cicchetti, P. Merle, A. Chaintreau, Quantitation in gas chromatography: usual practices and performances of a response factor database, *Flavour Fragr. J.*, **23**, 450–459 (2008).
74. *(a)* Eurachem Guide, The fitness for purpose of analytical methods. A laboratory guide to method validation and related topics, 1st English Ed., 1998; *(b)* CITAC/Eurachem Guide to quality in analytical chemistry, an aid to accreditation, 2002; *(c)* V. Gold, K. L. Loening, A. D. McNaught, *Compendium of Chemical Terminology; IUPAC Recommendations*, P. Sehmi (Ed.), Blackwell Scientific Publications, Oxford, 1987.
75. D. Sybilska, T. Koscielski, α-Cyclodextrin as a selective agent for the separation of *o*-, *m*- and *p*-xylene and ethylbenzene mixtures in gas–liquid chromatography, *J. Chromatogr.*, **261**, 357–362 (1983).
76. Z. Juvancz, G. Alexander, J. Szejtli, Permethylated β-cyclodextrin as stationary phase in capillary gas chromatography, *J. High Resolut. Chromatogr.*, **10**, 105–107 (1987).
77. V. Schurig, H.-P. Nowotny, Separation of enantiomers on diluted permethylated β-cyclodextrin by high-resolution gas chromatography, *J. Chromatogr.*, **441**, 155–163 (1988).
78. H.-P. Nowotny, D. Schmalzing, D. Wistuba, V. Schurig, Extending the scope of enantiomer separation on diluted methylated β-cyclodextrin derivatives by high-resolution gas chromatography, *J. High Resolut. Chromatogr.*, **12**, 383–392 (1989).
79. *(a)* C. Bicchi, V. Manzin, A. D'Amato, P. Rubiolo, Cyclodextrin derivatives in GC separation of enantiomers of essential oil, aroma and flavour compounds, *Flavour Fragr. J.*, **10**, 127–137 (1995); *(b)* C. Bicchi, A. D'Amato, V. Manzin, Derivatized cyclodextrins in enantiomer GC separation of volatiles, in *Flavours and Fragrances*, K. Swift (Ed.), The Royal Society of Chemistry, Cambridge, 1997, pp. 57–69; *(c)* C. Bicchi, A. D'Amato, P. Rubiolo, Cyclodextrin derivatives as chiral selector for direct GC separation of enantiomers in essential oil, aroma and flavour fields, *J. Chromatogr. A*, **843**, 99–121 (1999); *(d)* W. A. König, D. H. Hochmuth, Enantioselective gas chromatography in flavor and fragrance analysis: strategies for the identification of known and unknown plant volatiles, *J. Chromatogr. Sci.*, **42**, 423–439 (2004).
80. *(a)* M. Jung, D. Schmalzing, V. Schurig, Theoretical approach to the gas chromatographic separation of enantiomers on dissolved cyclodextrin derivatives, *J. Chromatogr.*, **552**, 43–57 (1991); *(b)* V. Schurig, H.-P. Nowotny, Gas chromatographic separation of enantiomers on cyclodextrin derivatives, *Angew. Chem.*, **102**, 969–986 (1990); *Angew. Chem. Int. Ed.*, **29**, 939–957 (1990).
81. A. Dietrich, B. Maas, V. Karl, P. Kreis, D. Lehmann, B. Weber, A. Mosandl, Stereoisomeric flavour compound. Part LV: stereodifferentiation of some chiral volatiles on heptakis-(2,3-di-*O*-acetyl-6-*O*-*tert*-butyldimethylsilyl)-β-cyclodextrin, *J. High Resolut. Chromatogr.*, **15**, 176–179 (1992).
82. C. Bicchi, G. Artuffo, A. D'Amato, A. Galli, M. Galli, Cyclodextrin derivatives in GC separation of racemic mixtures of volatiles: part I, *J. High Resolut. Chromatogr.*, **14**, 301–305 (1991).
83. *(a)* G. Schomburg, H. Husmann, E. Hübinger, W. A. König, Multidimensional capillary gas chromatography–enantiomeric separations of selected cuts using a chiral second column, *J. High Resolut. Chromatogr.*, **7**, 404–410 (1984); *(b)* L. Mondello, M. Catalfamo, G. Dugo, P. Dugo, Multidimensional tandem capillary gas chromatography system for the analysis of real complex samples. Part I: development of a fully automated tandem gas chromatography system, *J. Chromatogr. Sci.*, **36**, 201–209 (1998).
84. *(a)* R. Shellie, P. Marriott, Comprehensive two-dimensional gas chromatography with fast enantioseparation, *Anal. Chem.*, **74**, 5426–5430 (2002); *(b)* R. Shellie, L. Mondello, G. Dugo,

P. Marriot, Enantioselective gas chromatographic analysis of monoterpenes in essential oils of the family Myrtaceae, *Flavour Fragr. J.*, **19**, 582–585 (2004).
85. E. Liberto, C. Cagliero, B. Sgorbini, C. Bicchi, D. Sciarrone, B. D'Acampora Zellner, L. Mondello, P. Rubiolo, Enantiomer identification in the flavour and fragrance fields by 'interactive' combination of linear retention indices from enantioselective GC and mass spectrometry, *J. Chromatogr. A*, **1195**, 117–126 (2008).
86. C. Bicchi, E. Liberto, C. Cagliero, C. Cordero, B. Sgorbini, P. Rubiolo, Conventional and narrow bore short capillary columns with cyclodextrin derivatives as chiral selectors to speed-up enantioselective-gas chromatography and enantioselective-gas chromatography mass spectrometry analyses, *J. Chromatogr. A*, **1212**, 114–123 (2008).
87. I. Hardt, W. A. Koenig, Diluted versus undiluted cyclodextrin derivates in capillary gas chromatography and the effect of linear carrier gas velocity, column temperature, and length on enantiomer separation, *J. Microcolumn Sep.*, **5**, 35–40 (1993).
88. A. Mosandl, Enantioselective capillary gas chromatography and stable isotope mass spectrometry in the authenticity of flavours and essential oils, *Food Rev. Int.*, **7**, 597–664 (1995), and references cited therein.
89. J. C. Giddings, Concepts and comparisons in multidimensional separation, *J. High Resolut. Chromatogr.*, **10**, 319–323 (1987).
90. *(a)* D. R. Deans, A new technique for heart cutting in gas chromatography, *Chromatographia*, **1**, 18–21 (1968); *(b)* G. Schomburg, F. Weeke, F. Müller, M. Oréans, Multidimensional gas chromatography (MDC) in capillary columns using double oven instruments and a newly designed coupling piece for monitoring detection after pre-separation, *Chromatographia*, **16**, 87–91 (1982); *(c)* L. Mondello, M. Catalfamo, P. Dugo, G. Dugo, Multidimensional capillary GC-GC for the analysis of real complex samples. Part II. Enantiomeric distribution of monoterpene hydrocarbons and monoterpene alcohols of cold-pressed and distilled lime oils, *J. Microcolumn Sep.*, **10**, 203–212 (1998).
91. *(a)* A. Bernreuther, N. Christoph, P. Schreier, Determination of the enantiomeric composition of β-lactones in complex natural matrices using multidimensional capillary gas-chromatography, *J. Chromatogr.*, **481**, 363–367 (1989); *(b)* A. Mosandl, U. Hener, U. Hagenauer-Hener, A. Kustermann, Direct enantiomer separation of chiral γ-lactones from food and beverages by multidimensional gas chromatography, *J. High Resolut. Chromatogr.*, **12**, 532–536 (1989); *(c)* P. Herion, G. Full, P. Winterhalter, P. Schreier, C. Bicchi, Enantiodifferentiation of isomeric vitispiranes, *Phytochem. Anal.*, **4**, 235–239 (1993).
92. Z. Liu, J. B. Phillips, Comprehensive 2-dimensional gas-chromatography using an on-column thermal modulator interface, *J. Chromatogr. Sci.*, **29**, 227–231 (1991).
93. *(a)* J. Dalluge, J. Beens, U. A. T. Brinkman, Comprehensive two-dimensional gas chromatography: a powerful and versatile analytical tool, *J. Chromatogr. A*, **1000**, 69–108 (2003); *(b)* M. Adahchour, J. Beens, R. J. J. Vreuls, U. A. T. Brinkman, Recent developments in comprehensive two-dimensional gas chromatography (GC $\times$ GC): I. Introduction and instrumental set-up, *Trends Anal. Chem.*, **25**, 438–454 (2006); *(c)* M. Adahchour, J. Beens, U. A. T. Brinkman, Recent developments in the application of comprehensive two-dimensional gas chromatography, *J. Chromatogr. A*, **1186**, 67–108 (2008).
94. W. Khummueng, J. Harynuk, P. J. Marriott, Modulation ratio in comprehensive two-dimensional gas chromatography, *Anal. Chem.*, **78**, 4578–4587 (2006).
95. D. Ryan, P. Morrison, P. Marriott, Orthogonality considerations in comprehensive two-dimensional gas chromatography, *J. Chromatogr. A.*, **1071**, 47–53 (2005).
96. C. Cordero, C. Bicchi, P. Rubiolo, Group-type and fingerprint analysis of roasted (coffee and hazelnut samples) food matrices by comprehensive two-dimensional gas chromatography, *J. Agric. Food Chem.*, **56**, 7655–7666 (2008), and references cited therein.
97. R. G. Brereton, *Applied Chemometrics for Scientists*, Wiley, Chichester, 2007, pp. 147–167.
98. *(a)* I. Marcos Lorenzo, J. L. Pérez Pavón, M. E. Fernández Laespada, C. García Pinto, B. Moreno Cordero, L. R. Henriques, M. F. Peres, M. P. Simões, P. S. Lopes, Application of headspace–mass spectrometry for differentiating sources of olive oil, *Anal. Bioanal. Chem.*, **374**, 1205–1211 (2002); *(b)* V. Kinton, E. Pfannkoch, J. Whitecavage, Discrimination of soft

drinks using a chemical sensor and principal component analysis, Gerstel Application Note 2, 2002.

99. A. J. Taylor, R. S. T. Linforth, B. A. Harvey, A. Blake, Atmospheric pressure chemical ionization mass spectrometry for *in vivo* analysis of volatile flavour release, *Food Chem.*, **71**, 327–338 (2000).
100. *(a)* W. J. Soeting, J. Heidema, A mass spectrometric method for measuring flavour concentration/time profiles in human breath, *Chem. Senses*, **13**, 607–617 (1988); *(b)* A. J. Taylor, R. S. T. Linforth,Techniques for measuring volatile release *in vivo* during consumption of food, in *Flavor Release*, D. D. Roberts, A. J. Taylor (Eds), ACS Symp. Ser., Vol. 763, American Chemical Society, Washington D. C., 2000, pp. 8–21, and references cited therein; *(c)* I. Baek, R. S. T. Linforth, A. Blake, A. J. Taylor, Sensory perception is related to the rate of change of volatile concentration in-nose during eating of model gels, *Chem. Senses*, **24**, 155–160 (1999).
101. D. Mayra, T. Märka, W. Lindinger, H. Brevard, C. Yeretzian, Breath-by-breath analysis of banana aroma by proton-transfer-reaction mass spectrometry, *Int. J. Mass Spectrom.*, **223/224**, 743–756 (2003).
102. W. Lindinger, A. Hansel, A. Jordan, On-line monitoring of volatile organic compounds at pptv levels by means of proton-transfer-reaction mass spectrometry (PTR-MS). Medical applications, food control and environmental research, *Int. J. Mass Spectrom. Ion Processes*, **173**, 191–241 (1998).
103. *(a)* M. Ni, S. E. Reichenbach, A. Visvanathan, J. TerMaat, E. B. Ledford Jr., Peak pattern variations related to comprehensive two-dimensional gas chromatography acquisition, *J. Chromatogr. A*, **1086**, 165–170 (2005); *(b)* B. V. Hollingsworth, S. E. Reichenbach, Q. Tao, A. Visvanathan, Comparative visualization for comprehensive two-dimensional gas chromatography, *J. Chromatogr. A*, **1105**, 51–58 (2006).

4

Plant Volatile Signalling: Multitrophic Interactions in the Headspace

André Kessler and Kimberly Morrell*

Cornell University, Department of Ecology and Evolutionary Biology, E445 Corson Hall, Ithaca, NY 14853, USA

4.1 Introduction

Plant scents have influenced cultures and myths since ancient times. In Hindu legend, the Astomi had no need to eat or drink and survived by smelling roots, apples and flowers.[1] They died from smelling unpleasant odours. In Greek mythology, the water nymph Mintha was metamorphosed into a mint plant and emitted an aromatic mint plume when trampled underfoot by her lover's jealous wife, the goddess Persephone. An ecological function of plant odours for a plant's survival and fitness was long denied. The German botanist Sprengel was the first to suggest that characteristics of flowers, such as colour, shape and smell, influence how they reproduce.[2] Darwin extended this hypothesis in his book '*On the Various Contrivances by which British and Foreign Orchids are Fertilised by Insects, and on the Good Effects of Intercrossing*' when he suggested that plants and their pollinators co-evolve and that floral scents evolved to attract pollinators and vector pollen.[3]

Since Darwin, scientists accepted that floral odours attract pollinator insects; however, it was only after discovering how to collect and analyse volatile organic compounds (VOCs) that scientists were able to study the multiple roles VOCs play in mediating interactions between organisms.[4]

A number of principles have emerged from research on floral VOCs that built the basis for our understanding of the signalling function of VOCs in general. First, floral VOC bouquets can contain tens to hundreds of different compounds,[5,6] but not all compounds in the complex

*Note: corresponding author.

The Chemistry and Biology of Volatiles Edited by Andreas Herrmann

mixture may have a biological function. Pollinators seem to use only a subset of VOCs in a flower's scent to guide their foraging choices. Moreover, the coevolution of plants and their pollinators and the influence of plant-pollinator interactions on plant fitness has resulted in some remarkable adaptations. For example, nursery pollination systems, as in *Yucca filamentosa*,[7] *Ficus* spp.,[8,9] *Silene latifolia*,[10] *Macrozamia lucida*[11] or *Glochidion* spp.[12] exhibit interdependent mutualistic interactions with their pollinators. In contrast, plants such as *Chiloglottis* and *Ophrys* orchids[13,14] attract pollinators by mimicking food and mates without providing a reward.[15] In these specialized and highly co-evolved interactions, pollinator insects seem to use single compounds or distinct combinations of compounds from floral bouquets to differentiate hosts from nonhosts.[7,8,13]

Second, some compounds in the VOC mixture may have multiple functions including attracting pollinators,[6] repelling nectar and pollen robbers[16] or increasing antimicrobial properties of the nectar.[17]

Third, while floral scent varies widely among related plant species,[18] unrelated taxa can show convergence in their floral VOC bouquets. This pattern seems to be driven by positive selection by major groups of pollinators.[5,15,19]

Fourth, floral VOCs contain context-dependent information, so a certain compound might result in different responses from different flower visitors.[20] For example, ground-blooming carrion-mimicking flowers attract blowflies by producing the foul-smelling dimethyl disulfide (**1**, Figure 4.1) in combination with heat, CO_2 emission and hair-like trichomes.[15] The same compound presented by flowers with high nectar production in neotropical vines and trees, attracts bats as pollinators.[21] Similarly, the ubiquitous floral and vegetative VOC methyl salicylate (**2**, MeSA) can attenuate honey bee visits to a flower,[22] attract orchid bees from long distances,[23] attract predatory mites when induced by herbivore damage[24] and even function as a pheromone that repels copulation attempts in *Pieris* butterflies.[25]

VOC emission in the headspace of vegetative tissues (vegetative VOCs) has been one of the major foci of the research on chemical plant defences in the past few decades.[26] Some of the volatile secondary metabolites are concentrated in storage organs (e.g. exudating trichomes, resin ducts) where they are released immediately upon herbivore attack and compromise insect development, fecundity and survival. Like floral VOCs, vegetative VOCs can mediate interactions between plants and many organisms in their biotic environment (tritrophic interactions). Many of the VOCs emitted from vegetative tissues are inducible in response to herbivore damage and influence herbivorous insects' ability to find and select a host plant. Moreover, because the specific induction of VOCs can reveal the whereabouts of herbivores to their natural enemies (e.g. predators and parasitoids), VOCs can function as indirect defences by facilitating host/prey finding of natural enemies, thereby reducing damage by herbivores.[27] In addition, herbivore-induced VOCs can provide signals for neighbouring undamaged plants to ready their inducible anti-herbivore defences.[28] The degree to which these observed ecological consequences of herbivore-

1 (Dimethyl disulfide) **2** (MeSA)

Figure 4.1 *Examples of VOCs containing context-dependent information* induced VOC emission are adaptive and drive the evolution of VOC signalling has been debated[29] because data on the fitness consequences, natural variation and specific signalling function of VOCs are limited.

Visitor attraction to floral VOCs directly affects a plant's fitness (positively by pollinators versus negatively by pollen/nectar robbers and florivores) by influencing seed yield. In contrast, herbivore-induced vegetative VOCs influence plant fitness indirectly by repelling herbivores and attracting predators and parasitoids, as these insects do not typically feed on seeds. Herbivory does not necessarily result in reduced seed production depending on the plant's ability to tolerate damage and to invest resources in direct defences, such as toxic, anti-nutritive and anti-digestive compounds.[30] The fitness effect of attracting natural enemies of herbivores to the plant depends on the degree and timing of the reduction in herbivory.[29] This makes measurements of fitness effects and evaluations of costs and benefits of VOC-mediated tritrophic interactions difficult and demands a deeper understanding of the underlying mechanisms.

In order to assume an adaptive function of herbivore-induced VOCs in multitrophic interactions it needs to be demonstrated that: (*a*) there is natural genotypic variation in the production and/or inducibility of VOC emission to allow for natural selection, (*b*) VOC emission provides a specific and reliable signal, (*c*) this signal can be perceived, processed and responded to by another organism (e.g. herbivore, predator, plant) and (*d*) this VOC-mediated interaction with other organisms has significant fitness benefits for the plant. Recent advances in collecting and analysing VOCs,[4] along with an explosive growth in our understanding of the biosynthetic and signalling pathways underlying VOC production[17,31] and the physiological mechanism that allow organisms to perceive VOCs[32] allow us to tackle critical questions in VOC signalling co-evolution. Moreover, these advances generated an awareness of the parallels between plant–pollinator and plant–herbivore interactions – enabling us to compare the ecological and evolutionary roles played by VOCs in both of these arenas.[33]

Here we review recent studies on the mechanisms and ecological functions of herbivore-induced VOC emission. In particular, we focus on the specificity of herbivore-induced VOC production and the potential costs and benefits of VOC emission. We also evaluate their fitness consequences of VOC-mediated interactions with herbivores, natural enemies of herbivores and neighbouring plants. In light of the behavioural ecology and evolutionary implications of herbivore-induced VOC emission we discuss the potential applications of VOC-mediated indirect defences in agriculture.

4.2 The Specificity and Complexity of Herbivore-Induced VOC Production

When attacked by herbivores, plants respond with a bewildering array of changes in primary and secondary metabolism.[34] In general, plants downregulate primary and upregulate secondary metabolic processes.[35] These responses are thought to alleviate the impact of additional physiological stresses that are associated with the damaged leaf tissue, such as oxidative stress, photoinhibition or drought stress. Herbivore-induced changes frequently reduce the quality of the leaf tissue as a food source for herbivores by reducing the availability of nutritious primary metabolites and by accumulating toxic, anti-nutritive and anti-digestive secondary metabolites.[34,36] The latter are referred to as induced direct defences because of their direct impact on herbivore performance. VOCs can also function as indirect defences – plant traits that facilitate the residence, foraging and host-finding behaviour of natural enemies of herbivores (including predators, parasitoids, parasites, pathogens).

Interestingly, many plant responses, particularly herbivore-induced VOC emissions, are specific to the attacking organism and/or the genotype of the attacked plant, suggesting that both plant genotype and attacker phenotype play a crucial role in the elicitation of VOC production. There are remarkable species-specific differences in plant scents that result from a large diversity of odour compounds. So far about 1700 VOCs from nearly 1000 plant species in 100 families have been described,[6,17] and the VOC blend emitted by a single herbivore-attacked plant contains tens to hundreds of compounds.[37–39] This complex blend includes compounds from multiple compound classes, such as green leaf volatiles (e.g. **3**, **4**), terpenoids (e.g. **5**, **6**) and aromatic compounds (e.g. **2**, **7**, Figure 4.2). A study of the volatile blends of 31 inbred maize (*Zea*

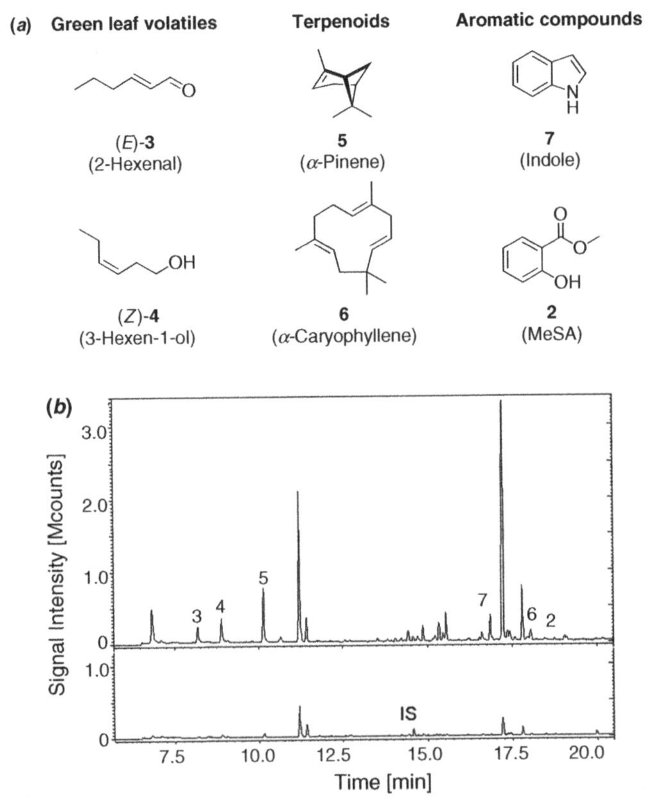

Figure 4.2 *Herbivore-induced volatile organic compound emission. (a) Commonly induced volatile organic compounds of three compound classes. (b) Representative gas chromatograms of goldenrod, Solidago altissima, headspace volatile organic compound emission from plants damaged by larval Trirhabda virgata (upper chromatogram) and undamaged plants (lower chromatogram). (c) Chrysomelidae beetle larva, Trirhabda virgata. Numbers on peaks in (b) refer to compound structures in (a), the major peaks are yet unidentified peaks*

Figure 4.2 (continued)

mays) lines induced with caterpillar regurgitant showed enormous variation both in the total amount of volatiles emitted and in the composition of volatile blends.[40] A comparable diversity in volatile blends was found when comparing different wild tobacco (*Nicotiana attenuata*) accessions,[41] wild accessions of *Datura wrightii*[42] and a number of Mexican maize varieties.[43] Although data on the natural genetic diversity of herbivore-induced VOC emission are still limited, these examples clearly illustrate that volatile blends are distinctive for particular genotypes.

The phenotypic plasticity of herbivore emissions and the mechanisms that cause plants to alter VOC emission in response to herbivory have been more comprehensively studied. Plant responses to herbivory (including VOC emission) result from specific transcriptional and metabolic reconfigurations of the plant[34] that are triggered by plant endogenous and herbivore-specific elicitation processes.

4.2.1 Plant Endogenous Wound Signalling

Recent studies show that a number of plant endogenous signalling pathways are activated in response to tissue damage (Scheme 4.1). Cross-talk between these pathways determines the downstream activation of genes involved in a metabolic reconfiguration of the plant.[34] Herbivore-induced VOC emission is dependent on and influenced by the functioning of at least three key phytohormones and their derivatives, jasmonic acid (**8**, JA, octadecanoid pathway), salicylic acid (**9**, SA, shikimate pathway) and ethylene (**10**, ET, Figure 4.3).[34,44,45] Plants deficient in different aspects of the octadecanoid pathway usually

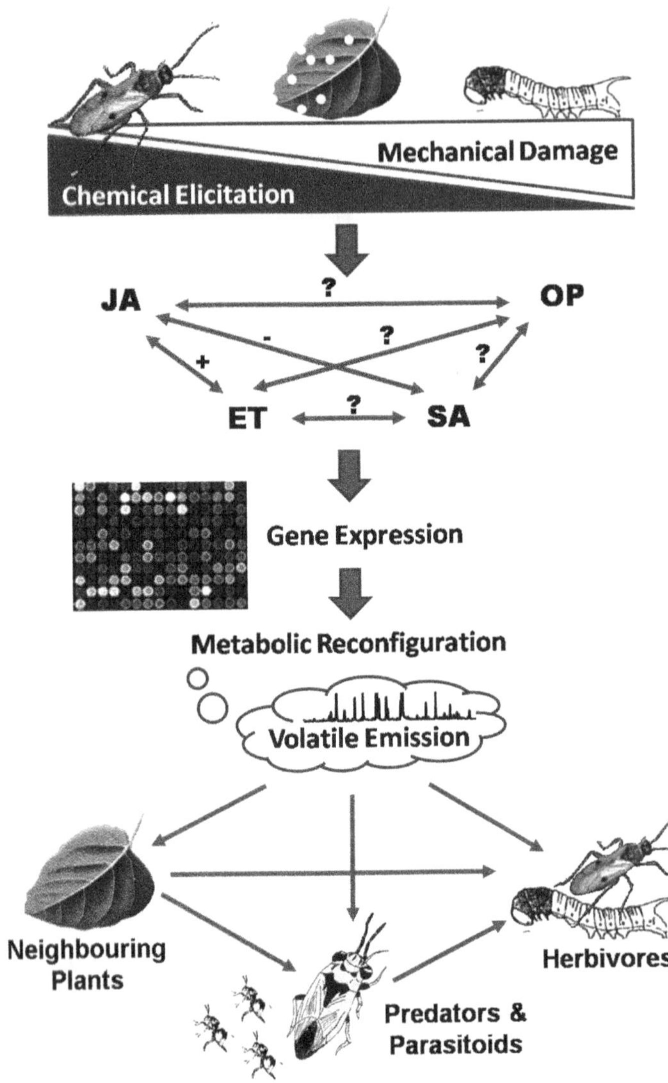

Scheme 4.1 The elicitation and ecological consequences of herbivore-induced volatile organic compound emission. In response to damage and elicitors from herbivores, plants trigger phytohormone-mediated pathways, transcribe defence-related genes and produce toxins, anti-digestive and anti-nutritive compounds. During these metabolic changes, plants emit volatile organic compounds (VOCs) into the headspace that are perceived by neighbouring plants, herbivores, predators and parasitoids and mediate complex multitrophic interactions. JA = jasmonic acid; ET = ethylene; SA = salicylic acid; OP = other phytohormones

show dramatically altered induced metabolic changes in response to herbivore damage and do not show herbivore-induced VOC emission. For example, the silencing of the lipoxygenase 3 (LOX3) gene, a specific wound and herbivore-induced lipoxygenase isoform that catalyses a crucial step in JA biosynthesis in wild tobacco, *Nicotiana attenuata*, inhibits

(±)-**8** (JA) | **9** (SA) | **10** (ET) | (±)-**11** (MeJA) | **12** (ACC)

Figure 4.3 *Structures of three key phytohormones **8–10** influencing herbivore-induced VOC emission and their derivatives or precursors (**11**, **12**)*

herbivore-induced responses, including VOC production.[46] Similarly, the application of JA or its methyl ester methyl jasmonate (**11**, MeJA) usually results in an induction of VOC emission and reasonably mimics herbivore-induced VOC emission.[41,47]

However, the seemingly small differences between JA-induced and herbivore-induced VOC bouquets can have significant impacts on their signalling function. The VOC blend induced in JA-treated lima bean plants lacks methyl salicylate (MeSA, **2**), which is produced by plants attacked by spider mites. Without MeSA in the blend, predatory mites are less attracted to spider mite-infested plants, as compared to JA-treated plants.[47] In contrast, predatory *Geocoris pallens* bugs are equally attracted to MeJA-treated and herbivore-attacked wild tobacco plants in the field because they hone in on single VOCs that are produced in response to both treatments.[37,48]

Despite the fact that JA is necessary and sufficient to trigger plant VOC responses, repetitive mechanical damage that induces endogenous JA bursts and herbivory still induces differential VOC emission in most plant species tested.[41,47] Moreover, emissions in response to wounding frequently lack the increased production of terpenoids that are found in response to herbivore damage, while exogenous application of JA or MeJA results in increased accumulation of terpenoids in the headspace. Differential plant responses to herbivory and JA application have been attributed to differential activation of signalling molecules and phytohormones, as well as to signalling pathway cross-talk.[49,50] For example, the exposure of lima bean plants to ET (**10**) through its precursor 1-aminocyclopropane-1-carboxylic acid (ACC, **12**, Figure 4.3) enhances VOC induction by JA and results in an increased attraction of predatory mites to the plants.[45]

What mechanisms cause the differential activation and thus allow the complex cross-talk of plant signalling pathways and, as a consequence, high specificity of the VOC response? The early events in plant wound recognition were reviewed recently[51] and include ion imbalances along membranes and the modulation of ion channels that differentially trigger downstream signalling networks of kinases and phytohormones.[52] Two major mechanisms influence the specificity of plant responses and VOC emission – the way herbivores rupture the plant tissue and the application of herbivore-derived chemical elicitors.[49]

First, herbivores tend to feed in bouts over long periods of time and feed on plant tissue in species-specific patterns, both of which may result in differential endogenous signal activation compared to standard mechanical damage. Support for this hypothesis comes from recent experiments using a mechanic caterpillar robot (MecWorm), which allows manipulation of the physical appearance and duration of wounding to better resemble herbivore-inflicted tissue damage.[53] Mimicking herbivore damage duration and area with the MecWorm resulted in the induction of herbivore-specific VOC emissions from lima bean plants that were qualitatively similar to VOC emissions induced by *Spodoptera littoralis* caterpillars but showed some quantitative differences. These results suggest that the way herbivores damage a plant affects the induced VOC emissions.[53]

Second, while feeding on the plant, herbivores apply chemical elicitor compounds that alter the endogenous wound response and contribute to the differential VOC emission we observe in many plant species after herbivore attack.[54] Regurgitant from lepidopteran larvae applied to wounds elicits VOC emission that is comparable to the emission from plants that are attacked by caterpillars.[41,55,56] This finding marked a breakthrough in the understanding of the induction mechanisms of volatile compounds and started a search for herbivore-derived elicitor compounds.

4.2.2 Herbivore-Derived Elicitors of VOC Emission

Lytic enzymes, such as β-glucosidase from *Pieris brassicae* caterpillar salivary secretions,[57] were the first class of elicitor compounds identified. Additional lytic enzymes have been isolated from the saliva of other lepidopteran species, such as glucose oxidase in *Helicoverpa zea*,[58] alkaline phosphatase from piercing/sucking insects such as whiteflies, *Bemisia tabaci*,[59] and numerous watery digestive enzymes from aphid saliva.[60] So far, only salivary enzymes from lepidopteran caterpillars have been shown to induce VOC emission similar to actively feeding herbivores.[57]

However, bioassay-driven fractionations of lepidopteran and orthopteran regurgitants and coleopteran oviposition fluids resulted in the discovery of four nonenzymatic elicitor compound classes (Figure 4.4).

Volicitin [*N*-(17-hydroxylinolenoyl)-L-glutamine, **13**] was the first in a series of fatty acid–amino acid conjugates (FACs, such as **13**, **14**) isolated from lepidopteran larval regurgitant. Extracted from *Spodoptera exigua* larvae, volicitin (**13**) induced excised maize seedlings to release the same odour blend of volatile terpenoids and indole that is released during caterpillar feeding.[61] Additional FACs with similar effects on VOC induction have been isolated from regurgitant of larval Sphingidae,[62] Noctuidae and Geometridae.[61,63] Interestingly, Tumlinson and Lait demonstrated that FACs are synthesized in the larvae by enzymes embedded in the membranes of the crop and anterior midgut tissue.[64] Moreover, the fatty acid moiety of the elicitors is obtained from the plant on which the caterpillar feeds, whereas the amino acid precursor derives from stores within the insect. FAC elicitation explains most of the specificity of herbivore-induced responses (particularly induced VOC emissions) in maize and wild tobacco, *N. attenuata*,[37,39,41,61] but is not sufficient to explain the responses of lima beans (*Phaseolus lunatus*) and cotton (*Gossypium hirsutum*) to attacking lepidopteran caterpillars.[65] Therefore it was hypothesized that any given herbivore species may produce a variety of elicitor compounds, each of which may only trigger specific responses in a subset of host plant species.[66]

Cowpea plants (*Vigna unguiculata*) emit specific blends of terpenoids when attacked by *Spodoptera frugiperda* caterpillars; however, the FACs present in *S. frugiperda* regurgitant do not explain the induction of specific VOCs in the bean plants.[66] The search for the active compounds in the caterpillars' regurgitant resulted in the isolation and identification of a new class of elicitor compounds – inceptins. Inceptins (e.g. **15**) specifically increase ET, JA and SA concentrations and elicit direct (cinnamic acid) and indirect defence (terpenoid) compound production. Inceptins are disulfide-bridged peptides that are proteolytic fragments of the chloroplastic adenosine-5′-triphosphate (ATP) synthase γ-subunit regulatory region.[66]

Inceptins

+ICDINGVCVDA-

15

Proteolytic fragments of chloroplastic ATP synthase from *Spodoptera frugiperda* larval oral secretions

Fatty acid-amino acid conjugates (FACs)

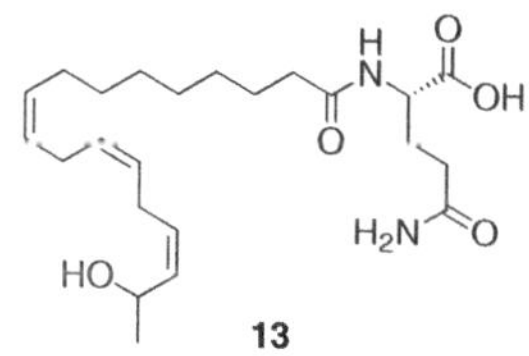

13
(Volicitin)

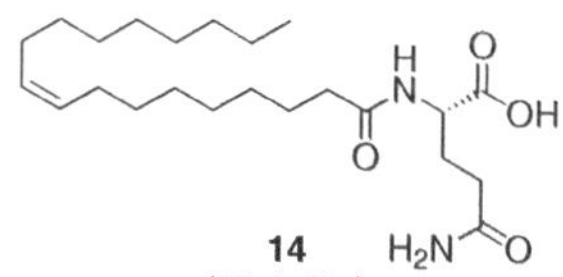

14
(18:1-Gln)

Lepidopteran caterpillar (*e.g. Spodoptera ssp. Manduca ssp.*) regurgitant

Caeliferins

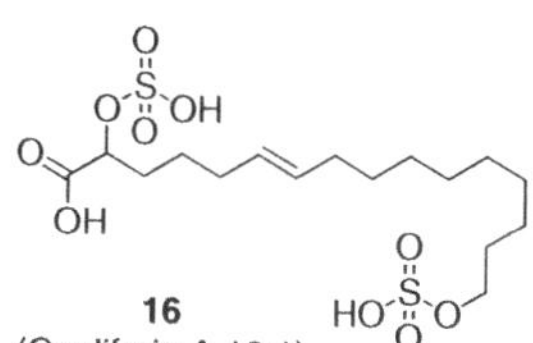

16
(Caeliferin A 16:1)

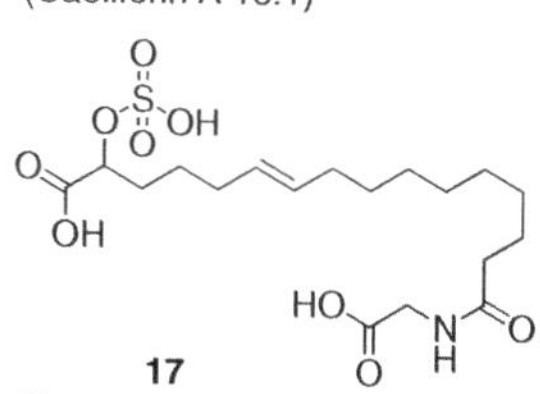

17
(Caeliferin B 16:1)

American bird grasshopper (*Schistocerca americana*) regurgitant

Bruchins

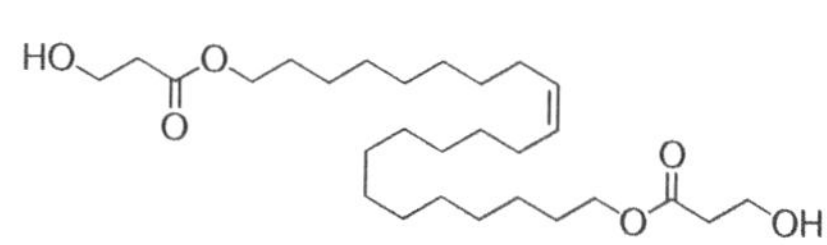

18
(Bruchin B)

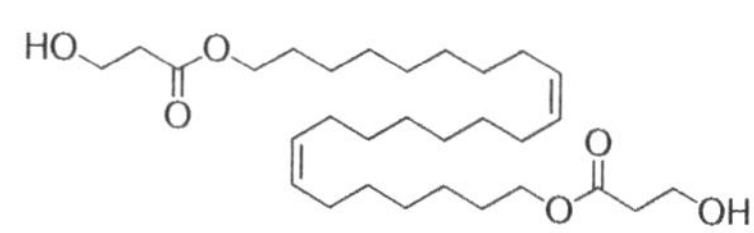

19
(Bruchin D)

Pea weevil (*Bruchus pisorum*) and cowpea weevil (*Callosobruchus maculatus*) oviposition fluid

Figure 4.4 *Herbivore-derived elicitor compounds, their chemical structures and origins*

Another class of elicitors was isolated from the regurgitant of American birdwing grasshoppers, *Schistocera americana*. The compounds, termed 'caeliferins', contain saturated and monounsaturated sulfated α-hydroxy fatty acids in which the ω-carbon is decorated with either a sulfated hydroxyl (**16**) or carboxyl functional group conjugated to glycine via an amide bond (**17**).[56] The grasshopper regurgitant contains a series of these caeliferin compounds with fatty acid chains varying in length between 15 and 20 carbon atoms. The most common and active one has 16 carbon atoms. Crude grasshopper regurgitant or isolated caeliferins applied to damaged leaves of maize seedlings induce the release of VOCs. The analysis of the regurgitant of several other orthopteran species indicated that most grasshoppers probably produce caeliferins. In contrast to the fatty acid moiety in FACs, the fatty acids in caeliferins with 16 carbons do not likely come from the host plant. The fatty acid composition of caeliferins in grasshopper regurgitant does not change with diet.[56]

Like herbivore feeding, insect oviposition triggers specific plant responses,[67] including VOC emission.[68–70] For example, the oviposition of egg masses of the pine sawfly *Diprion pini* into a slit in pine (*Pinus sylvestris*) needles results in VOC emission from the whole twig. The VOCs attract the eulophid egg parasitoid *Chrysonotomyia ruforum*, allowing the plant to defend itself against herbivore attack prior to being damaged by the feeding larvae.[71] The elicitor, which is likely a peptide or protein, is present in the oviduct secretion and coats the surface of the *D. pini* eggs.[72] Similarly, compounds in the oviposition fluids of elm leaf beetles, *Xanthogaleruca luteola*, induce specific VOC emission in held elm, *Ulmus minor*, which attracts the egg parasitoid wasp, *Oomyzus gallerucae*.[73] The identity and characteristics of elicitor compounds in oviposition fluids remain largely unknown.[67] However, a new compound class of elicitors has been isolated from pea weevil (*Brochus pisorum*) oviposition fluids. The elicitor compounds were named 'bruchins' and are long-chain α,ω-dioles, esterified at one or both oxygen atoms with 3-hydroxypropanoic acid (e.g. **18**, **19**, Figure 4.4). These compounds are potent plant growth regulators and elicit a tumour-like growth of undifferentiated cells (neoplasm) at the oviposition site, beneath the egg.[74]

The transcriptional and metabolic reconfiguration of the plant resulting from the interplay between wound-induced plant endogenous signalling and its alteration by herbivore-derived elicitors can induce a highly specific VOC signal that includes information about the plant species attacked and the species and age of the attacking herbivore.[37,75–77] An investigation of the phylogenetic distribution of elicitor activity revealed that all tested plant families responded at least to one of the tested elicitor compound classes with significant increases in ET, JA and sometimes SA production.[78] However, the timing and strength of the phytohormonal responses to similar elicitor compounds were highly variable among different plant species. Although the mechanisms are unknown, these findings suggest that specific receptor–ligand interactions mediate herbivore recognition and plant responses. Moreover, to induce particular VOC blends in response to herbivory, plants require functional biosynthetic pathways and their regulation under herbivore-inducible promoters.[17] That means that natural genetic variation in VOC induction depends on the biosynthetic properties of the plant and the signalling pathways regulating the activation of certain pathways in response to herbivore damage.

In conclusion, VOCs can provide complex and highly specific signals that can potentially spread information within a plant or inform other organisms about host quality. In the following sections, we review how emitting plants and other organisms use this information, and we focus on the ecological functions of herbivore-induced VOCs (within and between plant signal transduction, herbivore repellence, natural enemy attraction).

4.3 Ecological Consequences of VOC Emission

4.3.1 Within-Plant Defence Signalling

In order to mount successful defences against attacking herbivores, plants must elicit whole-plant defences and produce toxins, anti-digestive and anti-microbial compounds in most of the foliar or root tissue.[79] To do this, plants use different signalling pathways to spread the information of an attack to undamaged plant tissues – enabling them to express systemic resistance before herbivores reach other parts of the plant. These defence signals include chemical signals transmitted internally through the xylem and phloem (vasculature) and volatiles (airborne chemical signals), although it has long been debated whether vascular or volatile signals are the dominant method of signal transfer.

Scientists have shown that the transport of defence signals can follow the transport of sugars, nutrients and water through the vasculature, where it will reliably reach its destination.[80,81] However, vascular connectivity is not uniform throughout a plant. Rather, it is limited by phyllotaxy, particularly in plants with complex architecture.[81] Vascular connectivity is high in orthostichous leaves, intermediate in adjacent leaves and absent in opposite leaves.[82,83] Moreover, vascular connections are absent between branches in shrubs with long branches emanating from a basal crown.[84] Thus, vascular signals elicit patchy defences in neighbouring leaves and branches. Moreover, vascular signal transfer is relatively slow over long distances and may take hours to days to elicit systemic induced resistance.[81] Thus, vascular signalling transport is not sufficient to explain the faster systemic wound responses that are frequently observed. There is increasing evidence that plants can use VOC signals to elicit defences in neighbouring leaves and branches – even when they are not physically connected.[85,86]

Unlike vascular signals, all leaves within air contact of a damaged leaf can detect volatile signals, and volatiles travel rapidly – providing the opportunity for advanced warning in distal undamaged plant parts. For instance, volatiles from herbivore-damaged lima bean leaves elicit extrafloral nectar secretion in neighbouring undamaged leaves.[87] Volatiles also facilitate signal transfer between tomato leaves. When air contact is eliminated, only leaves with direct vascular connections to the damaged leaf show significant changes in defence chemistry,[82] suggesting that volatiles mediate the interaction. Similarly, volatiles have been shown to elicit defence compound production in neighbouring branches of cottonwood,[88] sagebrush[89] and blueberry plants[84] with long, distal branches and limited vascular connectivity.

In cottonwood (*Populus deltoids*), systemic defences deterred feeding by leaf beetles (*Plagiodera versicolora*) in orthostichous but not adjacent leaves.[90] Similarly, feeding by *Lema trilinea* on *Solanum dulcamara* resulted in leaf position-dependent induction of systemic defences and larval performance.[91] The induction of systemic defences and larval performance decreased significantly on leaves connected to the damaged leaf, yet remained high on leaves lacking vascular connectivity. These experiments show that spatial control of induction is ecologically relevant and that, to ward off herbivores completely, uniform systemic responses are needed. By overcoming limitations in vascular connectivity, herbivory-induced volatiles could provide a signal that makes uniform systemic responses possible.

4.3.2 Herbivore-Induced VOC Emission as Part of a Metabolic Reconfiguration of the Plant

The herbivore-induced production of VOCs is only part of a metabolic reconfiguration that includes the increased production of secondary metabolites and defensive proteins. To understand the role of herbivore-induced VOCs as signals for other organisms, one needs to understand the correlation between VOC emission and the production of other, nonvolatile secondary metabolites and the plant's overall physiology. Direct defence compound production, herbivore resistance and VOC signalling are frequently correlated, and the potential physiological and ecological interaction between direct and indirect defences have been investigated for potential trade-offs.[34] Plant defence theory predicts trade-offs among defensive traits because it assumes either that plants have limited resources to allocate to growth, defences and other vital functions or that there are phenotypic and ecological pleiotropic effects that link metabolic processes. A plant should optimize the allocation of resources to maximize fitness, which often involves trade-offs (i.e. between direct and indirect defences).[30] For example, the production of exudating trichomes as direct defensive traits frequently compromises top-down control of herbivory by predators and parasitoids.[92] Although trade-offs between different types of plant defences have repeatedly been reported,[93] there seem to be only a few examples of a physiological trade-off between the production of direct defensive compounds and VOC emission. For example, different accessions of lima bean (*Phaseolus lunatus*) differ in the amounts of toxic hydrogen cyanide (HCN) that they produce in response to damage – high cyanogenic accessions release low amounts of VOCs and vice versa.[94] Quantitative trade-offs between cyanogenesis and VOC releases were evident between accessions and ontogenetically, which suggests differential resource allocation to the two types of defences. The general lack of evidence for trade-offs between different types of plant defences likely suggests a more complicated relationship of defensive traits as currently perceived and may go beyond simple resource allocation models.[95]

For example, there are studies on herbivore-induced metabolic changes in plants that find a positive correlation between direct defence compound production, resistance and VOC emission.[48] One likely reason for a positive correlation is a common biochemical origin and thus a simultaneous regulation of volatile and nonvolatile compounds (e.g. terpenoids, phenylpropanoids). Another potential reason is that selective forces acting on direct and indirect defensive chemical traits are similar, which would result from a synergistic processes of bottom-up and top-down effects on herbivore performance. Such a synergistic effect of direct and indirect defences has been observed in the wild tobacco *Nicotiana attenuata*, where herbivore damage resulted in increased production of toxic secondary metabolites and VOC emission. The toxic compounds decreased the growth rate of the voracious pest *Manduca quinquemaculata* but did not increase mortality.[48] Increased VOC emission resulted in increased attraction of predatory bugs, *Geocoris pallens*, to the plant. Because the toxins make *Manduca* caterpillars grow more slowly, they remain vulnerable to predators for longer periods of time – thereby increasing their mortality and the fitness of the plant.[37,48] Synergistic effects of chemical direct and indirect defences on herbivore performance can be viewed as a special case of the slow growth–high mortality principle and are likely to occur when the direct defence compound production does not compromise the efficacy of predators and parasitoids.

Many specialized herbivore species can sequester plant toxins and use them for their own defences. As a result, these herbivores can only be attacked by specialized natural enemies that themselves can tolerate high amounts of the plant toxins. Moreover, combined physical and chemical direct defences, such as exudating trichomes, can reduce the recruitment of bodyguards to the plant.[96] It is noteworthy that these negative effects of plant defences on natural enemies come predominantly from constitutively produced direct defences. The regulation of defence gene expression under a herbivore-inducible promoter could allow the

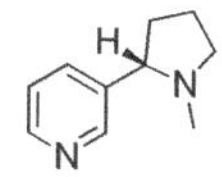

20
(Nicotine)

Figure 4.5 *Structure of nicotine (**20**), a toxin to many insect species*

plant to specifically orchestrate direct and indirect defences and overcome the compromising effects of direct defences on top-down control of herbivory. For example, elicitor compounds in the regurgitant of *Manduca sexta* hornworms attenuate nicotine production in wild tobacco (*N. attenuata*). Nicotine (**20**, Figure 4.5) is potentially toxic to attacking parasitoid wasps.[97] The herbivore-derived compounds specifically induce an endogenous ethylene burst that results in the attenuation of gene expression of a key enzyme in nicotine biosynthesis – putrescine methyl transferase.[98] While the toxic nicotine production is attenuated, VOCs are emitted and other growth-slowing compounds, such as phenolics and proteinase inhibitors, are strongly induced in response to *Manduca* damage, allowing the plant to avoid compromising the positive interaction with parasitoids and predators.[99]

The links between induced VOC emission and herbivory, and between induced VOC emission and plant quality, as food for herbivores have some basic implications for how different organisms respond to VOCs. For example, herbivores can inform their host plant choice and/or find mating partners on plants damaged by conspecifics. Natural enemies can potentially extract information about the presence and identity of a potential host/prey species. Plants can eavesdrop on the VOC signal from a damaged neighbour and gain information about potential future herbivory.

4.3.3 Herbivores Use VOCs to Select Host Plants

Herbivores consider many qualities when selecting a host plant. Are other herbivores around?[100] Was the plant previously damaged?[101] Is the plant surrounded by conspecifics or heterospecifics?[102] Does the plant have sufficient light, water and nutrients; and is it young or old?[103] Some studies have shown that VOC blends influence herbivores' ability to differentiate between hosts of different food quality. Abiotic factors like photoperiod, nutrients, soil moisture and temperature influence the intensity and composition of volatile blends. When conditions are ideal (as in young plants growing in high-nutrient, well-lit environments), plants release larger quantities of volatiles and are more attractive to herbivores. Volatiles not only reflect environmental suitability, but they also reflect the genetic quality of a plant.[103] Silencing defence genes has been shown to alter herbivore preference. Jassbi *et al.* showed how silencing the geranylgeranyl diphosphate synthase gene in *Nicotiana attenuata* improves host plant quality, causing tobacco hornworm larvae to gain weight three times faster on transgenic compared to control plants.[104] Similarly, oviposition by *Pieris rapae* decreases on *Arabidopsis thaliana* with knocked-down indole glucosinolate production, suggesting that it is an important host-finding cue.[105] In this way, VOCs are reliable indicators of host plant quality.

Studies have shown that many herbivores use VOCs to find quality plant hosts.[106] One key indicator of quality is prior damage. Damaged plants release different VOC blends than undamaged plants, and this often deters herbivory.[37,107] In cotton, VOCs from plants with high damage deter feeding by *Spodoptera littoralis* after just three days, and the effects persist for at least 14 days after damage.[108] Furthermore, the effects are strongest in the youngest leaves, which are preferred by herbivores. Damage by the chrysomelid beetle *Agelastica alni* increases with increasing distance from damaged *Alnus glutinosa* (black alder), and this effect is the result of plant communication, not predator recruitment.[109] Although damage aboveground is not necessarily a predictor of belowground herbivory, damage in one area of the plant may influence the suitability of the entire plant for future herbivory. For example, in experiments on the rice water weevil (*Lissorhoptrus oryzophilus*, a root herbivore) and fall armyworm (*Spodoptera frugiperda*, a foliage-feeding herbivore) on rice, severe damage by one herbivore decreases densities and growth rates of the other herbivore by up to 40%.[110]

In some cases, VOCs from damaged plants actually increase herbivore recruitment. This specific herbivore behaviour could help to find mates[111] or host plants in environments where preferred hosts are sparse (since a bad host is better than no host at all). Though a number of scientists have commented on plant palatability and the apparency theory,[112] only recent studies showed evidence that VOCs could mediate this interaction. Studies on wild tobacco plants (*N. attenuata*) genetically modified to silence green leaf volatile production in response to damage showed that flea beetles, *Epitrix hirtipennis*, accumulate on plants with high damage-induced green leaf volatile emission.[113] Similarly, the attraction of neonate fall armyworms (*Spodoptera frugiperda*) to VOCs from damaged cowpea plants facilitates their host plant search.[114] As discussed in the next section, VOCs from damaged plants do make them more apparent to insects of higher trophic levels – enabling them to attract beneficial predators and parasitoids. In these cases, apparency is advantageous for the plant and one should expect plants (and possibly plant communities) to benefit from apparency to natural enemies. Nonetheless, apparency to herbivores may represent an ecological cost of VOC emission.

4.3.4 VOCs as Indirect Defences Against Herbivores

VOCs are inducible indirect defences if they function as attractive signals for natural enemies of herbivores.[30] By announcing the presence of prey or host organisms to predators or parasitoids, respectively, VOCs can facilitate prey and host search behaviour of natural enemies – effectively employing them as 'bodyguards'.[115] Herbivore-induced volatile emission, popularly referred to as a 'plant's cry for help',[116] has been shown in at least 13 plant families[117] and is one of the most efficient ways plants cope with insect attackers.[118] What is the evidence for the defensive function of herbivore-induced VOCs and what are the fitness effects for the plant and the community consequences of VOC signalling?

A number of carnivorous insect species respond to specific herbivore-induced VOC emissions, including entomophagous nematodes,[119] predacious mites,[120] Heteroptera (Figure 4.6)[37,48,113] and Coleoptera, as well as a wealth of parasitoid wasps.[39,57,121] Classic examples of VOC-mediated indirect defences are: (*a*) attracting predatory mites to spider mite-attacked lima beans,[122] and (*b*) attracting the parasitoid braconid wasp, *Cotesia marginiventris*, to maize seedlings that are attacked by the beet armyworm, *Spodoptera exigua*.[39] Meanwhile a whole range of folivorous mites[115,123] and insects,[37,39,41,124,125] stem borers,[126,127] seed feeders[128] and root feeders[119,129] have been shown to induce VOCs and attract predators and parasitoids. Moreover, the microinjuries caused by lepidopteran

Figure 4.6 *Big-eyed bug, Geocoris pallens, attacking a mirid bug Tupiocoris notatus. G. pallens is a generalist predator that is attracted by herbivore-induced plant VOCs*

caterpillars moving across the plant surfaces[130] and oviposition by female beetles[67] and sawflies[131] increase VOC emissions and induce plant responses. Birds are also attracted to herbivore-induced plant VOCs and use them to find their prey[132] – broadening the spectrum of potential bodyguards responding to plants' 'calls for help'.

Most of the examples of VOC-mediated indirect defences come from crop plant systems studied in laboratory or agricultural settings. Although this may limit scientists' ability to apply these results to more complex natural systems, they illustrate the basic mechanism of predator and parasitoid attraction to herbivore-damaged plants. Moreover, recent experiments on wild systems in their native habitats suggest that the findings from crop plant systems may be generalisable[37] and some common characteristics of VOC-mediated defences can be identified.

First, VOC emission has to be inducible to sustainably function as an attractive signal for natural enemies. Natural enemies have to associate the VOC signal with their phytophagous prey/hosts, which is impossible when VOC emissions are constitutive and not inducible in response to herbivory. Second, herbivore-induced VOC emission increases the signal intensity associated with a feeding herbivore and thus the distances over which bodyguard carnivores can be recruited. By inducing systemic VOC emission in response to herbivore attack, plants provide a strong, reliable and highly specific signal that is strongly correlated with the presence of the herbivore and thus potential prey or hosts.

More importantly, the cue is assumed to be significantly stronger than any cue directly coming from the herbivore. Herbivores are under strong selective pressure to minimize signalling that can be associated with their presence and attract natural enemies. Yet herbivore feeding activity elicits strong VOC emissions, which suggests that plants are selected, in turn, to utilize cues that are specific yet inevitable for the attacking herbivore. It remains unknown if VOC-inducing herbivore-derived elicitors, such as FACs, inceptins, or caeliferins (Figure 4.4), play vital functions for the herbivores themselves or if their biosynthesis and application to the plant wound simply cannot be avoided. However, recent examples suggest that herbivores can choose to feed on plants or plant parts that are deficient in VOC signalling and defence elicitation[48,133] or actively inhibit signalling that results in induced VOC emission.[134]

Third, another characteristic of VOC-mediated indirect defences is the high specificity of induced VOC bouquets and the correspondingly specific attraction of natural enemies. The specificity is evident in both the de novo production of certain compounds and/or the relative changes of volume ratios in the entire scent bouquet.[37] Carnivorous mites and insects can often discriminate between VOC blends induced by different herbivore species.[75,117,125] In particular, specialist parasitoid wasps utilize specific compounds to find their appropriate hosts. For example, the parasitoid wasp *Cotesia kariyai* exclusively attacks first and second instar *Mythimna separate* caterpillars. Older instar caterpillars have physical defences that successfully repel the parasitoid. It seems that specific plant VOCs allow the parasitoid to differentiate between the earlier and later instar *Mythimna* larvae feeding on the plants. Interestingly, this response can be elicited by applying regurgitant collected from caterpillars of different instars to the plant.[77] Moreover, carnivores can learn to discriminate and associate a particular blend with prey/host organisms, which allows them to respond only to the most reliable VOC signals.[135,136]

In contrast to specialists, which seem to rely on a few specific components in VOC blends,[77,137] generalist predators seem to be attracted by basic VOC signalling or even single compounds out of the VOC bouquet. Generalist big-eyed bugs, *Geocoris pallens* (Figure 4.6), are attracted by singly offered terpenoid and green leaf compounds, all of which are generally part of damaged plants' VOC blends.[37,113] Moreover, experiments with wild tobacco plants (*N. attenuata*) deficient in green leaf VOC or terpenoid production in response to hornworm (*Manduca sexta*) damage demonstrated that *G. pallens* can use both green leaf VOCs and terpenoids independently for prey location.[113] Utilizing general compounds in the VOC blend could be advantageous for generalist predators by allowing them to find multiple prey species on a large number of plant species efficiently. However, this behaviour may also come with the cost of being attracted to a nonsuitable prey. Karban found *G. pallens* attracted to sagebrush plants that were primarily attacked by grasshoppers, which are too large for them to eat.[138] In this case, herbivore-induced VOC emission attracts natural enemies but does not reduce herbivory. Even if not all predators visiting a plant significantly reduce its herbivore population, plants may still benefit from an increased predator visitation rate by multiplying the negative effects of induced defences on the herbivore population and thus the positive effects on plant fitness.[48] Moreover, the mere presence of predatory insects can reduce herbivorous insect performance nonlethally by increasing stress-related behaviours – thereby decreasing the impact of herbivores on the plant.[139]

Although there are multiple accounts for the attractiveness of herbivore-induced VOCs to natural enemies of herbivores and the damage-reducing effects of their attraction, we still know relatively little about how often VOC-mediated plant–insect interactions actually improve plant fitness and whether VOCs, as indirect defence traits, truly coevolve with predatory insects and herbivores. This led to the suggestion of alternative, non-adaptive hypotheses for why plants produce VOCs.[140] However, the interest in the evolutionary aspects of indirect defences is high and a growing number of studies use natural variation[40,43] or genetic manipulation of herbivore-induced VOC emissions for a functional analysis.[113,136] Such comparative and integrative experimental approaches will eventually allow scientists to assess the Darwinian fitness effects of VOC-mediated indirect defences for the plant.

Moreover, recent field studies illustrate the importance of considering whole-community effects when assessing the adaptive function of herbivore-induced VOC emission. Plant-induced responses (VOC emission, defensive chemistry) can mediate interactions between all organisms associated with the plant and can link the fate of organisms that do not directly interact in space and time. For example, belowground herbivory may influence shoot secondary metabolism and, by extension, the performance of aboveground herbivores and the associated VOC signalling.[141,142] VOC-mediated attraction of predators can be elicited by multiple herbivore species on a plant, but differential predator preferences may alter the herbivore community composition and thus the fitness effects of herbivory for the plant.[48] The presence of multiple herbivore or pathogen attackers on one plant can significantly alter plant VOC signalling, and this has complex consequences for tritrophic interactions and the plant's indirect defences.[141,143] Thus, induced plant responses to herbivory in general and VOC signalling in particular increase the arena in which plant insect interactions are played out. There are promising new approaches in analysing the role of herbivore-induced VOC emissions on entire community interaction networks.[144]

4.3.5 VOCs in Plant–Plant Interactions

In addition to the multiple interactions that VOCs mediate within the plant arthropod community, VOCs can elicit defensive responses within and between plants.[28,87,89,145] At present, we know that volatile blends vary depending on the plant species damaged, the type and the developmental stage of an herbivore – providing the opportunity for plants to gain specific information about herbivores in their vicinity from volatile blends.[146] Although plants derive fitness benefits from repelling herbivores and attracting predators and parasitoids (see above), the ecological relevance of VOCs in plant–plant signalling is still debated, and the extent to which signal propagation between plants influences plant fitness, arthropod community dynamics and plant community structure remains unclear.[28]

Although plants that emit VOCs ('emitters') derive clear fitness benefits from inducing systemic direct and indirect defences, weighing the overall costs and benefits of volatile emission for an emitter is complicated. This is partially because of the complex interaction that induced VOC emission can mediate within the arthropod community of the plant. However, it is also because VOCs are not private signals, so downwind plants ('receivers') may eavesdrop on the signal and benefit from advanced information about their risk of herbivory – potentially altering competitive interactions between plants. The identity and relatedness of the receiver likely influences whether VOCs represent a net cost or benefit to an emitter.[49] If VOCs are reliable indicators of impending herbivory and neighbouring plants share herbivores, then receivers could prime or elicit their defences prematurely – gaining a competitive advantage.

The fitness of an emitter may be positively affected when kin eavesdrop on the VOC signal. For example, sagebrush individuals respond more strongly to VOCs from damaged clones than damaged nonclones, suggesting that VOCs allow self/nonself discrimination and possibly kin recognition.[147] This could provide a fitness advantage to an emitter, through inclusive fitness, by amplifying collective defences and increasing a clone's chances of repelling herbivores or attracting predators or parasitoids. Premature defence elicitation after VOC perception could be a key competitive strategy in plant populations that are clonal and grow in dense monocultures.[87]

However, in complex ecological systems, it is also likely that neighbouring conspecifics may benefit from responding to VOCs if they share herbivores. For example, when one lima bean branch is damaged by herbivores, VOCs are perceived by undamaged lima bean plants that then induce extrafloral nectar production.[85] Similarly, sagebrush (*Artemisia tridentata*) becomes more resistant to herbivory after exposure to VOCs from damaged neighbours.[148] Plants within 60 cm of a damaged sagebrush experienced this benefit. Moreover, heterospecifics in the vicinity of a damaged emitter can also benefit from responding to VOCs. For example, wild tobacco (*Nicotiana attenuata*) responds to VOCs from damaged sagebrush (*Artemisia tridentata*) by priming its defences.[149] 'Priming' allows a plant to respond more quickly and strongly to subsequent feeding by an herbivore.[150]

When the predictability of future herbivory via herbivore-induced VOCs is low, then eliciting defences is a net cost, but if they are attacked and their defences are ready, then they may experience fitness benefits by minimizing herbivore damage. It is likely that priming is actually less costly to a plant than defence elicitation, since defence proteins and metabolites are not produced until a primed plant is attacked.[149,150] If receivers gain a competitive advantage, emitters might experience a reduction in fitness through altered competitive abilities. Finally, although scientists have begun to examine the ecological relevance of VOC signalling between plants, through behavioural assays, it remains to be seen how eavesdropping shapes the succession of plant communities and herbivore movement through the population. Does eavesdropping provide net fitness benefits or costs to emitters and receivers? If eavesdropping benefits receivers, do they experience enhanced competitive ability and therefore greater abundance in the population over time?

4.4 Conclusion

Although vegetative and floral VOCs and the way they structure plant–insect and plant–pollinator interactions are often considered separately, recognizing the parallels between these two fields can advance both and possibly facilitate agricultural applications. As we discussed at the outset of this chapter, there are key parallels between vegetative and floral VOC blends and their use by arthropods.

First, vegetative and floral VOC blends are complex – they each contain tens to hundreds of different compounds. Experiments on herbivores, predators/parasitoids and pollinators show that insects typically use only a subset of the volatiles to identify and evaluate a potential host plant or find potential prey on it. There seem to be distinct differences in the proportion of a VOC blend generalists and specialists respond to – while generalist carnivores are able to use a broad range of wound response VOCs singly to locate hosts, specialists tend to utilize a smaller subset of compounds in the blend.

Second, many of the components of vegetative and floral VOC blends have multiple functions, whose relative fitness costs or benefits depend on the ecological context of the plant–insect interaction. This is evident in genetically transformed wild tobacco (*N. attenuata*) plants that cannot synthesize green leaf volatiles and are simultaneously more attractive to herbivorous flea beetles (which damage it more heavily) and to the generalist predator *Geocoris pallens*.[113] The outcome of this interaction depends on the relative damage that flea beetles do to a plant, which is in turn a function of how many new herbivores and predators are recruited. This is also true for floral VOCs which provide fitness benefits for the plant by attracting pollinators, but may result in fitness costs when pollen and nectar robbers are attracted to the plant by the same signal.

As VOC signals from damaged plants are often correlated with defence compound production, they enable herbivores to extract information about host plant quality and the presence of mating partners or conspecific and heterospecific competitors. Predators gain information about the presence and identity of potential prey/hosts. The diverse functions of herbivore-induced VOCs increase the complexity of potential interactions and make their fitness effects context-dependent.

The fact that plants, themselves, can respond to VOC signals from damaged neighbours adds to the complexity and requires a similarly careful cost–benefit analysis. Within-plant signalling may benefit plants by overcoming limitations in vascular connectivity. However, the effect of eavesdropping on the fitness of the emitter likely depends on whether the competing neighbour is kin or not.

Our knowledge about fitness consequences of herbivore-induced VOC emission is still very limited but will grow with the increasing use of modern molecular tools in field studies,[113,151] the use of comparative approaches[40,43] and the consideration of community networks interacting with the plant.[48,144] These new developments will also allow us to consider the possibility of harnessing VOC-mediated induced plant defences in agriculture.[152] Genetically manipulating the expression of plant VOC traits facilitates studies on the function of induced VOC emission and enables scientists to maximize the effects of VOC-mediated tritrophic interactions.[136,152] It seems that breeding programmes have inadvertently selected against induced defences in some crops. For example, North American maize (*Zea mays*) lines no longer emit (*E*)-*β*-caryophyllene (**21**, Figure 4.7) – a compound that attracts beneficial predatory nematodes and controls western corn rootworm (*Diabrotica virgifera*). Restoring this trait resulted in 60% fewer adult beetles emerging[153] and could presumably enhance yields dramatically as well. Similarly, a comparison of herbivore performance on wild and cultivated *Brassica oleracea* revealed distinct differences in primary and secondary metabolite levels that affected herbivore performance.[154]

One of the most applied examples of utilizing VOC-mediated tritrophic interactions in agriculture comes from a study of maize (*Zea mays*) grown in a 'push–pull' intercropping system in Kenya.[127,155] Maize plants are grown together with an intercrop species (the 'push plants' *Desmodium uncinatum*, *Melinis miutiflora*) that constitutively emits VOCs that would usually be emitted by herbivore-attacked maize plants. This odour repels ovipositing adults of two stem borer species (*Busseola fusca*, *Chilo partellus*), the major insect pests in the system, and attracts parasitoids into the field. The emission of damage-indicating VOCs protects crops by reducing herbivore abundance directly through decreased herbivore attraction and indirectly by increasing predation pressure. The herbivore repellent effect of

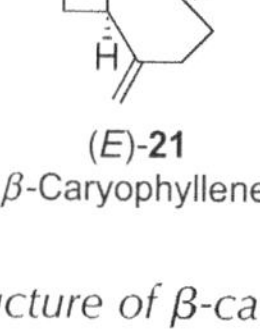

(*E*)-**21**
(*β*-Caryophyllene)

Figure 4.7 *Structure of β-caryophyllene (**21**)*

Desmodium is multiplied by growing hyper-attractive plants, the 'pull plants' (e.g. Napier grass, *Pennisetum purpureum*) in the circumference of the maize fields to attract the stem borers away from the fields.[154]

The Kenyan push–pull system is a promising example of how the utilization of natural VOC-mediated plant defences could help to make agriculture more sustainable. However, applications like this require a detailed understanding of the community and chemical ecology of the system as well as detailed assessment of the plant fitness consequences of herbivore-induced VOC emission. This challenge hopefully drives research in the entire field that studies multitrophic interactions in the headspace.

Acknowledgements

We thank Anurag Agrawal, Danny Kessler and Rayko Halitschke for inspirational conversations and helpful comments on this manuscript. Because of the brevity and the focus of this chapter we have not been able to refer to all the relevant literature and we apologize to those authors whose work we have not had space to cite. The authors were supported with funds from the National Science Foundation (DEB-0717139) and Cornell University.

References

1. E. A. Schwanbeck, *Megasthenis Indica,* Sumptibus Pleimesii, Bibliopolae, Bonn, 1846.
2. C. K. Sprengel, *Das entdeckte Geheimnis der Natur im Bau und in der Befruchtung der Blumen*, Meyer & Müller, Berlin, 1793.
3. C. R. Darwin, *On the Various Contrivances by which British and Foreign Orchids are Fertilised by Insects, and on the Good Effects of Intercrossing*, John Murray, London, 1862.
4. D. Tholl, U. S. R. Roese, Detection and identification of floral scent compounds, in *The Biology of Floral Scent,* N. Dudareva, E. Pichersky (Eds), CRC Press, Boca Raton, 2006, pp. 3–26.
5. H. E. M. Dobson, Relationship between floral fragrance and type of pollinator, in *The Biology of Floral Scent*, N. Dudareva, E. Pichersky (Eds), CRC Press, Boca Raton, 2006, pp. 147–198.
6. J. T. Knudsen, R. Eriksson, J. Gershenzon, B. Stahl, Diversity and distribution of floral scent, *Bot. Rev.*, **72**, 1–120 (2006).
7. G. P. Svensson, M. O. Hickman, S. Bartram, W. Boland, O. Pellmyr, R. A. Raguso, Chemistry and geographic variation of floral scent in *Yucca filamentosa* (Agavaceae), *Am. J. Bot.*, **92**, 1624–1631 (2005).
8. (*a*) C. Chen, Q. Song, Responses of the pollinating wasp *Ceratosolen solmsi marchali* to odor variation between two floral stages of *Ficus hispida*, *J. Chem. Ecol.*, **34**, 1534–1544 (2008); (*b*) C. Chen, Q. Song, M. Proffit, J.-M. Bessiere, Z. Li, M. Hossaert-McKey, Private channel: a single unusual compound assures specific pollinator attraction in *Ficus semicordata*, *Funct. Ecol.*, **23**, 941–950 (2009).
9. M. Proffit, B. Schatz, R. M. Borges, M. Hossaert-McKey, Chemical mediation and niche partitioning in non-pollinating fig-wasp communities, *J. Anim. Ecol.*, **76**, 296–303 (2007).
10. S. Doetterl, A. Juergens, K. Seifert, T. Laube, B. Weissbecker, S. Schueltz, Nursery pollination by a moth in *Silene latifolia*: the role of odours in eliciting antennal and behavioural responses, *New Phytol.*, **169**, 707–718 (2006).
11. I. Terry, G. H. Walter, C. Moore, R. Roemer, C. Hull, Odor-mediated push–pull pollination in cycads, *Science*, **318**, 70 (2007).
12. T. Okamoto, A. Kawakita, M. Kato, Interspecific variation of floral scent composition in *Glochidion* and its association with host-specific pollinating seed parasite (*Epicephala*), *J. Chem. Ecol.*, **33**, 1065–1081 (2007).

13. M. Ayasse, F. P. Schiestl, H. F. Paulus, C. Loefstedt, B. Hansson, F. Ibarra, W. Francke, Evolution of reproductive strategies in the sexually deceptive orchid *Ophrys sphegodes*: how does flower-specific variation of odor signals influence reproductive success? *Evolution*, **54**, 1995–2006 (2000).
14. F. P. Schiestl, M. Ayasse, H. F. Paulus, C. Loefstedt, B. S. Hansson, F. Ibarra, W. Francke, Orchid pollination by sexual swindle, *Nature*, **399**, 421–422 (1999).
15. A. Jürgens, S. Dötterl, S. Doetterl, U. Meve, The chemical nature of fetid floral odours in stapeliads (Apocynaceae–Asclepiadoideae–Ceropegieae), *New Phytol.*, **172**, 452–468 (2006).
16. (*a*) L. S. Adler, R. Karban, S. Y. Strauss, Direct and indirect effects of alkaloids on plant fitness via herbivory and pollination, *Ecology*, **82**, 2032–2044 (2001); (*b*) D. Kessler, K. Gase, I. T. Baldwin, Field experiments with transformed plants reveal the sense of floral scents, *Science*, **321**, 1200–1202 (2008); (*c*) R. A. Raguso, Why are some floral nectars scented? *Ecology*, **85**, 1486–1494 (2004).
17. *The Biology of Floral Scent*, N. Dudareva, E. Pichersky (Eds), CRC Press, Boca Raton, 2006.
18. P. Goldblatt, J. C. Manning, Radiation of pollination systems in the Iridaceae of sub-Saharan africa, *Ann. Bot.*, **97**, 317–344 (2006).
19. J. T. Knudsen, L. Tollsten, Floral scent in bat-pollinated plants: a case of convergent evolution, *Bot. J. Linn. Soc.*, **119**, 45–57 (1995).
20. R. A. Raguso, Wake up and smell the roses: the ecology and evolution of floral scent, *Annu. Rev. Ecol. Syst.*, **39**, 549–569 (2008).
21. O. von Helversen, L. Winkler, H. J. Bestmann, Sulphur-containing ‘perfumes’ attract flower-visiting bats, *J. Comp. Physiol., A*, **186**, 143–153 (2000).
22. J. A. Henning, Y.-S. Peng, M. A. Montague, L. R. Teuber, Honey bee (Hymenoptera: Apidae) behavioral response to primary alfalfa (Rosales: Fabaceae) floral volatiles, *J. Econ. Entomol.*, **85**, 233–239 (1992).
23. T. Eltz, K. Lunau, Antennal response to fragrance compounds in male orchid bees, *Chemoecology*, **15**, 135–138 (2005).
24. J. G. de Boer, M. A. Posthumus, M. Dicke, Identification of volatiles that are used in discrimination between plants infested with prey or nonprey herbivores by a predatory mite, *J. Chem. Ecol.*, **30**, 2215–2230 (2004).
25. J. Andersson, A. K. Borg-Karlson, C. Wiklund, Sexual cooperation and conflict in butterflies: a male-transferred anti-aphrodisiac reduces harassment of recently mated females, *Proc. R. Soc. Lond., Ser. B*, **267**, 1271–1275 (2000).
26. (*a*) M. Bruinsma, H. Ijdema, J. J. A. van Loon, M. Dicke, Differential effects of jasmonic acid treatment of *Brassica nigra* on the attraction of pollinators, parasitoids, and butterflies, *Entomol. Exp. Appl.*, **128**, 109-116 (2008); (*b*) E. E. Farmer, Surface to air signals, *Science*, **411**, 854–856 (2001).
27. M. Dicke, Chemical ecology of host-plant selection by herbivorous arthropods: a multitrophic perspective, *Biochem. Syst. Ecol.*, **28**, 601–617 (2000).
28. I. T. Baldwin, R. Halitschke, A. Paschold, C. C. von Dahl, C. A. Preston, Volatille signaling in plant–plant interactions: ‘talking trees’ in the genomic era, *Science*, **311**, 812–815 (2006).
29. J. D. Allison, J. D. Hare, Learned and naïve natural enemy responses and the interpretation of volatile organic compounds as cues or signals, *New Phytol.*, **184**, 768–782 (2009).
30. R. Karban, I. T. Baldwin, *Induced Responses to Herbivory*, Chicago University Press, Chicago, 1997.
31. D. R. Gang, Evolution of flavors and scents, *Annu. Rev. Plant Biol.*, **56**, 301–325 (2005).
32. M. A. Carlsson, B. S. Hansson, Detection and coding of flower volatiles in nectar-foraging insects, *Biol. Floral Scent*, 243–261 (2006).
33. (*a*) N. Dudareva, E. Pichersky, J. Gershenzon, Biochemistry of plant volatiles, *Plant Physiol.*, **135**, 1893–1902 (2004); (*b*) E. Pichersky, T. D. Sharkey, J. Gershenzon, Plant volatiles: a lack of function or a lack of knowledge? *Trends Plant Sci.*, **11**, 421–421 (2006); (*c*) R. A. Raguso, Floral scent in a whole-plant context: moving beyond pollinator attraction, *Funct. Ecol.*, **23**, 837–840 (2009).
34. A. Kessler, I. T. Baldwin, Plant responses to insect herbivory: the emerging molecular analysis, *Annu. Rev. Plant Biol.*, **53**, 299–328 (2002).

35. S. C. Welter, E. A. Bernays, Arthropod impact on plant gas exchange, in *Insect–Plant Interactions, Vol. 1*, E. A. Bernays (Ed.), CRC Press, Boca Raton, 1989, pp. 135–150.
36. E. E. Farmer, C. A. Ryan, Interplant communication: airborne methyl jasmonate induces synthesis of proteinase inhibitors in plant leaves, *Proc. Natl. Acad. Sci. USA*, **87**, 7713–7716 (1990).
37. A. Kessler, I. T. Baldwin, Defensive function of herbivore-induced plant volatile emissions in nature, *Science*, **291**, 2141–2144 (2001).
38. O. E. Krips, P. E. L. Willems, R. Gols, M. A. Posthumus, G. Gort, M. Dicke, Comparison of cultivars of ornamental crop *Gerbera jamesonii* on production of spider mite-induced volatiles, and their attractiveness to the predator *Phytoseiulus persimilis*, *J. Chem. Ecol.*, **27**, 1355–1372 (2001).
39. T. C. J. Turlings, J. H. Tumlinson, W. J. Lewis, Exploitation of herbivore-induced plant odors by host-seeking parasitic wasps, *Science*, **250**, 1251–1253 (1990).
40. T. Degen, C. Dillmann, F. Marion-Poll, T. C. J. Turlings, High genetic variability of herbivore-induced volatile emission within a broad range of maize inbred lines, *Plant Physiol.*, **135**, 1928–1938 (2004).
41. R. Halitschke, A. Kessler, J. Kahl, A. Lorenz, I. T. Baldwin, Ecophysiological comparison of direct and indirect defenses in *Nicotiana attenuata*, *Oecologia*, **124**, 408–417 (2000).
42. J. D. Hare, Variation in herbivore and methyl jasmonate-induced volatiles among genetic lines of *Datura wrightii*, *J. Chem. Ecol.*, **33**, 2028–2043 (2007).
43. M. E. F. Hoballah, C. Tamo, T. C. J. Turlings, Differential attractiveness of induced odors emitted by eight maize varieties for the parasitoid *Cotesia marginiventris*: is quality or quantity important? *J. Chem. Ecol.*, **28**, 951–968 (2002).
44. (*a*) W. Boland, J. Hopke, J. Donath, J. Nüske, F. Bublitz, Jasmonic acid and coronatin induce odor production in plants, *Angew. Chem.*, **107**, 1715–1717 (1995); *Angew. Chem. Int. Ed.*, **34**, 1600–1602 (1995); (*b*) R. Ozawa, G. Arimura, J. Takabayashi, T. Shimoda, T. Nishioka, Involvement of jasmonate- and salicylate-related signaling pathway for the production of specific herbivore-induced volatiles in plants, *Plant Cell Physiol.*, **41**, 391–398 (2000); (*c*) R. Ozawa, K. Shiojiri, M. W. Sabelis, G. I. Arimura, T. Nishioka, J. Takabayashi, Corn plants treated with jasmonic acid attract more specialist parasitoids, thereby increasing parasitization of the common armyworm, *J. Chem. Ecol.*, **30**, 1797–1808 (2004); (*d*) R. M. P. van Poecke, M. A. Posthumus, M. Dicke, Herbivore-induced volatile production by *Arabidopsis thaliana* leads to attraction of the parasitoid *Cotesia rubecula*: chemical, behavioral, and gene-expression analysis, *J. Chem. Ecol.*, **27**, 1911–1928 (2001); (*e*) K. Ament, M. R. Kant, M. W. Sabelis, M. A. Haring, R. C. Schuurink, Jasmonic acid is a key regulator of spider mite-induced volatile terpenoid and methyl salicylate emission in tomato, *Plant Physiol.*, **135**, 2025–2037 (2004).
45. J. Horiuchi, G. Arimura, R. Ozawa, T. Shimoda, J. Takabayashi, T. Nishioka, Exogenous acc enhances volatiles production mediated by jasmonic acid in lima bean leaves, *FEBS Lett.*, **509**, 332–336 (2001).
46. R. Halitschke, I. T. Baldwin, Antisense lox expression increases herbivore performance by decreasing defense responses and inhibiting growth-related transcriptional reorganization in *Nicotiana attenuata*, *Plant J.*, **36**, 794–807 (2003).
47. M. Dicke, R. Gols, D. Ludeking, M. A. Posthumus, Jasmonic acid and herbivory differentially induce carnivore-attracting plant volatiles in lima bean plants, *J. Chem. Ecol.*, **25**, 1907–1922 (1999).
48. A. Kessler, I. T. Baldwin, Herbivore-induced plant vaccination. Part I. The orchestration of plant defenses in nature and their fitness consequences in the wild tobacco *Nicotiana attenuata*, *Plant J.*, **38**, 639–649 (2004).
49. A. Kessler, R. Halitschke, Specificity and complexity: the impact of herbivore-induced plant responses on arthropod community structure, *Curr. Opin. Plant Biol.*, **10**, 409–414 (2007).
50. (*a*) S. J. Park, Y. H. Huang, P. Ayoubi, Identification of expression profiles of sorghum genes in response to greenbug phloem-feeding using cDNA subtraction and microarray analysis, *Planta*, **223**, 932–947 (2006); (*b*) J. S. Thaler, R. M. Bostock, Interactions between abscisic-acid-mediated responses and plant resistance to pathogens and insects, *Ecology*, **85**, 48–58 (2004);

(*c*) C. C. von Dahl, M. Havecker, R. Schlogl, I. T. Baldwin, Caterpillar-elicited methanol emission: a new signal in plant–herbivore interactions? *Plant J.*, **46**, 948–960 (2006).
51. M. E. Maffei, A. Mithofer, W. Boland, Before gene expression: early events in plant–insect interaction, *Trends Plant Sci.*, **12**, 310–316 (2007).
52. M. E. Maffei, A. Mithofer, W. Boland, Insects feeding on plants: rapid signals and responses preceding the induction of phytochemical release, *Phytochemistry*, **68**, 2946–2959 (2007).
53. A. Mithofer, G. Wanner, W. Boland, Effects of feeding *Spodoptera littoralis* on lima bean leaves. II. Continuous mechanical wounding resembling insect feeding is sufficient to elicit herbivory-related volatile emission, *Plant Physiol.*, **137**, 1160–1168 (2005).
54. P. W. Pare, M. A. Farag, V. Krishnamachari, H. M. Zhang, C. M. Ryu, J. W. Kloepper, Elicitors and priming agents initiate plant defense responses, *Photosynth. Res.*, **85**, 149–159 (2005).
55. H. T. Alborn, T. H. Jones, G. S. Stenhagen, J. H. Tumlinson, Identification and synthesis of volicitin and related components from beet armyworm oral secretions, *J. Chem. Ecol.*, **26**, 203–220 (2000).
56. H. T. Alborn, T. V. Hansen, T. H. Jones, D. C. Bennett, J. H. Tumlinson, E. A. Schmelz, P. E. A. Teal, Disulfooxy fatty acids from the American bird grasshopper *Schistocerca americana,* elicitors of plant volatiles, *Proc. Natl. Acad. Sci. USA*, **104**, 12976–12981 (2007).
57. L. Mattiacci, M. Dicke, M. A. Posthumus, Beta-glucosidase – an elicitor of herbivore-induced plant odor that attracts host-searching parasitic wasps, *Proc. Natl. Acad. Sci. USA*, **92**, 2036–2040 (1995).
58. H. Eichenseer, M. C. Mathews, J. L. Bi, J. B. Murphy, G. W. Felton, Salivary glucose oxidase: multifunctional roles for helicoverpa zea? *Arch. Insect Biochem. Physiol.*, **42**, 99–109 (1999).
59. C. J. Funk, Alkaline phosphatase activity in whitefly salivary glands and saliva, *Arch. Insect Biochem. Physiol.*, **46**, 165–174 (2001).
60. P. W. Miles, Aphid saliva, *Biol. Rev.*, **74**, 41–85 (1999).
61. H. T. Alborn, T. C. J. Turlings, T. H. Jones, G. Stenhagen, J. H. Loughrin, J. H. Tumlinson, An elicitor of plant volatiles from beet armyworm oral secretion, *Science*, **276**, 945–949 (1997).
62. R. Halitschke, U. Schittko, G. Pohnert, W. Boland, I. T. Baldwin, Molecular interactions between the specialist herbivore *Manduca sexta* (Lepidoptera, Sphingidae) and its natural host *Nicotiana attenuata.* III. Fatty acid–amino acid conjugates in herbivore oral secretions are necessary and sufficient for herbivore-specific plant responses, *Plant Physiol.*, **125**, 711–717 (2001).
63. G. Pohnert, V. Jung, E. Haukioja, K. Lempa, W. Boland, New fatty acid amides from regurgitant of Lepidopteran (Noctuidae, Geometridae) caterpillars, *Tetrahedron*, **55**, 11275–11280 (1999).
64. J. H. Tumlinson, C. G. Lait, Biosynthesis of fatty acid amide elicitors of plant volatiles by insect herbivores, *Arch. Insect Biochem. Physiol.*, **58**, 58–64 (2005).
65. D. Spiteller, G. Pohnert, W. Boland, Absolute configuration of volicitin, an elicitor of plant volatile biosynthesis from Lepidopteran larvae, *Tetrahedron Lett.*, **42**, 1483–1485 (2001).
66. E. A. Schmelz, M. J. Carroll, S. LeClere, S. M. Phipps, J. Meredith, P. S. Chourey, H. T. Alborn, P. E. A. Teal, Fragments of ATP synthase mediate plant perception of insect attack, *Proc. Natl. Acad. Sci. USA*, **103**, 8894–8899 (2006).
67. M. Hilker, T. Meiners, Early herbivore alert: insect eggs induce plant defense, *J. Chem. Ecol.*, **32**, 1379–1397 (2006).
68. (*a*) S. Colazza, A. Fucarino, E. Peri, G. Salerno, E. Conti, F. Bin, Insect oviposition induces volatile emission in herbaceous plants that attracts egg parasitoids, *J. Exp. Biol.*, **207**, 47–53 (2004); (*b*) T. Meiners, N. K. Hacker, P. Anderson, M. Hilker, Response of the elm leaf beetle to host plants induced by oviposition and feeding: the infestation rate matters, *Entomol. Exp. Appl.*, **115**, 171–177 (2005); (*c*) T. Meiners, A. Köpf, C. Stein, M. Hilker, Chemical signals mediating interactions between *Galeruca tanaceti L.* (Coleoptera, Chrysomelidae) and its egg parasitoid *Oomyzus galerucivorus* (Hedqvits) (Hymenoptera, Eulophidae), *J. Insect Behav.*, **10**, 523–539 (1997).
69. M. Hilker, T. Meiners, Induction of plant responses to oviposition and feeding by herbivorous arthropods: a comparison, *Entomol. Exp. Appl.*, **104**, 181–192 (2002).
70. T. Meiners, M. Hilker, Induction of plant synomones by oviposition of a phytophagous insect, *J. Chem. Ecol.*, **26**, 221–232 (2000).

71. M. Hilker, C. Kobs, M. Varma, K. Schrank, Insect egg deposition induces *Pinus sylvestris* to attract egg parasitoids, *J. Exp. Biol.*, **205**, 455–461 (2002).
72. M. Hilker, C. Stein, R. Schroder, M. Varama, R. Mumm, Insect egg deposition induces defence responses in *Pinus sylvestris*: characterisation of the elicitor, *J. Exp. Biol.*, **208**, 1849–1854 (2005).
73. (*a*) T. Meiners, M. Hilker, Host location in *Oomyzus gallerucae* (Hymenoptera: Eulophidae), an egg parasitoid of the elm leaf beetle *Xanthogaleruca luteola* (Coleoptera: Chrysomelidae), *Oecologia*, **112**, 87–93 (1997); (*b*) T. Meiners, C. Westerhaus, M. Hilker, Specificity of chemical cues used by a specialist egg parasitoid during host location, *Entomol. Exp. Appl.*, **95**, 151–159 (2000).
74. R. P. Doss, J. E. Oliver, W. M. Proebsting, S. W. Potter, S. R. Kuy, S. L. Clement, R. T. Williamson, J. R. Carney, E. D. DeVilbiss, Bruchins: insect-derived plant regulators that stimulate neoplasm formation, *Proc. Natl. Acad. Sci. USA*, **97**, 6218–6223 (2000).
75. S. Rasmann, T. C. J. Turlings, First insights into specificity of belowground tritrophic interactions, *Oikos*, **117**, 362–369 (2008).
76. (*a*) J. Takabayashi, M. Dicke, M. A. Posthumus, Variation in composition of predator-attracting allelochemical emitted by herbivore-infested plants: relative influence of plant and herbivore, *Chemoecology*, **2**, 1–6 (1991); (*b*) J. Takabayashi, T. Noda, S. Takahashi, Plants produce attractants for *Apanteles kariyai*, a parasitoid of *Pseudaletia separata* – cases of communication and misunderstanding in parasitoid–plant interactions, *Appl. Entomol. Zool.*, **26**, 237–243 (1991); (*c*) T. C. J. Turlings, P. J. McCall, H. T. Alborn, J. H. Tumlinson, An elicitor in caterpillar oral secretions that induces corn seedlings to emit chemical signals attractive to parasitic wasps, *J. Chem. Ecol.*, **19**, 411–425 (1993).
77. J. Takabayashi, S. Takahashi, M. Dicke, M. A. Posthumus, Developmental stage of herbivore *Pseudaletia separata* affects production of herbivore-induced synomone by corn plants, *J. Chem. Ecol.*, **21**, 273–287 (1995).
78. E. A. Schmelz, J. Engelberth, H. T. Alborn, J. H. Tumlinson, P. E. A. Teal, Phytohormone-based activity mapping of insect herbivore-produced elicitors, *Proc. Natl. Acad. Sci. USA*, **106**, 653–657 (2009).
79. S. S. Duffey, M. J. Stout, Antinutritive and toxic components of plant defense against insects, *Arch. Insect Biochem. Physiol.*, **32**, 3–37 (1996).
80. U. Schittko, I. T. Baldwin, Constraints to herbivore-induced systemic responses: bidirectional signaling along orthostichies in nicotiana attenuata, *J. Chem. Ecol.*, **29**, 763–770 (2003).
81. C. Orians, Herbivores, vascular pathways, and systemic induction: facts and artifacts, *J. Chem. Ecol.*, **31**, 2231–2242 (2005).
82. C. M. Orians, S. D. P. Smith, L. Sack, How are leaves plumbed inside a branch? Differences in leaf-to-leaf hydraulic sectoriality among six temperate tree species, *J. Exp. Bot.*, **56**, 2267–2273 (2005).
83. M. A. Watson, B. B. Casper, Morphogenetic constraints on patterns of carbon distribution in plants, *Annu. Rev. Ecol. Syst.*, **15**, 233–258 (1984).
84. C. R. Rodriguez-Saona, L. E. Rodriguez-Saona, C. J. Frost, Herbivore-induced volatiles in the perennial shrub, *Vaccinium corymbosum,* and their role in inter-branch signaling, *J. Chem. Ecol.*, **35**, 163–175 (2009).
85. M. Heil, J. C. Silva Bueno, Herbivore-induced volatiles as rapid signals in systemic plant responses: how to quickly move the information, *Plant Signal Behav.*, **2**, 191–193 (2007).
86. J. F. Stuefer, S. Gomez, T. van Molken, Clonal integration beyond resource sharing: implications for defence signalling and disease transmission in clonal plant networks, *Evol. Ecol.*, **18**, 647–667 (2004).
87. M. Heil, J. C. Silva Bueno, Within-plant signaling by volatiles leads to induction and priming of an indirect plant defense in nature, *Proc. Natl. Acad. Sci. USA*, **104**, 5467–5472 (2007).
88. Z. Hu, Y. Shen, F. Shen, Y. Luo, X. Su, Evidence for the signaling role of methyl jasmonate, methyl salicylate and benzothiazole between poplar (*Populus simonii x P. pyramidalis* 'Opera 8277') cuttings, *Trees*, **23**, 1003–1011 (2009).
89. R. Karban, I. T. Baldwin, K. J. Baxter, G. Laue, G. W. Felton, Communication between plants: induced resistance in wild tobacco plants following clipping of neighboring sagebrush, *Oecologia*, **125**, 66–71 (2000).

90. C. G. Jones, R. F. Hopper, J. S. Coleman, V. A. Krischik, Control of systemically induced herbivore resistance by plant vascular architecture, *Oecologia*, **93**, 452–456 (1993).
91. D. V. Viswanathan, J. S. Thaler, Plant vascular architecture and within-plant spatial patterns in resource quality following herbivory, *J. Chem. Ecol.*, **30**, 531–543 (2004).
92. (*a*) A. J. Gassmann, J. D. Hare, Indirect cost of a defensive trait: variation in trichome type affects the natural enemies of herbivorous insects on *Datura wrightii*, *Oecologia*, **144**, 62–71 (2005); (*b*) J. A. Rudgers, S. Y. Strauss, J. E. Wendel, Trade-offs among anti-herbivore resistance traits: insights from Gossypieae (Malvaceae) *Am. J. Bot.*, **91**, 871–880 (2004).
93. (*a*) A. A. Agrawal, A. Janssen, J. Bruin, M. A. Posthumus, M. W. Sabelis, An ecological cost of plant defence: attractiveness of bitter cucumber plants to natural enemies of herbivores, *Ecol. Lett.*, **5**, 377–385 (2002); (*b*) A. Biere, H. B. Marak, J. M. M. van Damme, Plant chemical defense against herbivores and pathogens: generalized defense or trade-offs? *Oecologia*, **140**, 430–441 (2004); (*c*) M. Heil, R. M. Bostock, Induced systemic resistance (isr) against pathogens in the context of induced plant defences, *Ann. Bot.*, **89**, 503–512 (2002); (*d*) R. Karban, J. H. Myers, Induced plant-responses to herbivory, *Annu. Rev. Ecol. Syst.*, **20**, 331–348 (1989); (*e*) H. B. Marak, A. Biere, J. M. M. van Damme, Two herbivore-deterrent iridoid glycosides reduce the in-vitro growth of a specialist but not of a generalist pathogenic fungus of *Plantago lanceolata* L., *Chemoecology*, **12**, 185–192 (2002); (*f*) J. L. Steward, K. H. Keeler, Are there trade-offs among antiherbivore defenses in *Ipomoea* (Convolvulaceae), *Oikos*, **53**, 79–86 (1988).
94. D. J. Ballhorn, S. Kautz, U. Lion, M. Heil, Trade-offs between direct and indirect defences of lima bean (*Phaseolus lunatus*), *J. Ecol.*, **96**, 971–980 (2008).
95. A. Agrawal, Macroevolution of plant defense strategies, *Trends Ecol. Evol.* **22**, 103–109 (2007).
96. A. M. Lambert, Effects of prey availability, facultative plant feeding, and plant defenses on a generalist predator, *Arthropod Plant Interact.*, **1**, 167–173 (2007).
97. K. M. Kester, P. Barbosa, Behavioral-responses to host foodplants of 2 populations of the insect parasitoid *Cotesia congregata* (Say), *Oecologia*, **99**, 151–157 (1994).
98. R. A. Winz, I. T. Baldwin, Molecular interactions between the specialist herbivore *Manduca sexta* (Lepidoptera, Sphingidae) and its natural host *Nicotiana attenuata.* IV. Insect-induced ethylene reduces jasmonate-induced nicotine accumulation by regulating putrescine *n*-methyltransferase transcripts, *Plant Physiol.*, **125**, 2189–2202 (2001).
99. C. Voelckel, T. Krugel, K. Gase, N. Heidrich, N. M. van Dam, R. Winz, I. T. Baldwin, Anti-sense expression of putrescine *n*-methyltransferase confirms defensive role of nicotine in *Nicotiana sylvestris* against *Manduca sexta*, *Chemoecology*, **11**, 121–126 (2001).
100. A. Magro, J. N. Tene, N. Bastin, A. F. G. Dixon, J. L. Hemptinne, Assessment of patch quality by ladybirds: relative response to conspecific and heterospecific larval tracks a consequence of habitat similarity? *Chemoecology*, **17**, 37–45 (2007).
101. R. J. McGuire, M. T. J. Johnson, Plant genotype and induced responses affect resistance to herbivores on evening primrose (*Oenothera biennis*), *Ecol. Entomol.*, **31**, 20–31 (2006).
102. B. Randlkofer, E. Obermaier, T. Meiners, Mother's choice of the oviposition site: balancing risk of egg parasitism and need of food supply for the progeny with an infochemical shelter? *Chemoecology*, **17**, 177–186 (2007).
103. S. P. Gouinguene, T. C. J. Turlings, The effects of abiotic factors on induced volatile emissions in corn plants, *Plant Physiol.*, **129**, 1296–1307 (2002).
104. A. R. Jassbi, K. Gase, C. Hettenhausen, A. Schmidt, I. T. Baldwin, Silencing geranylgeranyl diphosphate synthase in *Nicotiana attenuata* dramatically impairs resistance to tobacco hornworm, *Plant Physiol.*, **146**, 974–986 (2008).
105. M. de Vos, K. L. Kriksunov, G. Jander, Indole-3-acetonitrile production from indole glucosinolates deters oviposition by *Pieris rapae*, *Plant Physiol.*, **146**, 916–926 (2008).
106. S. Finch, R. H. Collier, Host-plant selection by insects – a theory based on 'appropriate/inappropriate landings' by pest insects of cruciferous plants, *Entomol. Exp. Appl.*, **96**, 91–102 (2000).
107. C. M. de Moraes, M. C. Mescher, J. H. Tumlinson, Caterpillar-induced nocturnal plant volatiles repel conspecific females, *Nature*, **410**, 577–580 (2001).

108. P. Anderson, M. Jonsson, U. Morte, Variation in damage to cotton affecting larval feeding preference of *Spodoptera littoralis*, *Entomol. Exp. Appl.*, **101**, 191–198 (2001).
109. T. Tscharntke, S. Thiessen, R. Dolch, W. Boland, Herbivory, induced resistance, and interplant signal transfer in *Alnus glutinosa*, *Biochem. Syst. Ecol.*, **29**, 1025–1047 (2001).
110. K. V. Tindall, M. J. Stout, Plant-mediated interactions between the rice water weevil and fall armyworm in rice, *Entomol. Exp. Appl.*, **101**, 9–17 (2001).
111. (*a*) A. R. Harari, D. Ben-Yakir, D. Rosen, Male pioneering as a mating strategy: the case of the beetle *Maladera matrida*, *Ecol. Entomol.*, **25**, 387–394 (2000); (*b*) J. H. Loughrin, D. H. Potter, T. R. Hamiltonkemp, Volatile compounds induced by herbivory act as aggregation kairomones for the Japanese-beetle (*Popillia japonica,* Newman), *J. Chem. Ecol.*, **21**, 1457–1467 (1995).
112. P. M. Reader, T. R. E. Southwood, The relationship between palatability to invertebrates and the successional status of a plant, *Oecologia*, **51**, 271–275 (1981).
113. R. Halitschke, J. A. Stenberg, D. Kessler, A. Kessler, I. T. Baldwin, Shared signals – 'alarm calls' from plants increase apparency to herbivores and their enemies in nature, *Ecol. Lett.*, **11**, 24–34 (2008).
114. M. J. Carroll, E. A. Schmelz, P. E. A. Teal, The attraction of *Spodoptera frugiperda* neonates to cowpea seedlings is mediated by volatiles induced by conspecific herbivory and the elicitor inceptin, *J. Chem. Ecol.*, **34**, 291–300 (2008).
115. M. Dicke, M. W. Sabelis, How plants obtain predatory mites as bodyguards, *Neth. J. Zool.*, **38**, 148–165 (1988).
116. (*a*) M. Dicke, M. W. Sabelis, Does it pay plants to advertise for bodyguards towards a cost–benefit analysis of induced synomone production? in *Causes and Consequences of Variation in Growth Rate and Productivity of Higher Plants*, H. Lambers, M. L. Cambridge, H. Konings, T. L. Pons (Eds), SPB Academic Publishing, The Hague, 1990, pp. 341–358; (*b*) J. Takabayashi, M. W. Sabelis, A. Janssen, K. Shiojiri, M. van Wijk, Can plants betray the presence of multiple herbivore species to predators and parasitoids? The role of learning in phytochemical information networks, *Ecol. Res.*, **21**, 3–8 (2006).
117. M. Dicke, Are herbivore-induced plant volatiles reliable indicators of herbivore identity to foraging carnivorous arthropods? *Entomol. Exp. Appl.*, **91**, 131–142 (1999).
118. M. Dicke, Behavioural and community ecology of plants that cry for help, *Plant Cell Environ.*, **32**, 654–665 (2009).
119. S. Rasmann, T. G. Kollner, J. Degenhardt, I. Hiltpold, S. Toepfer, U. Kuhlmann, J. Gershenzon, T. C. J. Turlings, Recruitment of entomopathogenic nematodes by insect-damaged maize roots, *Nature*, **434**, 732–737 (2005).
120. M. Dicke, K. J. van der Maas, J. Takabayashi, L. E. M. Vet, Learning affects response to volatile allelochemicals by predatory mites, *Proc. Sect. Exp. Appl. Entomol., Neth. Entomol. Soc.*, 31–36 (1990).
121. C. M. de Moraes, W. J. Lewis, P. W. Pare, H. T. Alborn, J. H. Tumlinson, Herbivore-infested plants selectively attract parasitoids, *Nature*, **393**, 570–573 (1998).
122. M. Dicke, M. W. Sabelis, Origin and function of semiochemicals in a tritrophic system of host plant, spider mites and predatory mites, *Colloq. – Inst. Nat. Rech. Agron.*, 17–18 (1988).
123. M. Dicke, T. A. Vanbeek, M. A. Posthumus, N. Bendom, H. Vanbokhoven, A. E. Degroot, Isolation and identification of volatile kairomone that affects acarine predator–prey interactions – involvement of host plant in its production, *J. Chem. Ecol.*, **16**, 381–396 (1990).
124. B. Drukker, P. Scutareanu, M. W. Sabelis, Do anthocorid predators respond to synomones from psylla-infested pear trees under field conditions? *Entomol. Exp. Appl.*, **77**, 193–203 (1995).
125. Y. J. Du, G. M. Poppy, W. Powell, J. A. Pickett, L. J. Wadhams, C. M. Woodcock, Identification of semiochemicals released during aphid feeding that attract parasitoid *Aphidius ervi*, *J. Chem. Ecol.*, **24**, 1355–1368 (1998).
126. R. P. J. Potting, L. E. M. Vet, M. Dicke, Host microhabitat location by stem-borer parasitoid *Cotesia flavipes* – the role of herbivore volatiles and locally and systemically induced plant volatiles, *J. Chem. Ecol.*, **21**, 525–539 (1995).
127. Z. R. Khan, D. G. James, C. A. O. Midega, J. Pickett, Chemical ecology and consevation biological control, *Biol. Control*, **45**, 210–224 (2008).

128. J. L. M. Steidle, A. Fischer, C. Gantert, Do grains whisper for help? Evidence for herbivore-induced synomones in wheat grains, *Entomol. Exp. Appl.*, **115**, 239–245 (2005).
129. R. W. H. M. van Tol, A. T. C. van der Sommen, M. I. C. Boff, J. van Bezooijen, M. W. Sabelis, P. H. Smits, Plants protect their roots by alerting the enemies of grubs, *Ecol. Lett.*, **4**, 292–294 (2001).
130. M. Pfeiffer, J. F. Tooker, D. S. Luthe, G. W. Felton, Plants on early alert: glandular trichomes as sensors for insect herbivores, *New Phytol.*, **184**, 644–656 (2009).
131. D. Kopke, R. Schroder, H. M. Fischer, J. Gershenzon, M. Hilker, A. Schmidt, Does egg deposition by herbivorous pine sawflies affect transcription of sesquiterpene synthases in pine? *Planta*, **228**, 427–438 (2008).
132. E. Mantyla, G. A. Alessio, J. D. Blande, J. Heijari, J. K. Holopainen, T. Laaksonen, P. Piirtola, T. Klemola, From plants to birds: higher avian predation rates in trees responding to insect herbivory, *PLoS One*, **3**, e2832, 2831–2838 (2008).
133. C. M. de Moraes, M. C. Mescher, Biochemical crypsis in the avoidance of natural enemies by an insect herbivore, *Proc. Natl. Acad. Sci. USA*, **101**, 8993–8997 (2004).
134. J. F. Tooker, J. R. Rohr, W. G. Abrahamson, C. M. de Moraes, Gall insects can avoid and alter indirect plant defenses, *New Phytol.*, **178**, 657–671 (2008).
135. (*a*) J. G. de Boer, M. Dicke, Olfactory learning by predatory arthropods, *Anim. Biol.*, **56**, 143–155 (2006); (*b*) B. Drukker, J. Bruin, G. Jacobs, A. Kroon, M. W. Sabelis, How predatory mites learn to cope with variability in volatile plant signals in the environment of their herbivorous prey, *Exp. Appl. Acarol.*, **24**, 881–895 (2000); (*c*) R. Schroder, L. Wurm, M. Varama, T. Meiners, M. Hilker, Unusual mechanisms involved in learning of oviposition-induced host plant odours in an egg parasitoid, *Anim. Behav.*, **75**, 1423–1430 (2008); (*d*) H. M. Smid, L. E. M. Vet, Learning in insects: from behaviour to brain, *Anim. Biol.*, **56**, 121–124 (2006); (*e*) H. M. Smid, G. H. Wang, T. Bukovinszky, J. L. M. Steidle, M. A. K. Bleeker, J. J. A. van Loon, L. E. M. Vet, Species-specific acquisition and consolidation of long-term memory in parasitic wasps, *Proc. R. Soc. Lond., Ser. B*, **274**, 1539–1546 (2007); (*f*) L. E. M. Vet, W. J. Lewis, R. T. Carde, Parasitoid foraging and learning, in *Chemical Ecology of Insects*, R. T. Carde, W. J. Bell (Eds), Chapman & Hall, New York, 1995, pp. 65–104.
136. C. Schnee, T. G. Koellner, M. Held, T. C. J. Turlings, J. Gershenzon, J. Degenhardt, The products of a single maize sesquiterpene synthase form a volatile defense signal that attracts natural enemies of herbivores, *Proc. Natl. Acad. Sci. USA*, **103**, 1129–1134 (2006).
137. (*a*) J. G. de Boer, M. Dicke, Experience with methyl salicylate affects behavioural responses of a predatory mite to blends of herbivore-induced plant volatiles, *Entomol. Exp. Appl.*, **110**, 181–189 (2004); (*b*) J. G. de Boer, M. Dicke, The role of methyl salicylate in prey searching behavior of the predatory mite *Phytoseiulus persimilis*, *J. Chem. Ecol.*, **30**, 255–271 (2004).
138. R. Karban, Damage to sagebrush attracts predators but this does not reduce herbivory, *Entomol. Exp. Appl.*, **125**, 71–80 (2007).
139. J. S. Thaler, C. A. M. Griffin, Relative importance of consumptive and non-consumptive effects of predators on prey and plant damage: the influence of herbivore ontogeny, *Entomol. Exp. Appl.*, **128**, 34–40 (2008).
140. E. van der Meijden, P. G. L. Klinkhammer, Conflicting interests of plants and the natural enemies of herbivores, *Oikos*, **89**, 202–208 (2000).
141. E. Guerrieri, G. Lingua, M. C. Digilio, N. Massa, G. Berta, Do interactions between plant roots and the rhizosphere affect parasitoid behaviour? *Ecol. Entomol.*, **29**, 753–756 (2004).
142. (*a*) I. Kaplan, R. Halitschke, A. Kessler, B. J. Rehill, S. Sardanelli, R. F. Denno, Physiological integration of roots and shoots in plant defense strategies links above- and belowground herbivory, *Ecol. Lett.*, **11**, 841–851 (2008); (*b*) N. M. van Dam, M. W. Oomen, Root and shoot jasmonic acid applications differentially affect leaf chemistry and herbivore growth, *Plant Signal Behav.*, **3**, 91–98 (2008).
143. (*a*) Y. J. Cardoza, J. H. Tumlinson, Compatible and incompatible xanthomonas infections differentially affect herbivore-induced volatile emission by pepper plants, *J. Chem. Ecol.*, **32**, 1755–1768 (2006); (*b*) J. G. de Boer, C. A. Hordijk, M. A. Posthumus, M. Dicke, Prey and non-prey arthropods sharing a host plant: effects on induced volatile emission and predator

attraction, *J. Chem. Ecol.*, **34**, 281–290 (2008); (*c*) C. M. Delphia, M. C. Mescher, C. M. de Moraes, Induction of plant volatiles by herbivores with different feeding habits and the effects of induced defenses on host-plant selection by thrips, *J. Chem. Ecol.*, **33**, 997–1012 (2007); (*d*) C. Rodriguez-Saona, J. A. Chalmers, S. Raj, J. S. Thaler, Induced plant responses to multiple damagers: differential effects on an herbivore and its parasitoid, *Oecologia*, **143**, 566–577 (2005); (*e*) M. Rostas, T. C. J. Turlings, Induction of systemic acquired resistance in zea mays also enhances the plant's attractiveness to parasitoids, *Biol. Control*, **46**, 178–186 (2008); (*f*) R. Soler, J. A. Harvey, T. M. Bezemer, Foraging efficiency of a parasitoid of a leaf herbivore is influenced by root herbivory on neighbouring plants, *Funct. Ecol.*, **21**, 969–974 (2007).

144. (*a*) E. H. Poelman, C. Broekgaarden, J. J. A. van Loon, M. Dicke, Early season herbivore differentially affects plant defence responses to subsequently colonizing herbivores and their abundance in the field, *Mol. Ecol.*, **17**, 3352–3365, (2008); (*b*) E. H. Poelman, J. J. A. van Loon, M. Dicke, Consequences of variation in plant defense for biodiversity at higher trophic levels, *Trends Plant Sci.*, **13**, 534–541 (2008).
145. (*a*) I. T. Baldwin, J. C. Schultz, Rapid changes in tree leaf chemistry induced by damage – evidence for communication between plants, *Science*, **221**, 277–279 (1983); (*b*) M. Dicke, J. Bruin, Chemical information transfer between plants: back to the future, *Biochem. Syst. Ecol.*, **29**, 981–994 (2001).
146. (*a*) G. Arimura, K. Matsui, J. Takabayashi, Chemical and molecular ecology of herbivore-induced plant volatiles: proximate factors and their ultimate functions, *Plant Cell Physiol.*, **50**, 911–923 (2009); (*b*) M. W. Sabelis, J. Takabayashi, A. Janssen, M. R. Kant, M. van Wijk, B. Sznajder, N. S. Aratchige, I. Lesna, B. Belliure, R. C. Schuurink, T. Ohgushi, T. P. Craig, P. W. Price, Ecology meets plant physiology: herbivore-induced plant responses and their indirect effects on arthropod communities, in *Ecological Communities: Plant Mediation in Indirect Interaction Webs*, Cambridge University Press, Cambridge, 2007, pp. 188–217.
147. R. Karban, K. Shiojiri, Self-recognition affects plant communication and defense, *Ecol. Lett.*, **12**, 502–506 (2009).
148. (*a*) R. Karban, M. Huntzinger, A. C. McCall, The specificity of eavesdropping on sagebrush by other plants, *Ecology*, **85**, 1846–1852 (2004); (*b*) R. Karban, K. Shiojiri, M. Huntzinger, A. C. McCall, Damage-induced resistance in sagebrush: volatiles are key to intra- and interplant communication, *Ecology*, **87**, 922–930 (2006).
149. A. Kessler, R. Halitschke, C. Diezel, I. T. Baldwin, Priming of plant defense responses in nature by airborne signaling between *Artemisia tridentata* and *Nicotiana attenuata*, *Oecologia*, **148**, 280–292 (2006).
150. J. Engelberth, D. H. Alborn, E. A. Schmelz, J. H. Tumlinson, Airborne signals prime plants against insect herbivore attack, *Proc. Natl. Acad. Sci. USA*, **101**, 1781–1785 (2004).
151. A. Kessler, R. Halitschke, K. Gase, I. T. Baldwin, Silencing the jasmonate cascade: induced plant defenses and insect populations, *Science*, **305**, 665–668 (2004).
152. J. Degenhardt, J. Gershenzon, I. T. Baldwin, A. Kessler, Attracting friends to feast on foes: engineering terpene emission to make crop plants more attractive to herbivore enemies, *Curr. Opin. Biotechnol.*, **14**, 169–176 (2003).
153. J. Degenhart, I. Hiltpold, T. G. Köllner, M. Frey, A. Gierl, J. Gershenzon, B. E. Hibbard, M. R. Ellersieck, T. C. J. Turlings, Restoring a maize root signal that attracts insect-killing nematodes to control a major pest, *Proc. Natl. Acad. Sci. USA*, **106**, 13213–13218 (2009).
154. R. Gols, L. M. A. Witjes, J. J. A. van Loon, M. A. Posthumus, M. Dicke, J. A. Harvey, The effect of direct and indirect defenses in two wild brassicaceous plant species on a specialist herbivore and its gregarious endoparasitoid, *Entomol. Exp. Appl.*, **128**, 99–108 (2008).
155. Z. R. Khan, J. A. Pickett, J. van den Berg, L. J. Wadhams, C. M. Woodcock, Exploiting chemical ecology and species diversity: stem borer and striga control for maize and sorghum in Africa, *Pest Manage. Sci.*, **56**, 957–962 (2000).

5

Pheromones in Chemical Communication

Kenji Mori

The University of Tokyo, 1-20-6-1309, Mukogaoka, Bunkyo-ku, Tokyo 113-0023, Japan

5.1 Introduction

5.1.1 Definition of Pheromones

After the discovery of the first insect sex attractant bombykol (**1**) by Butenandt and his coworkers,[1] the term 'pheromone' was defined by Karlson and Lüscher in 1959.[2] The term is derived from the Greek *pherein* (to carry or transfer) and *horman* (to stir up or excite). Pheromones are substances which are secreted to the outside by an individual and received by a second individual of the same species, in which they release a specific reaction, for example, a definite behaviour or a developmental process.

As shown in Figure 5.1, pheromones are thought to be one of two groups of biofunctional molecules to spread information among individuals. Those molecules which spread information are called semiochemicals. They are divided into pheromones and allelochemicals, the latter of which are further divided into allomones, kairomones and synomones. Definitions of these terms are summarized in Figure 5.1. Pheromones are the topic of this chapter. Since pheromones are often used as kairomones by the predators of pheromone-releasing insects, kairomones are also treated briefly.

5.1.2 Classification of Pheromones

Pheromones are classified into two categories: releaser pheromones and primer pheromones (Figure 5.2). A releaser (or signaller) pheromone causes a change of behaviour in the receiver, while a primer pheromone has a physiological impact on the receiver.

The Chemistry and Biology of Volatiles Edited by Andreas Herrmann

Semiochemicals = Biofunctional molecules which spread information among individuals. Derived from *semio* (Greek) = sign.

- **Pheromones** = Biofunctional molecules which are used for communication between individuals within the same species. Derived from *pherein* (Greek) = carry, transfer + *horman* (Greek) = excite, stir up.
- **Allelochemicals** = Biofunctional molecules which are used for communication between individuals belonging to different species. Derived from *allelon* (Greek) = of each other.

Allelochemicals

- **Allomones** = Biofunctional molecules which evoke advantageous reactions for their producers. Derived from *allos* (Greek) = other.
- **Kairomones** = Biofunctional molecules which evoke advantageous reactions for their receivers. Derived from *kairo* (Greek) = opportune.
- **Synomones** = Biofunctional molecules which evoke advantageous reactions for both their producers and receivers. Derived from *syn* (Greek) = together with.

Figure 5.1 *Classification and definition of semiochemicals, especially pheromones and kairomones*

Releaser pheromones can be further classified as sex pheromones, aggregation pheromones, trail pheromones, etc., according to the type of behavioural change. Bombykol (**1**), the female-produced sex pheromone of the silkworm moth *Bombyx mori*, attracts male *B. mori*. (1*R*,5*S*,7*R*)-*exo*-Brevicomin (**2**), the female-produced aggregation pheromone of the western pine beetle *Dendroctonus brevicomis*, attracts both males and females of *D. brevicomis* in cooperation with (1*S*,5*R*)-frontalin (**3**). (*S*)-4-Methyl-3-heptanone (**4**) is

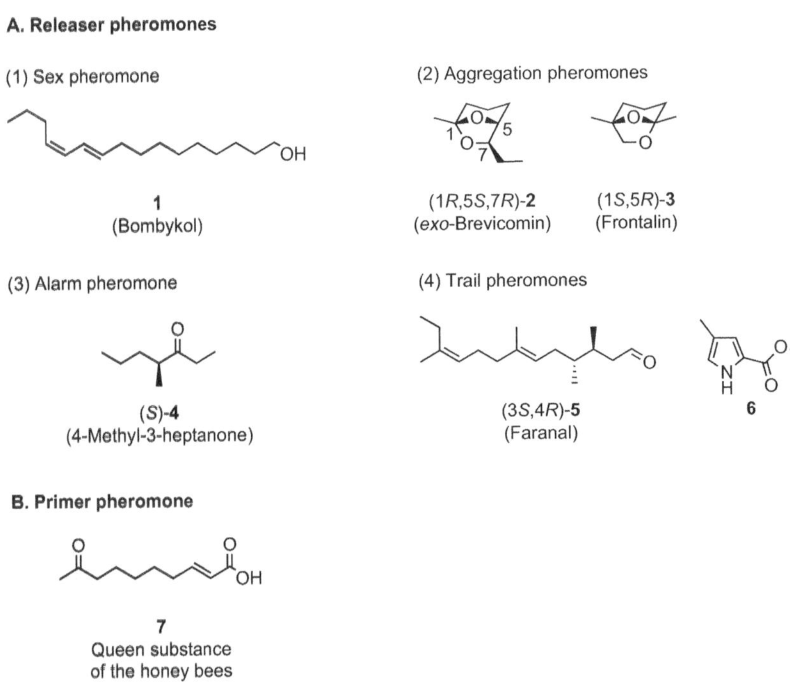

Figure 5.2 *Typical examples of volatile pheromones*

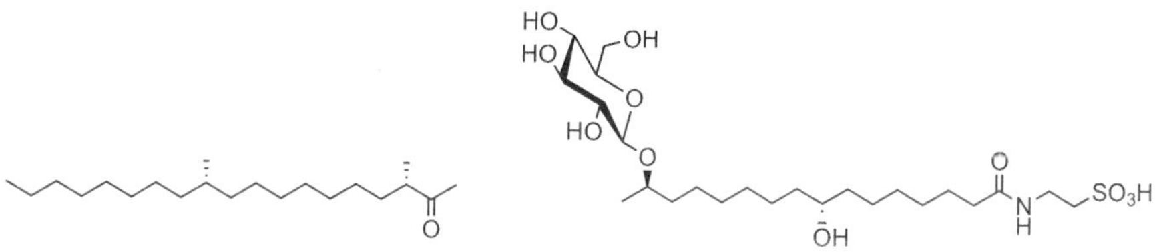

(3*S*,11*S*)-**8**
Sex pheromone of female German cockroach

9
Oviposition deterring pheromone of the European cherry fruit fly

Figure 5.3 *Examples of nonvolatile pheromones*

the principal alarm pheromone of the leaf cutting ant *Atta texana* to communicate danger to its neighbours. Trail pheromones inform neighbours about the trails to follow. (3*S*,4*R*)-Faranal (**5**) is the trail pheromone of the workers of the Pharaoh's ant *Monomorium pharaonis*. Its detection threshold is about 1 pg cm^{-1} of a trail. Methyl 4-methylpyrrole-2-carboxylate (**6**) is the trail pheromone of the leaf-cutting ant *A. texana*, a well known farming ant which cultivates mushrooms. Its threshold of detection is about 0.08 pg cm^{-1} or 3.48×10^8 molecules cm^{-1}. Accordingly, only 0.33 mg of **6** can draw a trail by which *A. texana* could travel around the world! The queen substance (**7**) of the honeybee *Apis mellifera* is an example of primer pheromones. It is secreted by the mandibular gland of a queen, inhibits the development of ovaries in worker ants and suppresses queen rearing in queenless colonies. The queen substance (**7**) is therefore known as the 'honeybee retinue pheromone'.

The structures in Figure 5.2 show that pheromones are small and volatile molecules with diverse functionalities. Only volatile pheromones are discussed in this chapter. Within the context of pheromones, 'volatile' means that a pheromone compound can evaporate and reach its receptor sites by diffusion. In the case of insects this is via the antennae. In quantities from < 100 molecules to several hundred, pheromone molecules can cause signal transduction from the receptor sites in the antennae of insects.[3] Due to the extremely high sensitivity of the receptor sites, a less volatile or an almost nonvolatile compound can serve as a pheromone, although we human beings cannot detect it by our own olfactory system. As shown in Figure 5.3, there are some pheromone molecules big enough to be nonvolatile. The German cockroach *Blattella germanica* employs nonvolatile **8** as the female-produced sex stimulant pheromone. A male *B. germanica* perceives **8** by touching the body surface of a female with his antennae. The ketone **8** belongs to the so-called contact pheromones. Glucosidic sulfonic acid (**9**) is the oviposition-deterring pheromone secreted by the females of the European cherry fruit fly, *Rhagoletis cerasi*. The female fly lays only a single egg into half-ripe cherries. As a glucoside with a sulfonic acid functionality, **9** is obviously nonvolatile. One should not assume that all the pheromones are small and volatile compounds.

5.2 History of Pheromone Research

The French naturalist Fabre (1823–1915) was the first to fully describe the attraction of male peacock moths, *Saturnia pyri*, by females. Fabre also noticed that the attractiveness of the females could not be masked by odoriferous substances such as naphthalene, but when the antennae of the males were covered with lacquer, they were not attracted to the females.

Accordingly, the antennae were thought to be the organ responsible for the attraction. As for the cause of the attraction, various kinds of radiation, such as infrared waves, were once assumed to be the reason. In 1925, however, Bloor in the United States showed the sex attractant of the gypsy moth *Lymantria dispar* to be a relatively stable and unsaponifiable substance soluble in organic solvents. Butenandt (1903–1995; Nobel Prize in 1939 for his hormone works) in Germany was the pioneer who clarified the nature of the female-produced sex pheromone of the silkworm moth *Bombyx mori*.

Scheme 5.1 summarizes the isolation and structure determination of the silkworm moth pheromone, bombykol (**1**).[3] Butenandt and his associates worked for 22 years (1939–1961) to clarify its structure. By extracting half a million scent glands from virgin *B. mori*, bombykol derivative **A** (12 mg) was then obtained as red crystals. The silkworms

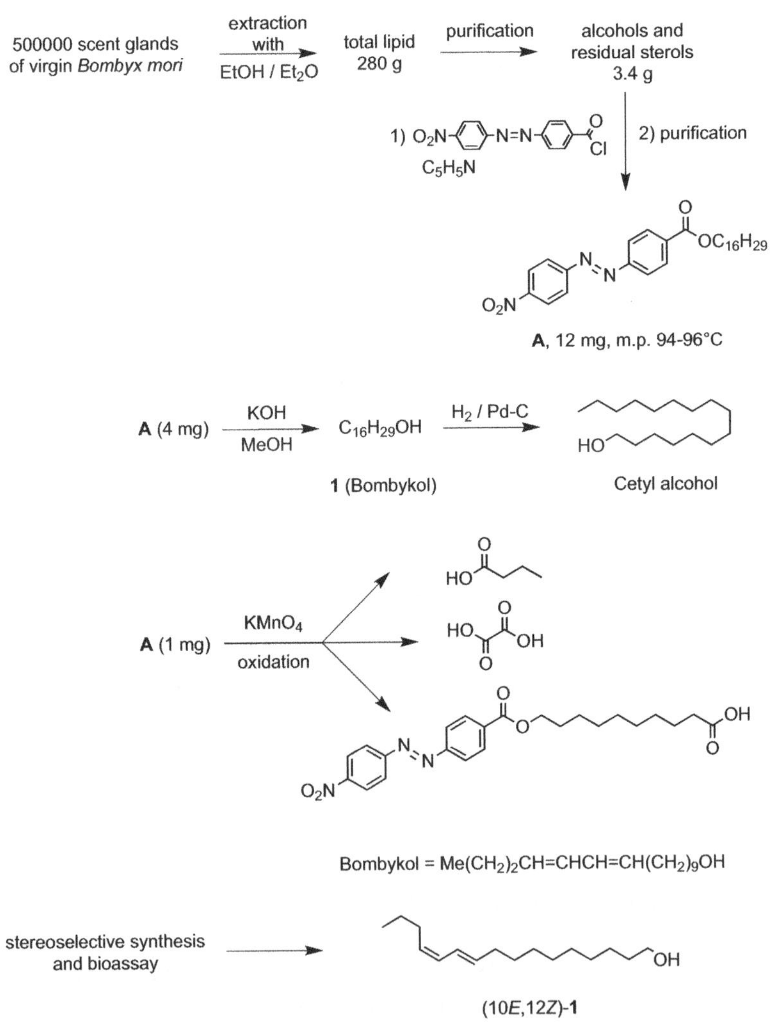

Scheme 5.1 *Structure determination of bombykol (**1**)*

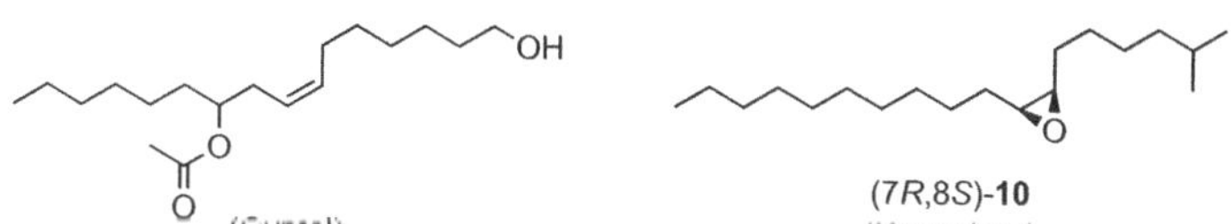

(Gyptol)

(7*R*,8*S*)-**10**
(Disparlure)

Figure 5.4 *Structure of gypsy moth pheromone* ***10***

were commercially available in Germany, Yugoslavia, Italy and Japan. A quantitative bioassay method to observe the sexual behaviour of male *B. mori* was developed and employed to monitor the active principle during the isolation. The availability of a large number of silkworms was the key to the success of Butenandt's work.

The carbon skeleton of bombykol (**1**) was determined by using 4 mg of **A** (Scheme 5.1). Alkaline hydrolysis of **A** afforded an unsaturated alcohol, $C_{16}H_{29}OH$, whose hydrogenation gave 1-hexadecanol (cetyl alcohol). Thus, bombykol possesses a C_{16} straight-chain carbon skeleton. The positions of the two double bonds were determined in 1959 by $KMnO_4$ oxidation of 1 mg of **A** to give *n*-butyric acid, oxalic acid and the ester of 10-hydroxydecanoic acid. Thus, bombykol must be 10,12-hexadecadien-1-ol (**1**). Finally, all the possible *cis*/*trans*-isomers of **1** were synthesized, and their pheromone activities were bioassayed in 1961 on male *B. mori* to show (10*E*,12*Z*)-**1** to be as potent as the natural pheromone at concentrations as low as 10^{-12} μg cm^{-3} in petroleum ether.

In the structure determination of pheromones, one must manipulate extremely minute amounts of samples, and therefore erroneous results were reported occasionally. Figure 5.4 shows the erroneous and correct structures of the female-produced sex pheromone of the gypsy moth *Lymantria dispar*. In 1960, Jacobson reported gyptol as the structure of the gypsy moth pheromone,[4] but its true structure was shown to be disparlure (**10**) by Beroza in 1970.[5] Although this was an extreme case of incorrect structural proposal, small mistakes took place frequently, which could be corrected by synthetic works.[6] Thanks to the development of microanalytical techniques, a small amount of a pheromone sample is sufficient for the structure determination. Compounds **1, 11–13** are depicted as examples in Table 5.1.

5.3 Research Techniques in Pheromone Science

Two useful monographs are available concerning the techniques for isolation, purification and identification of pheromones.[10,11]

5.3.1 The Collecting of Pheromones

In the past, pheromones were collected by extracting the whole bodies or abdominal tips of insects, which contained the pheromonal glands, using organic solvents such as hydrocarbons, diethyl ether or dichloromethane. This method, however, gives a mixture containing a large amount of fats and sterols, whose removal is always problematic. It is now common to use adsorbents to collect volatile pheromones in the air. Porapak Q, an ethylvinylbenzene-divinylbenzene copolymer, is often employed for the adsorption. Trapped volatiles are recovered from Porapak Q by elution with pentane or diethyl ether. Solid phase microextraction (SPME) with special adsorbents is also a popular method for pheromone collection.[11]

Table 5.1 Samples (and amounts) employed for the structure elucidation of pheromones

Researchers and year of their work	Name of compound	Structure[a]	Amount of sample [mg]
Butenandt *et al.*,[1] 1959	Bombykol (pheromone of *Bombyx mori*)	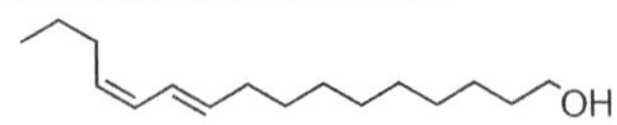 **1**	12.0 (as a derivative)
Persoons *et al.*,[7] 1976	Periplanone-B (pheromone of *Periplaneta americana*)	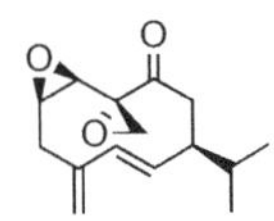 **11**	0.2
Oliver *et al.*,[8] 1992	Pheromone of *Biprorulus bibax*	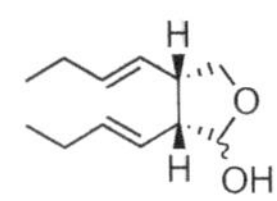 **12**	0.075
Wakamura *et al.*,[9] 2001	Posticlure (pheromone of *Orgyia postica*)	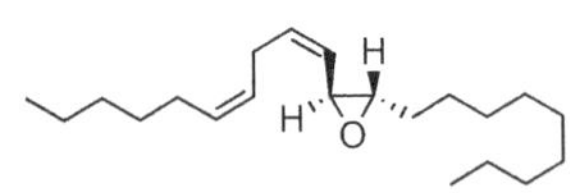 **13**	0.01

Note: [a] The stereostructures (including *cis/trans*-isomerism) were determined later by synthesis.

5.3.2 Bioassay-Guided Purification

Crude pheromone extracts are usually purified by various chromatographic techniques such as high performance liquid chromatography (HPLC) and gas chromatography (GC). The course of purification must be checked by proper bioassays. The traditional way is to observe the behaviour of the organism caused by chromatographic fractions, and another method is to carry out electroantennographic assay. Schneider in Germany was the first to use the electroantennogram (EAG) to study pheromone perception by *B. mori*. His study revealed that slow olfactory receptor potentials could be recorded from an isolated antenna placed between two glass capillary microelectrodes connected to an amplifier and a recording instrument. When GC separation of a crude extract is coupled with EAG apparatus, one can readily find any pheromonally active fractions. Every pheromone component elicits an EAG response. In rare occasions, however, nonpheromonal compounds such as inhibitors may also evoke EAG responses. By using a GC apparatus connected with EAG and mass spectrometry (MS), one can identify the known pheromone components by referring to the MS library.

5.3.3 Structure Determination and Synthesis

In the case of a new pheromone, MS is not a sufficient tool to clarify its structure; a nuclear magnetic resonance (NMR) study will reveal its structure. In the case of a pheromone with possible stereoisomers, its stereoselective synthesis will be the final proof of the proposed structure. The absolute configuration of a chiral and nonracemic pheromone can usually be elucidated by its enantioselective synthesis.

5.3.4 Field Bioassay

Field bioassay of the synthetic pheromone of a pest insect gives us information about whether it can be used practically as a population-monitoring agent or even as a mass-trapping agent. Communication disruption between the males and females of pest insects can be effected by synthetic pheromones, and this is a particularly useful method for pest control. For the practical use of pheromones in forests and crop fields, substantial amounts of pheromones must be synthesized first of all to enable a large-scale test and also to check the toxicity and environmental fate of the pheromones.

5.3.5 Structure Elucidation of the Male-Produced Aggregation Pheromone of the Stink Bug *Eysarcoris lewisi* – A Case Study

Nibbled rice, or rice damaged by pest insects, is a serious economic problem in Japan, because consumers do not accept such damaged rice. A stink bug *Eysarcoris lewisi* is known as one of the major species of rice bugs that cause nibbled rice in northern Japan. *E. lewisi* usually lives in meadows and fields and comes into the rice paddy fields where it attacks rice plants at the time of their grain formation. Its emergence can hardly be surveyed by conventional means, especially because it cannot be attracted by light traps. The possibility of using its pheromone for monitoring its population was therefore examined by Takita and her coworkers. In 2005 Takita proposed the structure of the male-produced aggregation pheromone of *E. lewisi* as (*E*)-2-methyl-6-(4′-methylenebicyclo[3.1.0]hexyl)hept-2-en-1-ol (**14**, Scheme 5.2).[12]

The collection, purification and structure elucidation of the *E. lewisi* pheromone was executed as summarized below.[13] Adult males reared for ten days after emergence were aerated ($1.8\,l\,min^{-1}$), and the airborne materials were adsorbed on Porapak Q, which was eluted with acetone. The acetone solution was concentrated *in vacuo* at room temperature. A total of 60 200 male equivalents (ME; volatiles obtained from a single male per day; about 0.51 μg) of the crude extract was purified by HPLC (twice) and then by preparative GC on two different columns to give a fraction with strong pheromone activity, as checked by the capture of *E. lewisi* in pheromone traps. Mass spectral analysis revealed the molecular weight of the pheromone to be 220.1840 ($C_{15}H_{24}O$). After catalytic hydrogenation, the hydrogenated pheromone possessed the molecular weight of 224, indicating the presence of two double bonds in the pheromone. NMR studies (both ^{1}H- and ^{13}C-NMR, together with proton COSY, HMQC, NOESY and HMBC analysis) caused Takita to propose **14** as the structure of the pheromone.[12]

Synthesis of the pheromone candidate **14** was achieved by Mori in 2007.[14] As shown in Scheme 5.2, (6*R*)- and (6*S*)-**14** were synthesized from the enantiomers of citronellal. However, the synthetic **14** showed ^{1}H- and ^{13}C-NMR data different from those of the natural pheromone. The (Z)-isomers (6*R*)- and (6*S*)-**15** were then synthesized, whose NMR data were in accord with those of the natural pheromone. Hence, bioassay proved the activity of (6*R*)-**15** as the aggregation pheromone of *E. lewisi.*

The next stage was to determine the absolute configuration at the positions of the ring fusion, C(1′) and C(5′), of the natural pheromone.[15] For this purpose, the diastereomeric mixture of ketone (6*R*)-**16** (Scheme 5.3) was reduced with L-Selectride to give a mixture

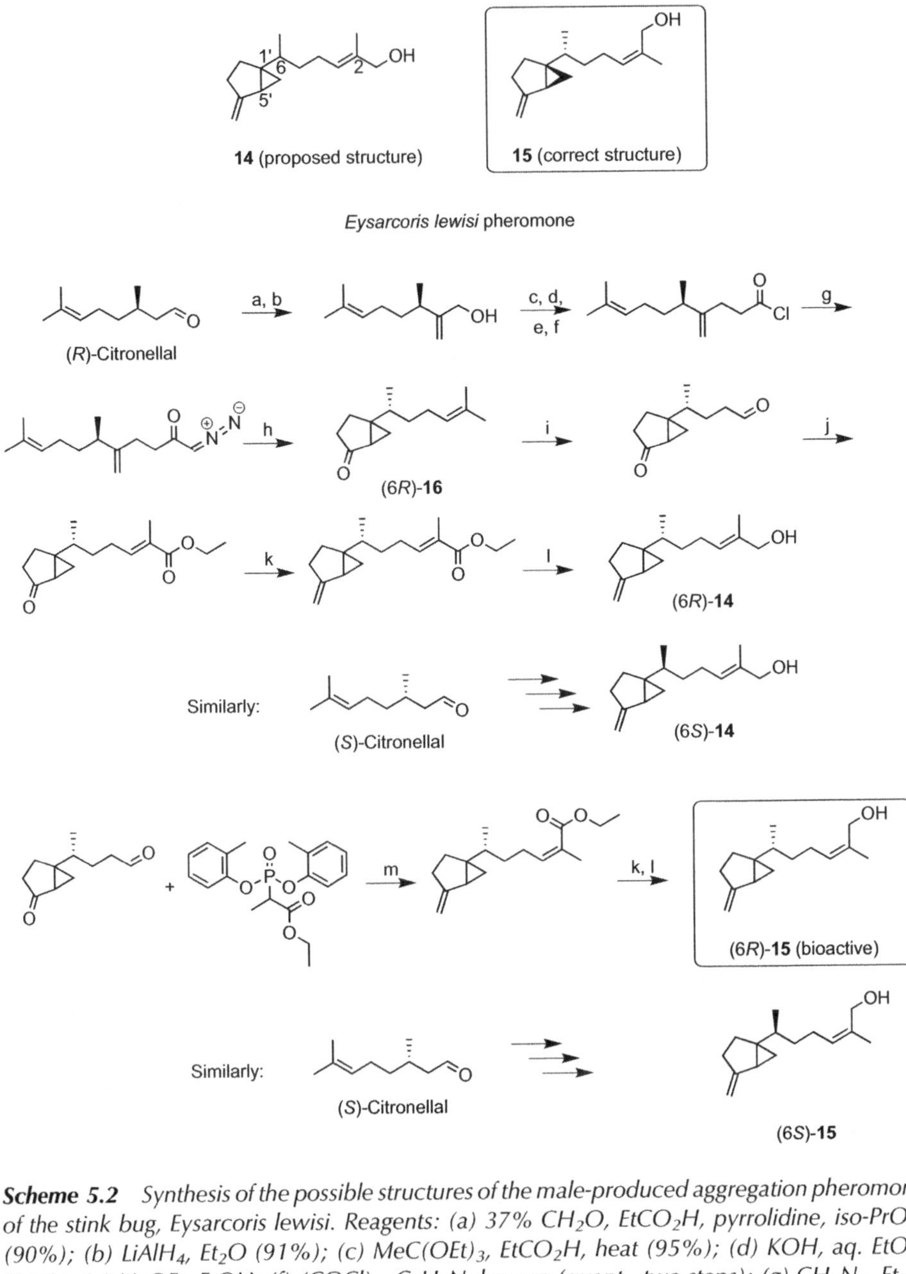

Scheme 5.2 *Synthesis of the possible structures of the male-produced aggregation pheromone of the stink bug, Eysarcoris lewisi. Reagents: (a) 37% CH_2O, $EtCO_2H$, pyrrolidine, iso-PrOH (90%); (b) $LiAlH_4$, Et_2O (91%); (c) $MeC(OEt)_3$, $EtCO_2H$, heat (95%); (d) KOH, aq. EtOH (83%); (e) NaOEt, EtOH; (f) $(COCl)_2$, C_5H_5N, hexane (quant., two steps); (g) CH_2N_2, Et_2O (quant.); (h) Cu, $CuSO_4$, cyclohexane, heat (58%); (i) OsO_4, $NaIO_4$, THF, tert-BuOH, H_2O (quant.); (j) $Ph_3P{=}C(Me)CO_2Et$, THF, CH_2Cl_2 (57%); (k) $Ph_3P(Me)Br$, n-BuLi, THF (96%); (l) $(iso\text{-}Bu)_2AlH$, toluene (51–55%); (m) NaH, THF (quant.)*

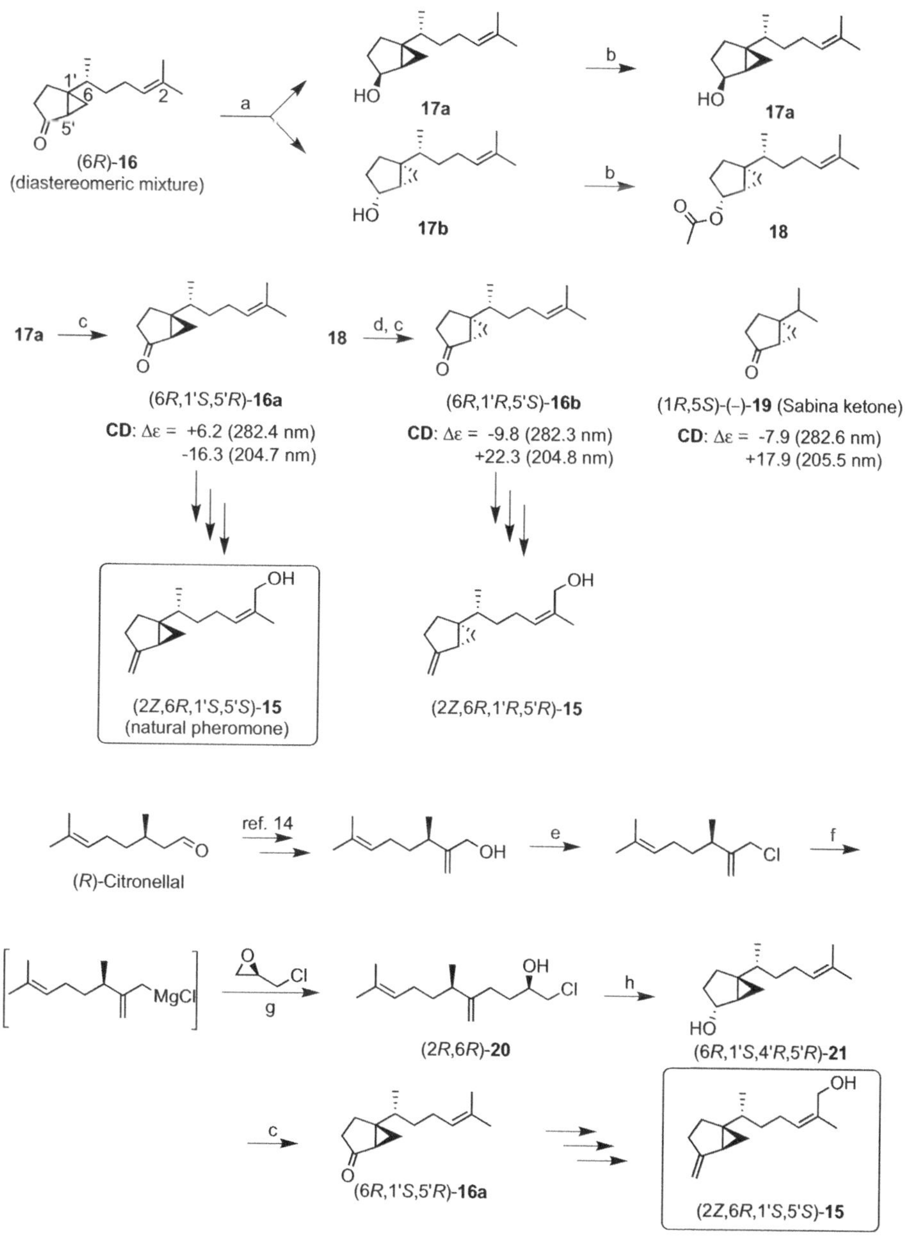

Scheme 5.3 *Synthesis of the male-produced aggregation pheromone of the stink bug, Eysarcoris lewis. Reagents: (a) [1] LiB(sec-Bu)$_3$H(L-Selectride), THF; [2] 30% H_2O_2, dil. NaOH (94%); (b) [1] lipase PS-D (Amano), CH_2=CHOAc, room temperature, 10–13 h, repeated three times; [2] SiO_2 chromatography; (c) (n-Pr)$_4$NRuO$_4$, NMO, molecular sieves (4Å), CH_2Cl_2, room temperature, (95% quant.); (d) K_2CO_3, MeOH (quant.); (e) MsCl, pentane, –5 °C, 0.5 h; then pyridine, room temperature, 20 h (85%); (f) Mg, $(CH_2Br)_2$, THF, room temperature; (g) CuI, THF, –78 °C to 0 °C, 1 h (50%); (h) n-BuLi, 2,2,6,6-tetramethylpiperidine, tert-BuOMe, –78 °C to room temperature, 12 h (95%)*

of two diastereomeric alcohols, **17a** and **17b**. Asymmetric acetylation of the mixture of **17a** and **17b** with vinyl acetate in the presence of lipase PS-D (Amano) was followed by chromatographic purification to give the recovered alcohol **17a** and the acetate **18**. The alcohol **17a** was oxidized with tetra(*n*-butyl)ammonium perruthenate and *N*-methylmorpholine *N*-oxide to give (6*R*,1′*S*,5′*R*)-**16a**. The acetate **18** was also converted to the ketone (6*R*,1′*R*,5′*S*)-**16b**. The circular dichroism (CD) spectra of these two diastereomers of **16** were then compared with that of the known (1*R*,5*S*)-(–)-sabina ketone (**19**). The ketone (6*R*,1′*S*,5′*R*)-**16a**, which showed the CD spectrum antipodal to that of (1*R*,5*S*)-sabina ketone (**19**), yielded the bioactive pheromone with its ^{1}H- and ^{13}C-NMR spectra identical to those of the natural pheromone. Accordingly, the natural pheromone of *E. lewisi* was shown to be (2*Z*,6*R*,1′*S*,5′*S*)-**15**. The other stereoisomers of **15** were biologically inactive.[16]

The final work for synthetic chemists was to carry out an enantioselective synthesis of the natural pheromone (2*Z*,6*R*,1′*S*,5′*S*)-**15**. The lower half of Scheme 5.3 summarizes our recent synthesis of the pheromone.[15] Hodgson's intramolecular cyclopropanation procedure (**20** → **21**) could be employed successfully.[17] The synthetic pheromone amounted to about 400 mg, which was sufficient to carry out further biological studies. The acetate of (2*Z*,6*R*,1′*S*,5′*S*)-**15** turned out to be the acetate isolated from an African plant *Haplocarpha scaposa*, and was identified by Bohlmann and Wallmeyer in 1982.[18] It sometimes happens that the same compound is produced by different organisms in different parts of the world.

5.4 Structural Diversity Among Pheromones

Organisms utilize many different kinds of volatile compounds as pheromones (Figures 5.5–5.8), which have evolved over the course of millions of years. Their stereochemistry is usually very important for their bioactivity.

Hydrocarbons serve as the pheromones of various insects (Figure 5.5). Alkanes with methyl branchings are common structures, as can be seen in the cases of (5*S*,9*S*)-**22** and **23**. In the case of the fly pheromone muscalure (**24**), (*Z*)-geometry of the double bond is important for its bioactivity. The peach miner moth pheromone (*S*)-**25** is used practically in Japan for pest control. Terpenes are frequently employed as pheromones. Acyclic tetraene **26**, monocyclic triene **27**, bicyclic diene **28**, spiro-diene **29** and macrocyclic tetraene **30** are all known as pheromones.

Epoxides, generated by the epoxidation of alkenes, are also common structures among pheromones (Figure 5.5). The epoxide **31** is practically useful in monitoring the population of the fall webworm moth, a pest of ornamental trees. Leucomalure (**32**) is a rare *bis*-epoxide. The green stink bug *Nezara viridula* uses bisabolene epoxide (**33**) as its pheromone.

Aliphatic alcohols as well as terpene alcohols are frequently employed components of the pheromone (Figure 5.5). The codling moth pheromone (**34**) is widely used in the United States to decrease the moth population in apple orchards. Beetles and weevils employ terpene alcohols **35**, **36** and **37** as pheromones. The use of **35** and **37** was studied for the control of forest pests.

A. Hydrocarbons

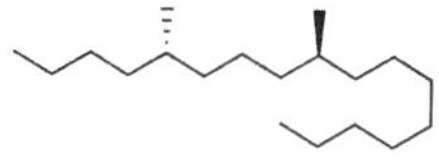

(5*S*,9*S*)-**22**
Leucoptera scitella (♀)
pest moth against apples

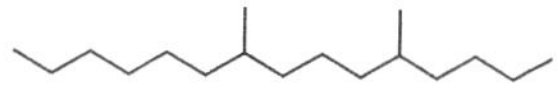

23 (unknown stereochemistry)
Leucoptera coffeella (♀)
coffee leaf miner moth

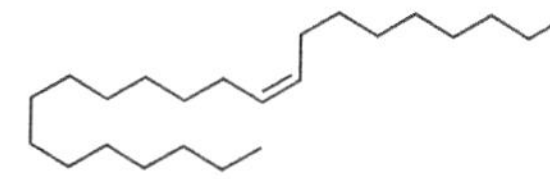

(*Z*)-**24**
Musca domestica (♀)
sex stimulant for house fly

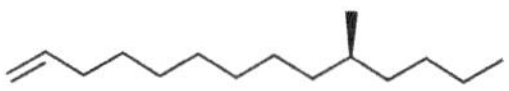

(*S*)-**25**
Lyonetia clerkella (♀)
peach miner moth

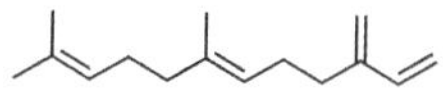

(*E*)-**26** (β-Farnesene)
Aphis gossypii
alarm signal of the cotton aphid

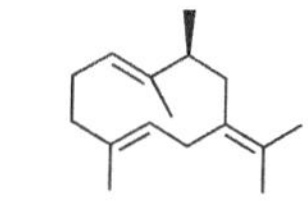

(*S*)-**27** (9-Methylgermacrene-B)
Lutzomyia longipalpis (♂)
sandfly in Lapinha, Brazil

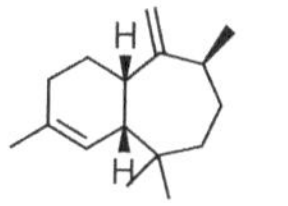

(1*S*,3*S*,7*R*)-**28** (3-Methyl-α-himachalene)
Lutzomyia longipalpis (♂)
sandfly in Jacobina, Brazil

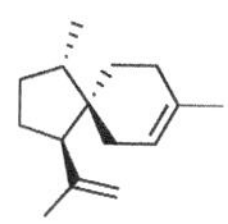

(1*R*,4*R*,5*S*)-**29** (Acoradiene)
Gnatocerus cornutus (♂)
broad-horned flour beetle

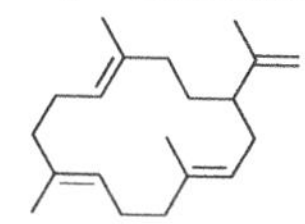

(*E*,*E*,*E*)-**30** (Neocembrene)
Nasutitermes exitosus
termite trail pheromone

B. Epoxides

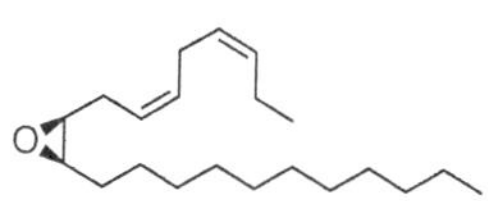

(3*Z*,6*Z*,9*S*,10*R*)-**31**
Hyphantria cunea (♀)
fall webworm moth

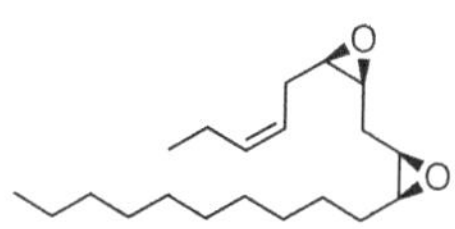

(3*E*,6*R*,7*S*,9*R*,10*S*)-**32** (Leucomalure)
Leucoma salicis (♀)
satin moth

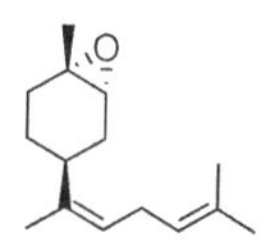

(1'*S*,3'*R*,4'*S*,*Z*)-**33**
Nezara viridula (♂)
green stink bug

C. Alcohols and their Esters

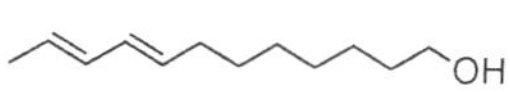

(8*E*,10*E*)-**34**
Cydia pomonella (♀)
codling moth

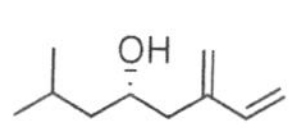

(*S*)-**35** (Ipsenol)
Ips paraconfusus (♂)
California five-spined ips

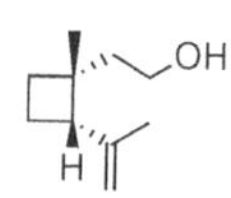

(1*R*,2*S*)-**36** (Grandisol)
Anthonomus grandis (♂)
cotton boll weevil

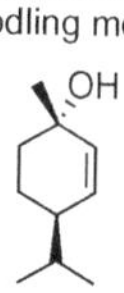

(1*S*,4*R*)-**37** (Quercivorol)
Platypus quercivorus (♂)
ambrosia beetle

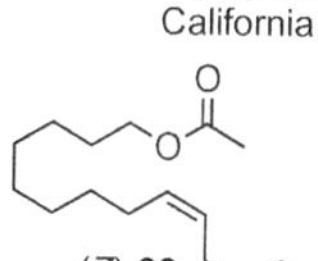

(*Z*)-**38**

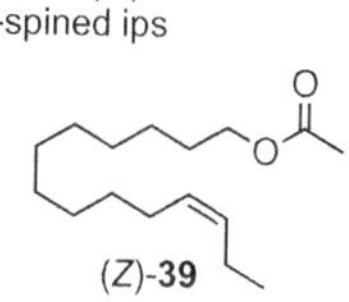

(*Z*)-**39**

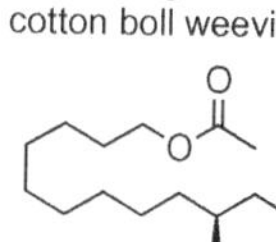

(*R*)-**40**

Adoxophyes honmai (♀)
smaller tea tortrix moth

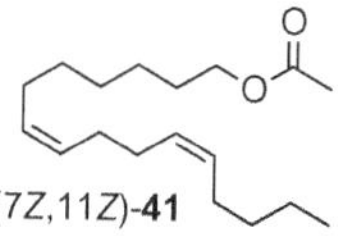

(7*Z*,11*Z*)-**41**

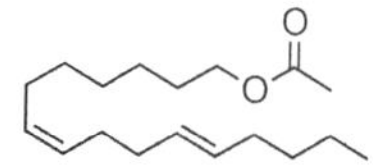

(7*Z*,11*E*)-**42**

Pectinophora gossypiella (♀)
pink bollworm moth (cotton pest)

Figure 5.5 *Structural diversity of pheromones (part 1 of 4)*

C. Alcohols and their Esters (continued)

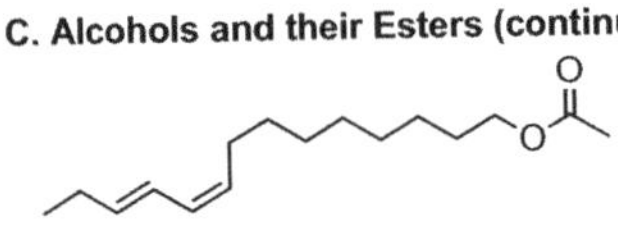

(9*Z*,11*E*)-**43**
Spodoptera litura (♀)
cotton leafworm

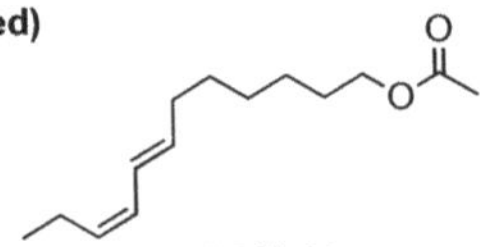

(7*E*,9*Z*)-**44**
Lobesia botrana (♀)
grape vine moth

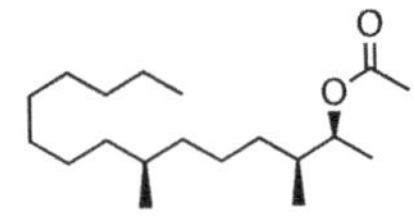

(1*S*,2*S*,6*S*)-**45**
Neodiprion pinetum (♀)
white pine sawfly

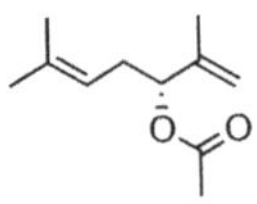

(*R*)-**46**
Pseudococcus comstockii (♀)
comstock mealybug

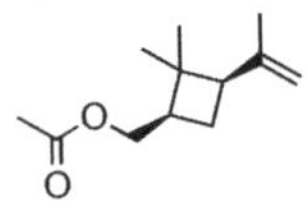

(1*R*,3*R*)-**47**
Plamococcus citri (♀)
citrus mealybug

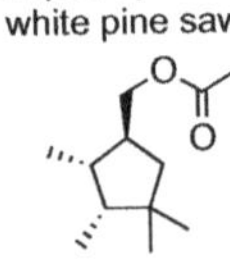

(1*S*,2*S*,3*R*)-**48**
Pseudococcus viburni (♀)
obscure mealybug

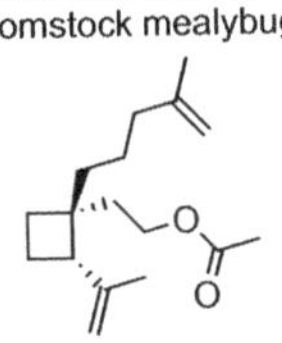

(1*R*,2*S*)-**49**
Aspidiotus nerii (♀)
oleander scale

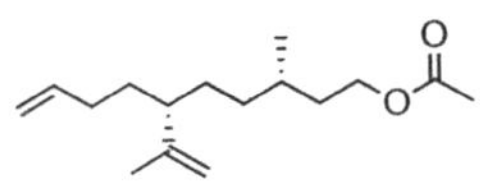

(3*S*,6*R*)-**50**

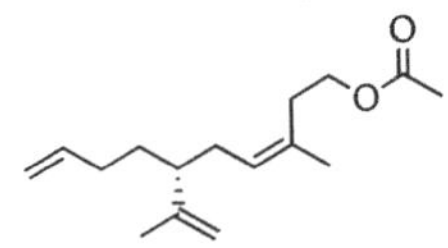

(3*Z*,6*R*)-**51**

Aonidiella auranti (♀)
California red scale

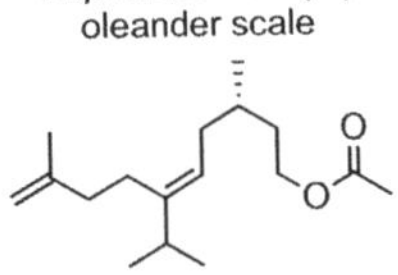

(3*S*,5*E*)-**52**
Aonidiella citrina (♀)
yellow scale

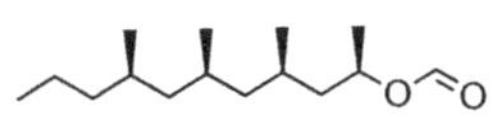

(1*R*,3*R*,5*R*,7*R*)-**53** (Lardolure)
Lardoglyphus konoi
acarid mite

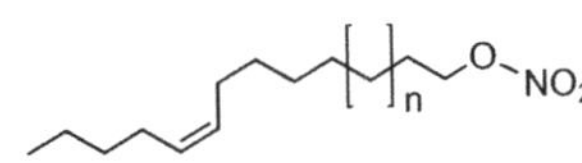

54 (n = 1)
55 (n = 2)
54/**55** = 50:1
Bucculatrix thurberiella (♀)
cotton leaf perforator

D. Aldehydes, Ketones and Esters

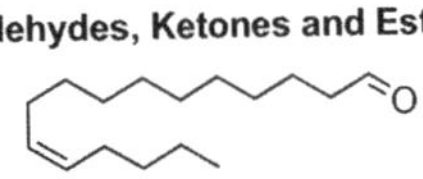

(*Z*)-**56**

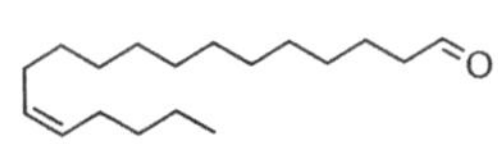

(*Z*)-**57**

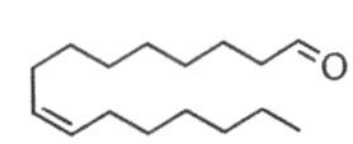

(*Z*)-**58**

Chilo suppressalis (♀)
rice stem borer

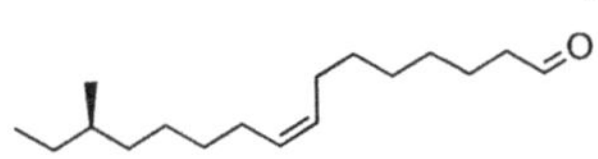

(8*Z*,14*R*)-**59**

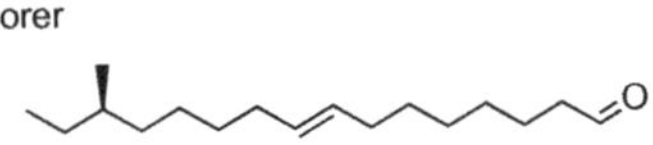

(8*E*,14*R*)-**60**

Trogoderma granarium (♀)
khapra beetle

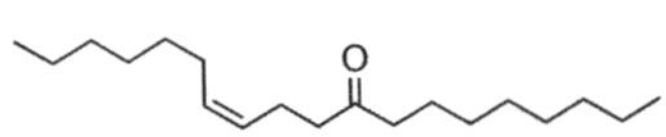

(*Z*)-**61**

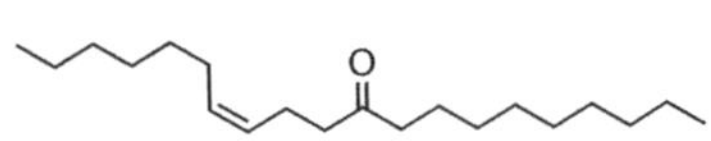

(*Z*)-**62**

Carposia niponensis (♀)
peach fruit moth

Figure 5.6 *Structural diversity of pheromones (part 2 of 4)*

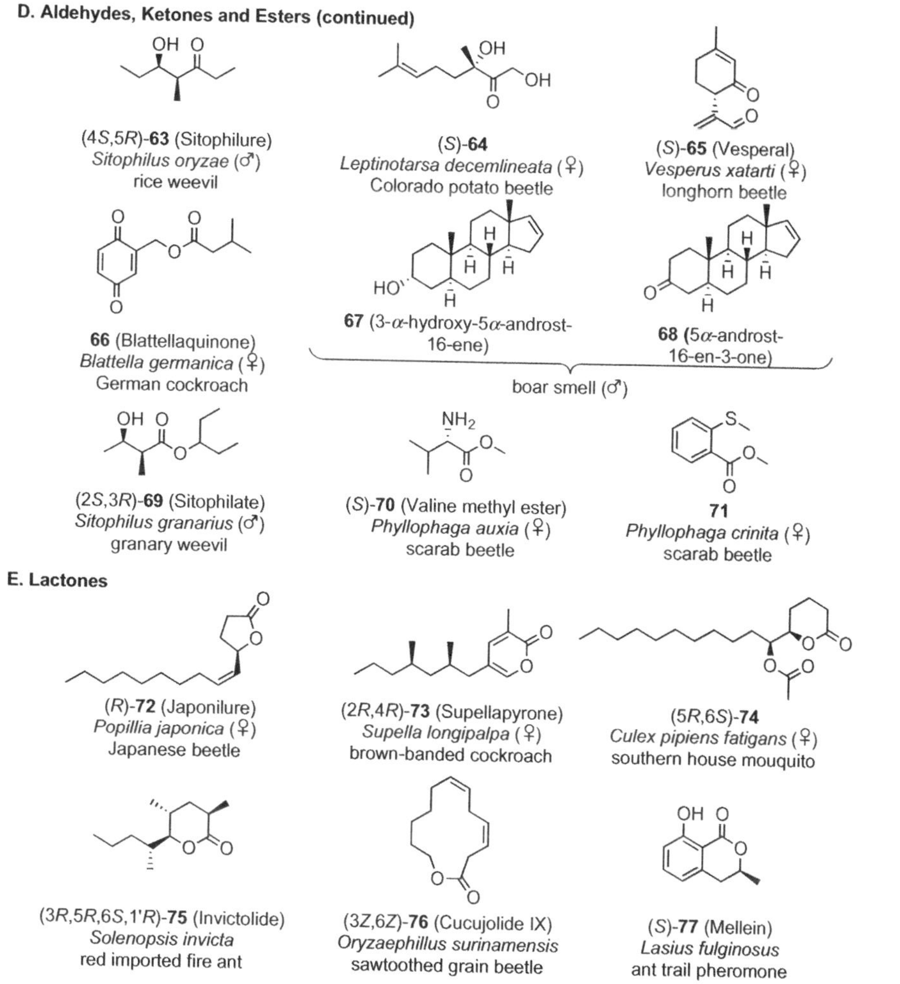

Figure 5.7 Structural diversity of pheromones (part 3 of 4)

Figures 5.5 and 5.6 show representative pheromones as esters of alcohols. Acetates are most common as the pheromones of lepidopteran insects (butterflies, moths). There are a number of pest species among them, and therefore the acetate pheromones have been studied extensively. The smaller tea tortrix moth, a pest in tea orchards, uses a mixture of **38**, **39** and **40** as its pheromone. This pheromone is widely used in Japan to reduce the population of that moth. In American cotton fields, the pink bollworm moth pheromone, a blend of (7*Z*,11*Z*)-**41** and (7*Z*,11*E*)-**42**, is employed to prevent cotton boll infestation by the moth. Another cotton pest, the cotton leafworm moth, employs **43** as its pheromone. The grape vine moth pheromone (**44**) is successfully used in Italy to decrease the moth population. Pine sawfly pheromones such as **45** have been extensively studied to monitor the population of the sawflies, which are harmful to the pine forests in northern Europe. Mealybugs employ the acetates **46–48** of terpene alcohols as their pheromones, and scale

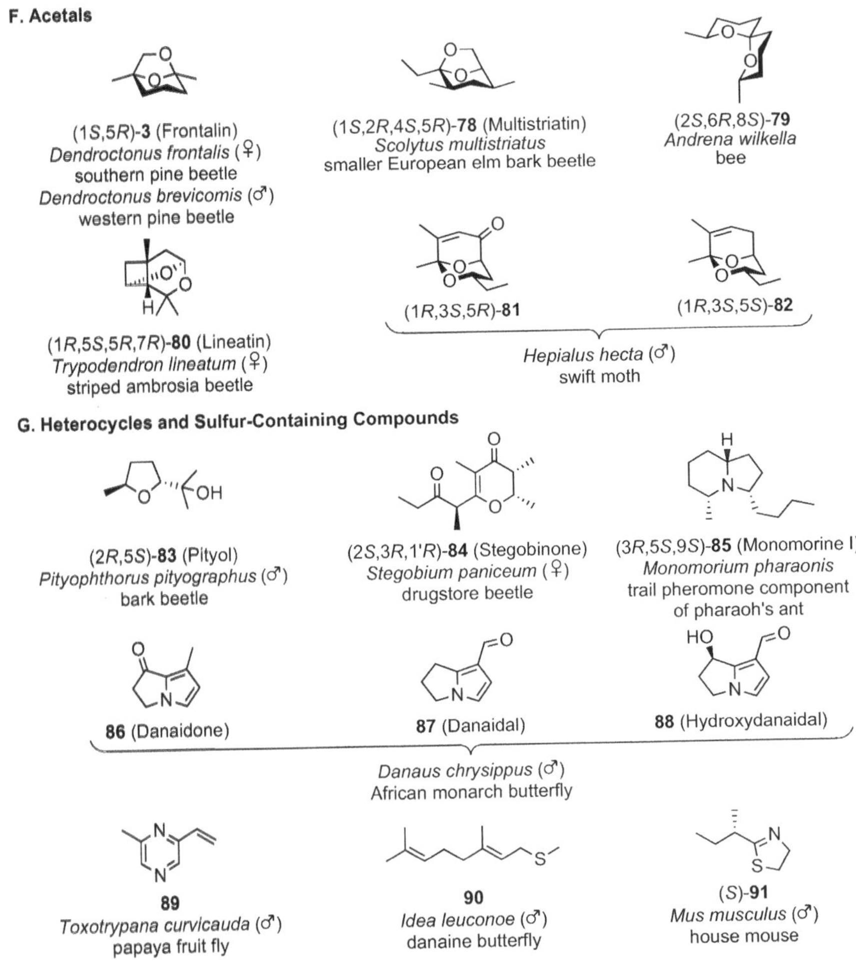

Figure 5.8 *Structural diversity of pheromones (part 4 of 4)*

insects use the acetates **49–52** of sesquiterpene alcohols as their pheromones. Mealybugs and scale insects are pests which are hardly detectable by ordinary methods, and their pheromones are considered excellent agents for monitoring their populations. The formate **53** is the pheromone of a mite. The cotton leaf perforator moth uses a mixture of nitrate esters **54** and **55** as its pheromone.

Aldehydes, ketones and esters are also common members of the pheromones group, as shown in Figure 5.6. The pheromone of the rice stem borer, which is a severe pest in Asian countries, is a mixture of three aldehydes: **56**, **57** and **58**. A mixture of two aldehydes (*Z*)-**59** and (*E*)-**60** provides the pheromone of the khapra beetle. The peach fruit moth uses a mixture of **61** and **62** as its pheromone. The rice weevil employs **63** (Figure 5.7) as its pheromone. The pheromone of the Colorado potato beetle is a unique dihydroxy ketone (**64**). The longhorn beetle, a pest in French vineyards, uses vesperal (**65**) as its pheromone.

Blattellaquinone (**66**) is the long-range sex attractant of female German cockroaches. The boar smell of *Sus scrofa* produced by males works on receptive females (sows) as a sex attractant, and consists of **67** and **68**. These steroids presumably affect the females of other higher animals, too. The granary weevil employs ester **69** (sitophilate) as its pheromone, while the pheromone of a scarab beetle is (*S*)-valine methyl ester (**70**). Another scarab beetle uses a sulfur-containing ester (**71**) as its pheromone.

Lactones also work as pheromones. Japonilure (**72**) is the pheromone of the Japanese beetle. The pheromone of the brown-banded cockroach is supellapyrone (**73**). Acetoxylactone **74** is the major component of the oviposition attractant pheromone from the eggs of the southern house mosquito. Invictolide (**75**) is the queen-recognition pheromone of the red imported fire ant. Macrolide **76** is the aggregation pheromone of the sawtoothed grain beetle. (*S*)-Mellein (**77**) is an ant trail pheromone.

Acetals are pheromones used by beetles, bees and higher animals. Frontalin (**3**, Figure 5.8) is the aggregation pheromone of bark beetles and the sex pheromone of the male Asian elephant. α-Multistriatin (**78**) is the pheromone of the smaller European elm bark beetle. Several bees and wasps, as well as fruit flies, use **79** as their pheromone. Lineatin (**80**), a unique cage-shaped molecule, is the pheromone of the striped ambrosia beetle. The swift moth employs **81** and **82** as its pheromone.

Oxygen heterocycles, nitrogen heterocycles and sulfur-containing compounds are also used as pheromones. Pityol (**83**) is a bark beetle pheromone, while stegobinone (**84**) is the pheromone of the drugstore beetle. Monomorine I (**85**) is a component of the trail pheromone of the Pharaoh's ant. African monarch butterflies use **86**–**88** as pheromones. The papaya fruit fly employs **89** as its sex pheromone. A danaine butterfly uses geranyl methyl sulfide (**90**) as its pheromone. The male house mouse employs (*S*)-**91** as its pheromone component.

As we have seen in Figures 5.5–5.8, diverse small molecules with different functionalities are used as pheromones. Their common feature is their volatility. Nonvolatile compounds such as sugars cannot be long-distance pheromones.

5.5 Complexity of Multicomponent Pheromones

In 1959 when the first pheromone bombykol [(10*E*,12*Z*)-10,12-hexadecadien-1-ol (**1**)] was identified as the female-produced sex attractant of the silkworm moth *B. mori*, the pheromonal activity was thought to be totally due to that single compound.[1] Now, it is generally believed that a pheromone is composed of plural pheromone components. This resembles the present situation in flavour and fragrance chemistry to regard a mixture as a whole as being responsible for the particular sense of smell. Some examples of multicomponent pheromones were already given in Figures 5.5–5.8. Two examples are examined in detail in this section.

The fall webworm moth (*Hyphantria cunea*) is a troublesome pest in Eastern Europe and Japan, where its larvae attack fruit trees and ornamental trees such as grape, peach, pear, apple, cherry, poplar and platan. Its female-produced sex pheromone was first studied by Roelofs and coworkers, who identified three pheromone components (**31**, **92**, **93**; Figure 5.9).[19] A blend of **31**, **92** and **93**, however, was pheromonally inactive when tested against *H. cunea*. Two additional pheromone components (**94**, **95**) were subsequently

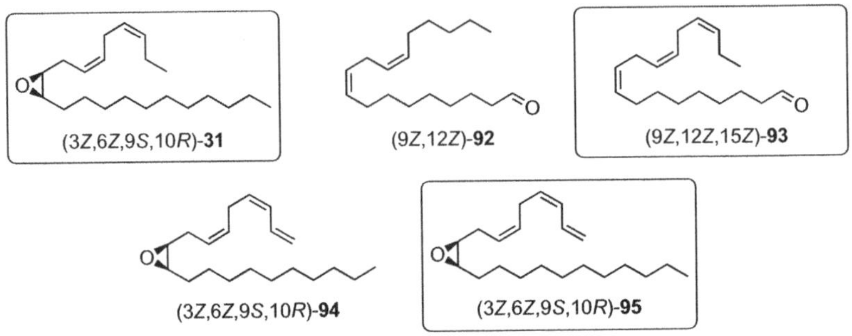

Figure 5.9 *Pheromone components of the female fall webworm moth, Hyphantria cunea*

identified by Tóth *et al.*, and a mixture of these five components (**31**, **92–95**) attracted *H. cunea* males.[20] Later, Senda *et al.* found that a blend of **31**, **93** and **95** could attract the males, and this blend was commercially developed as a lure to monitor the population of *H. cunea*.[21] In the practical use of pheromones, it is essential to clarify the minimal requirement of pheromone components for attractiveness.

Recently, in 2008, Lacey *et al.* reported a typical example to show the complexity of a multicomponent pheromone.[22] A male-produced aggregation pheromone of the cerambycid beetle *Megacyllene caryae* contained as many as eight male-specific compounds, as shown in Figure 5.10. These are (2*R*,3*S*)-2,3-hexanediol and its enantiomer (2*S*,3*R*)-2,3-hexanediol (**96**), (*S*)-limonene (**97**), 2-phenylethanol (**98**), (*S*)-α-terpineol (**99**), nerol (**100**), neral (**101**) and geranial (**102**).[22] None of these were attractive as a single component. Both sexes of *M. caryae* were attracted to the complete blend of these eight compounds, but the elimination of anyone of them resulted in a decreased trap capture. Blends that were missing (2*S*,3*R*)-**96**, (2*R*,3*S*)-**96**, or a mixture of **101** and **102** (1 : 1) were pheromonally inactive. Modern studies on semiochemicals revealed the importance of a proper mixture to evoke a biological reaction.

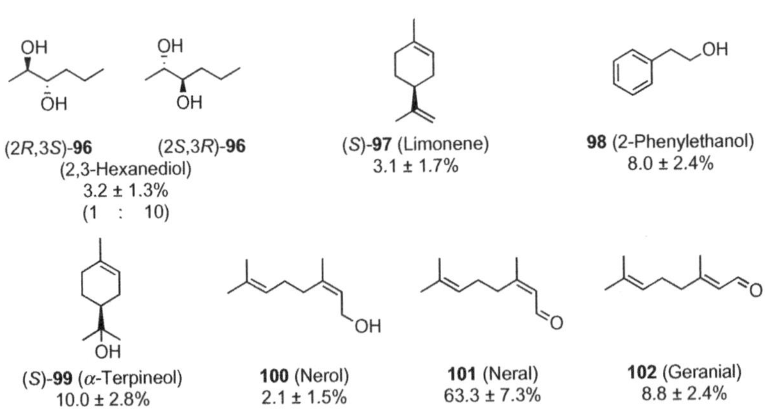

Figure 5.10 *Pheromone components of the male cerambycid beetle, Megacyllene caryae*

5.6 Stereochemistry and Pheromone Activity

The relationships between stereochemistry and bioactivity have been studied extensively with regard to pheromones since the early 1970s, when their pure enantiomers became available by synthesis. A detailed review with full references is available on this subject.[23] These relationships are far from straightforward. Organisms utilize chirality to enrich and diversify their communication system. The stereochemistry–pheromone activity relationships are classified into ten categories, as explained in Figure 5.11. It must be emphasized that these ten categories were found only through experiments using pure pheromone stereoisomers of synthetic origin.

5.6.1 Only a Single Enantiomer is Bioactive and its Opposite Enantiomer Does Not Inhibit the Response to the Active Isomer

This is the most common relationship, and the majority (about 60%) of the chiral pheromones belong to this category. (1*R*,5*S*,7*R*)-*exo*-Brevicomin (**2**), the aggregation pheromone of the western pine beetle, and (3*S*,4*R*)-faranal (**5**), the trail pheromone of the Pharaoh's ant, are typical members of this group. Pheromones in this category can be used as their racemates in practical applications.

5.6.2 Only One Enantiomer is Bioactive, and its Opposite Enantiomer Inhibits the Response to the Pheromone

Electroantennographic and behavioural responses of the gypsy moth to the enantiomers of disparlure (**10**), the gypsy moth pheromone, showed that (7*R*,8*S*)-**10** was bioactive, while (7*S*,8*R*)-**10** inhibited the activity of the (7*R*,8*S*)-isomer. Very strong inhibitory action of the opposite (*S*)-isomer of japonilure (*R*)-**72**, the sex pheromone of the Japanese beetle, is remarkable: (*R*)-**72** of 99% enantiomeric excess (*ee*) was about two-thirds as active as pure (*R*)-**72**, that of 90% *ee* was about one-third as active, that of 80% *ee* was about one-fifth as active, and both (*R*)-**72** of 60% *ee* and (±)-**72** were inactive. In the practical application of these pheromones, pure enantiomers of the pheromones have to be synthesized. Indeed, pure (7*R*,8*S*)-disparlure (**10**) and pure (*R*)-japonilure (**72**) are manufactured and used practically in pest control.

5.6.3 Only One Enantiomer is Bioactive, and its Diastereomer Inhibits the Response to the Pheromone

(4*S*,6*S*,7*S*)-Serricornin (**103**) is commercially important for the population-monitoring and mass-trapping of the cigarette beetle, *Lasioderma serricorne*. Since its (4*S*,6*S*,7*R*)-isomer is inhibitory, the commercial pheromone lure must be manufactured without contamination of the (4*S*,6*S*,7*R*)-isomer.

(2*S*,3*R*,1′*R*)-Stegobinone (**84**) is the sex pheromone component of the female drugstore beetle, *Stegobium paniceum*. A synthetic sample of **84** as a racemic and diastereomeric mixture showed only very weak activity. It was later shown that the addition of the (2*S*,3*R*,1′*S*)-isomer of **84** significantly reduces the response of the male drugstore beetle.

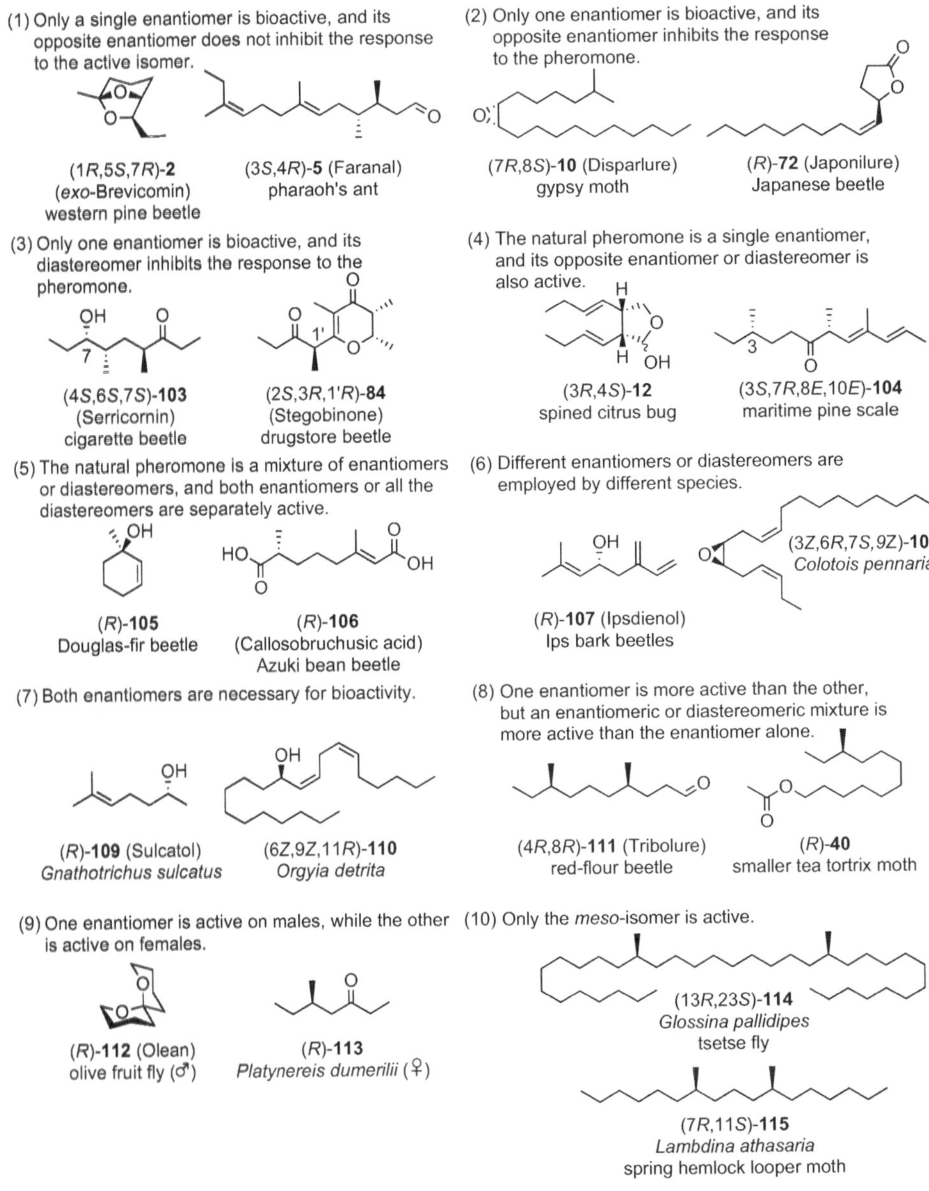

Figure 5.11 *Stereochemistry–pheromone activity relationships*

5.6.4 The Natural Pheromone is a Single Enantiomer, and its Opposite Enantiomer or Diastereomer is Also Active

The male-produced pheromone of the spined citrus bug, *Biprorulus bibax*, is (3*R*,4*S*)-**12**, but its opposite enantiomer (3*S*,4*R*)-**12** is as active as the pheromone itself.

Females of the maritime pine scale, *Matsucoccus feytaudi*, use (3*S*,7*R*)-**104** as a sex pheromone. Its (3*R*,7*R*)-isomer also showed bioactivity similar to the natural pheromone, while *M. feytaudi* males responded very weakly to the other two stereoisomers. It therefore seems that only the stereochemistry at C(7) is important for the expression of bioactivity.

5.6.5 The Natural Pheromone is a Mixture of Enantiomers or Diastereomers, and Both of the Enantiomers, or All of the Diastereomers are Separately Active

Females of the Douglas fir beetle, *Dendroctonus pseudotsugae*, produce an average of a 55:45 mixture of (*R*)- and (*S*)-**105**. The combined effect of the enantiomers is additive rather than synergistic, and both enantiomers are required for maximum response.

The adzuki bean beetle, *Callosobruchus chinensis*, uses callosobruchusic acid (**106**) as its pheromone. Although (*R*)-**106** is the major component of the natural pheromone (*R*/*S* = 3.1–3.4 : 1), (*R*)-**106** is only half as active as (*S*)-**106**.

5.6.6 Different Enantiomers or Diastereomers are Employed by Different Species

Ipsdienol (**107**) plays a pivotal role in the communication systems of bark beetles belonging to the genus *Ips*. (*R*)-Ipsdienol (**107**) is the aggregation pheromone of the bark beetles *Ips calligraphus* and *Ips avulses*, and (*S*)-**107** is the pheromone component of the California fivespined ips, *Ips paraconfusus*. *Ips pini* in New York employs a mixture of (*R*)-**107**/(*S*)-**107** at 32 : 68 to 56 : 44, while the same species in California uses a mixture of (*R*)-**107**/(*S*)-**107** at 89 : 11 to 98 : 2.

The chirality of pheromones is important to discriminate between two species of the winter-flying geometrid moths in central Europe. Thus (6*R*,7*S*)-**108** is the pheromone of *Colotois pennaria*, while *Erannis defoliaria* uses (6*S*,7*R*)-**108** as its pheromone.

5.6.7 Both Enantiomers are Necessary for Bioactivity

Sulcatol (**109**) is the aggregation pheromone of male *Gnathotrichus sulcatus*, an economically important ambrosia beetle in the Pacific coast of North America. The natural pheromone was known to be a 35 : 65 mixture of (*R*)-**109** and (*S*)-**109**. When Mori's synthetic enantiomers of **109** were bioassayed in Canada in 1976, neither (*R*)- nor (*S*)-**109** was bioactive. The maximum response of the beetle was to a racemic mixture (50 : 50) of the enantiomers, and the response to (±)-**109** was significantly greater than that to a 35 : 65 mixture. It thus became clear that the beetles must produce a mixture of enantiomers of **109** if they are to communicate with each other.

The tussock moth *Orgyia detrita* uses a 1.0:3.5 mixture of (*R*)-**110** and (*S*)-**110** as its pheromone. It is also known that (±)-**110** is more bioactive than the naturally occurring 1.0 : 3.5 mixture.

5.6.8 One Enantiomer is More Active Than the Other, but an Enantiomeric or Diastereomeric Mixture is More Active Than the Enantiomer Alone

Tribolure [(4*R*,8*R*)-**111**] is the male-produced aggregation pheromone of the red flour beetle, *Tribolium castaneum*. It was found that (4*R*,8*R*)-**111** was as active as the natural pheromone, while a mixture of (4*R*,8*R*)-**111** and its (4*R*,8*S*)-isomer in a ratio of 4:1 was about ten times more active than (4*R*,8*R*)-**111** alone.

The smaller tea tortrix moth, *Adoxophyes honmai*, uses (*R*)-**40** as a minor component of its pheromone bouquet (Figure 5.5), and (*R*)-**40** was found to be slightly more active than (*S*)-**40**. Further field tests suggest that there is an optimum *R*/*S* ratio of 95 : 5 for the trapping of males.

5.6.9 One Enantiomer is Active on Males, While the Other is Active on Females

In the case of olean (**112**), its (*R*)-isomer is active against the male olive fruit fly, *Bactrocera oleae*, while (*S*)-**112** activates the female, which produces (±)-**112** as the sex pheromone.

Another example is 5-methyl-3-heptanone (**113**), which has been isolated as a pheromone in the coelomic fluid of gravid specimens of nereid marine polychaetes. It is responsible for the induction of nuptial dance behaviour prior to the release of gametes in *Platynereis dumerilii*, and the female-produced (*S*)-**113** attracts the males, while the male-produced (*R*)-**113** is active on females.

5.6.10 Only the *meso*-Isomer is Active

There are some alkane pheromones with methyl branchings, whose *meso*-isomers are bioactive. (13*R*,23*S*)-13,23-Dimethylpentatriacontane (**114**) is the sex stimulant pheromone of the female tsetse fly, *Glossina pallidipes*. Neither its (13*R*,23*R*)- nor (13*S*,23*S*)-isomer is bioactive.

Female-produced sex pheromone components of the spring hemlock looper moth, *Lambdina athasaria*, are 7-methylheptadecane and 7,11-dimethylheptadecane (**115**). After the synthesis of all of their stereoisomers, a mixture of (*S*)-7-methylheptadecane and (7*R*,11*S*)-**115** (*meso*-isomer) was found to be bioactive.

Extensive joint work by biologists and chemists has revealed that diversity is the key of pheromone response. At present this diversity can be clarified only through experiments. It is therefore a prerequisite to study the relationship between stereochemistry (including *cis*/*trans*-isomerism) and bioactivity if one wants to use pheromones practically.

5.7 Pheromones With Kairomonal Activities

It has been believed that pheromones act only as pheromones to certain species. Recent studies have revealed that pheromones act also as kairomones for predators of the pheromone releasers. Two examples are given below.

The scarab beetle *Osmoderma eremita* and its larval predator, the click beetle *Elater ferrugineus*, are known as indicators of the species richness of the insect fauna of hollow deciduous trees in northern Europe. (*R*)-4-Decanolide (**116**, Figure 5.12) is the male-produced sex pheromone of *O. eremita*.[24] This lactone **116** is also employed by *E. ferrugineus* as a kairomone for the detection of tree hollows containing the larvae of *O. eremita*.[24]

Dunkelblum and coworkers, in cooperation with Mori's group, studied both the pheromonal and kairomonal activities of three female-produced sex pheromones (**117, 104, 118**) of the pine bast scales of *Matsucoccus* species together with their analogues (**119, 120**).[25,26] The Israeli pine bast scale, *M. josephi*, employs (2*E*,5*R*,6*E*,8*E*)-**117** as its sex pheromone, whereas (3*S*,7*R*,8*E*,10*E*)-**104** is used by *M. feytaudi* and (2*E*,4*E*,6*R*,10*R*)-**118** by *M.*

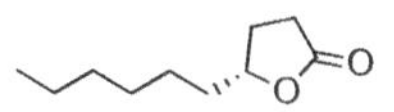

(*R*)-**116**
pheromone of the scarab beetle *Osmoderma eremita* and kairomone for its predator *Elater ferrugineus*

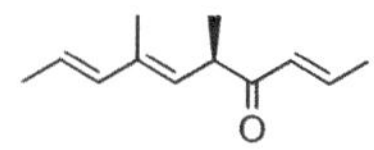

(2*E*,5*R*,6*E*,8*E*)-**117**
pheromone of the Israeli pine scale *Matsucoccus josephi* and kairomone for its predator *Elatophilus hebraicus*

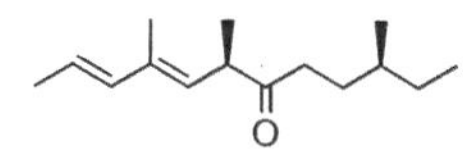

(3*S*,7*R*,8*E*,10*E*)-**104**
pheromone of the maritime pine scale *M. feytaudi* with kairomonal activities for *E. hebraicus*, *E. crassicornis*, *E. nigricornis* and *Hemerobius stigma*

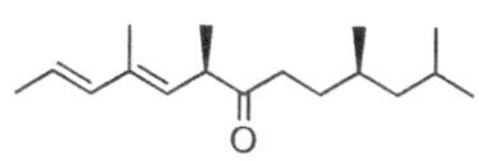

(2*E*,4*E*,6*R*,10*R*)-**118**
pheromone of the Japanese pine scale *M. matsumurae* with kairomonal activity for *E. hebraicus*

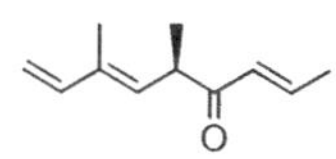

(2*E*,5*R*,6*E*)-**119**
pheromone mimic for *M. josephi* with no kairomonal activity for *E. hebraicus*

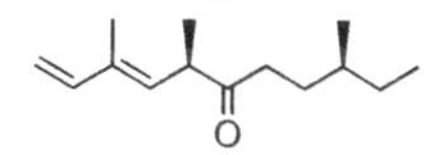

(3*E*,5*R*,9*S*)-**120**
pheromone mimic for *M. feytaudi* with no kairomonal activity for *E. crassicornis* and *H. stigma*

Figure 5.12 *Pheromones with kairomonal activities*

matsumurae. The pheromone **117** was found to be a potent kairomone for *Elatophilus hebraicus* adults, the major predatory bug against *M. josephi*. Strangely enough, the Israeli predator *E. hebraicus* was also attracted by **104** and **118**, although both *M. feytaudi* and *M. matsumurae* are absent in Israel. It seems as if the presence of the diene moiety of **117**, **104** and **118** is sufficient to attract the adult *E. hebraicus*. The kairomonal activity of **117**, **104** and **118** is a general phenomenon, and the *M. feytaudi* pheromone attracts such predators as *Elatophilus crassicornis* and *Hemerobius stigma* in Portugal and *E. nigricornis* and *H. stigma* in Italy and France.

The pheromone analogues **119** and **120** showed interesting biological properties. The nor-analogue **119** of *M. josephi* pheromone **117** preserved the pheromonal activity but eliminated its kairomonal activity.[25] Similarly, removal of the terminal methyl group from the diene moiety of **104**, producing the nor-analogue **120**, again preserved its pheromonal activity but eliminated the kairomonal activity.[26] Thus, subtle and designed alterations in the structure of the diene system change the mode of the kairomonal activity markedly. The two mimics **119** and **120** may be useful to capture only the pest pine bast scales without disturbing their predators.

5.8 Mammalian Pheromones

Since insect pheromone science has become a mature field, studies on the pheromones of mammals and aquatic organisms are now becoming attractive areas to work in. In 1996 Rasmussen *et al.* found that female Asian elephants, *Elephas maximus*, use a 97:3 mixture of (*Z*)- and (*E*)-7-dodecenyl acetates (**121**, **122**; Figure 5.13) as their sex pheromone to show their readiness to mate. The acetates **121** and **122** are employed by the females of 126 species of insects, including the moth *Trichoplusia ni*.[27] Thus, as different as elephants are from moths, they share a mating ritual.

According to Rasmussen's recent study, male Asian elephants *Elephas maximus* release frontalin (**3**) from the temporal gland on the face during musth, which is an annual period of

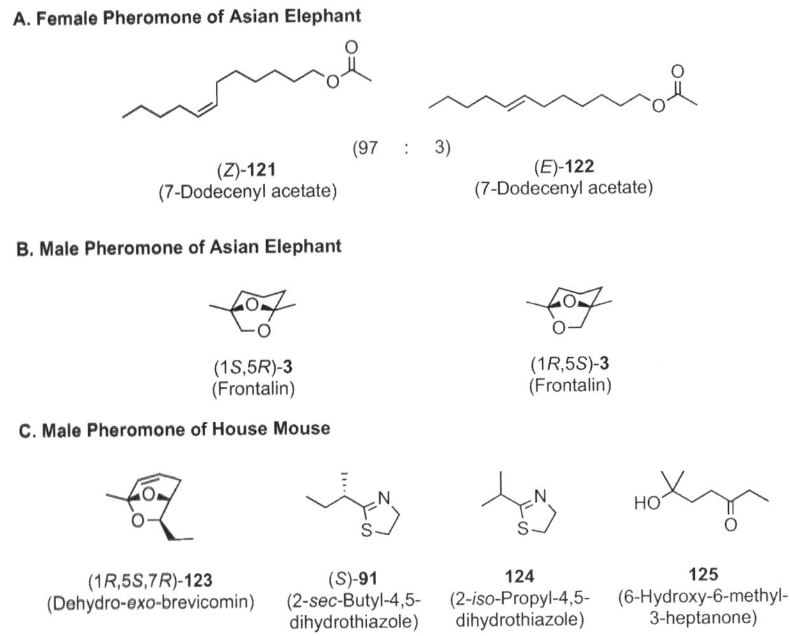

Figure 5.13 *Mammalian pheromones*

sexual activity and aggression.[28] The ratio of frontalin enantiomers enables other elephants to distinguish both the maturity of male elephants in musth and the phase of musth. In young males, significantly more (1*R*,5*S*)-(+)-**3** than (1*S*,5*R*)-(–)-**3** is released. As the elephant matures, the ratio becomes almost equal to emit (±)-frontalin. Musth periods get longer as males age. Secretions containing a high concentration of **3** at racemic ratios attract follicular phase females, whereas the secretions repulse males, as well as luteal phase and pregnant females. The importance of the enantiomeric composition of frontalin (**3**) in the behaviour of Asian elephants could be noticed only after the advent of enantioselective (chiral) GC. It must be added that bark beetles employ (1*S*,5*R*)-(–)-frontalin (**3**) as their pheromone component.

Genes in the major histocompatibility complex (MHC), known for their role in immune recognition and transplantation success, are involved in modulating mate choice in mice and perhaps also in humans.[29] Volatile body odours of mice are regulated by MHC genes, and it is these odour differences that underline mate choice and familial recognition. An individual's olfactory identity is coded in part by a pattern of volatile semiochemicals, which is regulated by genes in MHC.[30] In this connection, effects of mouse pheromone components on attractiveness of a male mouse evolved as an interesting research subject. For this study, previously synthesized mouse pheromone components (**123**, **91**),[31,32] as well as newly synthesized ones (**124**, **125**),[33] were employed.

In many wild animals, older males are often preferred by females, because they carry 'good' genes that account for their viability. In the case of the house mouse *Mus musculus*, higher levels of (1*R*,5*S*,7*R*)-dehydro-*exo*-brevicomin (**123**; *exo*-brevicomin is a bark beetle pheromone), (*S*)-2-*sec*-butyl-4,5-dihydrothiazole (**91**) and 2-isopropyl-4,5-dihydrothiazole (**124**) were detected in the urine of aged male mice than in that of normal adult males, while a lower level of 6-hydroxy-6-methyl-3-heptanone (**125**) was observed.[34] When **123**, **91** and **124** were added to the urine of normal adult males, their urine showed an enhanced attractiveness for female mice. The addition of **125** had no effect at all. Accordingly, it is established in the case of the house mouse that semiochemicals control the mate-selection process.[34] The search to clarify the roles of semiochemicals in higher animals including humans will continue to be an interesting area of pheromone science, with potential impacts on perfume industries.

5.9 Invention of Pheromone Mimics

There are practical demands for the invention of pheromone mimics, because pheromones are often too labile to be used in pest control. Various mimics have been prepared to date, and some of them are described in this section.

Tacke *et al.* synthesized the enantiomers of sila-linalool (**127**) by the method shown in Scheme 5.4.[35] The starting material (**126**) was converted to (±)-**127**, which was resolved by GC to give both (+)-**127** and (–)-**127**. They were both bioactive as tested by electro-antennographic detection (EAD) on the males of the vernal solitary bee, *Colletes cunicularius*. The bioactivity of the sila-pheromone (**127**) was not significantly different with that of the natural linalool, although **127** must be slightly less volatile than linalool due to the atomic weight of silicon being heavier than that of oxygen. A silicon atom can substitute a carbon atom. This is a good example of bioisosterism.

(1*S*,5*R*)-Frontalin (**3**) is the aggregation pheromone of bark beetles such as *Dendroctonus brevicomis* and *D. frontalis*. Strunz *et al.* synthesized its isomer **128**, which was shown to be pheromonally active.[36] Bravo *et al.* synthesized the trifluoro analogue of frontalin (**129**, Figure 5.14).[37] Its bioactivity, however, was not reported.

(4*S*,5*R*)-Eldanolide (**130**) is the male-produced sex pheromone of the African sugarcane borer, *Eldana saccharina*. Itoh *et al.* reported the synthesis and pheromonal activity of its fluorinated analogues (**131**–**133**).[38,39] Both enantiomers of **131** were bioactive, while the other four analogues showed no activity, as revealed by EAD.

(7*R*,8*S*)-Disparlure (**10**) is the female-produced sex pheromone of the gypsy moth, *Lymantria dispar*. Plettner and her coworkers synthesized and bioassayed its 5-oxa analogues (**134**).[40] GC-EAD bioassay revealed both enantiomers of **134** to be bioactive. The dose–response for (7*R*,8*S*)-**134** and that for (7*S*,8*R*)-**134** were similar. Interestingly,

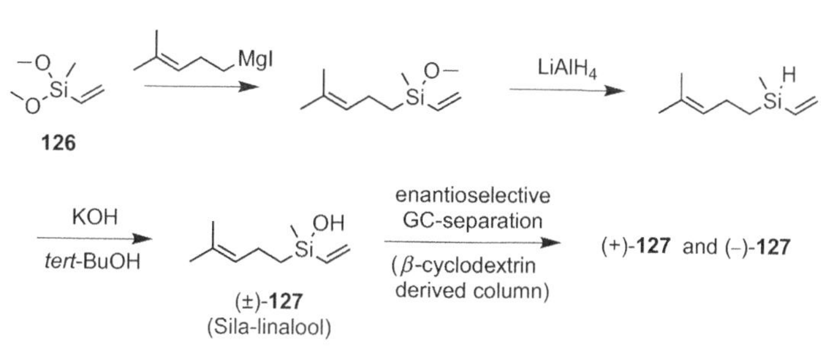

Scheme 5.4 Synthesis of sila-linalool **127** as pheromone mimic

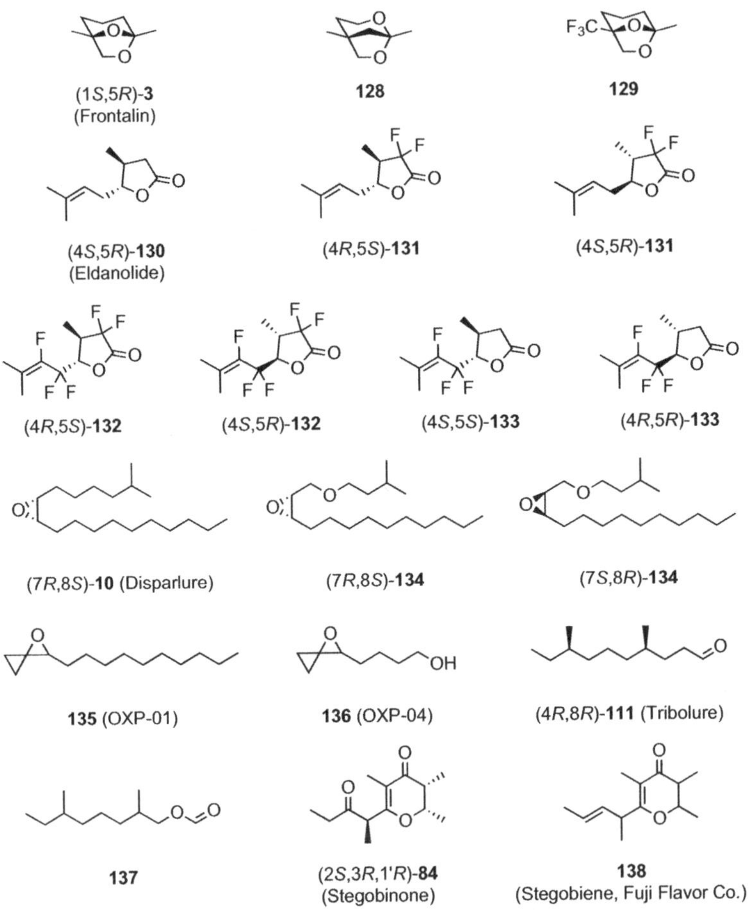

Figure 5.14 *Pheromones and their mimics*

pheromone-binding protein 1 (PBP1), which binds (7*S*,8*R*)-**10** strongly, binds the two enantiomers of **134** with nearly the same affinity as (7*S*,8*R*)-**10**. The affinity of PBP1 for naturally occurring (7*R*,8*S*)-**10** is known to be much weaker than for its enantiomer. Neither (7*R*,8*S*)-**134** nor (7*S*,8*R*)-**134** worked as a pheromone inhibitor. The concept of bioisosterism works in this case, too, although with a subtle difference. Two disparlure mimics with unusual structures were reported by Solari *et al.*[41] 2-Decyl-1-oxaspiro[2.2]pentane (OXP-01, **135**) works as an inhibitor against the action of disparlure (**10**), and 2-(4-hydroxybutyl)-1-oxaspiro[2.2]pentane (OXP-04, **136**) is more pheromonally active than **10** and has an additive effect in the blend with **10**.[41]

(4*R*,8*R*)-4,8-Dimethyldecanal (**111**, tribolure) is the aggregation pheromone of the flour beetle, *Tribolium castaneum* and *T. confusum*. Due to the air-sensitivity of **111** as an aldehyde, the more stable formate ester (**137**) was synthesized, found to be bioactive and practically used for commercial pheromone traps.[42] This is another example of bioisosterism to replace a carbon atom with an oxygen atom.

(2*S*,3*R*,1′*R*)-Stegobinone (**84**) is the female sex pheromone of the drugstore beetle, *Stegobium paniceum*. Its (2*S*,3*R*,1′*S*)-isomer is a strong inhibitor of the pheromone action. The methyl group at C(1′) of **84** is so readily epimerisable that the natural **84** soon becomes biologically inactive, and **84** cannot be used practically. Scientists at Fuji Flavor Co. synthesized stegobiene (**138**), which showed pheromone activity and could be used commercially to monitor the population of the drugstore beetle.

5.10 Conclusion

Recent progress in pheromone science has been so rapid that it is rather difficult to summarize all the topics of pheromone research in this chapter. I did not discuss biological issues such as the molecular biology of pheromone perception and the biosynthesis of pheromones. Those who want to learn the biological aspects should refer to the two-volume reference work of Schulz.[43] All the known pheromones are listed on the internet web sites 'Pherolist'[44] and 'The Pherobase'.[45] Pheromones discovered prior to 1990 are listed in the encyclopaedic work by Mayer and McLaughlin.[46] Hardie and Minks edited an excellent monograph on the pheromones of nonlepidopteran agricultural pests.[47] Pheromone chapters by Francke and Schulz, and also by others, are good sources of information in pheromone science.[48]

In order to preserve our global ecological system, we need a knowledge of chemical ecology to understand more about the roles of bioactive natural products, including pheromones, in the environment. We must both discover and solve many problems which still remain unexplained. Pheromone science is an interdisciplinary science between chemistry and biology. No one can be an expert in both of these areas unless he or she is satisfied to remain with a superficial understanding. Thus we have to remember the following words of Paul the Apostle: 'The person who thinks he knows something really does not know as he ought to know' (I Corinthians 8 : 2).

Acknowledgements

I wish to thank Dr. T. Tashiro for his help.

References

1. A. Butenandt, R. Beckmann, D. Stamm, E. Hecker, On the sex attractant of the silk moth *Bombyx mori*. Purification and structure determination. (Ueber den Sexuallockstoff des Seidenspinners *Bombyx mori*. Reindarstellung und Konstitutionsermittlung), *Z. Naturforsch.*, **14b**, 283–284 (1959).
2. P. Karlson, M. Lüscher, 'Pheromones': a new term for a class of biologically active substances, *Nature*, **183**, 55–56 (1959).
3. E. Hecker, A. Butenandt, Bombykol revisited – reflections on a pioneering period and on some of its consequences, in *Techniques in Pheromone Research*, H. E. Hummel, T. A. Miller (Eds), Springer-Verlag, New York, 1984.
4. M. Jacobson, M. Beroza, W. A. Jones, Insect sex attractants. I. The isolation, identification, and synthesis of the sex attractant of the gypsy moth, *J. Am. Chem. Soc.*, **83**, 4819–4824 (1961).

5. B. A. Bierl, M. Beroza, C. W. Collier, Potent sex attractant of the gypsy moth: its isolation, identification, and synthesis, *Science*, **170**, 87–89 (1970).
6. K. Mori, Synthetic examination of incorrectly proposed structures of biomolecules, *Chem. Rec.*, **5**, 1–16 (2005).
7. C. J. Persoons, P. E. J. Verwiel, F. J. Ritter, E. Talman, P. J. F. Nooijen, W. J. Nooijen, Sex pheromones of the American cockroach, *Periplaneta americana*: a tentative structure of periplanone-B, *Tetrahedron Lett.*, 2055–2058 (1976).
8. J. E. Oliver, J. R. Aldrich, W. R. Lusby, R. M. Waters, D. G. James, A male-produced pheromone of the spined citrus bug, *Tetrahedron Lett.*, **33**, 891–894 (1992).
9. S. Wakamura, N. Arakaki, M. Yamamoto, S. Hiradate, H. Yasui, T. Yasuda, T. Ando, Posticlure: a novel *trans*-epoxide as a sex pheromone component of the tussock moth, *Orgyia postica* (Walker), *Tetrahedron Lett.*, **42**, 687–689 (2001).
10. H. E. Hummel, T. A. Miller (Eds), *Techniques in Pheromone Research*, Springer-Verlag, New York, 1984.
11. J. G. Millar, K. F. Haynes (Eds), *Methods in Chemical Ecology*, Vol. 1 and 2, Kluwer, Dordrecht, 1998.
12. M. Takita, Isolation of the aggregation pheromone of *Eysarcoris lewisi*, *Tohoku Nogyo Seika Joho*, **19**, 50–51 (2005).
13. M. Takita, H. Sugie, J. Tabata, S. Ishii, S. Hiradate, Isolation and estimation of the aggregation pheromone from *Eysarcoris lewisi* (Distant) (Heteroptera: Pentatomidae), *Appl. Entomol. Zool.*, **43**, 11–17 (2008).
14. K. Mori, Synthetic studies aimed at the elucidation of the stereostructure of the aggregation pheromone, 2-methyl-6-(4′-methylenebicyclo[3.1.0]hexyl)hept-2-en-1-ol, produced by the male stink bug *Eysarcoris lewisi*, *Tetrahedron: Asymmetry*, **18**, 838–846 (2007).
15. T. Tashiro, K. Mori, Synthesis and absolute configuration of the male aggregation pheromone of the stink bug *Eysarcoris lewisi* (Distant), (2*Z*,6*R*,1′*S*,5′*S*)-2-methyl-6-(4′-methylenebicyclo [3.1.0]hexyl)hept-2-en-1-ol, *Tetrahedron: Asymmetry*, **19**, 1215–1223 (2008).
16. K. Mori, T. Tashiro, T. Yoshimura, M. Takita, J. Tabata, S. Hiradate, H. Sugie, Determination of the absolute configuration of the male aggregation pheromone, 2-methyl-6-(4′-methylenebicyclo [3.1.0]hexyl)hept-2-en-1-ol, of the stink bug *Eysarcoris lewisi* (Distant) as 2*Z*,6*R*,1′*S*,5′*S* by its synthesis, *Tetrahedron Lett.*, **49**, 354–357 (2008).
17. D. M. Hodgson, Y. K. Chung, I. Nuzzo, G. Freixas, K. K. Kulikiewicz, E. Cleator, J.-M. Paris, Intramolecular cyclopropanation of unsaturated terminal epoxides and chlorohydrins, *J. Am. Chem. Soc.*, **129**, 4456–4462 (2007).
18. F. Bohlmann, M. Wallmeyer, Naturally occurring terpene derivatives. Part 392. Sesquisabinene derivatives from *Haplocarpha* species, *Phytochemistry*, **21**, 1157–1158 (1982).
19. A. S. Hill, B. G. Kovalev, L. N. Nikolaeva, W. L. Roelofs, Sex pheromone of the fall webworm moth, *Hyphantria cunea*, *J. Chem. Ecol.*, **8**, 383–396 (1982).
20. M. Tóth, H. R. Buser, A. Peña, H. Arn, K. Mori, T. Takeuchi, L. N. Nicolaeva, B. G. Kovalev, Identification of (3*Z*,6*Z*)-1,3,6-9,10-epoxyheneicosatriene and (3*Z*,6*Z*)-1,3,6-9,10-epoxyeicosatriene in the sex pheromone of *Hyphantria cunea*, *Tetrahedron Lett.*, **30**, 3405–3408 (1989).
21. S. Senda, T. Saika, K. Furumori, Pheromone control of fall webworm, *Hyphantria cunea* Drury (Lepidoptera: Arctiidae), *Nitto Giho*, **91**, 97–102 (1993).
22. E. S. Lacey, J. A. Moreira, J. G. Millar, L. M. Hanks, A Male-produced aggregation pheromone blend consisting of alkanediols, terpenoids, and an aromatic alcohol from the cerambycid beetle *Megacyllene caryae*, *J. Chem. Ecol.*, **34**, 408–417 (2008).
23. K. Mori, Significance of chirality in pheromone science, *Bioorg. Med. Chem.*, **15**, 7505–7523 (2007).
24. G. P. Svensson, M. C. Larsson, Enantiomeric specificity in a pheromone–kairomone system of two threatened saproxylic beetles, *Osmoderma eremita* and *Elater ferrugineus*, *J. Chem. Ecol.*, **34**, 189–197 (2008).
25. S. Kurosawa, M. Takenaka, E. Dunkelblum, Z. Mendel, K. Mori, Pheromonal and kairomonal activities can be separated: synthesis and bioactivity studies of pine bast scale sex pheromones and their analogues, *ChemBioChem*, **1**, 56–66 (2000).

26. Z. Mendel, E. Dunkelblum, M. Branco, J. C. Franco, S. Kurosawa, K. Mori, Synthesis and structure–activity relationship of diene modified analogs of *Matsucoccus* sex pheromones, *Naturwissenschaften*, **90**, 313–317 (2003).
27. L. E. L. Rasmussen, T. D. Lee, W. L. Roelofs, A. Zhang, G. D. Daves Jr., Insect pheromone in elephants, *Nature*, **379**, 684 (1996).
28. D. R. Greenwood, D. Comeskey, M. B. Hunt, L. E. L. Rasmussen, Chemical communication: chirality in elephant pheromones, *Nature*, **438**, 1097–1098 (2005).
29. K. Yamazaki, G. K. Beauchamp, Genetic basis for MHC-dependent mate choice, *Adv. Genetics*, **59**, 129–145 (2007).
30. G. K. Beauchamp, K. Yamazaki, Chemical signaling in mice, *Biochem. Soc. Trans.*, **31**, 147–151 (2003).
31. M. Novotny, S. Harvey, B. Jemiolo, J. Alberts, Synthetic pheromones that promote inter-male aggression in mice, *Proc. Natl. Acad. Sci. USA*, **82**, 2059–2061 (1985).
32. T. Tashiro, K. Mori, Pheromone synthesis. Part 197. Synthesis of the enantiomers of 2-*sec*-butyl-4,5-dihydrothiazole and (1*R*,5*S*,7*R*)-3,4-dehydro-*exo*-brevicomin, pheromone components of the male mouse, *Mus musculus*, *Eur. J. Org. Chem.*, 2167–2173 (1999).
33. T. Tashiro, K. Osada, K. Mori, Pheromone synthesis. Part 238. Syntheses of 2-isopropyl-4,5-dihydrothiazole and 6-hydroxy-6-methyl-3-heptanone, pheromone components of the male mouse, *Mus musculus*, *Biosci. Biotechnol. Biochem.*, **72**, 2398–2402 (2008).
34. K. Osada, T. Tashiro, K. Mori, H. Izumi, The identification of attractive volatiles in aged male mouse urine, *Chem. Senses*, **33**, 815–823 (2008).
35. R. Tacke, T. Schmid, M. Hofmann, T. Tolasch, W. Francke, Sila-linalool as a pheromone analogue: a study on C/Si bioisosterism, *Organometallics*, **22**, 370–372 (2003).
36. G. M. Strunz, C. M. Yu, L. Ya, P. S. White, E. A. Dixon, Synthesis of a biologically active isomer of frontalin and some new products from acylative decarboxylation of 3-methyltricarballylic acid, *Can. J. Chem.*, **68**, 782–786 (1990).
37. P. Bravo, E. Corradi, M. Frigerio, S. V. Meille, W. Panzeri, C. Pesenti, F. Viani, Synthesis of (−)-(1*S*,5*R*)- and (+)-(1*R*,5*S*)-trifluoro analogues of frontalin, *Tetrahedron Lett.*, **40**, 6317–6320 (1999).
38. T. Itoh, K. Sakabe, K. Kudo, P. Zagatti, M. Renou, Asymmetric synthesis of both enantiomers of α,α-difluoroeldanolide: an interesting property of their biological activity, *Tetrahedron Lett.*, **39**, 4071–4074 (1998).
39. T. Itoh, K. Kudo, K. Yokota, N. Tanaka, S. Hayase, M. Renou, Synthesis and biological activity of point-fluorinated pheromone analogs of *Eldana saccharina*, *Eur. J. Org. Chem.*, 406–412 (2004).
40. J. A. H. Inkster, I. Ling, N. S. Honson, L. Jacquet, R. Gries, E. Plettner, Synthesis of disparlure analogues, using resolution on microcrystalline cellulose triacetate-I, *Tetrahedron: Asymmetry*, **16**, 3773–3784 (2005).
41. P. Solari, R. Crnjar, A. Frongia, G. Sollai, F. Secci, M. Spiga, C. Masala, A. Liscia, Oxaspiropentane derivatives as effective sex pheromone analogues in the gypsy moth: electrophysiological and behavioral evidence, *Chem. Senses*, **32**, 755–763 (2007).
42. K. Mori, S. Kuwahara, M. Fujiwhara, Pheromone synthesis: Part CVII – synthesis of 2,6-dimethyloctyl formate, a potent mimic of the aggregation pheromone of the flour beetles, *Proc. Ind. Acad. Sci., Chem. Sci.*, **100**, 113–117 (1988).
43. S. Schulz (Ed.), *The Chemistry of Pheromones and Other Semiochemicals*, Topics in Current Chemistry, Vols 239 and 240, Springer, Berlin, 2004, 2005.
44. http://www-pherolist.slu.se/.
45. http://www-pherobase.com/.
46. M. S. Mayer, J. R. McLaughlin, *Handbook of Insect Pheromones and Sex Attractants*, CRC Press, Boca Raton, 1991.
47. J. Hardie, A. K. Minks (Eds), *Pheromones of Non-Lepidopteran Insects Associated with Agricultural Plants*, CABI Publishing, Wallingford, 1999.
48. L. Mander, H.-W. Liu (Eds), *Comprehensive Natural Products II: Chemistry and Biology*, Vol. 4, Chemical Ecology, Chapters 4–7, Elsevier, Oxford, 2010.

6

Use of Volatiles in Pest Control

J. Richard M. Thacker and Margaret R. Train

Biological Sciences, University of the West of Scotland, Paisley, PA1 2BE, UK

6.1 Introduction

Volatile compounds that are used in the management and/or control of pest species are extensive in their chemistry, formulation and mode of action. They include vapour active synthetic insecticides that are directly toxic to target organisms, repellents that deter feeding by pests, pheromones that alter the behaviour of pests and various botanical volatiles that can be used to manipulate the behaviour of beneficial species that can also contribute to pest control, for example predators and parasitoids. Many of these volatile compounds are used as synthetic versions of naturally occurring botanical products while others exist solely as synthetic chemicals. In this chapter we attempt to provide a basic review concerning the use of volatile compounds for the management of pest species. We begin by considering the use of volatile compounds as pest deterrents.

6.2 Repellents (DEET, Neem, Essential Oils)

Insect repellents are volatile compounds that cause insects to make orientated movements away from the source.[1] Historically, smoke, plants, plant oils applied to skin and even camel urine, have all been used to repel pest species.[2] Even today, many individuals throughout Asia plant a neem tree in their gardens to deter biting arthropods. Examples of volatile compounds that are now known to repel crop pest species include monoterpenes from *Chrysanthemum* that repel diamondback moths[3] and isoprenes from tobacco that deter the tobacco hornworm.[4] Many other examples of such deterrence now exist, although in the vast majority of cases the exact mechanisms involved remain to be elucidated.[5]

The Chemistry and Biology of Volatiles Edited by Andreas Herrmann

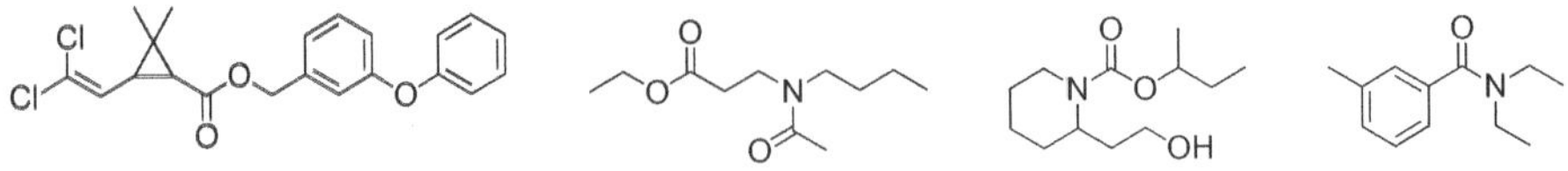

1 (Permethrin) 2 (IR3535®) 3 (Picaridin) 4 (DEET)

Figure 6.1 *Commercially available compounds for personal protection against insect bites*

In the case of human exposure to biting arthropods and our attempts to deter such attacks research is more advanced. Commercially, there are at present a number of over the counter products available for personal protection against insect bites. These include the insecticide permethrin (**1**), IR3535 (ethyl *N*-butyl-*N*-acetylaminoproprionate, **2**), picaridin (icaridin, **3**) and the most extensively used volatile for this purpose DEET (*N,N*-diethyl-2-toluamide, **4**; Figure 6.1).[2,6] DEET has been in use since 1956 and is the product of choice for more than 30% of the United States population for protection from biting insects.[7,8]

It is known that biting arthropods, such as mosquitoes, are attracted to the mammalian epidermis by the presence of carbon dioxide, lactic acid, perspiration, moisture, odour, warmth, hormones including oestrogen and kairomones such as ammonia.[9] Topical repellents provide the wearer with a vaporous chemical layer that is offensive to pest species. For example, olfactory receptors in antennal sensilla of mosquitoes are activated by DEET (**4**) as well as by other known insect repellants such as eucalyptol (**5**, Figure 6.2) and linalool (**6**). This activation of olfactory receptors deters feeding/biting by insects.[2,7]

The importance of using volatile compounds to protect against arthropod bites should not be underestimated. Although in so-called developed countries biting arthropods are mostly a nuisance, in developing countries biting arthropods vector the pathogens that cause major diseases such as dengue fever, yellow fever and malaria. The World Health Organization estimates that such pathogens pose a threat to more than 2.4 billion people in 92 countries. Malaria alone is responsible for more than 1 million deaths per year.[10]

In addition to synthetic products there are a number of volatile plant extracts that are commercially available and are marketed as products that are able to repel biting insects. Many of these extracts are essential oils with the most well known compound being citronella, an essential oil extracted from the grass *Cymbopognon nardus*.[2] The main active ingredients in these extracts appear to be (3*R*)-(+)-(β)-citronellol (**7**) and (3*R*)-(+)-(β)-citronellal (**8**).[11] Although citronella based products are widely available the utility of these extracts has been questioned. For example, research has indicated that the oil of *C. nardus* is only effective as a repellent for 40 min, compared with a maximum duration of

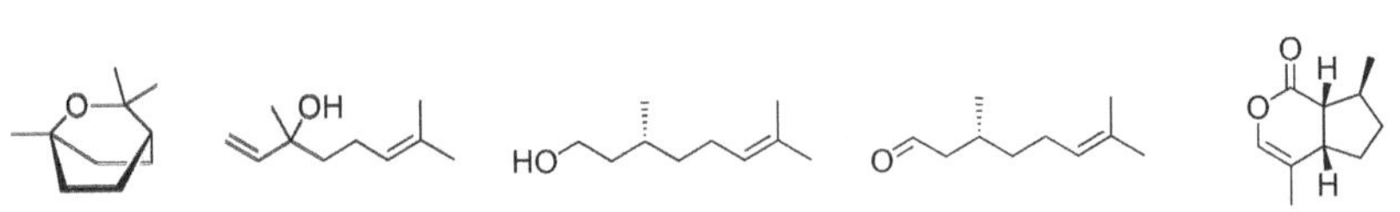

5 (Eucalyptol) 6 (Linalool) (*R*)-7 (Citronellol) (*R*)-8 (Citronellal) 9 (Nepetalactone)

Figure 6.2 *Structures of typical insect repellents in essential oils*

16–20 h for DEET (**4**). While other research on burning candles containing 3% citronella oil indicated that there was very little effect in decreasing the landing rate of *Culex* mosquitoes.[7]

In addition to citronella extracts there are a number of other plants with essential oils which have been identified as possessing insect repellent properties. These include basil, bog myrtle, catnip, cedarwood, eucalyptus, geranium, keffir lime, lavender, mountain mint, mugwort, neem osage orange, pennyroyal, peppermint, rosemary, sweet orange, syringa tree, thyme, turmeric, turpentine, vetiver and wintergreen.[2,12,13]

Of these, there are numerous studies indicating that it is the essential oils of Lamiaceae species that possess the most effective arthropod-deterring properties.[14] Examples of common plants in this family include catnip, lavender, peppermint and spearmint, all of which are known to have varying degrees of insect toxicity. For example, the oil of *Nepeta cataria* or catnip as it is more commonly known has a well documented effect on feline species and research has shown that it is nepetalactone (**9**) that is the active component of catnip, with the *cis,trans*- and *trans,cis*-isomers being present in its essential oil.[15,16] Hot water extracts of *Nepeta cataria* repel flea beetles and fresh catnip repels black ants. Other studies have shown that nepetalactone obtained from *Nepeta cataria* may be ten times more effective than DEET at repelling mosquitoes.[15]

The fresh and dried leaves of *Eucalyptus globulus* are commonly used across the African continent to reduce arthropod biting activity[12,17] and to reduce feeding on crops. For example, eucalyptus leaves are frequently used in grain stores in Brazil to deter feeding by *Sitophilus zeamais* and *Rhyzopertha dominica*.[18] Indeed, research has indicated that this and other closely related species displayed high levels of repellency towards a variety of invertebrates. For example, in a study of the repellency, toxicity and chemical composition of extracts of *Eucalyptus saligna* the eight major constituents were identified as 1,8-cineol (eucalyptol, **5**), α-pinene (**10**), limonene (**11**), terpinene (**12**, α-isomer shown), *p*-cymene (cymol, **13**), terpinen-4-ol (**14**), terpineol (**15**, α-isomer shown) and carvacrol (**16**). Overall, crude eucalyptus extracts were the most effective against the grain pests *Sitophilus zeamais* and *Tribolium confusum* however individual constituents were still both repellent and toxic in their own right (Figure 6.3).[19]

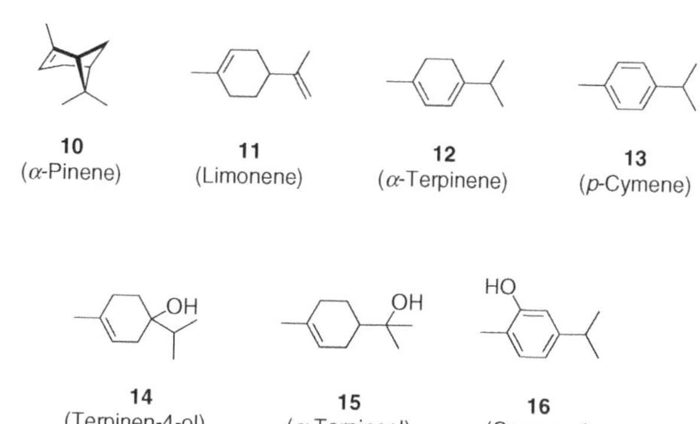

Figure 6.3 *Major components of Eucalyptus globulus essential oil*

In addition to the use of essential oils from plants as insect repellents there are a number of plant species that produce volatile compounds that are insecticidal and/or behaviour-modifying in nature. Many of these compounds were discovered as a result of European explorations of the New World in the fifteenth to seventeenth centuries. They include various alkaloids, terpenoids and phenolics and, collectively, are phytochemicals that are often regarded as examples of secondary plant substances. Perhaps the most well known examples today comprises extracts from neem the tree *Azadirachta indica* and extracts of natural pyrethoids from the chrysanthemum *Chrysanthemum cinerariaefolium* but there are others including nicotine and various isobutylamides (both of which can be volatilized and used as fumigants).

Research into the insecticidal activity of extracts from the neem tree *Azadiracta indica* has been wide ranging over the past 40 years.[20] By 2009, it had been reported that over 500 arthropod species were susceptible to extracts from this tree and to extracts from the related plants *Azadirachta excelsa* and *Melia azaderach*.[21] The effects of exposure to extracts from these species appear to be related to their hormone like activity and more specifically to antifeedant effects. The exact mode of action of the principle active ingredient, azadirachtin, is still to be determined but it seems clear that some of these effects derive from the volatile components released by these plants. In the Indian subcontinent it is common for a tree to be planted outside a house to reduce problems associated with dipterous pests.

Overall, DEET (**4**) remains the most widely used repellent of arthropods in developed countries. In developing countries however, there is still substantial reliance on extracts from various plants, notably the neem tree. DEET of course is a synthetic product. In the next section of this review we consider another type of volatile chemical used to manage pest species: the use of volatile synthetic pesticides.

6.3 Volatile Synthetic Chemicals and Fumigants

As of 2009 there were over 450 different insecticidal formulations approved for pest control use in the United Kingdom (UK). Many of these approvals were for the use of volatile compounds for the management of pest species. For example, one of the simplest uses of a synthetic volatile compound for pest control comprises the use of naphthalene (**17**, Figure 6.4) or more commonly dichlorobenzene (**18**) in mothballs. Both of these compounds transition directly from a solid to a gas and they are meant to be used in enclosed situations where the gas can build up and kill pest species.

17
(Naphthalene)

18
(Dichlorobenzene)

Figure 6.4 *Structures of synthetic volatiles naphthalene (**17**) and dichlorobenzene (**18**) used in mothballs*

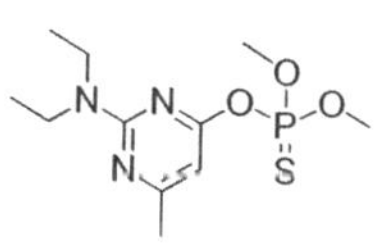

19
(Pirimiphos-methyl)

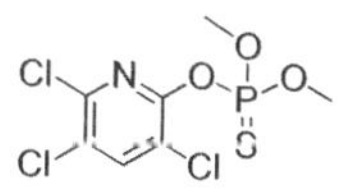

20
(Chlorpyrifos-methyl)

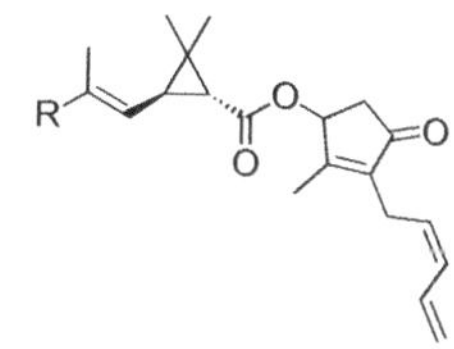

21: R = CH_3 (Pyrethrin I)
22: R = $COOCH_3$ (Pyrethrin II)

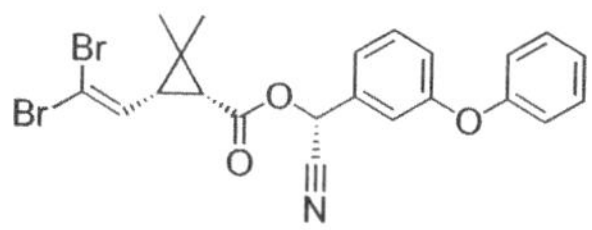

(Deltamethrin)

23
(Methyl bromide)

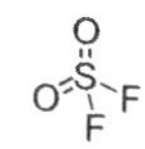

(Sulfuryl fluoride)

Figure 6.5 *Structures of compounds for pest control approved for use in food storage*

The most extensive use of volatile compounds for pest control in the UK however comprises the control of storage pests. Compounds that are approved for the control of storage pests include pirimiphos methyl (**19**) and chlorpyrifos methyl (**20**; organophosphates), pyrethrins (**21**, **22**) and aluminium or magnesium phosphide (AlP or Mg_3P_2, phosphine; Figure 6.5). Table 6.1 lists the compounds which were approved for use in UK food storage in 2009.[22] Of these, phosphine is widely used to control pests within grain

Table 6.1 *Compounds approved for use in UK food storage 2009. Data collated from the website of the Pesticides Safety Directorate (UK),[22] (see also Figure 6.5)*

Active ingredient	Use	Trade names
Pirimiphos methyl (**19**)	Food and grain storage areas	1. Actellic Smoke Generator No 10 2. Actellic Smoke Generator No 20
Chlorpyrifos methyl (**20**)	Crop handling and storage structures	1. Reldan 22
Aluminium phosphide	Space fumigation	1. Degesch Fumigation Pellets 2. Detia Gas Ex-P 3. Detia Gas Ex-T 4. Phostoxin I
Aluminium phosphide	Crop handling and storage structures	1. Degesch Fumigation Tablets 2. Detia Gas-Ex-B 3. Detia Gas-Ex-B-Forte
Deltamethrin	Crop handling and storage structures	1. K-Obiol EC25
Methyl bromide	Food storage areas	1. Mebrom 100 2. Methyl Bromide 100% 3. Methyl Bromide 98%
Sulfuryl fluoride	Crop handling and storage structures	1. ProFume

stores because, while it is highly toxic to organisms that are respiring (i.e. pests in grain stores), it has no impact on the viability of dormant grain. There is now evidence of resistance development in storage pests to phosphine but the exact mechanism of this resistance is as yet unknown.

Worldwide, fumigation is used either in bulk grain to control pests of stored products or for fabric treatment of buildings and empty bins to remove residual insect populations. Fumigants are gaseous pesticides that (at a given temperature, concentration and long enough exposure) kill all insect life stages.[23,24] Fumigants are most effective when fumigation follows a hygiene practice that has removed the potential sources and routes of reinfestation (e.g. grain residues and harbourage sites such as birds' nests, animal feed, grain spillages, debris).[25] The gas can be distributed using an air circulation system (i.e. closed-loop fumigation) to ensure even distribution throughout the grain and store.[24] Solid formulations (aluminium phosphide) can be added by hand, on trays under fumigation sheets, using a probe or with an automatic dispenser delivering a designated dosage as grain is loaded into the bins.[23]

A survey of pesticide use on UK farms ($n = 335$) during the period 2002 to 2003 indicated 60% of grain was treated by fumigation. The most commonly used pesticide by farmers for fabric treatment (82% by weight) was pirimiphos methyl (**19**). The main grain fumigants used and applied by contractors were aluminium phosphide and methyl bromide (**23**).[26] The latter compound however is currently being phased out as it depletes ozone, and it will be withdrawn from use worldwide by 2015 in agreement with the Montreal protocol.[27]

Historically, the use of fumigants for pest management has often produced resistant strains of pests. As a result, a longer period of fumigation is usually needed under gas-tight conditions to prevent leakage (leading to under-dosing), to achieve effective control. The exposure time varies with species and whether they are susceptible or resistant strains. For example, laboratory tests indicate a resistant strain of *Rhyzopertha dominica*, originally from Bangladesh, survived phosphine at a dosage of 1.45 mg l^{-1}, with death only occurring after 72 h exposure.[28]

Over-exposure to fumigants can lead to death and these products should only be applied by certified licensed personnel trained in the use of respirators, gas detectors and first aid.[23] For health and safety reasons fumigation should be carried out by a minimum of two people, in sealed leakproof environments and the area has to be well ventilated after the designated period of fumigation.

The most effective fumigant used in the world today is phosphine (PH_3). This volatile has been used since the 1930s.[28] The gas is produced from pellets, sachets or tablets of aluminium or magnesium phosphide when in contact with moisture in the air.[26] It is a colourless, odourless, toxic gas (impurities produce a garlic-like smell) that attacks the respiratory system.[28] It reacts with copper, copper salts and precious metals found in electrical equipment so these need to be removed or covered during fumigation.[23] Phosphine is most effective when used at low concentrations over long periods (minimum 5–16 days) at 15–20 °C to ensure all life stages are killed. Three grams of aluminium phosphide releases 1 g of phosphine gas and a dosage of 2 g gas t^{-1} produce or 1.5 g m^{-3} has been recommended.[25]

Phosphine spontaneously combusts when in contact with liquid water due to a rapid release of gas, at temperatures above 100 °C, reduced pressures or when concentrations exceed 25 mg l^{-1} in air.[25,28] Death due to phosphine poisoning of humans, through ingestion and inhalation, occurs at a concentration of 2.8 mg l^{-1}. A full recovery from mild poisoning is obtained after exposure to fresh air, complete rest and abstinence from alcohol consumption for 48 h. A maximum threshold level of 0.3 ppm day^{-1} for individuals working 8-h daily shifts during a 5-day week is recommended.[23] The United States Food Quality Protection Act (FQPA) of 1996 indicates a tolerance limit of 0.01 ppm phosphine residues in food.[29] When well aerated, little residue (aluminium hydroxide, nondecomposed aluminium phosphide) is left on the grain, even after repeated fumigations.[23]

When phosphine is used frequently, in poorly sealed structures and at a lower exposure time, resistance builds up in the insect population. Insect resistance to phosphine has occurred since the 1970s and has spread throughout the world, including the UK. Of the 849 strains of insect collected from 82 countries, 20% were recorded as resistant to phosphine.[23] Resistant insects include *Cadra cautella* (almond moth), *Cryptolestes ferrugineus* (flat grain beetle), *Oryzaephilus surinamensis* (saw-toothed grain beetle), *Plodia interpunctella* (Indian meal moth), *Rhyzopertha dominica* (lesser grain borer), *Sitophilus granarius* (granary weevil), *Sitophilus oryzae* (rice weevil), *Tribolium castaneum* (rust-red flour beetle), *Tribolium confusum* (confused flour beetle) and *Trogderma granarium* (Khapra beetle).[28,30–32] Eggs and pupae are more resistant to phosphine, although the former become less so with age.[28] Resistance relates to a low level uptake of phosphine, detoxification processes within resistant strains and the presence of resistance genes. The current recommendation to deal with resistance development is to expose pests for a longer time at a lower concentration, as resistant strains can survive high concentrations (due to a protective narcosis effect) when exposed for too short a period.[23,28,30] Methyl phosphine, a phosphine analogue, is being developed as a fumigant to control phosphine-resistant insects.[30]

Pirimiphos methyl (**19**) is another volatile that can be used as a spray, fog, smoke and dust for fabric treatment and as a grain protectant.[26,32] It is typically sprayed on grain as it is loaded into storage facilities and is effective against immature stages of insects.[25] Some resistance has been recorded in Brazil for *Sitophilus oryzae* (L.), *Oryzaephilus surinamensis*, *Rhyzopertha dominica*, *Tribolium castaneum* and *Cryptolestes species*.[32] In New South Wales, Australia, populations of *S. granarius* (L.), *O. surinamensis*, *R. dominica* and *T. castaneum* all contained resistant individuals.[33]

Chlorpyrifos methyl (**20**) is a broad-spectrum organophosphate used as a spray to treat empty bins and as a grain protectant. It is most effective against insects when farm-stored wheat is cooled by aeration compared to uncooled grain. However this is an expensive treatment which has little effect on *Rhyzopertha dominica*, a major wheat pest in many countries, and resistant strains of *Oryzaephilus surinamensis* have been recorded in Australia and the UK.[34]

Many insecticides that are not naturally volatile can also be made to volatilize and can then be applied for the purpose of fumigation. Typically the equipment involved in the application of these compounds would be some type of fogging machine. Fumigants are used in enclosed spaces such as greenhouses and storage facilities (e.g. grain stores; discussed above). They are also used at ports and for treatment of ships and aeroplanes. With fogging machines thermal energy nozzles use a stream of hot gas to vaporize a liquid pesticide which then condenses as it leaves the sprayer, forming a cloud of aerosol droplets.

When these droplets are less than 15 μm in diameter a fog is produced. The hot gas is the exhaust from a petrol engine that is expelled through a long pipe (0.5–1.0 m in length). The liquid pesticide is fed into the hot gas stream at a rate controlled by a restrictor. Only pesticides that are heat-stable are suitable for use with fogging machines. These machines have found widespread use for pesticide application in enclosed environments like greenhouses, food storage facilities and buildings. They have also been used successfully in plantation crops with very dense canopies.

Aerosols are widely used in household pest control. The active ingredient is dissolved in a petroleum solvent and pressurized in a container with a propellant gas such as carbon dioxide. When sprayed, the petroleum solvent quickly evaporates to leave a fog of microscopic droplets that are suspended in air. The droplets produced are typically less than 10 μm in diameter and because there is a danger associated with inhaling these droplets, most aerosols have a very low concentration of active ingredient. This makes aerosols expensive to use. In agricultural situations, fogging machines can be used to produce aerosols, by using thermal energy to vaporize the active ingredients. As stated above, for these to work successfully the active ingredient must obviously be heat-stable.

Although there are a significant number of synthetic insecticides that are vapour-active and have been, and continue to be used in pest management programmes, many of these compounds are not without problems. Indeed, many active ingredients continue to be withdrawn from the market as a result of tougher regulatory regimes. For example, in March 2009 the European Commission adopted a new approach to the registration of products for pest management, moving from a risk-based assessment to a hazard-based system. The upshot of this change is that many products currently used in pest management in Europe will no longer be available in the future. There is clearly a will at the political level to move towards the use of far more environmentally friendly products for pest control. One type of product that falls within this area comprises pheromones. These are discussed in the next section. Pheromones, of course, are extremely volatile organic molecules.

6.4 Pheromones

Pheromones are volatile chemicals that mediate intraspecific interactions within species.[35] The very first pheromone to be identified was the sex pheromone of the silkworm moth *Bombyx mori*. This identification of bombykol [(10*E*,12*Z*)-hexadecadienol (**24**), Figure 6.6] was undertaken in the late 1950s and involved the dissection of ca. 500 000 female moths. The first trials that involved assaying pheromones for pest control occurred in the mid-1960s. Since then, hundreds of other pheromones for over 1500 insect species have been identified[36] and there are now more than 50 available on a commercial basis.[37] In relation to arthropod pest control, pheromones have been developed that can be used in the control of Coleoptera, Diptera, Lepidoptera and Acarina. Compounds **25–61** in Figure 6.6 and Table 6.2 show the great diversity of pheromones that have been registered for use in pest management in the United States.[38] The figure also makes reference to volatile compounds that act as kairomones. These are discussed later in the chapter.

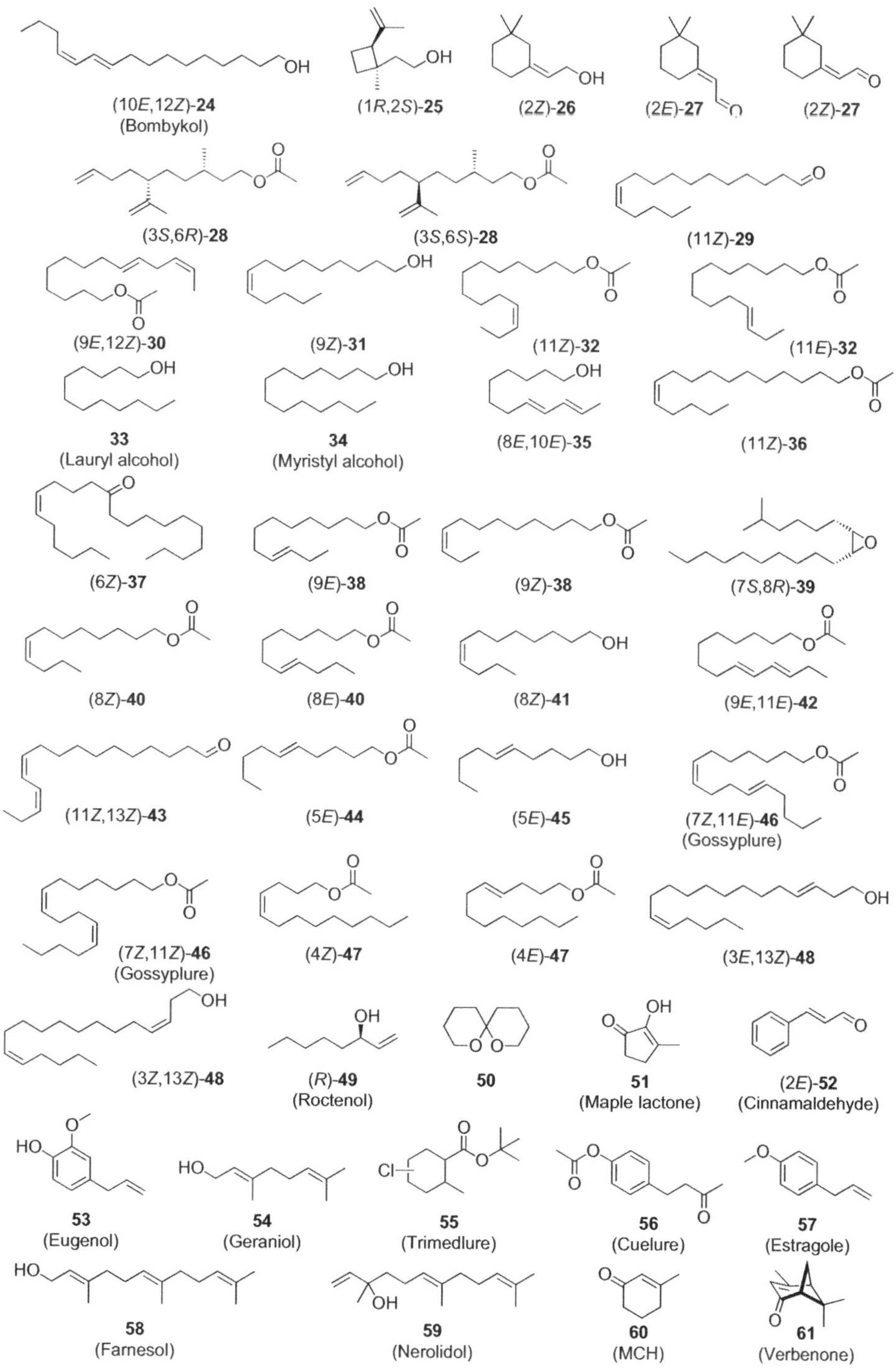

Figure 6.6 *Structures of selected pheromones and kairomones*

Table 6.2 *Selected pheromones and kairomones (volatile compounds) approved for use by the Environmental Protection Agency (EPA) in the United States*[38]

Active ingredients	Type of pheromone and use[a]
2-[(1*R*,2*S*)-1-Methyl-2-(1-methylethenyl) cyclobutyl] ethanol (**25**), (2*Z*)-(3,3-dimethylcyclohexylidene) ethanol (**26**), (2*E*)- and (2*Z*)-(3,3-dimethylcyclohexylidene) acetaldehyde (**27**)	Boll weevil aggregation pheromone. Used in lure and kill traps with the insecticide malathion.
(3*S*,6*R*)- and (3*S*,6*S*)-3-methyl-6-isopropenyl-9-decen-1-yl acetate (**28**)	California red scale sex attractant. Used for mating disruption.
(11*Z*)-Hexadecenal (**29**)	Artichoke plume moth sex pheromone.
(9*E*,12*Z*)-Tetradecadienyl acetate (**30**), (9*Z*)-tetradecen-1-ol (**31**)	Beet armyworm sex pheromone.
(11*Z*)-Tetradecenyl acetate (**32**)	Black headed fire worm moth sex pheromone and leaf rollers (various).
(11*E*)-Tetradecenyl acetate (**32**)	*Sparganothis* fruitworm moth sex pheromone.
(11*E*)- and (11*Z*)-tetradecenyl acetate (**32**)	Omnivorous leaf roller moth sex pheromone.
(11*Z*)-Tetradecenyl acetate (**32**), lauryl alcohol (**33**), myristyl alcohol (**34**), (8*E*,10*E*)-dodecadien-1-ol (**35**)	Codling moth sex pheromone, hickory shuckworm moth sex pheromone, oblique-banded leaf roller moth sex pheromone and pandemic leaf roller moth sex pheromone.
(11*Z*)-Hexadecenyl acetate (**36**)	Diamondback moth sex pheromone.
(6*Z*)-Heneicosen-11-one (**37**)	Douglas fir tussock moth sex pheromone.
(9*E*)-Dodecenyl acetate (**38**)	Eastern pine shoot borer moth sex pheromone.
(9*Z*)-Dodecenyl acetate (**38**)	Grapeberry moth sex pheromone.
(7*S*,8*R*)-Epoxy-2-methyloctadecane (**39**)	Gypsy moth sex pheromone.
(8*Z*)- and (8*E*)-dodecenyl acetate (**40**), (8*Z*)-dodecen-1-ol (**41**)	Koa seed worm moth sex pheromone and macadamia nut borer moth sex pheromone.
(9*E*,11*E*)-Tetradecadienyl acetate (**42**)	Light brown apple moth sex pheromone.
(11*Z*,13*Z*)-Hexadecadienal (**43**)	Navel orangeworm moth sex pheromone.
(8*Z*)- and (8*E*)-dodecenyl acetate (**40**), (8*Z*)-dodecen-1-ol (**41**), (5*E*)-decenyl acetate (**44**), (5*E*)-decen-1-ol (**45**)	Oriental fruit moth sex pheromone and peach twig borer moth sex pheromone.
(7*Z*,11*E*)- and (7*Z*,11*Z*)-hexadecadienyl acetate (gossyplure; **46**)	Pink bollworm moth sex pheromone.

(4*E*)- and (4*Z*)-tridecenyl acetate (**47**)	Tomato pinworm moth sex pheromone.
(3E,13Z)- and (3Z,13Z)-octadecadien-1-ol (**48**)	Western polar clearwing moth sex pheromone.
1-Octen-3-ol (octenol) and (*R*)-(–)-1-octen-3-ol (roctenol; **49**)	Used as a kairomone in traps to attract mosquitoes and other biting flies.
1,7-Dioxaspiro-[5,5]-undecane (**50**)	Olive fly sex pheromone.
2-Hydroxy-3-methyl-2-cyclopenten-1-one (**51**)	Kairomone attractant for cockroach management.
3-Phenylpropenal (cinnamaldehyde; **52**)	Kairomone attractant for corn rootworms.
2-Methoxy-4-(2-propen-1-yl)-phenol (Eugenol; **53**)	Kairomone attractant for Japanese beetles.
(*E*)-3,7-Dimethyl-2,6-octadien-1-ol (geraniol; **54**)	Kairomone attractant for Japanese beetles.
1,1-Dimethylethyl 4-(or 5-)chloro-2-methylcyclohexanecarboxylate (trimedlure; **55**)	Mediterranean fruit fly sex pheromone.
4-[4-(Acetyloxy)phenyl]-2-butanone (cuelure; **56**)	Melon fly sex pheromone.
4-Allyl anisole (estragole; **57**)	Bark beetle anti-attractant for use in forestry.
3,7,11-Trimethyl-2,6,10-dodecatriene-1-ol (farnesol; **58**), 3,7,11-trimethyl-1,6,10-dodecatriene-3-ol (nerolidol; **59**)	Spider mite sex pheromones that are used in combination with insecticides for mite control.
3-Methyl-2-cyclohexene-1-one (MCH; **60**)	Bark beetle anti-attractant for use in forestry.
4,6,6-Trimethylbicyclo[3.1.1]hept-3-en-2-one (Verbenone) (**61**)	Bark beetle anti-attractant for use in forestry.

Note: [a]Used for mating disruption and for lure and kill if not stated otherwise.

Pheromones are used by arthropods to locate mates, for dispersion, for trail formation, to signal alarm and for maturation.[39] Of these, the pheromones that are used for mate location have probably received the most attention in relation to pest control. The pheromones that fall into this category comprise both sex pheromones and aggregation pheromones. The best studied sex pheromones are those produced by Lepidoptera. In most cases it is females who release the pheromone in order to attract males. The pheromone disperses downwind and receptive males then respond via a process known as chemoanemotaxis (i.e. chemically mediated, wind-mediated movement). The concentration of pheromone required to elicit a response varies with circumstances and species. Laboratory studies have shown that just one molecule of pheromone may be enough to elicit a neurological response (a receptor to fire) in a receiving species while field experiments with bombykol (**24**, sex pheromone of *B. mori*) have shown that 200 molecules ml^{-1} air are enough to elicit a behavioural response in a downwind, receptive male. Pheromones are therefore active at very low concentrations. In pest control, these pheromones have been used extensively to monitor pest species in order to improve the timing of other control measures such as pesticide applications. Aggregation pheromones also bring individuals together to mate. However, they may also serve to overcome host defences. They have been most extensively used in pest control with various species of bark beetle. Alarm pheromones are produced in response to attack by predators. They therefore serve as a warning and generally elicit an escape response in receiving and responsive individuals.

Chemically, most pheromones are comprised of single or multiple, low molecular weight, volatile molecules. This makes sense because pheromones function to elicit specific responses that are time-limited. To date, most of the pheromones that have been chemically characterized are produced by lepidopterous pests. Typically, these pheromones are C_{10}–C_{21} unsaturated aliphatic alcohol, acetate, aldehyde or ketone derivatives. Most insects synthesize pheromones via specific biochemical pathways in which geometrical and optical isomerisms are strictly controlled. The result of this is that most insects only respond to specific isomers and many insects also require specific ratios of different compounds within a pheromone formulation. For example, the moths *Clepsis spectrana* and *Adoxophyes orana* have an identical sex pheromone that is composed of (11*Z*)- and (9*Z*)-tetradecenyl acetates (**32** and **62**, respectively, Figure 6.7). However, it is the precise ratio of these two components that determines which species will respond.

Because pheromones are volatile compounds specific formulations have been developed for their release. Collectively, these are called controlled release formulations. The aim is to control the release rate per se, the uniformity of the release rate and the period of time during which the pheromone is active. Where pheromones are blends of more than one compound, it is also important that the pheromone dispenser releases each compound in the correct

(11*Z*)-**32** (9*E*)-**62**

Figure 6.7 *Sex pheromone of* Clepsis spectrana *and* Adoxophyes orana*, both composed of (11Z)-***32*** and (9Z)-***62*** at different ratios*

proportions. The main formulations developed over the past 40 years are plastic fibres, laminate flakes, microcapsules and pheromone-impregnated rubber septa. If these specific formulations were not used then a pheromone applied across a crop would rapidly evaporate.

Pheromones can be used in a number of different ways in pest control. First, they can be used to monitor pest species numbers. Second, they can be used to mass-trap pest species. Third, they can be used to disrupt mating. Fourth, they can be used to attract species to a poisoned bait. This approach is known as 'lure and kill'. Finally, pheromones can be used to deter pest species from attacking host plants (i.e. as anti-attractants). Of these uses, by far the most successful application of pheromones has been in monitoring.

In monitoring systems the number of individuals trapped (usually males) is used as an indicator of pest presence. Such information is then used to guide the targeting of other control measures, such as a pesticide application. However, such data can also be useful for detecting changes in long-term population dynamics and/or in recording the spread of a novel pest imported from elsewhere. It is this approach that has been used to track the spread of the gypsy moth *Lymantria dispar* across North America.[40] Monitoring of pest species can also be used to assay the presence of exotic species. For example, such an approach was taken during 2002–2005 in a national survey in New Zealand to determine the presence of beetles that might pose a threat to the countries' trees and forests. The approach taken was to use traps baited with combinations of the volatiles α-pinene (**10**), β-pinene (**63**), ethanol (**64**), frontalin (**65**) and ipsdienol (**66**, Figure 6.8). The latter two volatiles are bark beetle aggregation pheromones, while the former are kairomones that beetles use to locate suitable host trees. This study trapped over 27 000 beetles and highlighted the usefulness of such an approach in detecting the presence of potentially inimical species.[41]

Mass trapping involves population reduction by collection of a large proportion of the pest population. Because sex pheromones generally only attract males this technique has been exploited most efficiently with aggregation pheromones that attract both sexes. The opposite approach to mass trapping is to use pheromones that deter further colonization of host plants. Within this context one of the most well studied compounds comprises the anti-attractant pheromone verbenone (**61**, Figure 6.6). This pheromone deters bark beetles from colonizing coniferous trees. This compound has been used both for individual tree protection and for area-wide bark beetle control over a number of years.[42]

Mating disruption involves swamping an area with a sex pheromone so that males are unable to locate females because of sensory adaptation (habituation) or competition. It is an approach that has been successfully commercialized for a number of major lepidopteran pest species, such as the pink bollworm *Pectinophora gossypiella*, the rice stem borer *Chilo*

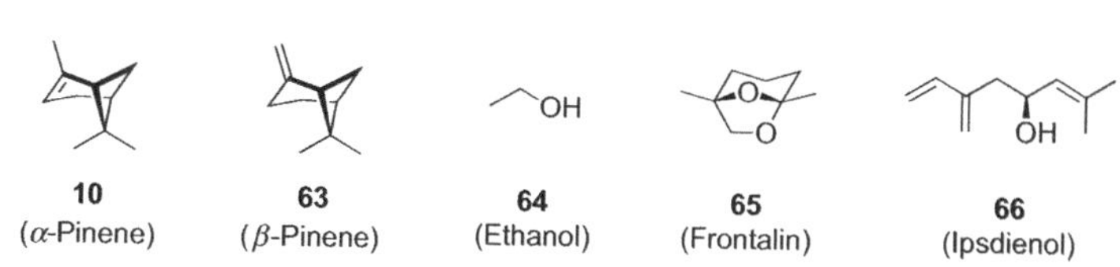

Figure 6.8 *Structures of volatiles used to determine the presence of beetles that might pose a threat to native trees and forests in New Zealand*

suppressalis, the codling moth *Cydia pomonella*, the gypsy moth *Lymantria dispar* and the diamondback moth *Plutella xylostella*.[43] In total, there are more than 30 different pheromone formulations that are currently marketed for the management of pest species via the process of mating disruption. In the case of the pink bollworm, the pheromone marketed as 'gossyplure' (**46**) has now been used for monitoring and mating disruption of this pest for more than 30 years.[44]

The 'lure and kill' approach involves attracting individuals to a source that is treated with a poison, usually a pesticide. The pest picks up a lethal dose and subsequently dies. This approach has been used successfully with a number of orchard pests, including the codling moth *Cydia pomonella*,[45] the light brown apple moth *Epiphyas postvittana*,[46] the olive fly *Bactrocera oleae*, the plum fruit moth *Cydia funebrana*, the tufted apple bud moth *Platynota idaeusalis* and the European elm bark beetle *Scolytis multistriatus*.[47]

In addition to the uses described above, attempts have also been made to exploit oviposition and alarm pheromones in pest control. For example, many pest species mark where they have laid eggs with a pheromone to deter other females of that species from laying eggs nearby. In theory, these pheromones could be used to reduce egg laying in a crop or to redirect egg laying elsewhere. For example, field trials with the oviposition-deterring pheromones of both the cherry fruit fly *Rhagoletis cerasi* and the apple maggot fly *Rhagoletis pomonella* indicated that these chemicals were effective in reducing egg laying on treated fruit. Alarm pheromones cause pest species to move. They are produced when pests become aware of predators. At present, there is one commercially available formulation: the spider mite *Tetranychus urticae* alarm pheromone. These pheromones are intended to be used as mixtures with pesticides. The aim is to get the pest species to increase its movement and so enhance its exposure to the pesticide. Increased movement is clearly of most relevance where contact pesticides are used. However, alarm pheromones can also increase the toxicity of systemic products, permitting reduced doses of pesticide to be applied. Trials with the aphid alarm pheromone farnesene (**67**, Figure 6.9) have shown that combinations of reduced-rate insecticide and pheromone are effective at controlling cereal aphids. This pheromone however is not yet commercially available.

One novel approach to the use of pheromones however that has been commercialized comprises the use of a predator aggregation pheromone. The idea is to improve the efficacy of a biological control agent. The aggregation pheromone of the spined soldier bug *Podisus maculiventris* is marketed in the United States as a product to enhance biological control in the home garden market.[48,49]

In addition to using volatile compounds that mediate intraspecific interactions (i.e. pheromones) there are also a number of volatile compounds that mediate interspecific interactions between species. These compounds and their use in pest management are discussed in the next two sections of this review.

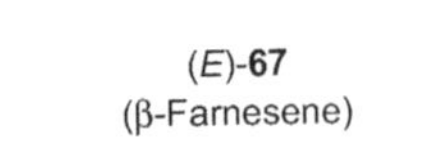

(*E*)-**67**
(β-Farnesene)

Figure 6.9 *Structure of aphid alarm pheromone farnesene (**67**)*

6.5 Volatile Allelochemicals

Chemicals that mediate interactions between different species are called allelochemicals. These interactions may benefit both equally, the emitting organism or the receiving organism. Technically, they are referred to as synomones, allomones and kairomones. Among these, it is kairomones that have received the most attention in relation to arthropod pest control. These are volatile chemicals that insects use as attractants (benefit to receiver). For example, recent trials with an attract and kill formulation based on host plant volatiles and an insecticide have looked at improving the management of the Colorado potato beetle *Leptinotarsa decemlineata*. In this work the volatile compounds (*S*)-(+)-linalool (**6**), (*Z*)-3-hexenyl acetate (**68**) and methyl salicylate (**69**, Figure 6.10) were used to attract adults and larvae to the pyrethroid insecticide permethrin. Overall the data indicated that substantial reductions in insecticide use may be possible, particularly in the management of early instars.[50]

It has been suggested that host kairomones could also be used to disrupt the location of food crops by pests. However, by far their greatest use has been in trapping programmes. For example, the essential oil constituents eugenol (**53**) and geraniol (**54**, Figure 6.6) have been used as lures to trap the Japanese beetle *Popilla japonica*, while methyl eugenol (**70**, Figure 6.10) is attractive to the oriental fruit fly *Dacus dorsalis*.[51]

At the agricultural level probably the single most extensive use of an attractant allelochemical comprises the use of 'trimedlure' (**55**) to attract the Mediterranean fruit fly *Ceratitis capitata*. This chemical compound is not a pheromone, but a synthetic attractant which can be used to lure insects either to sticky traps or to pesticide-treated traps. Baited traps were first developed in the 1970s and they are still in use in many countries today. The lure has proven particularly useful for monitoring pest presence within export/import crops at airports and seaports.

More recently, a number of chemicals that act as kairomones have been exploited to improve the efficacy of beneficial species. Examples include spraying crops with honeydew to attract lacewings[52] and the commercial development of 2-phenylethanol (**71**) as an attractant for a range of predatory species.[49,53]

One of the more novel uses of volatile kairomones to manage pest species that has been developed in the past decade comprises the development of machines called 'midge magnets'. Studies by Bhasin *et al.*[54] indicated that midges respond positively to the volatile compounds carbon dioxide (**72**, Figure 6.11), 1-octen-3-ol (**49**), acetone (**73**), butanone (**74**) and lactic acid (**75**) and that these volatiles are important in host location. Acetone and 1-octen-3-ol in particular appear to act synergistically. As of 2009, there are a variety of traps available on a commercial basis worldwide. All these traps use slightly different combina-

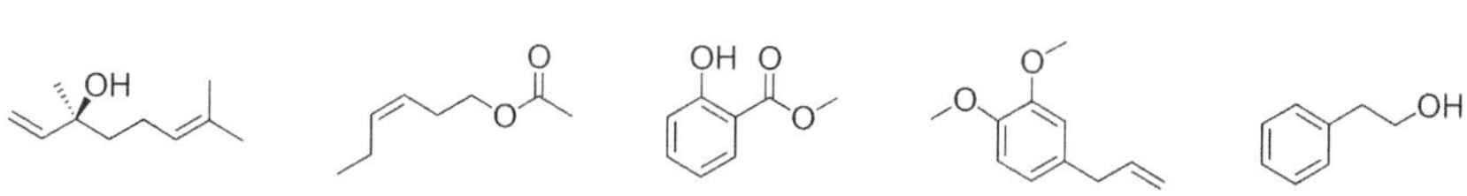

(*S*)-**6**
(Linalool)

(*Z*)-**68**

69

70
(Methyl eugenol)

71

Figure 6.10 *Volatile compounds used as attractants for different species*

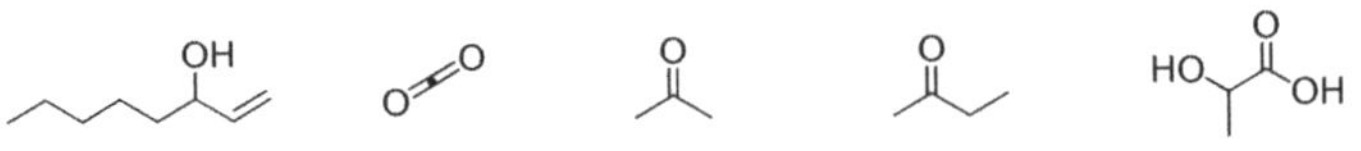

49 **72** **73** (Acetone) **74** (Butanone) **75** (Lactic acid)

Figure 6.11 *Volatile compounds used as midge attractants*

tions of the volatiles mentioned above to attract dipterous pest species – most often they are marketed for mosquito or midge control.

6.6 Plant Volatiles and Behavioural Modification of Beneficial Insects

It is known that many predators and parasitoids are attracted to host pests by pheromones and by volatiles released from damaged plants.[55,56] For example, bean plants attacked by phytophagous mites release volatiles to attract predatory mites while corn plants attacked by caterpillars and bean plants attacked by aphids both release compounds that are attractive to hymenopterous parasitoids.[57] More recently, it has been reported that such responses are not limited to invertebrate species. For example, the number of passerine birds feeding on caterpillars has been reported to be correlated with the emission of terpenoids from trees that were under attack.[5]

The volatile chemicals produced by plants when damaged have been called 'herbivore-induced plant volatiles' (HIPVs) and, at present, a great deal of research has therefore focussed on the manipulation of these volatiles to improve the efficacy of beneficial species within pest control.[49,58]

The production of HIPVs appears to be extremely complex. Compounds in the oral secretions of pest species appear to trigger the synthesis and release of HIPVs, as does egg deposition in some species and root feeding for others.[59] Furthermore, it is now clear that plants can perceive and respond to volatiles released by pest-infested neighbouring plants in a priming process.[60] One of the more important compounds in the process appears to be the plant hormone jasmonic acid (**76**, Figure 6.12),[61] although research indicates that this is not the only chemical involved in the priming process.[56] Studies at the molecular level are currently attempting to identify the genes involved in the priming process. Research with methyl jasmonate (**77**) and conifer trees has shown that this compound increases the activity of monoterpene synthetases, causing increased titres of terpenoid and phenolic defensive compounds.[62] As a result preliminary studies are being undertaken to investigate whether this defence-inducing plant hormone product could be used in combination with pheromones and kairomones to protect trees from attack by bark beetles.[63]

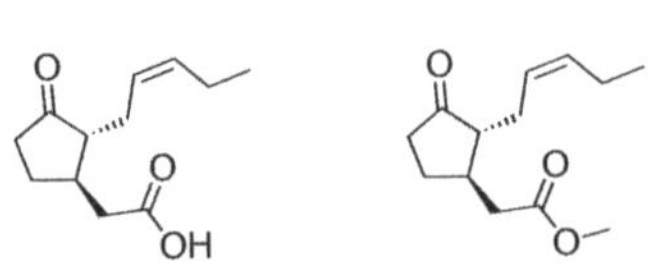

76 (Jasmonic acid) **77** (Methyl jasmonate)

Figure 6.12 *Structures of jasmonic acid (**76**) and methyl jasmonate (**77**)*

Chemically the HIPVs that attract predators and parasitoids include various terpenoids as well as products from the lipoxygenase, shikimate and tryptophan biochemical pathways. Two of the first synthetic HIPVs to be used in a field situation comprised (*Z*)-3-hexenyl acetate (**68**) and methyl salicylate (**69**).[64] These synthetic HIPVs attracted a range of predatory species, including predatory hemipteran, coleopteran, neuropteran and dipteran species.[65] Associated with this increased recruitment of predatory species were reductions in pest populations. Interestingly, predatory species do not appear to respond to these synthetic volatiles directly[65] and it has been suggested that these compounds act as elicitors that stimulate plants to produce further volatile compounds that then serve to recruit predators and parasitoids.[49]

Recent research has attempted to combine **68** and **69** with botanical insecticides. The idea is to use chemical products that target species directly in combination with biological control agents recruited by attraction to the HIPV. Evidence that such an approach may work is provided by Charleston *et al.*[66] and by Khan *et al.*[49] In the case of the latter, trials in hops and grapes in which plants were sprayed with botanical pesticides in combination with HIPVs clearly attracted greater populations of predatory and parasitic beneficial species.

Overall, the interactions that take place between host plants, pests and predatory species are complex and situation-specific.[67] For example, in laboratory studies Reddy reports that the predator *Chrysoperla carnea* responded positively to volatiles emitted by eggplant, okra and pepper infested with red spider mite.[68] However, such a behavioural response was not detected in tomato plants infested with the same pest species. The authors suggest that this may be associated with the impacts of trichomes and trichome exudates on the attractiveness of plants but this was not measured. Whatever the exact mechanism behind predator/parasitoid orientation to plants infested with pest species, it is clear that this is an area of chemical ecology that may be extremely fruitful in relation to the manipulation of volatile compounds for environmentally benign pest control strategies.

6.7 Concluding Comments

The aim of this review has been to highlight the extensive range of uses of volatile organic compounds in pest management programmes. Many of the applications discussed comprise commercially successful products (i.e. volatile synthetic insecticides, biorational pheromones). In contrast, other uses of volatile compounds are restricted to the manipulation of plant parts (and the associated volatiles) by people at the local level in many African, Asian and South American cultures. One of the greatest areas of research activity at present comprises developments in our understanding of plant phytochemistry and the manipulation of various volatiles for pest control purposes, whether it be to attract or deter pest species and/or their associated natural enemies. At the molecular level, studies are beginning to unravel the key genes involved and it is already the case that genetically engineered plants have been produced with altered abilities to express volatile compounds. One novel example of this comprises the production of genetically engineered crop plants that are able to produce pest species pheromones.[69] Such plants could be used as a source for pheromone extraction that may be economically superior to conventional pheromone production methods ('molecular farming') or they could be used directly in the management of pest species (e.g. if they were able to produce alarm pheromones).

The above said and given that arthropods are numerically the most successful group of species on our planet, it is highly likely that pest species would to be able to develop and evolve in response to anything we can use to suppress their populations. However, these new studies do point to novel ways to manage pest species while minimizing our reliance on more environmentally damaging products.

References

1. V. G. Dethier, L. B. Browne, C. N. Smith, The designation of chemicals in terms of the responses they elicit for insects, *J. Econ. Entomol.*, **53**, 134–136 (1960).
2. M. Brown, A. A. Hebert, Insect repellents: an overview, *J. Am. Acad. Dermatol.*, **36**, 243–249 (1997).
3. H. Wang, W.-F. Guo, P.-J. Zhang, Z.-Y. Wu, S.-S. Liu, Experience-induced habituation and preference towards non-host odours in ovipositing females of a moth, *J. Chem. Ecol.*, **34**, 330–338 (2008).
4. J. Laothawornkitkul, N. D. Paul, C. E. Vickers, M. Possell, J. E. Taylor, P. M. Mullineaux, C. N. Hewitt, Isoprene emissions influence herbivore feeding decisions, *Plant, Cell Environ.*, **31**, 1410–1415 (2008).
5. S. B. Unsicker, G. Kunert, J. Gershenzon, Protective perfumes: the role of vegetative volatiles in plant defence against herbivores, *Curr. Opin. Plant Biol.*, **12**, 479–485 (2009).
6. S. Licciardi, J. P. Herve, F. Darriet, J.-M. Hougard, V. Corbel, Lethal and behavioural effects of three synthetic repellents (DEET, IR3535 and KBR 3023) on *Aedes aegypti* mosquitoes in laboratory assays, *Med. Vet. Entomol.*, **20**, 288–293 (2006).
7. M. S. Fradin, Mosquitoes and mosquito repellents: a clinician's guide, *Ann. Intern. Med.* **128**, 931–940 (1998).
8. R. N'Guessan, M. Rowland, T.-L. Moumouni, N. B. Kesse, P. Carnevale, Evaluation of synthetic repellents on mosquito nets in experimental huts against insecticide-resistant *Anopheles gambiae* and *Culex quinquefasciatus* mosquitoes, *Trans. R. Soc. Trop. Med. Hyg.*, **100**, 1091–1097 (2006)
9. (*a*) Y. T. Qiu, R. C. Smallegange, J. J. A. van Loon, C. J. F. ter Braak, W. Takken, Interindividual variation in the attractiveness of human odours to the malaria mosquito *Anopheles gambiae*, *Med. Vet. Entomol.*, **20**, 280–287 (2006); (*b*) L. J. Zwiebel, W. Takken, Olfactory regulation of mosquito–host interactions, *Insect Biochem. Mol. Biol.*, **34**, 645–652 (2004).
10. W. Van der Goes van Naters, J. R. Carlson, Insects as chemosensors of humans and crops, *Nature*, **444**, 302–307 (2006).
11. T. Burfield, S. L. Reekie, Mosquitoes, malaria and essential oils, *Int. J. Aromather.*, **15**, 30–41 (2005).
12. K. Pålsson, T. G. T. Jaenson, Plant products used as mosquito repellents in Guinea Bissau, West Africa, *Acta Trop.*, **72**, 39–52 (1999).
13. (*a*) M. B. Isman, Botanical insecticides, deterrents and repellents in modern agriculture and an increasingly regulated world, *Annu. Rev. Entomol.*, **51**, 45–66 (2006); (*b*) D. A. Ukeh, M. A. Birkett, J. A. Pickett, A. S. Bowman, A. J. Mordue, Repellent activity of alligator pepper, *Aframomum melegueta,* and ginger, *Zingiber officinale,* against the maize weevil, *Sitophilus zeamais*, *Phytochemistry*, **70**, 751–758 (2009).
14. (*a*) F. Mansour, U. Ravid, E. Putievsky, Studies on the effects of essential oils isolated from 14 species of *Labiatae* on the carmine spider mite *Tetranychus cinnabarinus*, *Phytoparasitica*, **14**, 137–142 (1986); (*b*) S. N. Johnson, P. J. Gregory, Chemically-mediated host-plant location and selection by root-feeding insects, *Physiol. Entomol.*, **31**, 1–13 (2006).
15. C. Peterson, J. Coats, Insect repellents – past, present and future, *Pestic. Outl.*, **12**, 154–158 (2001).
16. R. Pavela, Insecticidal activity of some essential oils against larvae of *Spodoptera littoralis*, *Fitoterapia*, **76**, 691–696 (2005).
17. D. R. Batish, H. P. Singh, R. K. Kohli, S. Kaur, Eucalyptus essential oil as a natural pesticide, *For. Ecol. Manage.*, **256**, 2166–2174 (2008).

18. J. P. Santos, Alternatives to chemical control of stored product insects on small farms in the tropics, in *Proceedings of the Ninth International Working Conference on Stored Product Protection*, São Paulo, ABRAPOS, Brazil, 2006, pp. 663–673.
19. A. L. Tapondjou, C. Adler, D. A. Fontem, H. Bouda, C. Reichmuth, Bioactivities of cymol and essential oils of *Cupressus sempervirens* and *Eucalyptus saligna* against *Sitophilus zeamais* Motschulsky and *Tribolium confusum* du Val *J. Stored Prod. Res.*, **41**, 91–102 (2005).
20. A. J. Mordue, A. Blackwell, Azadirachtin: an update, *J. Insect Physiol.*, **39**, 903–924 (1993).
21. (*a*) C. Coria, W. Almiron, G. Valladeres, C. Carpinella, F. Luduena, M. Defago, S. Palacios, Larvicide and oviposition deterrent effects of fruit and leaf extracts from *Melia azedarach* on *Aedes aegypti* (L.) (Diptera: Culicidae), *Bioresour. Technol.*, **99**, 3066–3070 (2008); (*b*) S. M. Greenberg, A. T. Showler, T.-X. Liu, Effects of neem-based insecticides on beet armyworm (Lepidoptera: Noctuidae) *Insect Sci.*, **12**, 17–23 (2005); (*c*) H. Schmutterer, M. Doll, The marrango or Philippine neem tree, *Azadirachta excelsa*: a new source of insecticides with growth regulating properties, *Phytoparasitica*, **21**, 79–86 (1993); (*d*) M. U. Ventura, M. Ito, Antifeedant activity of *Melia azedarach* (L.) extracts to *Diabrotica speciosa* (Genn.) (Coleoptera: Chrysomelidae) beetles, *Braz. Arch. Biol. Technol.*, **43**, 215–219 (2000); (*e*) C. B. Wandscheer, J. E. Duque, M. A. N. Da Silva, Y. Fukuyama, J. L. Wohlke, J. Adelman, J. D. Fontana, Larvicidal action of ethanolic extracts from fruit endocarps of *Melia azedarach* and *Azadirachta indica* against the dengue mosquito *Aedes aegypti*, *Toxicon*, **44**, 829–835 (2004).
22. Pesticides Safety Directorate (UK): http://www.pesticides.gov.uk.
23. E. J. Bond, Manual of Fumigation for Insect Control – FAO Plant Production and Protection Paper 54, Food and Agriculture Organization of the United Nations, Rome, 1989.
24. D. K. Weaver, A. R. Petroff, *Pest Management for Grain Storage and Fumigation*, Montana Department of Agriculture, Montana, 2004.
25. D. J. Walker, G. Farrell (Eds), *Food Storage Manual*, Natural Resources Institute, University of Greenwich/World Food Programme, Chatham/Rome, 2003.
26. A. Dawson, D. G. Garthwaite, M. R. Thomas, *Pesticide Usage Survey Report* 192. Farm Grain Stores in Great Britain 2002/03, Department of Environment, Food and Rural Affairs and Scottish Executive Environment and Rural Affairs Department, London, 2003.
27. (*a*) C. H. Bell, Fumigation in the 21st century, *Crop Prot.*, **19**, 563–569 (2000); (*b*) UNEP, *Montreal protocol on substances that deplete the ozone layer*, 1994 Report of the Methyl Bromide Technical Options Committee, 1995 Assessment, UNEP, Nairobi, 1995, p. 304.
28. M. Q. Chaudhry, A review of the mechanisms involved in the action of phosphine as an insecticide and phosphine resistance in stored-product insects, *Pestic. Sci.*, **49**, 213–228 (1997).
29. E. J. Donahaye, Current status of non-residual control methods against stored product pests, *Crop Prot.*, **19**, 571–576 (2000).
30. M. Q. Chaudhry, Phosphine resistance, *Pestic. Outlook*, **11**, 88–91 (2000).
31. H. Benhalima, M. Q. Chaudhry, K. A. Mills, N. R. Price, Phosphine resistance in stored-product insects collected from various grain storage facilities in Morocco, *J. Stored Prod. Res.*, **40**, 241–249 (2004).
32. I. Lorini, A. Ferreira Filho, Integrated pest management strategies used in stored grain in Brazil to manage phosphine resistance, in *Proceedings of the International Conference on Controlled Atmosphere and Fumigation in Stored Products,* Gold-Coast Australia, 2004, pp. 293–300.
33. G. A. Herron, Resistance to grain protectants and phosphine in coleopterous pests of grain stored on farms in New South Wales, *Aust. J. Entomol.*, **29**, 183–189 (1990).
34. (*a*) F. H. Arthur, Grain protectants: current status and prospects for the future, *J. Stored Prod. Res.*, **32**, 293–302 (1996); (*b*) D. W. Hagstrum, C. Reed, P. Kenkel, Management of stored wheat insect pests in the USA, *Integr. Pest Manage. Rev.*, **4**, 127–142 (1999).
35. (*a*) P. Karlson, M. Lüscher, 'Pheromones': a new term for a class of biologically active substances, *Nature*, **183**, 55–56 (1959); (*b*) D. A. Nordlund, Semiochemicals: a review of terminology, in *Semiochemicals: Their Role in Pest Control*, John Wiley & Sons, Ltd, New York, 1981.
36. W. L. Roelofs, Chemistry of sex attraction, *Proc. Natl. Acad. Sci. USA*, **92**, 44–49 (1995).
37. L. G. Copping (Ed.), *Biopesticide Manual*, British Crop Protection Council Publications, 1998.
38. Full data set available at: http://www.epa.gov/pesticides/biopesticides/.

39. (*a*) T. D. Wyatt, *Pheromones and Animal Behaviour: Communication by Smell and Taste*, Cambridge University Press, Cambridge, 2003; (*b*) M. R. E. Symonds, M. A. Elgar, The evolution of pheromone diversity, *Trends Ecol. Evol.* **23**, 220–228, (2008).
40. A. A. Sharov, D. Leonard, A. M. Liebhold, E. A. Roberts, W. Dickerson, 'Slow the spread': a national program to contain the gypsy moth, *J. For.*, **100** (5), 30–36 (2002).
41. E. G. Brockerhoff, D. C. Jones, M. O. Kimberley, D. M. Suckling, T. Donaldson, Nationwide survey for invasive wood-boring and bark beetles (Coleoptera) using traps baited with pheromones and kairomones, *For. Ecol. Manage.*, **228**, 234–240 (2006).
42. (*a*) J. H. Borden, L. J. Chong, T. J. Earle, D. P. W. Huber, Protection of lodgepole pine from attack by the mountain pine beetle *Dendroctonus ponderosae* (Coleoptera: Scolytidae) using high doses of verbenone in combination with nonhost bark volatiles, *For. Chron.*, **79**, 685–691 (2003); (*b*) N. E. Gillette, N. Erbigin, J. N. Webster, L. Pederson, S. R. Mori, J. D. Stein, D. R. Owen, K. M. Bischel, D. L. Wood, Aerially applied verbenone-releasing laminated flakes protect *Pinus contorta* stands from attack by *Dendroctonus ponderosae* in California and Idaho, *For. Ecol. Manage.*, **257**, 1405–1412 (2009).
43. (*a*) R. T. Carde, A. K. Minks, Control of moth pests by mating disruption: successes and constraints, *Annu. Rev. Entomol.*, **40**, 559–585 (1995); (*b*) E. M. Hegazi, M. A. Konstantopoulou, P. Milonas, A. Herz, B. E. Mazomenos, W. E. Khafagi, A. Zaitun, S. M. Abdel-Rahman, I. Helal, S. El-Kemny, Mating disruption of the jasmine moth *Palpita unionalis* (Lepidoptera: Pyralidae) using a two pheromone component blend: a case study over three consecutive olive growing seasons in Egypt, *Crop Prot.*, **26**, 837–844 (2007).
44. (*a*) L. A. Bariola, J. C. Keller, D. L. Turley, J. R. Farris, Migration and population studies of the pink bollworm in the arid west, *Environ. Entomol.*, **2**, 205–208 (1973); (*b*) L. K. Gaston, R. S. Kaae, H. H. Shorey, D. Sellers, Controlling the pink bollworm by disrupting sex pheromone communication between adult moths, *Science*, **196**, 904–905 (1977).
45. P.-J. Charmillot, D. Hofer, D. Pasquier, Attract and kill: a new method for control of the codling moth *Cydia pomonella*, *Entomol. Exp. Appl.*, **94**, 211–216 (2000).
46. D. M. Suckling, E. G. Brockerhoff, Control of light brown apple moth (Lepidoptera: Tortricidae) using an attracticide, *J. Econ. Entomol.*, **92**, 367–372 (1999).
47. P. Howse, I. Stevens, O. Jones (Eds), *Insect Pheromones and their Use in Pest Management*, Chapman and Hall, London, 1998.
48. J. R. Aldrich, Predators, in *Pheromones of Non-Lepidopteran Insects Associated with Agricultural Plants*, J. Hardie, A. K. Minks (Eds), CABI Publishing, Wallingford, 1999, pp. 357–381.
49. Z. R. Khan, D. G. James, C. A. O. Midega, J. A. Pickett, Chemical ecology and conservation biological control, *Biol. Control*, **45**, 210–224 (2008).
50. J. W. Martel, A. R. Alford, J. C. Dickens, Evaluation of a novel host plant volatile-based attracticide for management of Colorado potato beetle, *Leptinotarsa decemlineata* (Say), *Crop Prot.*, **26**, 822–827 (2007).
51. R. I. Vargas, J. D. Stark, M. H. Kido, H. M. Ketter, L. C. Whitehand, Methyl eugenol and cue-lure traps for suppression of male oriental fruit flies and melon flies (Diptera: Tephritidae) in Hawaii: effects of lure mixtures and weathering, *J. Econ. Entomol.*, **93**, 81–87 (2000).
52. P. K. McEwan, M. A. Jervis, N. A. C. Kidd, Use of a sprayed L-tryptophan solution to concentrate numbers of the green lacewing, *Chrysoperla carnea* in olive tree canopy, *Entomol. Exp. Appl.*, **70**, 97–99 (1994).
53. J. Zhu, A. A. Cossé, J. J. Obrycki, K. S. Boo, T. C. Baker, Olfactory reactions of the twelve-spotted lady beetle, *Coleomegilla maculata* and the green lacewing, *Chrysoperla carnea* to semiochemicals released from their prey and host plant: electroantennogram and behavioral responses *J. Chem. Ecol.*, **25**, 1163–1177 (1999).
54. (*a*) A. Bhasin, A. J. Mordue (Luntz), W. Mordue, Responses of the biting midge *Culicoides impunctatus* to acetone, CO_2, and 1-octen-3-ol in a wind tunnel, *Med. Vet. Entomol.*, **14**, 300–307 (2000); (*b*) A. Bhasin, A. J. Mordue (Luntz), W. Mordue, Field studies on the efficacy of host odour baits for the biting midge *Culicoides impunctatus* in Scotland, *Med. Vet. Entomol.*, **15**, 147–156 (2001).

55. (*a*) D. W. Whitman, Allelochemical interactions among plants, herbivores and their predators, in *Novel Aspects of Insect–Plant Interactions*, P. Barbosa, D. K. Letourneau (Eds), John Wiley & Sons, Ltd, New York, 1988, pp. 11–64; (*b*) J. S. Thaler, Jasmonate-inducible plant defences cause increased parasitism of herbivores, *Nature*, **399**, 686–688 (1999); (*c*) G. V. P. Reddy, A. Guerrero, Interactions of insect pheromones and plant semiochemicals, *Trends Plant Sci.*, **9**, 253–261 (2004).
56. T. C. J. Turlings, J. Ton, Exploiting scents of distress: the prospect of manipulating herbivore-induced plant odours to enhance the control of agricultural pests, *Curr. Opin. Plant Biol.*, **9**, 421–427 (2006).
57. (*a*) M. W. Sabelis, H. E. Van de Baan, Location of distant spider mite colonies by phytoseiid predators: demonstration of specific kairomones emitted by *Tetranychus urticae* and *Panonychus ulmi* (Acari: Phytoseeidae, Tetranychidae), *Entomol. Exp. Appl.*, **33**, 303–314 (1983); (*b*) T. C. J. Turlings, F. L. Wäckers, L. E. M. Vet, W. J. Lewis, J. H. Tumlinson, Learning of host-finding cues by hymenopterous parasitoids, in *Insect Learning: Ecological and Evolutionary Perspectives*, D. R. Papaj, A. Lewis (Eds), Chapman and Hall, New York, 1993, pp. 51–78; (*c*) Y. Du, G. M. Poppy, W. Powell, J. A. Pickett, L. J. Wadhams, C. M. Woodcock, Identification of semiochemicals released during aphid feeding that attract the parasitoid *Aphidius ervi*, *J. Chem. Ecol.*, **24**, 1355–1368 (1998).
58. (*a*) R. Karban, I. T. Baldwin, *Induced Responses to Herbivory*, The University of Chicago Press, Chicago, 1997; (*b*) Y. Lou, X. Hua, T. C. J. Turlings, J. Cheng, X. Chen, G. Ye, Differences in induced volatile emissions among rice varieties result in differential attraction and parasitism of *Nilaparvata lugens* eggs by the parasitoid *Anagrus nilaparvatae* in the field, *J. Chem. Ecol.*, **32**, 2375–2387 (2006); (*c*) J. A. Pickett, M. A. Birkett, T. J. A. Bruce, K. Chamberlain, R. Gordon-Weeks, M. C. Matthes, J. A. Napier, L. E. Smart, C. M. Woodcock, Developments in aspects of ecological phytochemistry: the role of *cis*-jasmone in inducible defense systems in plants, *Phytochemistry*, **68**, 2937–2945 (2007).
59. (*a*) H. T. Alborn, T. C. J. Turlings, T. H. Jones, G. Stenhagen, J. H. Loughrin, J. H. Tumlinson, An elicitor of plant volatiles from beet armyworm oral secretion, *Science*, **276**, 945–949 (1997); (*b*) R. Halitschke, U. Schittko, G. Pohnert, W. Boland, I. T. Baldwin, Molecular interactions between the specialist herbivore *Manduca sexta* (Lepidoptera, Sphingidae) and its natural host *Nicotiniana attenuata*. III. Fatty acid–amino acid conjugates in herbivore oral secretions are necessary and sufficient for herbivore-specific plant responses, *Plant Physiol.*, **125**, 711–717 (2001); (*c*) P. W. Pare, J. H. Tumlinson, De Novo biosynthesis of volatiles induced by insect herbivory in cotton plants, *Plant Physiol.*, **114**, 1161–1167 (1997); (*d*) S. Rasmann, T. G. Köllner, J. Degenhardt, I. Hiltpold, S. Toepfer, U. Kuhlmann, J. Gershenzon, T. C. J. Turlings, Recruitment of entomopathogenic nematodes by insect-damaged maize roots, *Nature*, **434**, 732–737 (2005).
60. (*a*) I. T. Baldwin, R. Halitschke, A. Paschold, C. C. von Dahl, C. A. Preston, Volatile signaling in plant–plant interactions: 'talking trees' in the genomics era, *Science*, **311**, 812–815 (2006); (*b*) A. Kessler, R. Halitschke, C. Diezel, I. T. Baldwin, Priming of plant defense responses in nature by airborne signaling between *Artemesia tridentata* and *Nicotiana attenuata*, *Oecologia*, **148**, 280–292 (2006).
61. J. Engleberth, H. T. Alborn, E. A. Schmelz, J. H. Tumlinson, Airborne signals prime plants against insect herbivore attack, *Proc. Natl. Acad. Sci. USA*, **101**, 1781–1785 (2004).
62. D. P. W. Huber, R. N. Phillippe, L. L. Madilao, R. N. Sturrock, J. Bohlmann, Changes in anatomy and terpene chemistry in roots of Douglas-fir seedlings following treatement with methyl jasmonate, *Tree Physiol.*, **25**, 1075–1083 (2005).
63. A. D. Graves, E. H. Holsten, M. E. Ascerno, K. P. Zogas, J. S. Hard, D. P. W. Huber, R. A. Blanchette, S. J. Seybold, Protection of spruce from colonization by the bark beetle, *Ips perturbatus*, in Alaska, *For. Ecol. Manage.*, **256**, 1825–1839 (2008).
64. D. G. James, Field evaluation of herbivore-induced plant volatiles as attractants for beneficial insects: methyl salicylate and the green lacewing *Chrysopa nigricornis*, *J. Chem. Ecol.*, **29**, 1601–1609 (2003).
65. D. G. James, Further field evaluation of synthetic herbivore-induced plant volatiles as attractants for beneficial insects, *J. Chem. Ecol.*, **31**, 481–495 (2005).

66. D. S. Charleston, R. Gols, K. A. Hordijk, R. Kfir, L. E. M. Vet, M. Dicke, Impact of botanical pesticides derived from *Melia azedarach* and *Azadirachta indica* plants on the emission of volatiles that attract parasitoids of the diamondback moth to cabbage plants, *J. Chem. Ecol.*, **32**, 325–349 (2006).
67. M. Dicke, M. W. Sabelis, How plants obtain predatory mites as bodyguards, *Neth. J. Zool.*, **38**, 148–165 (1987).
68. G. V. P. Reddy, Plant volatiles mediate orientation and plant preference by the preadtor *Chrysoperla carnea* Stephens (Neuroptera: Chysopidae), *Biol. Control*, **25**, 49–55 (2002).
69. P. Nešněrová, P. Šebek, T. Macek, A. Svatoš, First semi-synthetic preparation of sex pheromones, *Green Chem.*, **6**, 305–307 (2004).

7

Challenges in the Synthesis of Natural and Non-Natural Volatiles

Anthony A. Birkbeck

Firmenich SA, Division Recherche et Développement, 1 Route des Jeunes, B. P. 239, 1211 Genève 8, Switzerland

7.1 Introduction – The Art of Organic Synthesis

Since the birth of synthetic organic chemistry, organic chemists have been intrigued by nature's ingenuity, evident in the wealth of interesting structural motifs found in natural products and have strived to replicate many of nature's most beautiful creations. In this quest, the synthesis of natural and non-natural volatile molecules presents a fascinating set of challenges for the fragrance chemist. Following the isolation and structural elucidation of a natural product, teams of organic chemists around the world start postulating synthetic approaches with the aim of confirming the often complicated structures, developing novel, exciting synthetic methodologies, or discovering new chemical reactions. The constant improvement in the selectivity of reagents, catalysts and organic synthesis techniques since the first synthesis of vitamin B12 in the 1970s by Woodward and Eschenmoser,[1] to more recent successes such as palytoxin by Kishi,[2] or azadirachtin by Ley,[3] has meant that the successful construction of extremely complex natural product skeletons is now merely a matter of time and resources.[4] The tremendous advances in spectrometers and purification techniques have also meant that structural elucidation of complex intermediates in pure form is now routine, which enables the unambiguous synthesis of larger, more complicated structures to be tackled by organic chemists. Nature herself is a formidable chemist, employing reactions that display exquisite selectivity under very mild conditions and whilst nature's chemistry may sometimes suffer from a finite number of biosynthetic pathways, this does not in any way limit the structural diversity present in natural products.[5]

The Chemistry and Biology of Volatiles Edited by Andreas Herrmann

This chapter concentrates on those volatile organic molecules, between six and 18 carbon atoms in size, which have a sufficient vapour pressure at ambient temperature and pressure to ensure the quantity of molecules in the vapour phase necessary to elicit an olfactory system response. The chapter does not treat inorganic volatiles like ammonia despite the considerable challenges in their synthesis. For evidence of the stimulation of the mammalian olfactory system by nonvolatile peptides and proteins in aerosol form the reader is directed to relevant publications.[6] Taste too, involves both volatile molecules via retronasal sampling and nonvolatile molecules in aqueous solution. Smell and taste are thus intimately linked. The focus of this chapter, however, is on volatile fragrance molecules rather than volatile taste molecules, although by virtue of their interrelatedness, some overlap may occur. Described below are two of the landmark syntheses of volatile natural products from more than 30 years ago, which illustrate the state of the art at the time and have served as inspiration for the next generation of fragrance chemists.

7.2 Overcoming Challenges in the Small-Scale Synthesis of Natural Volatile Compounds

7.2.1 D,L-Caryophyllene (1964)

In their synthesis of D,L-caryophyllene (**1**), a volatile sesquiterpene isolated from clove oil, Corey and coworkers employed state of the art synthetic methodology and an ingenious retrosynthetic route to overcome the considerable challenges.[7] The ring junction stereochemistry, the strained four-membered ring and the non-trivial nine-membered ring with the correct double bond geometry were all achieved by careful choice of the synthetic route (Scheme 7.1).[8] The four-membered ring was constructed in one step using a [2 + 2] photochemically mediated reaction between cyclohexenone and excess isobutene which gave the desired regiochemistry in the *cis*-fused cyclohexanone **2**. Although the *trans* ring junction is present in the natural product this was corrected in the penultimate step of the synthesis. Functional group manipulation proceeded in selective fashion to install the carbon skeleton ideally setup for the key fragmentation reaction to transform the fused 4,6,5-membered rings into the desired 4,9-fused system.[8]

Regioselective introduction of the carboxymethyl and methyl substituents followed by the nucleophilic addition of the protected propargylic aldehyde derivative gave compound **3** which had installed all of the carbon atoms present in the natural product (Scheme 7.1). Exhaustive hydrogenation, deprotection, cyclisation and oxidation gave the spirolactone **4**. Base mediated lactone opening followed by Dieckmann ester condensation gave the required ketoester **5** which, after hydrolysis and decarboxylation, furnished the ketone **6**. Reduction of **6** gave the secondary alcohol as a mixture of epimers, selective activation of which gave the required fragmentation precursor **7**. Grob fragmentation of hydroxy tosylate **7** cleaved the fused five- and six-membered rings to form the desired nine-membered ring and installed the *trans* double bond geometry.[8] Although the *cis* ring junction present in the starting material could be preserved, prolonged exposure to the basic reaction conditions effectively epimerised the α-hydrogen giving the *trans* junction present in the natural product. The conversion of the ketone via a Wittig reaction into the required exo-methylene group was straightforward and yielded D,L-caryophyllene (**1**). The relative

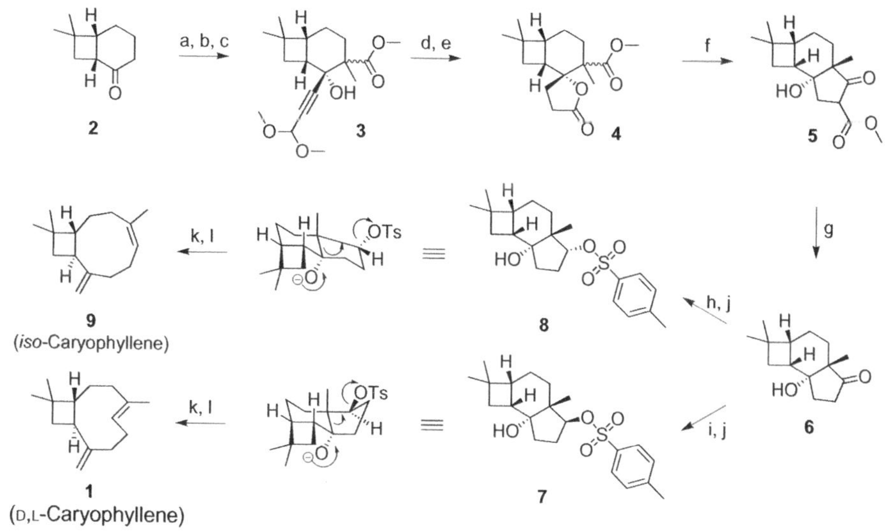

Scheme 7.1 *Corey synthesis of D,L-caryophyllene (**1**). Reagents and conditions: (a) NaH, dimethyl carbonate; (b) methylation; (c) propargyl aldehyde dimethyl acetal, base; (d) H_2, Pd/C (cat.); (e) chromic acid; (f) NaH, DMSO; (g) aqueous base, then pyridine ΔH; (h) $LiAlH_4$; (i) H_2, Raney-Nickel; (j) Tosyl chloride, pyridine; (k) NaH, DMSO, tert-butanol; (l) NaH, DMSO, triphenylphosphonium bromide*

stereochemistry of the alcohols, from the reduction of ketone **6**, determined the double bond geometry present in the nine-membered ring and similar treatment of the epimeric hydroxyl tosylate **8** furnished *iso*-caryophyllene (**9**).

7.2.2 β-Vetivone (1973)

Nearly a decade later, a concise stereoselective synthesis of β-vetivone (**10**), a key odour compound found in vetiver oil, was reported by Stork and coworkers (Scheme 7.2).[9] Applying his newly developed alkylation of kinetic enolates, this synthesis elegantly installed the extremely challenging quaternary spirocentre, relative to the methyl group, in a stereoselective way. This was achieved by the stepwise dialkylation of the cyclohexenone **11** relying on the greater reactivity of the allylic chloride moiety in compound **12** relative to the alkyl chloride portion to set up the correct relative stereochemistry between

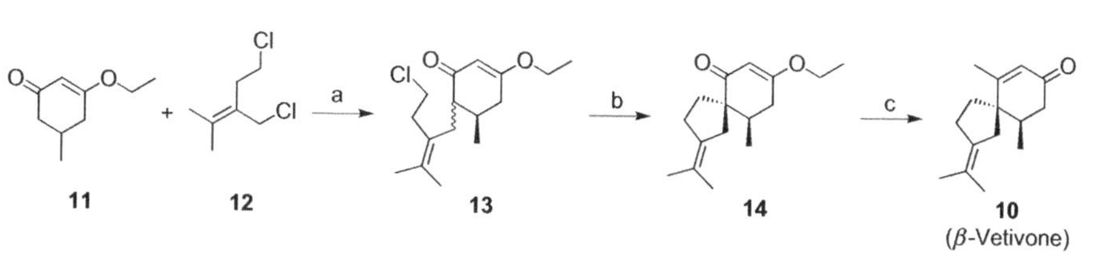

Scheme 7.2 *Stork synthesis of β-vetivone (**10**). Reagents and conditions: (a) LDA, HMPA, THF, –78 °C; (b) LDA, –78 °C, then room temperature (r.t.); (c) methyl lithium, 0 °C, then 1 N HCl, r.t., 60%*

the spirocyclic carbon and the adjacent methyl group during the second alkylation step of enone **13**.

The challenging exocyclic double bond in the five-membered ring was then already installed in intermediate enone **14**. Addition of methyl lithium to **14** followed by acidic enol ether deprotection and subsequent β-elimination furnished β-vetivone (**10**) in good overall yield.

These two landmark syntheses were capable of producing from milligrams to multigram quantities of material. The large-scale manufacture of either D,L-caryophyllene (**1**) or β-vetivone (**10**), employing the published synthetic routes, would however be highly unlikely due to the costs involved, the reagents used and the quantity of waste generated per kilogramme of product. The next four examples will examine the synthesis and manufacture of three natural volatile molecules, followed by one non-natural volatile molecule, displaying ever increasing complexity and posing ever increasing synthetic challenges.

7.3 Overcoming Challenges in the Large-Scale Synthesis of Nature Identical and Non-Natural Molecules

7.3.1 (Z)-3-Hexenol

The odour of (Z)-3-hexenol (**15**) is highly reminiscent of freshly cut grass and is released via the rapid enzymatic breakdown of unsaturated fatty acids in many flowers and fruits. The original synthesis of **15** by Bedoukian, in 1963, employed 1-butyne (**16**), a readily available petrochemical feedstock (Scheme 7.3).[10] Deprotonation with lithium amide in liquid ammonia followed by alkylation with ethylene oxide gave alkyne **17**. Selective reduction of **17** to the desired (Z)-3-hexenol (**15**) was achieved via hydrogenation over a Lindlar catalyst.[11]

The exponential advances in the catalytic technologies and the discovery of new reactions have meant that molecules essential to the perfumer's palette are increasingly being produced from nonpetrochemical feedstocks using cleaner and less waste-generating processes. The trimerisation of acetaldehyde (**18**) to give sorbic acid (**19**) and sorbol (**20**) derivatives is readily achieved under mildly basic conditions (Scheme 7.4). In pioneering work by Driessen-Hölscher and coworkers, the mild 1,4-hydrogenation of either acid **19** or alcohol **20** over a ruthenium cyclopentadienyl (Cp*) catalyst gave (Z)-3-hexenoic acid (**21**) in excellent yield and with excellent double bond isomer selectivities (*Z*/*E* > 98:2) or in the case of sorbol (**20**), 1,4-hydrogenation gave (Z)-3-hexenol (**15**) directly.[12] Reduction of the carboxylic acid functionality in the presence of the (Z) double bond in **21** could be readily achieved with a suitable hydride reducing agent. This synthesis of (Z)-3-hexenol (**15**) may soon become competitive on an industrial scale.[13]

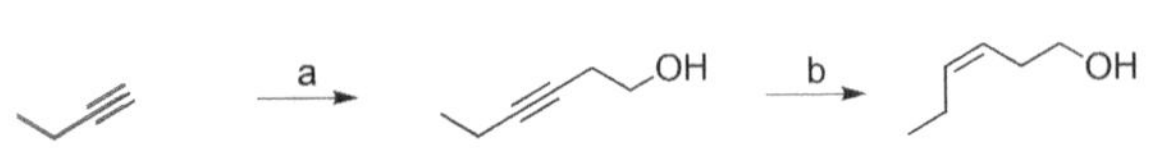

16 17 15

Scheme 7.3 *Synthesis of (Z)-3-hexenol (**15**). Reagents and conditions: (a) Li, NH_3 (l), ethylene oxide; (b) H_2, Lindlar cat.*

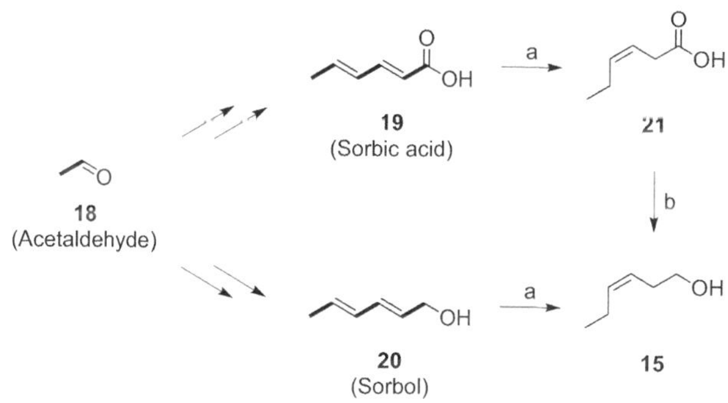

Scheme 7.4 *Selective 1,4-hydrogenation giving (Z)-3-hexenol (**15**). Reagents and conditions: (a) H_2, Ru Cp* cat.; (b) PMHS, Zn cat.*

7.3.2 Citral

Citral (**22**) is another large volume fragrance ingredient produced by state of the art catalytic technology, and BASF's continuous manufacture of **22** employs two consecutive [3,3]-sigmatropic rearrangements. The synthetic viability of the pivotal tandem [3,3]-Claisen–Cope rearrangement was amply demonstrated by Thomas of Firmenich in 1967 in his synthesis of α-sinensal (**23**, Scheme 7.5).[14]

Citral is the common name given to the mixture of geranial (**22a**) and its regioisomer neral (**22b**), which gives the characteristic zest odour to lemon. The BASF continuous synthesis of citral starts from isobutylene (**24**) and is widely regarded as the epitome of atom economy and engineering efficiency.[15] The key advantage of the BASF continuous process is the convergent manufacturing route. A Prins reaction between formaldehyde and isobutylene (**24**) yields isoprenol (**25**; Scheme 7.6). Isomerisation of **25** to prenol (**26**) and subsequent allylic oxidation yields prenal (**27**). Thus, the two five-carbon precursors are synthesised from the same feedstock. Combining prenol (**26**) and prenal (**27**) under acidic conditions to form the transient mixed acetal sets the chemistry in motion, elimination and recycling

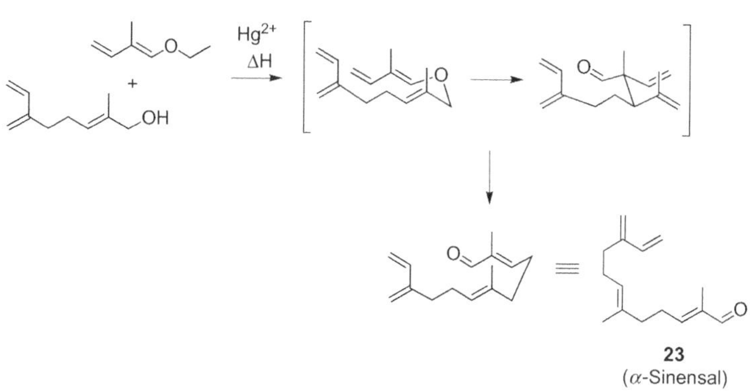

Scheme 7.5 *Synthesis of α-sinensal (**23**) via tandem [3,3]-Claisen then [3,3]-Cope rearrangement*

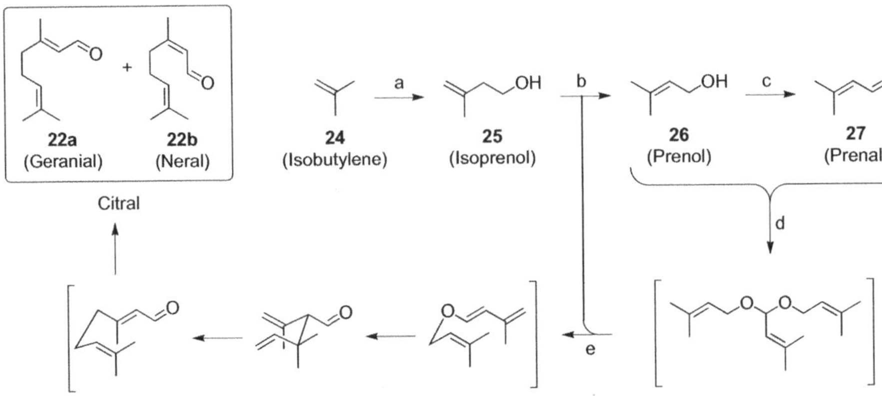

Scheme 7.6 *BASF continuous citral manufacture. Reagents and conditions: (a) formaldehyde, Prins reaction; (b) isomerisation; (c) Ag/SiO$_2$/O$_2$, 300–600 °C, 0.001 s^{-1}; (d) H$^+$, heat; (e) 100–200 °C, recycling of* ***26*** *by distillation*

of **26** followed by [3,3]-sigmatropic Claisen rearrangement gives an intermediate which then undergoes a further [3,3]-sigmatropic Cope rearrangement to yield citral (**22**). The fact that citral (**22**) itself is not stable under acidic conditions makes this synthesis even more remarkable.

With multithousand tonne volumes currently being manufactured, several challenges were elegantly overcome by clever engineering and an innovative synthesis. Citral (**22**) benefits from the advantage of being a key starting material for the synthesis of both large volume vitamins and ionones used in the fragrance industry. Whilst more than 1000 t year^{-1} of citral itself is consumed by the flavour and fragrance industry, approximately 40 times that quantity is consumed in the manufacture of vitamin E.[16] Having the necessary infrastructure in place and this quantity of citral available, BASF have subsequently extended the use of citral as a starting material towards the manufacture of several closely related terpenes. The enzymatic conjugate reduction of citral (**22**), yields optically active (*R*)-(+)-citronellal (**28**), which is the principal odour vector for lemongrass and a key precursor in the synthesis of (1*R*,2*S*,5*R*)-(–)-menthol (**29**, see below).[17] Alternatively, selective 1,4-hydrogenation of citral (**22**) yields citronellal (**28**) in racemic form.[18] Additionally the 1,2-reduction of citral (**22**) yields geraniol (**30**), which is an industrially important rose smelling alcohol ingredient in the perfumer's palette.[19] This alcohol (**30**) then becomes the starting material for a recently reported allylic rearrangement yielding another appreciated floral alcohol linalool (**31**; Scheme 7.7).[20] This transformation patented by BASF is the intriguing rearrangement of geraniol (**30**) into linalool (**31**) via the reactive distillation over a tungsten catalyst, which appears to be counter intuitive in the sense that primary allylic alcohols are transformed into a tertiary allylic alcohol (**31**; linalool). Linalool (**31**) is important in both lavender and bergamot reconstructions and is most associated with fresh and floral facets. The competitive strength that BASF have developed in their continuous citral process has now been further exploited in the synthesis of closely related terpene alcohols and aldehydes.

Both (*Z*)-3-hexenol (**15**) and citral (**22**) are challenging molecules to synthesise on industrial scale, and both contain double bonds able to display two isomeric forms (*E*/*Z* or

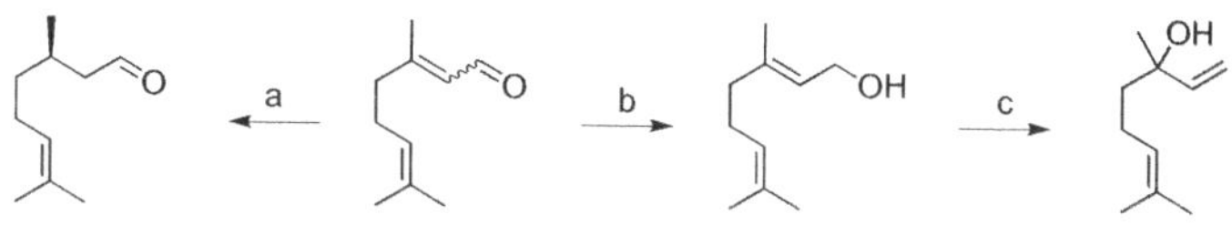

(*R*)-**28** (Citronellal) | **22** (Citral) | **30** (Geraniol) | **31** (Linalool)

Scheme 7.7 *Terpenoid synthesis from citral (**22**). Reagents and conditions: (a) enzymatic reduction; (b) H_2, cat.; (c) reactive distillation over tungsten cat.*

cis/*trans*) and different odour properties. In the next example we examine how the increased challenge, present in the form of stereoisomeric control, is overcome.

7.3.3 (–)-Menthol

(1*R*,2*S*,5*R*)-(–)-Menthol (**29**) gives the characteristic mint odour and cooling effect to peppermint oil (*Mentha piperita* sp.). Both the cooling effect and the odour are dependent on the absolute and relative configuration of the three stereocentres. Thus, for a feasible industrial synthesis the near complete control of both relative and absolute stereocentres is essential. Fortunately in the case of (1*R*,2*S*,5*R*)-menthol (**29**), once the absolute configuration of methyl group in (*R*)-citronellal (**28**) has been established, the Prins cyclisation occurs in highly diastereoselective fashion giving all of the substituents in the equatorial position around the cyclohexyl ring. The industrial synthesis by Takasago starts from myrcene (**32**) and employs the asymmetric isomerisation of the achiral allylic amine **33** to the enamine **34** (96–99% enantiomeric excess; *ee*) that has been performed on a 9 t scale {turnover number of the ruthenium 2,2′-bis(diphenylphosphino)-1,1′-binaphthyl [(*S*)-BINAP] catalyst approx. 8000, which with catalyst recycling reaches 400 000}.[21] Enamine **34** is subsequently hydrolyzed under acidic conditions to yield (*R*)-citronellal (**28**; 96–99% *ee*, Scheme 7.8).[21,22] Prins cyclisation of **28** under catalytic Lewis acidic conditions gives (1*R*,2*S*,5*R*)-(–)-isopulegol (**35**) in excellent stereoselectivity which, upon hydrogenation, yields **29** in excellent overall yield. This highly efficient catalytic isomerisation process results in the manufacture of 1000 t year^{-1} of (1*R*,2*S*,5*R*)-menthol (**29**), 1100 t year^{-1} of the precursor (1*R*,2*S*,5*R*)-isopulegol (**35**) and 1500 t year^{-1} of (*R*)-citronellal (**28**).[21]

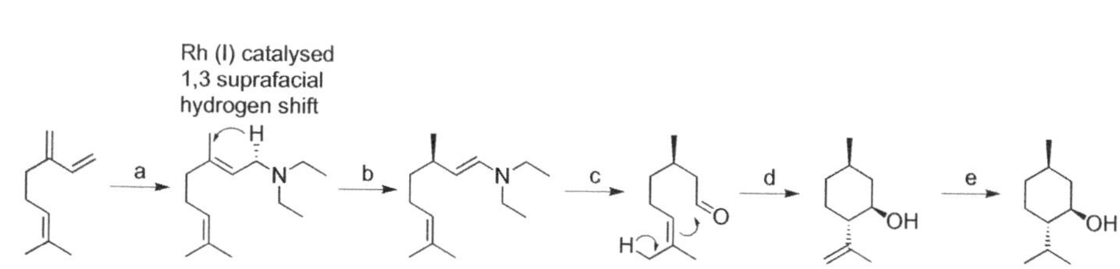

32 (Myrcene) | **33** | **34** | (*R*)-**28** (Citronellal) | (1*R*,2*S*,5*R*)-**35** (Isopulegol) | (1*R*,2*S*,5*R*)-**29** (Menthol)

Scheme 7.8 *Takasago (–)-menthol (**29**) process. Reagents and conditions: (a) lithium diethylamide; (b) (S)-BINAP Rh (I) cat., 96–99% ee; (c) H^+, H_2O; (d) $ZnBr_2$ cat.; (e) H_2, Raney-Nickel cat.*

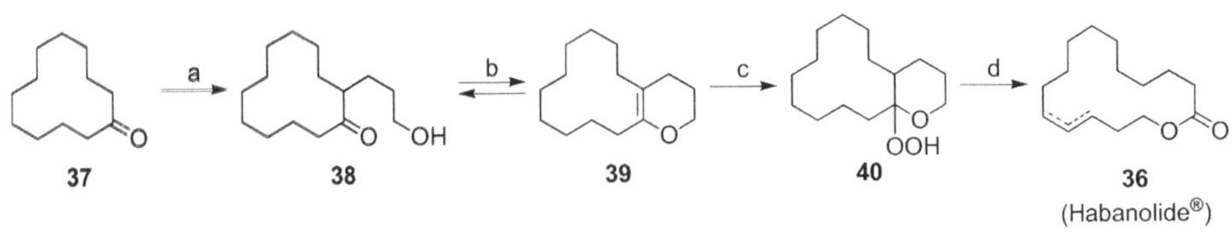

Scheme 7.9 *Industrial manufacture of habanolide (36). Reagents and conditions: (a) allyl alcohol, DTBP, ΔH; (b) H^+; (c) H_2O_2, acetic acid; (d) ΔH, Fe/Cu cat.*

Many non-natural menthol analogues are highly sought after as non-odorous physiological coolants.[23]

7.3.4 Habanolide

So far we have looked at examples of natural volatile molecules. Habanolide (**36**, Scheme 7.9) from Firmenich is a highly appreciated unsaturated macrocyclic musk lactone manufactured on a large scale that has not yet been found in nature.[24] The corresponding saturated analogue, however, has been isolated from Angelica root oil.[25] The challenge of constructing the entropically disfavoured 15-membered macrocyclic ring was overcome via the addition of a unit containing three carbon atoms to cyclododecanone (**37**), followed by cyclisation and ring expansion. The radical addition of cyclododecanone (**37**) to allyl alcohol yielded hydroxyl ketone **38**, which upon acidic treatment underwent ring closure and dehydration to afford dihydropyran **39**. Oxidation of the dihydropyran with acidic hydrogen peroxide gave the intermediate hydroperoxide **40**, which upon pyrolysis over a suitable mixed iron/copper catalyst underwent ring enlargement, producing the musk smelling unsaturated lactone **36** as a mixture of double bond isomers.

In the above four examples we have seen how, by overcoming considerable challenges, the efficient manufacture of natural or a non-natural volatile molecules has been achieved. These syntheses vary from two to five synthetic steps and each product costs less than 50 $ kg^{-1}, which is a tribute to the implementation of constant process improvements. In each case, the molecular structure was known, as was the financial and olfactory value of the odorant. The manufacture of such successful novel fragrance molecules is often undertaken by third parties following the expiry of the patent protecting them.

What if the structure and value of a fragrance molecule was not yet known, either because it had not been discovered or because it was a natural molecule of known structure but had not yet been synthesised via an industrially feasible route?

7.4 Remaining Challenges in the Large-Scale Synthesis of Natural and Non-Natural Volatiles

The exquisite sensitivity and power of discrimination present in the human olfactory system helps explain why no two identically smelling odorants have yet been discovered. The implication of this is that new molecules may display improved molecular properties or offer other advantages but their odour will always be non-identical to highly appreciated fragrance ingredients. The design of such novel fragrance molecules could typically follow two interrelated pathways.

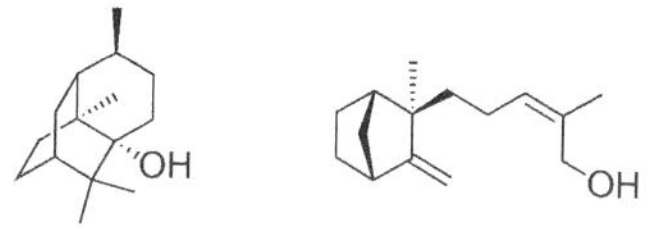

41 (Patchoulol) (*Z*)-**42** (*β*-Santalol)

Figure 7.1 *Structures of patchoulol (**41**) and β-santalol (**42**)*

First, one could envisage employing a readily available starting material, applying appropriate industrial catalytic technologies to make interesting structures in the hope that the odours will be sufficiently interesting to perfumers, the molecules will show good cost/performance, a favourable safety profile and the cost of production will allow the manufacture of such a molecule. This strategy would apply mainly to non-natural volatile molecules and could be seen as the industrial route already in place awaiting an appreciated odorant to be discovered.

An alternative strategy would be to design, then synthesise fragrance molecules with highly appreciated odours and favourable molecular properties, and then try to develop industrially feasible chemical reactions for their manufacture at an acceptable cost. In such cases, the molecule and odour could be described as already defined, but awaiting a manufacturing route. As we see in several examples below, this can apply to both natural, for example patchouli alcohol (patchoulol, **41**), *β*-santalol (**42**; Figure 7.1), or non-natural fragrance molecules (e.g. musks).

Every year, significant effort is dedicated to constantly improving catalytic methodologies and increasing the palette of reactions that can be utilised on an industrial scale. Despite this, many thousands of odorant molecules are synthesised every year and only a very small fraction of these are deemed worthy additions to the perfumer's palette and subsequently commercialised. Not surprisingly, tremendous effort has been dedicated to trying to correlate structure with odour and yet most attempts towards the accurate prediction of odour from molecular structure have met with disappointment.[26] Novel odorants of the future need to be designed, synthesised and evaluated, then undergo safety testing and be manufactured using a safe, cost-efficient and environmentally acceptable process, each of which poses interesting challenges.

In the initial phase of designing future novel non-natural volatile molecules as perfumery ingredients, chemists face the question of whether greater understanding of the olfactory system is an advantage. The interaction of volatile molecules with the olfactory system is extremely complex. Both the immune system and the olfactory system rely on a huge number of possible combinations of receptors to recognise and interact with both self and nonself molecules, and both play an essential role in the detection of molecules present in the environment surrounding the organism. We know, for instance, that the odour of an individual is related to their major histocompatibility complex (MHC) and that this cluster of genes plays a pivotal role in the immune system and has implications in the choice of mate.[27] The human olfactory system is able to discern odours from a wide variety of functional groups and molecular shapes. The environment immediately surrounding the binding domain in odour receptors or odour binding proteins is made up of L-amino acids linked via peptide bonds; thus, the binding sites are chiral environments and are capable of interacting differently with enantiomers and diastereoisomers that may be present in odorants.[28,29] We know that humans perceive odour in a combinatorial way; that is, an odorant molecule binds to multiple odorant receptors and odorant receptors bind to multiple odorants.[30] It is the electrical signal resulting from this pattern of binding interactions that is interpreted by the brain as an odour. Sandalwood may be an exception to this, as the structural prerequisites for sandalwood odour appear to be more specific.[31]

The sense of smell is able to discern functional groups in fragrance molecules. In addition, shape can be discerned via interaction with the chiral peptide backbone of the olfactive receptor (OR) protein. Indeed, systematic mutation studies have provided evidence of specificity being achieved by a single amino acid residue.[32] If we can discern the functional groups in fragrance molecules and we can discriminate shapes, where should we start when attempting to design a novel fragrance molecule? To help investigate this conundrum, the following example of cassis (or blackcurrant) is discussed.

7.5 Design and Synthesis of Novel Odorants and Potential Industrial Routes to a Natural Product

7.5.1 Cassis (Blackcurrant)

For many people thiol **43** (Figure 7.2) is instantly recognisable as cassis or blackcurrant and yet for others it is highly reminiscent of cat urine.[33] The closely related cyclic ketone **44** is the major odour contributor found in Buchu oil and also evokes a cassis type odour.[34] The hindered oxime Labienoxime (**45**, from Givaudan) and Buccoxime (**46**, from Symrise) also display cassis odours and are commercially successful fragrance materials.[35]

As we can see, many functional groups can evoke recognisable cassis odours. A further class of cassis odorants are the spiro ether class exemplified by Etaspirene (**47**), the molecule responsible for the fruity cassis twist in the perfume Pleasures (from Estée Lauder).[36] The synthesis of Etaspirene (**47**) employs the addition of the protected propargylic Grignard reagent to enone **48** which yields tertiary alcohol **49** (Scheme 7.10). Acidic deprotection followed by Lindlar reduction of **49** furnishes selectively the (*Z*)-alkene **50**, then subsequent cyclisation yields **47** as a mixture of diasteroisomers.

The structurally related unsaturated spiroether **51** and novel *seco* analogue (**52**, Figure 7.3) were prepared using similar chemistry and also display cassis odours.[37] The dihydrofuran ring in **47**, **51** and **52** is not a prerequisite for cassis odour, as the saturated ether **53**[37] and Cassifix (**54**, from IFF)[38] also display similar odours. The tetracyclic saturated ether **55**, isolated as a byproduct in less than 1% yield, was also claimed as an intense and substantive cassis odour.[39]

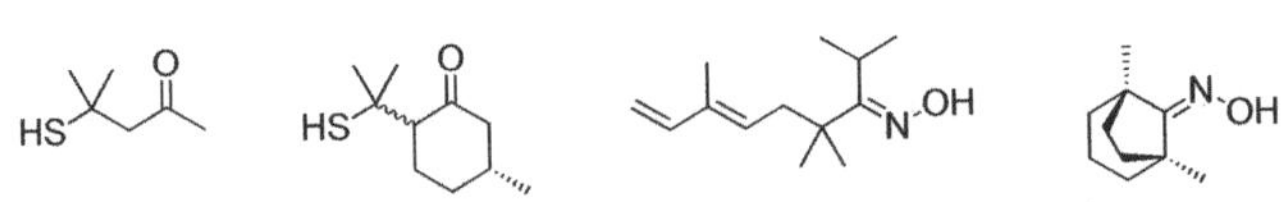

43 **44** **45** (Labienoxime®) **46** (Buccoxime®)

Figure 7.2 Structures of selected cassis odorants

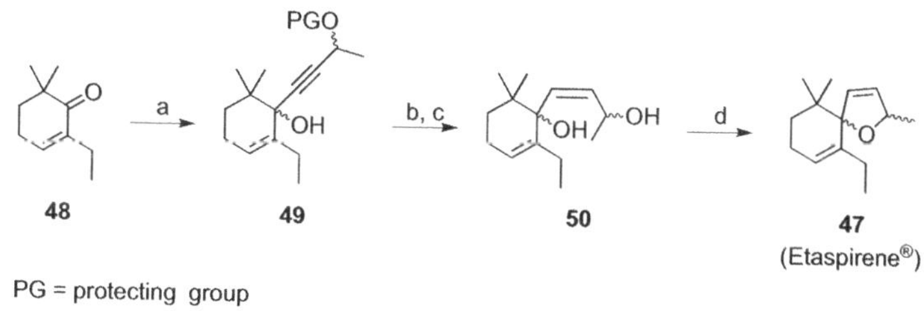

Scheme 7.10 Synthesis of etaspirene (**47**). Reagents and conditions: (a) protected propargylic alcohol, base; (b) deprotection; (c) H_2, Lindlar cat.; (d) acid cat.

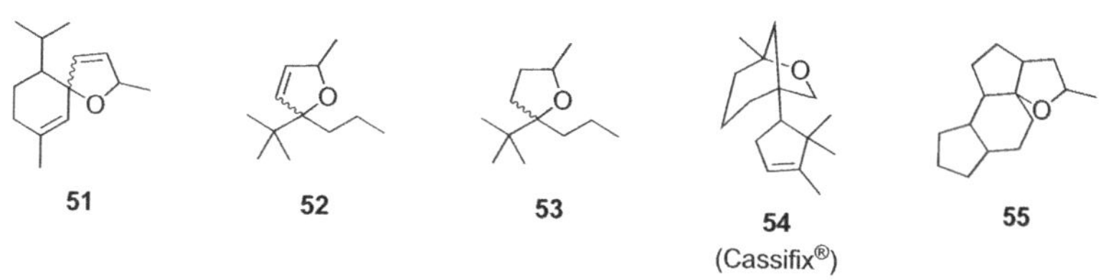

Figure 7.3 Structures of further cassis odorants

Further variation in the functional group with alcohols **56**[40] and **57**[41] also evokes the cassis odour. Alcohol **56**, claimed as a powerful cassis, passionfruit odorant, is readily synthesised from methyl pentadiene (**58**) and readily available enone **59** in a Diels–Alder reaction followed by hydride reduction (Scheme 7.11).[40]

Finally, the 2-substituted pyridine **60** (Scheme 7.11)[42] also displays a cassis, tomato leaf odour and indicates that a huge variety of functional groups are tolerated whilst maintaining a similar odour character. Although hindered thiols, oximes, spiroether, trisubstituted dihydrofurans, alcohols and even pyridine-containing molecules display cassis type odours, all are distinct cassis odorants with different facets; it seems that the steric odour environment is more important than the functional group to evoke the cassis odour. Cassis and patchouli odour classes both seem to require a very hindered functional group hidden within a small compact structure. For instance, the inclusion of a tertiary alcohol

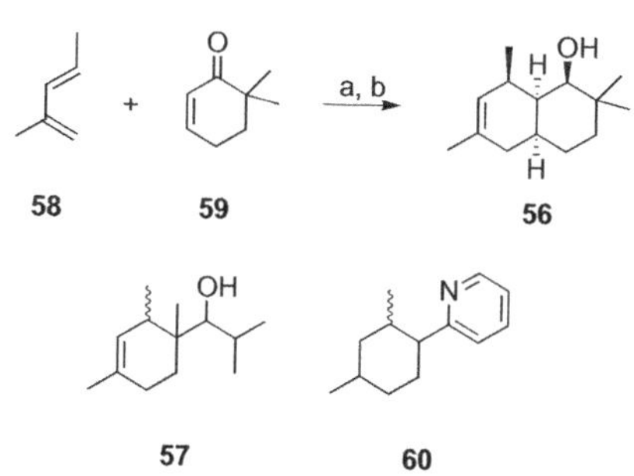

Scheme 7.11 Synthesis of decalin derivative **56** and structures of cassis odorants **57** and **60**. Reagents and conditions: (a) Lewis acid or heat; (b) $LiAlH_4$

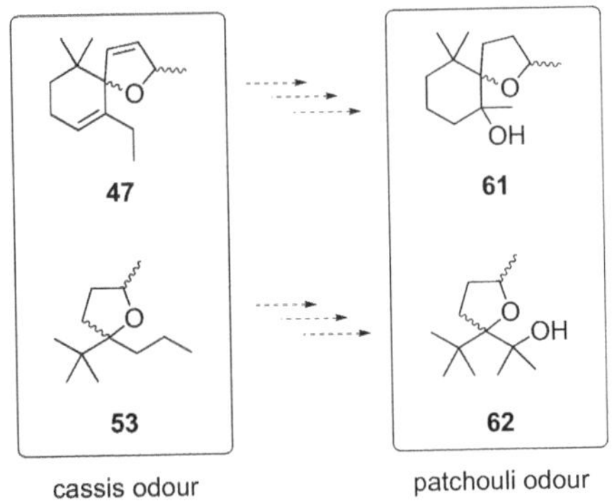

Scheme 7.12 Cassis versus patchouli odour

functionality to the cassis-like tetrahydrofuran skeletons of compounds **47** and **53** results in compounds **61** and **62** (Scheme 7.12), for which a shift from the cassis note to an earthy, woody and camphoraceous odour reminiscent of patchouli oil is observed. It is also interesting to note that both the ether and the tertiary functional groups are adjacent to each other.[37,43] However, not all hindered alcohols have cassis or patchouli odours, and the relationship between the two notes is unclear.

7.5.2 Patchouli

The steam distillation of the fermented leaves of *Pogostemon cablin* (Blanco) Benth. (Laminaceae), yields patchouli oil, a highly popular and substantive perfumery ingredient that finds widespread use in many product areas, particularly in laundry applications. This highly appreciated odour of patchouli oil has three important facets (woody/amber, earthy and camphoraceous odours) present together in complete harmony. The worldwide production of this oil is currently more than 1200 t year^{-1}, although seasonal supply variations often result in wide fluctuations in price (between 30 and 200 \$ kg^{-1}).[44] The structure of patchouli alcohol (patchoulol, **41**), that accounts for over 40% of the essential oil, was published erroneously in 1961.[45] Of the large number of syntheses of **41** that have been published to date, many employ either an intra- or intermolecular Diels–Alder reaction between a trimethyl cyclohexadiene derivative **63** and an alkene as the key retrosynthetic disconnection (Scheme 7.13).[46] A concise synthesis of racemic patchoulol (**41**) was published in 1974 by Naef and Ohloff, and the corresponding asymmetric version followed several years later (Scheme 7.13).[47]

The addition of the organolithium reagent derived from the optically active bromide **64** to the dienone **63** gave the required tertiary alcohol as a 1:1 mixture of diasteroisomers (**65**, **66**). Despite the directness of this approach the major disadvantage was the generation of two diastereoisomers of which only **65** undergoes the desired Diels–Alder reaction to yield (–)-patchoulol (**41**). Severe steric interactions between the sidechain methyl and the bridgehead methyl disfavour the cycloaddition in diastereoisomer **66**, thus effectively limiting the yield. The search for non-natural patchouli odorants has focused mainly on how to reduce complexity, and most studies have looked at sterically hindered tertiary alcohols buried within compact cyclic structures.[48] However, hindered tertiary alcohols such as

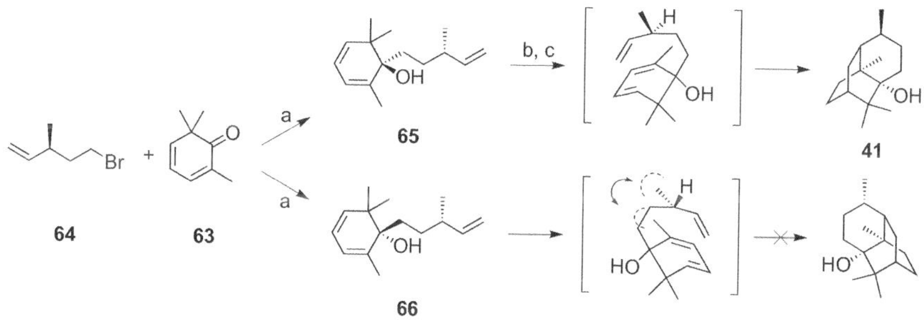

Scheme 7.13 Synthesis of (–)-patchoulol (**41**). Reagents and conditions: (a) Li and bromide **64** gave a 1:1 mixture of diastereoisomers; (b) 280–300 °C, 24 h, cat. base, 25% yield or 50% based on **65**; (c) H_2, Pd/C

geosmin (**67**, Scheme 7.14) also display very earthy odours and hindered secondary alcohols such as borneol (**68**) display very camphoraceous odours.[49] Recently, several interesting and industrially feasible spirocyclic patchouli-like odorants were reported.[50] Diels–Alder reaction of the readily available enone **69** with either butadiene or isoprene yielded ketones **70** and **71**, respectively. Ketone **70** itself was developed and launched due to its interesting herbal note. Reduction of ketone **70** gave the patchouli odorant **72** which is noteworthy due to the secondary alcohol functionality rather than the more typical tertiary alcohol functionality. Reduction of ketone **71** under similar conditions followed by acetylation furnished **73**, an extremely powerful and elegant amber odorant (Scheme 7.14).

Bifunctional patchouli alcohol-like odorants **74**, **75** and **76** (Scheme 7.15) were disclosed in 1987 by Barton and coworkers, following the oxidation, under '*gif*' conditions, of patchouli alcohol (**41**) in an attempt to mimic *in vivo* metabolism through the agency of cytochrome P450 enzymes.[51] Intriguingly, it was ketone **76** with the closely positioned ketone and alcohol functional groups that maintained a rich patchouli odour, whilst ketone **74** was deemed the most interesting odorant, with an orris concrete-like note.

Further bifunctional patchouli odorants **77**–**80** and the related tertiary alcohol **81** (Figure 7.4) were recently published by Kraft and coworkers, as part of detailed structure–odour relationship studies on *spiro* tertiary alcohols and hydroxyl ketones.[52,53] All four bifunctional molecules displayed a patchouli and or camphoraceous odour, often with

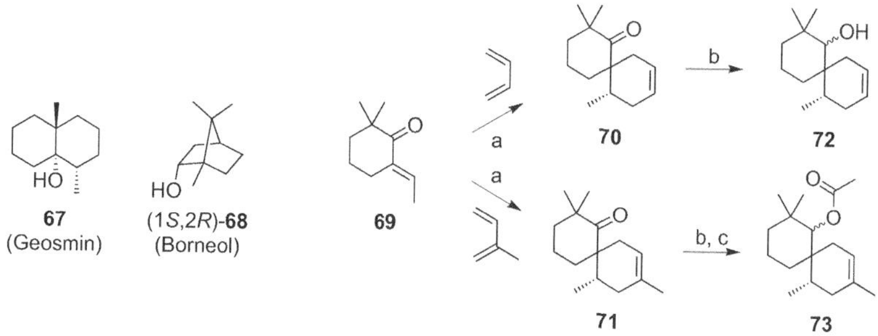

Scheme 7.14 Spirocyclic patchouli, amber and herbal odorants. Reagents and conditions: (a) Lewis acid or ΔH; (b) $LiAlH_4$; (c) acetic anhydride, cat., ΔH

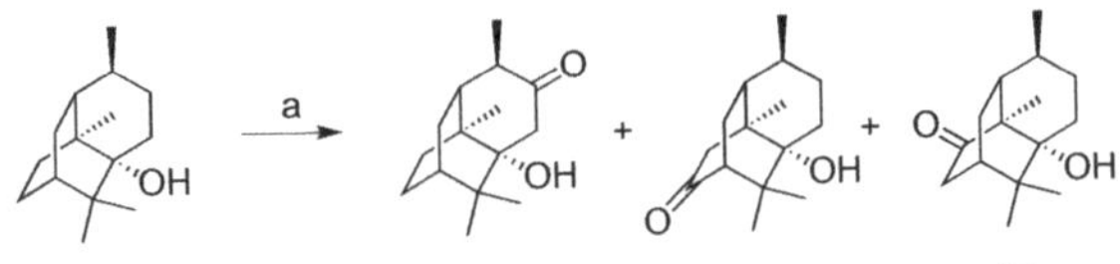

Scheme 7.15 *Oxidation of patchoulol (**41**) under 'gif' oxidation conditions. Reagents and conditions: (a) Fe(II), H_2O_2, pyridine, acetic acid*

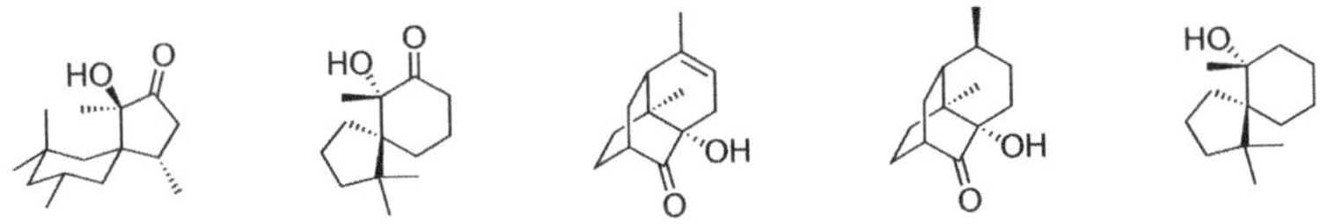

Figure 7.4 *Further spirocyclic patchouli odorants*

additional interesting tobacco or animalic facets with varying detection thresholds. Slight changes in structure gave large changes in odour, and the 'ring reversal' transition from the *spiro* hydroxyl ketone **77** (with a rich patchouli odour) to the related hydroxyl ketone **78** (which displayed a camphoraceous minty, slightly patchouli) reinforced the difficulties encountered in trying to predict odour from structure.[26]

The synthesis of analogues **79** and **80** employed an elegant intramolecular Prins cyclisation of decalin **82** to form the bicyclo[2.2.2]octane ring system (Scheme 7.16). The *gem* dimethyl motif in patchouli alcohol (**41**) was replaced with a carbonyl functional group and both **79** and **80** maintained appreciated patchouli odours.[53]

The secondary alcohol **83** was synthesised in three steps as part of a series of interesting and powerful woody odorants and patented for its appreciated patchouli odour with ambery nuances (Scheme 7.17).[54] Even more recently, alcohol **84** was patented for its extremely natural patchouli odour.[55]

A Diels–Alder reaction between methyl pentadiene (**85**) and methyl cyclohexenone (**86**) gave the fused bicyclic ketone **87** in good yield. Hydrogenation followed by carbonyl reduction furnished alcohol **84** in good overall yield.[55] From sterically hindered tertiary and secondary alcohols, we next investigate the wide variety of molecular structure that represent the musk odour class.

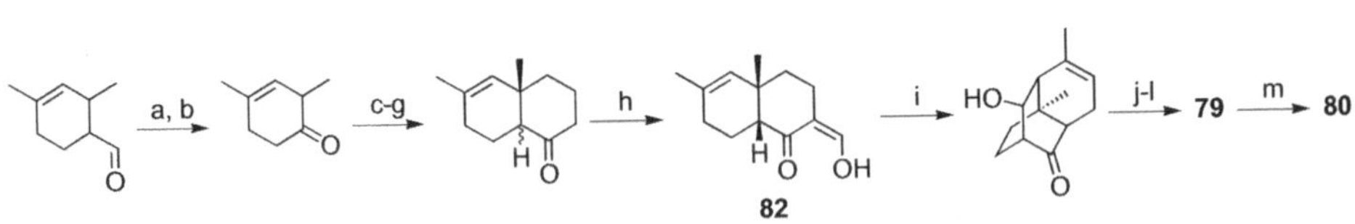

Scheme 7.16 *Prins cyclisation to form patchoulol skeleton. Reagents and conditions: (a) enamine formation; (b) oxygen, Cu(I); (c) base, 1,4-dimethoxy-2-butanone; (d) $LiAlH_4$; (e) acetic anhydride, base; (f) Li, NH_3; (g) oxalic acid, aq. methanol; (h) base, ethyl formate; (i) p-TSA, ΔH; (j) thiocarbamate formation; (k) radical deoxygenation; (l) base, oxidation; (m) H_2, Pd/C*

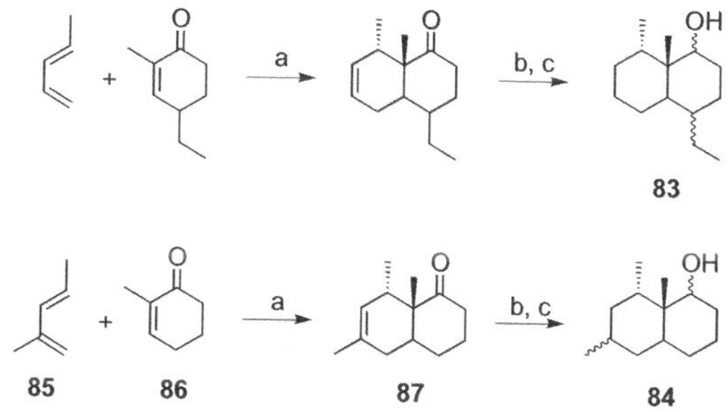

Scheme 7.17 Three-step syntheses of patchouli-like odorants **83** and **84**. Reagents and conditions: (a) Lewis acid or ΔH; (b) H_2, Pd/C cat.; (c) $LiAlH_4$

7.5.3 Musk

Many classes of musk odorants exist, including nitromusk, macrocyclic and polycyclic, containing many varied and interesting structures.[56] Macrocyclic musks in a pure form such as muscone (**88**) have been used in perfumery virtually since their structural determination by Ruzicka in 1926 and despite their high initial price.[56] The efficient synthesis of these large macrocyclic rings has been the subject of constant research and, in a novel three-step synthesis of racemic muscone (**88**), the radical mediated addition of acetone to 1,9-decadiene (**89**) gave the diketone **90** (Scheme 7.18).[57] Intramolecular aldol condensation of the diketone **90** followed by hydrogenation gave racemic muscone (**88**). The asymmetric hydrogenation of the pure (*E*)- and (*Z*)-isomers of enone **91** had previously been applied to the synthesis of (*R*)-muscone (**88**).[58]

Using another related aldol approach, non-natural unsaturated analogues such as (5*Z*)-muscenone (**92**) have been synthesised in stereoselective and enantioselective fashion (Scheme 7.19) confirming their highly appreciated nitro musk aspects.[59]

In a novel approach to the macrocylic musk **92**, Fehr and coworkers demonstrated the feasibility of the intramolecular aldol condensation of the *meso*-diketone **93** to furnish the optically active enone **94**, followed by Eschenmoser fragmentation giving alkyne **95** and Lindlar hydrogenation to yield (5*Z*)-**92**, as shown in Scheme 7.19.[59] In subsequent developments towards the synthesis of scalemic (5*Z*)-muscenone (**92**), Knopff and coworkers first reported the chiral aminoalkoxide-mediated condensation of the *meso*-diketone **93** and β-elimination of the transient aldol adduct, which gave enone **94** in 56–76% *ee*, and they more recently reported an amino acid catalyzed version of this aldol addition/dehydration sequence.[60]

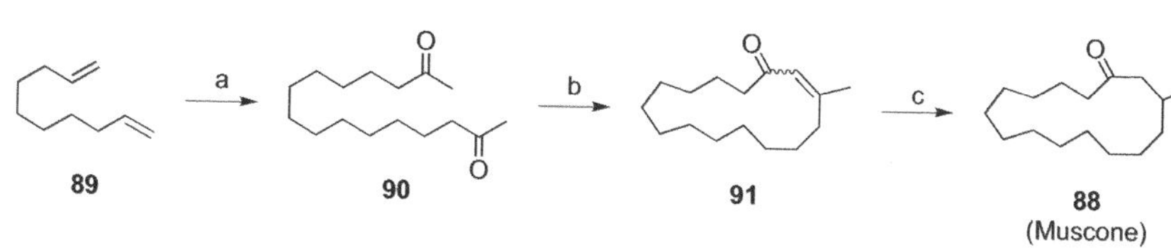

Scheme 7.18 A radical approach to muscone (**88**). Reagents and conditions: (a) Mn(II), acetone; (b) base, aldol; (c) H_2, cat.

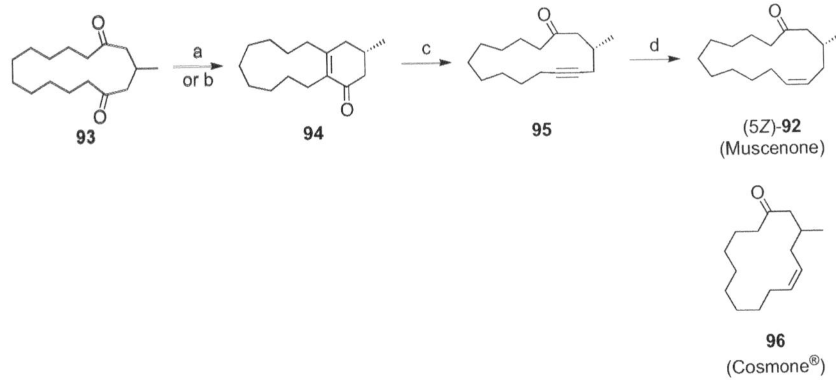

Scheme 7.19 *Asymmetric synthesis of (5Z)-muscenone (**92**) and structure of its lower homologue Cosmone (**96**). Reagents and conditions: (a) KOH, asymmetric reduction, separation; (b) ephedrine base, or amino acid cat.; (c) Eshenmoser fragmentation; (d) H_2, Lindlar cat.*

A lower homologue of (5*Z*)-muscenone (**92**) was recently released onto the market as Cosmone (**96**, from Givaudan)[61] and the synthesis of unnatural unsaturated, ether-containing and methyl-substituted macrocyclic ketones and lactones has been a very rewarding field of investigation[62–63], even 100 years after the isolation of the first macrocyclic musk.

Examples of yet another class of musks are shown in Figure 7.5. One of them, cyclomusk (**97**), was patented by BASF in 1976.[64] The structurally related Helvetolide (**98**) was patented in 1991 and has found widespread use as a top note musk in many commercially successful fragrances.[65] The replacement of the *gem* dimethyl group with a carbonyl functionality maintained the musk odour and gave rise to another successful odorant, Romandolide (**99**),[66] and subsequently the epoxyester **100**.[67] Further related members of the Helvetolide class of musk including **101–104** have been reported.[68,69]

Another interesting structural class of musk was discovered, patented and named Klauscenone (**105**, Figure 7.6).[70] Although Klauscenone (**105**) was never commercialised, during the synthesis of analogues, Moxalone (**106**) was discovered, developed and later launched on the market.[71] The related polymethylated bicyclic ketone Cashmeran (**107**, from IFF)[72], a commercially successful musk displaying woody aspects has been on the

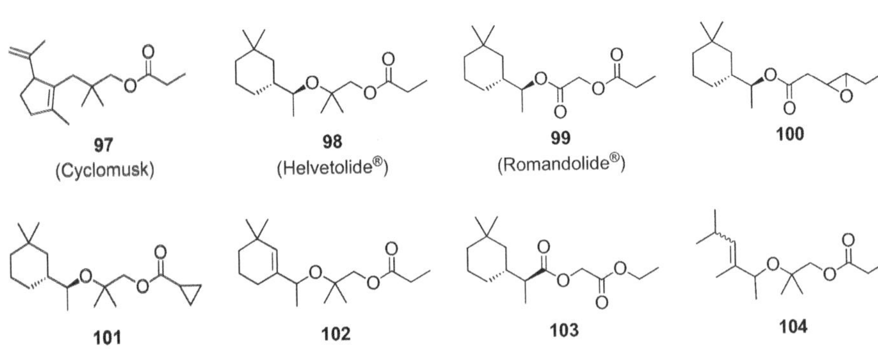

Figure 7.5 *A selection of cyclomusk analogues*

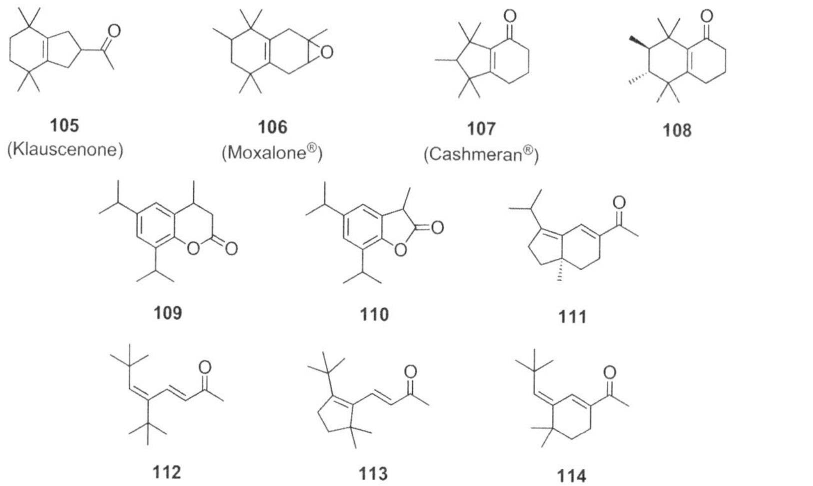

Figure 7.6 *Further novel musk skeletons*

market since the 1970s, and the more substantive and more powerful higher homologue **108** has been recently patented.[73]

Aromatic lactonic musks are yet another interesting structural class of musks, and six- (**109**)[74] and five-membered (**110**)[75] versions have been claimed. Novel indenone **111** displayed a strong musky odour and served as a novel lead for yet another structural class of musks.[76] In studying the interface between ionone (violet), woody and musk odour families, Kraft and coworkers synthesised both *seco* analogues typified by **112** and later monocyclic versions of five- (**113**) and six-membered rings (**114**), which further extended this exciting class of novel musk odorants (Figure 7.6).[77]

Many of these analogues show interesting nuances in addition to the musk odour descriptors. From this veritable wealth of structural diversity that evokes the musk odour as their primary descriptor, trying to accurately predict odour from structure is far from reliable. From an odour class where structural promiscuity is evident, we pass to an odorant class where the majority of members are either primary or secondary alcohols with a very well defined shape and functional group spacing: the sandalwood odorants.

7.5.4 Sandalwood

The slow growing and highly prized East Indian sandalwood (*Santalum album* L.), which is now under severe threat from overexploitation and made worse by its increasing scarcity, has always been subject to supply variations and therefore suffers from constant price fluctuations.[78] The current market price for the essential oil is approximately 1000 $\$\,kg^{-1}$ and represents a very attractive and extremely challenging target for fragrance chemists. The constant volatility of the East Indian sandalwood oil price has driven the search for synthetic sandalwood smelling compounds. Since the 1940s a number of these molecules have been manufactured and are available on the market. Two of the most recent additions to the synthetic sandalwood odorants are Firsantol (**115**)[79] and Javanol (**116**;[80] Figure 7.7).

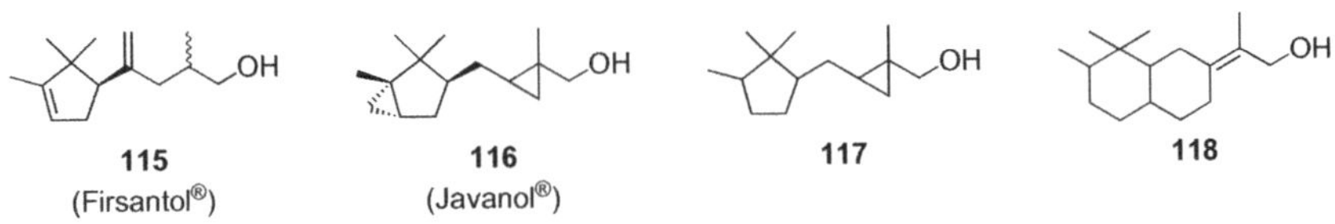

Figure 7.7 Structures of selected synthetic sandalwood odorants

Even more recently the saturated monocyclopropanated sandalwood odorant **117**[81] and the extremely natural and substantive sandalwood odorant alcohol **118**[82] were synthesised and patented.

As East Indian sandalwood oil itself is a highly appreciated perfumery ingredient already found in many prestigious perfumes, a synthetic replacement would be very desirable. (Z)-(−)-β-Santalol (**42**, Figure 7.8) is the key odour compound responsible for the creamy lactonic sandalwood character of the oil. The level of this compound in East Indian sandalwood oil is 20–25% and in the related Australian sandalwood oil (*S. spicatum*) is 3–8%. The closely related (Z)-(+)-α-santalol (**119**) which makes up 50–60% of the East Indian sandalwood oil has a less appreciated cedar-like sandalwood character.[83] The presence of the (Z)-configured allylic alcohol is also essential for the highly appreciated odour, as the (*E*)-(−)-β-santalol (**120**) has been described as weakly sandalwood, woody, medicinal and phenolic.[84] As the (+)-enantiomer of **42** has been shown to be odourless, clearly an industrially feasible asymmetric synthesis of the (Z)-(−)-β-santalol (**42**) would be even more desirable.[85] The corresponding *endo* isomer, *epi*-(Z)-(+)-β-santalol (**121**), has already been synthesised in racemic form and was found to be weak and uninteresting.[86] For the industrial synthesis of the more appreciated (Z)-(−)-β-santalol (**42**) several challenges must be overcome, including the selective synthesis of the (Z)-allylic alcohol, the installation of the *exo* sidechain and the *exo* methylene functional group, whilst avoiding the facile rearrangement of the norbornane skeleton under acidic conditions.

Despite many racemic, academic syntheses,[87] an asymmetric synthesis[85] and two theoretically industrially feasible racemic syntheses of (Z)-(−)-β-santalol (**42**) by Christenson and Willis (1979)[88] and Baumann and Hoffmann (1979),[89] neither synthetic (±)-(Z)-α-santalol (**119**) nor (±)-(Z)-β-santalol (**42**) are available commercially. Christenson and Willis described the racemic synthesis of **42** via buffered epoxidation of racemic camphene (**122**) followed by epoxide opening with the dianion of acetic acid and ring closure giving the spirolactone **123** (Scheme 7.20).[88] Sulfuric acid mediated rearrangement gave the fused lactone **124** containing the three carbon atom-containing sidechain with the required *exo*-configuration. Heating under acidic alcoholic conditions cleaved the lactone and generated the *exo*-methylene functionality in ester **125**. Reduction, followed by modified Wittig reaction gave racemic (Z)-β-santalol (**42**).

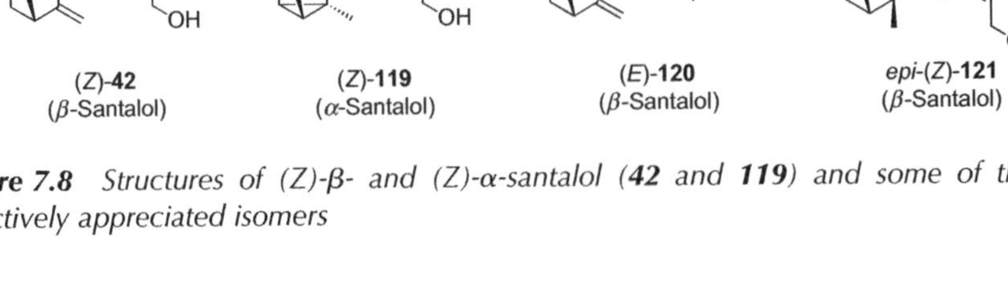

Figure 7.8 Structures of (Z)-β- and (Z)-α-santalol (**42** and **119**) and some of their less olfactively appreciated isomers

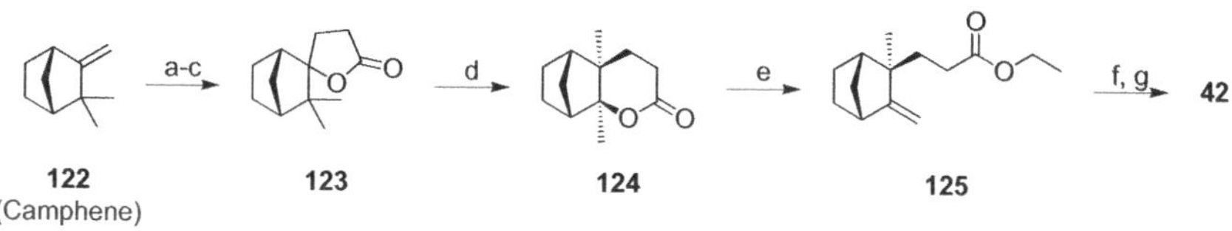

Scheme 7.20 Synthesis of (Z)-β-santalol (**42**). Reagents and conditions: (a) peracetic acid, sodium acetate; (b) LDA, acetonitrile; (c) KOH, ethanol, ΔH; (d) conc. H_2SO_4, –10°C; (e) p-TSA, ethanol, ΔH; (f) DIBAL-H; (g) n-BuLi, ethylidene triphenyl phosphorane, then formaldehyde

Around the same time, Baumann and Hoffmann, chemists at BASF, published a lengthy route to β-santalol (**42**) employing industrially feasible reactions, but unfortunately giving no control over the double bond geometry in the penultimate step and furnishing only (*E*)-β-santalol (**120**).[89]

Nearly 30 years after these syntheses,[88,89] the combination of two recent breakthroughs in catalytic technologies may result in the more efficient, industrially feasible synthesis of pure (Z)-(–)-β-santalol (**42**; Scheme 7.21). The first challenge presented by **42** is the efficient installation of the three carbon atom-containing sidechain to the bicyclo[2.2.1]heptane core in *exo*-fashion. The *exo* challenge has been elegantly overcome via a novel copper(I)-catalyzed fragmentation reaction observed initially as a side reaction by Fehr and coworkers.[90]

Preparation of the required olefin **126** was straightforward, by employing the recently reported asymmetric Diels–Alder reaction of crotonaldehyde and cyclopentadiene in water to give the aldehyde **127** with good *exo* selectivity and in 95% *ee* (Scheme 7.21).[91] Exhaustive hydrogenation of both the aldehyde carbonyl and the double bond afforded the corresponding alcohol, activation as the methyl carbonate and pyrolysis yielded the desired *endo*-methyl methylene [2.2.1]bicycloheptane **126**, with a slight erosion in *ee* (to 91%). Then the key '*ene*' reaction with a silyl-protected propargylic aldehyde derivative followed by deprotection produced alcohol **128**.[90] Alcohol **128** is an ideal substrate for a cyclisation/fragmentation reaction to give the desired unsaturated aldehyde **129** in excellent yield,

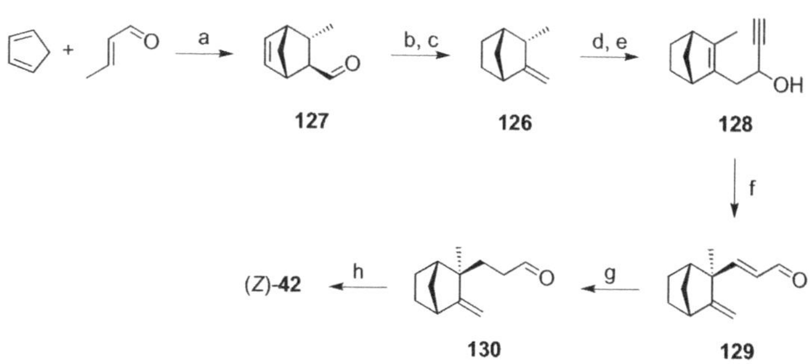

Scheme 7.21 Synthesis of (Z)-(–)-β-santalol (**42**) via a copper (I) catalysed fragmentation. Reagents and conditions: (a) exo/endo 72:28 (% ee ratio 95:76); (b) H_2, Raney-Nickel cat.; (c) activation, pyrolysis; (d) TMS propargylic aldehyde, dimethylaluminium chloride; (e) K_2CO_3, methanol; (f) Cu(I) cat., dichloroethane, 50°C, 98%; (g) H_2, Pd/$CaCO_3$ cat.; (h) n-BuLi, ethylidene triphenyl phosphorane, then formaldehyde

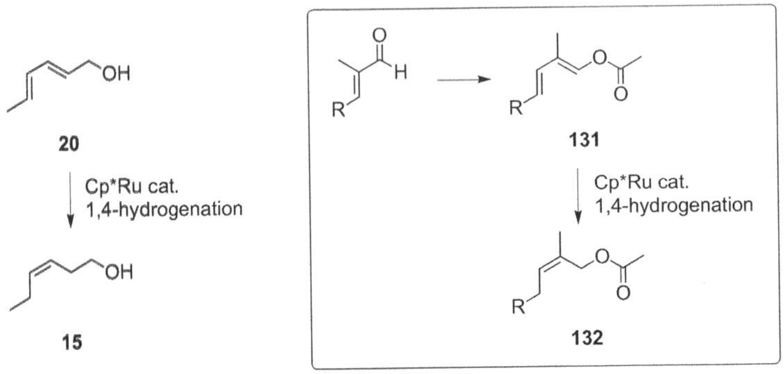

Scheme 7.22 *Extension of 1,4-hydrogenation of dienes to dienyl esters to give the (Z)-allylic alcohol motif in* ***132***

installing simultaneously the *exo*-methylene motif and the three carbon atom-containing sidechain in *exo*-fashion. One recrystallisation from pentane at low temperature improved the *ee* to 97%. Selective hydrogenation of the conjugated double bond followed by a Wittig reaction modified by Corey and Yamamoto[92] furnished (Z)-(–)-β-santalol (**42**; *Z*/*E* 97:3, 97% *ee*) in reasonable overall yield.[90] It was the (Z)-allylic alcohol motif[93] that had plagued the two previous industrially feasible syntheses. Although Wittig reactions are indeed utilised on an industrial scale in the multitonne manufacture of vitamins, the conversion of the aldehyde **130** into (Z)-**42** via a catalytic process would be required. It could be shown that the same 1,4-hydrogenation conditions developed for the synthesis of (Z)-3-hexenol (**15**)[13] when applied to a suitably substituted dienyl acetate such as **131** would result in the acetate protected form of the (Z)-allylic alcohol motif in **132** found in both (Z)-α-santalol (**119**) and (Z)-β-santalol (**42**; Scheme 7.22).[94] This has now been shown to be a viable approach for the large-scale preparation of (Z)-**42**.[90]

Aldol condensation of aldehyde **130** with propanal occurred smoothly to yield the enal **133**, requiring only double bond isomerisation and aldehyde reduction to yield (Z)-(–)-β-santalol (**42**; Scheme 7.23). Generation of dienyl acetate **134** occurred smoothly and the key transformation, the selective 1,4-hydrogenation, gave the (Z)-double bond with excellent selectivity (*Z*/*E* 98:2) to yield the allylic acetate **135**. Acetate **135**, after treatment under transesterification conditions, furnished (Z)-(–)-β-santalol (**42**) in good overall yield and in 97% *ee* from aldehyde **130**.[94]

As we can see, using a combination of innovative catalytic technologies may enable the efficient industrial synthesis of (Z)-β-santalol (**42**), one of the last classic synthetic challenges for fragrance chemists.

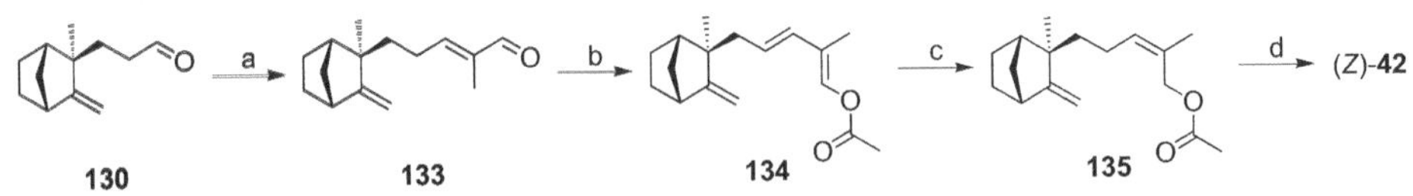

Scheme 7.23 *Synthesis of β-santalol (**42**) using (Z)-selective catalytic 1,4-hydrogenation. Reagents and conditions: (a) propanal, cat.; (b) acetic anhydride, triethylamine, ΔH; (c) H_2, RuCp* cat.; (d) K_2CO_3, methanol*

Table 7.1 *Challenges compared between pharmaceutical, agrochemical and fragrance industries*

Typical molecular properties	Pharmaceuticals[95]	Agrochemicals[96]	Fragrance guidelines
Molecular weight	300–500 amu	250–400 amu	100–300 amu
H-bond donors per molecule (OH, NH, etc.)	5	≤3	≤1
H-bond acceptors per molecule (C=O, etc.)	10	2–12	≤2
log $P_{o/w}$[a]	<5	<5	<5
Charged salts	Yes	Yes	No
Crystals	Yes	Yes	Rare
Purification processes	Crystallisation or chromatography	Crystallisation	Distillation
Protecting groups	Common	Rarely used	Rarely used
Typical cost [$ kg^{-1}]	10 000–150 000[b]	10–100	10–500
Length of synthesis	5–12 steps	<5 steps	2–7 steps
Chiral syntheses	Common	15–20% of market	Rare[97]

Note: [a]Logarithm of the octanol/water partition coefficient as a measure of hydrophobicity.
[b]Aspirin (Bayer), being a cheap pharmaceutical drug synthesized in one step from salicylic acid, would be an exception to this. Monoclonal antibodies costing many times this amount would also be exceptions. Lipitor (Pfizer) a representative large volume drug retails for 2 $ per 10 mg tablet, equivalent to approximately 200 000 $ kg^{-1}.

7.6 Other Challenges

The synthetic challenges posed by both natural and non-natural volatile molecules have occupied the world's synthetic chemists for many decades. Despite the significant advances that have been made in the areas of selective catalysis, the discovery of novel chemical reactions and the use of known technology in novel applications, formidable challenges still exist. Additional challenges are becoming significant along the pathway from the discovery of a novel fragrance molecule to its eventual launch onto the market. These additional challenges come in the form of ever tighter price constraints, together with a favourable ecological and safety profile. The ultimate challenge in synthesis of a novel fragrance molecule, however, is achieving the desired organoleptic purity in the final product. Organoleptic purity is sometimes very different to chemical purity, and odour quality is most often be achieved by distillation, which presents further unique challenges to the fragrance industry. Table 7.1 illustrates how these challenges vary across several chemical industries.[95–97]

7.7 Conclusion

This chapter focuses on a small selection of the very elegant solutions that chemists have devised to overcome considerable challenges in the synthesis of natural and non-natural volatile compounds. Whilst the processes involved in the olfactory system are more and more studied, their complexity became more and more evident, in particular when it comes to anticipating odours and hedonic effects. Therefore, serendipity and a certain touch of magic still play an important role in the design, synthesis and eventual production of novel fragrance molecules. We will continue to be dependent on the creativity and imagination of observant chemists. Despite many beautiful smelling molecules being isolated from natural sources or discovered by design (and by chance) every year within the research departments of fragrance companies, only a small fraction will be launched on the market to eventually find their way into a successful fragrance. Actually launching a successful fragrance ingredient on the market requires combining the cost-efficient synthesis of ingredients displaying desired molecular properties. These desirable molecular properties increasingly include considerations such as the metabolic fate of fragrance molecules within the olfactory system[98] and the environment.

To conclude this review, it is useful to set out some of the specific issues and challenges that arise for organic chemists from the growing concern for environmental biodegradability and interest in how the olfactory system recognises fragrance molecules. At present, high odour strength has been suggested to be due to odorant molecules inhibiting their own metabolism by being good inhibitors of the cytochrome P450 enzymes that occur within the nasal mucosa.[99] Both in vertebrates and in worms odorant receptors are G protein-coupled receptors, and it has been proposed that they are metalloproteins, implicating copper or zinc in odorant binding.[100] Insect odorant receptors however, can also act as gated ion channels.[101] As the olfactory system is able to recognise a huge variety of different functional groups from aldehydes, carboxylic acids, amines, thiols, alkenes, alkynes, ethers, aromatic rings through to alcohols and nitriles, a question for organic chemists is whether fragrance molecules are simply acting as Lewis basic ligands for a Lewis acidic metal(s) buried within an odorant receptor. In seminal work by Ohloff and coworkers of Firmenich, a distance of <3 Å between hydrogen bond acceptor and donor, in bi-functional odorants, was shown to be important for a muguet odour.[102] A further question for chemists is whether this is due to the proximity of both functional groups allowing their interaction, as a bidentate ligand, with a single metal ion? Could a larger distance between the hydrogen bond donor and acceptor imply more than one metal ion or simply a mixed message to the brain that is not interpreted as an odour? On this point, it has already been shown that odorants can act as both agonists and antagonists to odour receptors and that oxidation of odorants can produce antagonistic compounds.[103] Further, the metabolism of odorants is frequently mediated by oxidative cytochrome P450 enzymes which may be a closely related process to the antagonism of the interaction of odorants to odorant receptors. Greater understanding of these processes and answers to the questions outlined above will have important implications for how chemists approach to efficient design of future fragrance molecules.

Acknowledgements

The author wishes to thank the following colleagues and friends for their proof reading of this chapter, for their constructive suggestions and comments and mostly for their encouragement and enthusiasm: Christian Chapuis, Carolyn Deere, Cornelius Elie, Charles Fehr, Christian Margot, Fabrice Robvieux, Francesco Santoro, Lionel Saudan, Ben Smith and Roger Snowden.

Dedication

Dedicated to the memory of Beat Winter 1946–2009, discoverer of Hivernal and Lilyflore and so many other beautiful fragrant molecules.

References

1. (*a*) R. B. Woodward, Total synthesis of vitamin B12, *Pure Appl. Chem.*, **33**, 145–178 (1973); (*b*) R. B. Woodward, Recent advances in the chemistry of natural products, *Pure Appl. Chem.*, **25**, 283–304 (1971); (*c*) A. Eschenmoser, Current synthesis of natural organic substances. Example vitamin B12, *Naturwissenschaften*, **61**, 513–525 (1974); (*d*) A. Eschenmoser, Vitamin B_{12}: experiments concerning the origin of its molecular structure, *Angew. Chem.*, **100**, 5–40 (1988); *Angew. Chem. Int. Ed.*, **27**, 5–39 (1988).
2. R. W. Armstrong, J.-M. Beau, S. H. Cheon, W. J. Christ, H. Fujioka, W.-H. Ham, L. D. Hawkins, H. Jin, S. H. Kang, Y. Kishi, M. J. Martinelli, W. W. McWhorter Jr., M. Mizuno, M. Nakata, A. E. Stutz, F. X. Talamas, M. Taniguchi, J. A. Tino, K. Ueda, J. Uenishi, J. B. White, M. Yonaga, Total synthesis of palytoxin carboxylic acid and palytoxin amide, *J. Am. Chem. Soc.*, **111**, 7530–7533 (1989).
3. (*a*) G. E. Veitch, A. Boyer, S. V. Ley, The azadirachtin story, *Angew. Chem.*, **120**, 9542–9570 (2008); *Angew. Chem. Int. Ed.*, **47**, 9402–9429 (2008); (*b*) A. Boyer, G. E. Veitch, E. Beckmann, S. V. Ley, Second generation synthesis of azadirachtin: a concise preparation of the propargylic mesylate fragment, *Angew. Chem.*, **121**, 1343–1346 (2009); *Angew. Chem. Int. Ed.*, **48**, 1317–1320 (2009).
4. (*a*) K. C. Nicolaou, T. Montagnon, *Molecules that Changed the World*, Wiley-VCH, Weinheim, 2008; (*b*) M. A. Sierra, M. C. de la Torre, *Dead Ends and Detours: Direct Ways to Successful Total Synthesis*, Wiley-VCH, Weinheim, 2005; (*c*) K. C. Nicolaou, E. J. Sorensen, *Classics in Total Synthesis: Targets, Strategies, Methods*, Wiley-VCH, Weinheim, 1996; (*d*) K. C. Nicolaou, S. A. Snyder, *Classics in Total Synthesis II: More Targets, Strategies, Methods*, Wiley-VCH, Weinheim, 2003.
5. E. Pichersky, J. P. Noel, N. Dudareva, Biosynthesis of plant volatiles: nature's diversity and ingenuity, *Science*, **311**, 808–811 (2006).
6. (*a*) M. Spehr, K. R. Kelliher, X. H. Li, T. Boehm, T. Leinders-Zufall, F. Zufall, Essential role of the main olfactory system in social recognition of major histocompatibility complex peptide ligands, *J. Neurosci.*, **26**, 1961–1970 (2006); (*b*) S. D. Munger, T. Leinders-Zufall, F. Zufall, Subsystem organization of the mammalian sense of smell, *Annu. Rev. Physiol.*, **71**, 115–140 (2009); (*c*) J. Kwak, M. C. Opiekun, K. Matsamura, G. Preti, K. Yamasaki, G. K. Beauchamp, Major histocompatibility complex-regulated odortypes: peptide-free urinary volatile signals, *Physiol. Behav.*, **96**, 184–188 (2008).
7. E. J. Corey, R. B. Mitra, H. Uda, Total synthesis of D,L-caryophyllene and D,L-isocaryophyllene, *J. Am. Chem. Soc.*, **86**, 485–492 (1964).
8. P. S. Wharton, Stereospecific synthesis of 6-methyl-*trans*-cyclodecenone, *J. Org. Chem.*, **26**, 4781–4782 (1961).
9. G. Stork, R. L. Danheiser, B. Ganem, Spiroannelation of enol ethers of cyclic 1,3 diketones. Simple stereospecific synthesis of beta-vetivone, *J. Am. Chem. Soc.*, **95**, 3414–3415 (1973).
10. (*a*) P. Z. Bedoukian, The seven primary hexenols and their olfactory characteristics, *J. Agric. Food. Chem.*, **19**, 1111–1114 (1971); (*b*) H. Surburg, J. Panten, *Common Fragrance and Flavor Materials*, 5th Ed., Wiley-VCH, Weinheim, 2006.
11. As Lindlar catalyst is typically poisoned with lead and now subject to more stringent transport regulations, lead free palladium catalysts to allow the selective hydrogenation of alkynes to *cis*-alkenes are actively sought and BASF has been particularly active in this field. For a recent example see: P. T. Witte, Process for the preparation of an aqueous colloidal precious metal suspension (to BASF SE), EP 2 082 805 (2009); *Chem. Abstr.*, **151**, 200761 (2009).

12. S. Steines, U. Engelert, B. Driessen-Hölscher, Stereoselective catalytic hydrogenation of sorbic acid and sorbic alcohol with new Cp*Ru complexes, *Chem. Commun.*, 217–218 (2000).
13. (*a*) P. Dupau, L. Bonomo, 1,4-Hydrogenation of dienes with Ru complexes (to Firmenich SA), WO 2008/120175 (2008); *Chem. Abstr.*, **149**, 448584 (2008); (*b*) P. Dupau, L. Bonomo, 1,4-Hydrogenation of sorbol with Ru complexes (to Firmenich SA), WO 2008/120174 (2008); *Chem. Abstr.*, **149**, 425710 (2008); (*c*) L. Saudan, Hydrogenation processes in the synthesis of perfumery ingredients, *Acc. Chem. Res.*, **40**, 1309–1319 (2007).
14. A. F. Thomas, The synthesis and structure of α-sinensal, *Chem Commun.*, 947–949 (1967).
15. A. Nissen, W. Rebafka, W. Aquila, Verfahren zur Herstellung von Citral (to BASF AG), EP 0 021 074 (1981); *Chem. Abstr.*, **95**, 7513 (1981).
16. BASF website, Citral: a key building block for fine chemicals, http:\\www.basf.com.
17. (*a*) A. Schaedler, T. Friedrich, R. Stuemer, S. Rinck-Jahnke, Process for the enzymatic preparation of citronellal (to BASF AG), WO 2007/077121 (2007); *Chem. Abstr.*, **147**, 94103 (2007); (*b*) G. Heydrich, G. Gralla, M. Rauls, J. Schmidt-Leithoff, K. Ebel, W. Krause, S. Oehlenschäger, C. Jäkel, M. Friedrich, E. J. Bergner, N. Kashani-Shirazi, R. Paciello, Method for producing optically active, racemic menthol (to BASF SE), WO 2009/068444 (2009); *Chem. Abstr.*, **151**, 10749 (2009).
18. H.-G. Göbbel, G. Wegner, H. Fuchs, S. Unverricht, A. Salden, Method for the continuous hydrogenation of citral to give citronellal (to BASF AG), WO 2004/007414 (2004); *Chem. Abstr.*, **140**, 130115 (2004).
19. H.-G. Göbbel, T. Gerlach, F. Funke, K. Ebel, S. Unverricht, G. Kaibel, Selective hydrogenation of olefinically unsaturated carbonyl compounds (to BASF AG), US 6,916,964 (2005); *Chem. Abstr.*, **139**, 36649 (2003).
20. F. Haese, K. Ebel, Method for the isomerisation of allyl alcohols (to BASF AG), WO 02/24617 (2002); *Chem. Abstr.*, **136**, 281140 (2002).
21. R. Noyori, Asymmetric terpenoid synthesis, in *Recent Developments in Flavour and Fragrance Chemistry Proceedings of the Third International Haarmann & Reimer Symposium*, R. Hopp, K. Mori (Eds), VCH, Weinheim, 1993, pp. 3–11.
22. (*a*) H. Kumobayashi, Y. Okeda, T. Okazaki, S. Akutagawa, Process for the preparation of optically active enamines or imines (to Takasago Perfumery Co.), EP 0 170 470 (1986); *Chem. Abstr.*, **106**, 50506 (1987); (*b*) S. Otsuka, K. Tani, T. Yamagata, S. Akutagawa, H. Kumobayashi, M. Yagi, Process for the preparation of enamines or imines (to Takasago Perfumery Co.), EP 0 068 506 (1983); *Chem. Abstr.*, **98**, 215832 (1983).
23. M. Erman, New developments in physiological coolants, *Perfum. Flavor.*, **32** (October), 20–35 (2007).
24. P. Fankhauser, P. Fantini, Use of unsaturated macrocyclic ketones as perfuming ingredients (to Firmenich SA), EP 0 424 787 (1991); *Chem. Abstr.*, **115**, 239362 (1991).
25. (*a*) M. Kerschbaum, Über Lactone mit großen Ringen – die Träger des vegetabilischen Moschus-Duftes, *Ber. Dtsch. Chem. Ges.*, **60**, 902–909 (1927); (*b*) K. Kerrola, B. Galambosi, H. Kallio, Characterization of volatile composition and odour of Angelica (*Angelica archangelica* subsp. *archangelica* L.) root extracts, *J. Agric. Food. Chem.*, **42**, 1979–1988 (1994) and references cited therein.
26. (*a*) G. Woker, The relations between structure and smell in organic compounds, *J. Phys. Chem.*, **10**, 455–473 (1906); (*b*) K. J. Rossiter, Structure–odor relationships, *Chem. Rev.*, **96**, 3201–3240 (1996); (*c*) C. S. Sell, On the unpredictability of odour, *Angew. Chem.*, **118**, 6402–6410 (2006); *Angew. Chem. Int. Ed.*, **45**, 6245–6261 (2006); (*d*) A. Triller, E. A. Boulden, A. Churchill, H. Hatt, J. Englund, M. Spehr, C. S. Sell, Odorant–receptor interactions and odor percept: a chemical perspective, *Chem. Biodivers.*, **5**, 862–886 (2008); (*e*) C. S. Sell, Rational odorant design: fantasy or feasibility? *Perfum. Flavor.*, **33** (September), 48–52 (2008).
27. R. E. Brown, What is the role of the immune system in determining individually distinct body odors?, *Int. J. Immunopharm.*, **17**, 655–661 (1995).
28. R. Bentley, The nose as a stereochemist. Enantiomers and odour, *Chem. Rev.*, **106**, 4099–4112 (2006).

29. (*a*) C. S. Sell, Scent through the looking glass, *Chem. Biodivers.*, **1**, 1899–1920 (2004); (*b*) E. Brenna, C. Fuganti, S. Serra, Enantioselective perception of chiral odorants, *Tetrahedron: Asymmetry*, **14**, 1–42 (2003).
30. (*a*) R. Axel, Scents and sensibility: a molecular logic of olfactory perception (Nobel lecture) *Angew. Chem.*, **117**, 6264–6282 (2005); *Angew. Chem. Int. Ed.*, **44**, 6110–6127 (2005); (*b*) L. B. Buck, Unraveling the sense of smell (Nobel lecture) *Angew. Chem.*, **117**, 6283–6296 (2005); *Angew. Chem. Int. Ed.*, **44**, 6128–6140 (2005).
31. (*a*) C. Chapuis, In the quest for a virtual pseudo receptor for sandalwood-like odorants, Part I, *Chem. Biodivers.*, **7**, 980–1021 (2004) and references cited therein;. (*b*) G. Buchbauer, A. Hillisch, K. Mraz, P. Wolshann, Conformational parameters of sandalwood-odor activity: conformational calculations on sandalwood, *Helv. Chim. Acta*, **77**, 2286–2296 (1994); (*c*) S. Bieri, K. Monastyrskaia, B. Schilling, Olfactory receptor neuron profiling using sandalwood odorants, *Chem. Senses*, **29**, 483–487 (2004).
32. (*a*) A. Keller, H. Zhang, Q. Chi, L. B. Vosshall, H. Matsunami, Genetic variation in a human odorant receptor alters odour reception, *Nature*, **449**, 468–473 (2007); (*b*) K. Schmiedeberg, E. Shirokova, H.-P. Weber, B. Schilling, W. Meyerhof, D. Krautwurst, Structural determinants of odorant recognition by human olfactory receptors OR1A1 and OR1A2 *J. Struct. Biol.*, **159**, 400–412 (2007).
33. E. Polak, G. Fetison, A.-M. Fombon, A. Skalli, Structure–odor relationships for 'catty' smelling mercapto compounds in humans, *J. Agric. Food. Chem.*, **36**, 355–359 (1988) and references cited therein.
34. D. Lamparsky, P. Schudel, *p*-Menthane-8-thiol-3-one, a new component of Buchu leaf oil, *Tetrahedron Lett.*, **36**, 3323–3326 (1971).
35. (*a*) G. Ohloff, 75 Jahre Riechstoff- und Aroma-Chemie im Spiegel der Helvetica Chimica Acta. Teil I, *Helv. Chim. Acta*, **75**, 1341–1415 (1992); (*b*) G. Ohloff, 75 Jahre Riechstoff- und Aroma-Chemie im Spiegel der Helvetica Chimica Acta. Teil II, *Helv. Chim. Acta*, **75**, 2041–2107 (1992); (*c*) G. Fráter, J. A. Bajgrowicz, P. Kraft, Fragrance chemistry, *Tetrahedron*, **54**, 7633–7703 (1998); (*d*) P. Kraft, J. A. Bajgrowicz, C. Denis, G. Fráter, Odds and trends: recent developments in the chemistry of odorants, *Angew. Chem.*, **112**, 3106–3138 (2000); *Angew. Chem. Int. Ed.*, **39**, 2980–3010 (2000).
36. (*a*) P. Fankhauser, Composé spiranique insaturé, son utilisation dans les parfums et les aromes et procédé pour sa préparation (to Firmenich SA), EP 0 033 959 (1981); *Chem. Abstr.*, **95**, 203728 (1981); (*b*) C. Chapuis, 1-Oxaspiro(4,5)dec-3-ene derivatives as perfuming ingredients (to Firmenich SA), WO 2007/129236 (2007); *Chem. Abstr.*, **147**, 527764 (2007).
37. (*a*) A. J. A. van der Weerdt, H. Renes, Perfume and flavouring compositions, perfumed articles and materials, and flavoured food products containing an oxaspirodecadiene derivative as essential compound as well as the oxaspiro[4.5]decadiene derivative *per se* (to Naarden International BV), EP 0 094 722 (1983); *Chem. Abstr.*, **100**, 120864 (1984); (*b*) P. Kraft, K. Popaj, A. Abate, Design, synthesis and olfactory properties of 2-substituted 2-*tert*-butyl-5-methyl-2,5-dihydrofurans: *seco*-derivatives of the theaspirenes, *Synthesis*, **16**, 2798–2809 (2005).
38. A. P. S. Narula, J. J. De Vigilio, C. Benaim, A. V. Ouwerkerk, O. Gillotin, Substituted cyclopentenyl-oxabicyclooctanes, cyclopentenyl-formyl cyclohexenes, cyclopentenyl-hydroxymethyl cyclohexenes, processes for preparing the same and organoleptic uses thereof (to International Flavors and Fragrances Inc.), US 5,087,707 (1991); *Chem. Abstr.*, **116**, 152090 (1992).
39. D. Munro, C. S. Sell, Novel fragrance compound (to Quest International BV), WO 99/21847 (1999); *Chem. Abstr.*, **130**, 338013 (1999).
40. C. Tarchini, 1,2,3,4,4a,8a-Octahydro-2,2,6,8-tétraméthyl-1-naphthénol son utilisation en tant qu'ingredient parfumant et procédé pour sa préparation (to Firmenich SA), EP 0 167 709 (1986); *Chem. Abstr.*, **105**, 226083 (1986).
41. K. Palmer, Novel fragrance compounds (to Quest International BV), WO 01/00551 (2001); *Chem. Abstr.*, **134**, 71734 (2001).

42. D. Munro, Pyridine derivatives useful as fragrance materials (to Quest BV), WO 2004/048336 (2004); *Chem. Abstr.*, **141**, 38532 (2004).
43. W. Skorianetz, W. Renold, G. Ohloff, K.-H. Schulte-Elte, Esters of certain tetramethyl and pentamethyl-1-oxa-spiro[4.5]decan-6-ols with certain alkanoic acids (to Firmenich SA), US 4,014,905 (1977); *Chem. Abstr.*, **87**, 73246 (1977).
44. B. M. Lawrence, A preliminary report on the world production of some selected essential oils and countries, *Perfum. Flavor.*, **34** (January), 39–44 (2009).
45. (*a*) G. Büchi, R. E. Erickson, N. Wakabayashi, Terpenes XVI. Constitution of patchouli alcohol and absolute configuration of cedrene, *J. Am. Chem. Soc.*, **83**, 927–938 (1961); (*b*) G. Büchi, W. D. Macleod, Synthesis of patchouli alcohol, *J. Am. Chem. Soc.*, **84**, 3205–3206 (1962); (*c*) See also: D. H. R. Barton *et al.* (ref. 51) for a more detailed discussion of the patchouli alcohol story.
46. (*a*) T. V. Magee, G. Stork, P. Fludzinski, A total synthesis of *rac*-patchouli alcohol, *Tetrahedron Lett.*, **36**, 7607–7610 (1995); (*b*) K. Yamada, Y. Kyotani, S. Manabe, M. Suzuki, Total synthesis of (±)-patchouli alcohol and (±) seychellene via a common homoisotwistane intermediate, *Tetrahedron*, **35**, 293–298 (1979); (*c*) R. N. Mirrington, K. J. Schmalzl, Studies with bicyclo[2.2.2]octenes. V. The total synthesis of (±)-patchouli alcohol, *J. Org. Chem.*, **37**, 2871–2877 (1972); (*d*) A. Srikrishna, G. Satyanarayana, An enantiospecific total synthesis of (–)-patchouli alcohol, *Tetrahedron: Asymmetry*, **16**, 3992–3997 (2005); (*e*) H. Spreitzer, A study on the odor/structure relationship of patchoulol and norpatchoulenol, *Helv. Chim. Acta*, **73**, 1730–1733 (1990); (*f*) W. Oppolzer, R. L. Snowden, A short synthesis of norpatchoulenol, *Tetrahedron Lett.*, **37**, 3505–3506 (1978).
47. (*a*) F. Naef, G. Ohloff, A short stereoselective total synthesis of racemic patchouli alcohol, *Helv. Chim. Acta*, **57**, 1868–1870 (1974); (*b*) F. Naef, R. Decorzant, W. Giersch, G. Ohloff, A stereocontrolled access to (±)-, (–)- and (+)-patchouli alcohol, *Helv. Chim. Acta*, **64**, 1387–1397 (1981).
48. F. Naef, R. Decorzant, W. Thommen, An efficient approach to patchouli alcohol analogues, *Helv. Chim. Acta*, **60**, 1196–1202 (1977).
49. E. Napolitano, E. Giovani, N. Ceccarelli, P. Pelosi, Tertiary alcohols with earthy odour, *J. Agric. Food Chem.*, **44**, 2806–2809 (1996) and references cited therein.
50. C. Vial, R. Moretti, A. Charpilloz, P. Fankhauser, P. Fantini, *Spiro* compounds as perfuming ingredients (to Firmenich SA), US 2004/0072721 (2004); *Chem. Abstr.*, **140**, 326677 (2004).
51. D. H. R. Barton, J.-C. Beloeil, A. Billon, J. Boivin, J.-Y. Lallemand, S. Mergui, Functionalization of saturated hydrocarbons, oxidation of patchouli alcohol by the '*gif* system': isolation and organoleptic properties of three new ketonic derivatives, *Helv. Chim. Acta*, **70**, 272–280 (1987).
52. (*a*) P. Kraft, W. Eichenberger, D. Frech, From vetiver to patchouli: discovery of a new high-impact spirocyclic patchouli odorant, *Eur. J. Org. Chem.*, 3233–3245 (2005); (*b*) P. Kraft, A. Bruneau, Ring reversal of a spirocyclic patchouli odorant: molecular modelling, synthesis and odor of 6-hydroxy-1,1,6-trimethylspiro[4.5]deca-7-one, *Eur. J. Org. Chem.*, 2257–2267 (2007); (*c*) P. Kraft, Woody pretzels: spirocycles from vetiver to patchouli and georgywood, *Chem. Biodivers.*, **5**, 970–998 (2008); (*d*) A. Jahnke, C. Burschka, R. Tacke, P. Kraft, Synthesis and olfactory evaluation of (4a*R**,8a*R**)-1,1,8a-trimethyl-decahydronaphthalen-4a-ol: a *cis*-decanol intersection structure of (–)-patchoulol and (5*R**,6*S**)-1,1,6-trimethylspiro[4.5]decan-6-ol, *Synthesis*, 62–68 (2009); erratum: *Synthesis*, 512 (2009).
53. P. Kraft, C. Weymuth, C. Nussbaumer, Total synthesis and olfactory evaluation of (1*R**,3*S**,6*S**,7*S**,8*S**)-3-hydroxy-6,8-dimethyltricyclo[5.3.1.0]undecan-2-one: a new synthetic route to the patchoulol skeleton, *Eur. J. Org. Chem.*, 1403–1412 (2006).
54. R. Moretti, Patchouli odorant (to Firmenich SA), WO 2007/135582 (2007); *Chem. Abstr.*, **148**, 10969 (2008).
55. R. Moretti, O. Etter, Patchoulol odorant (to Firmenich SA), WO 2009/066193 (2009); *Chem. Abstr.*, **150**, 564007 (2009).
56. (*a*) A. S. Williams, The synthesis of macrocyclic musks, *Synthesis*, 1707–1723 (1999); (*b*) M. Gautschi, J. A. Bajgrowicz, P. Kraft, Fragrance chemistry – milestones and perspectives, *Chimia*, **55**, 379–387 (2001); (*c*) P. Kraft, G. Fráter, Enantioselectivity of musk odor sensation,

Chirality, **13**, 388–394 (2001); (*d*) P. Kraft, 'Brain aided' musk design, *Chem. Biodivers.*, **1**, 1957–1974 (2004).

57. D. Munro, K. Palmer, Preparation of macrocyclic ketones utilizing addition of ketones to alkenes (to Quest International BV), WO 00/01648 (2000); *Chem. Abstr.*, **132**, 78314 (2000).
58. T. Yamamoto, M. Ogura, T. Kanisawa, A novel synthetic method for (*R*)- and (*S*)-muscones by enantioseletive hydrogenation of (*E*)- and (*Z*)-3-methyl-2-cyclopentadecen-1-ones catalyzed by *p*-tolyl-BINAP-Ru(II) complexes, *Tetrahedron*, **58**, 9209–9212 (2002).
59. C. Fehr, J. Galindo, O. Etter, An efficient enantioselective synthesis of (+)-(*R,Z*)-5-muscenone and (–)-(*R*)-muscone – an example of a kinetic resolution and enantioconvergent transformation, *Eur. J. Org. Chem.*, 1953–1957 (2004).
60. (*a*) O. Knopff, J. Kuhne, C. Fehr, Enantioslective intramolecular aldol addition/dehydration reaction of a macrocyclic diketone: synthesis of the musk odorants (*R*)-muscone and (*R,Z*)-5-muscenone, *Angew. Chem.*, **119**, 1329–1332 (2007); *Angew. Chem. Int. Ed.*, **46**, 1307–1310 (2007); (*b*) O. Knopff, A process for the preparation of optically active cyclohexenones (to Firmenich SA), WO 2005/077875 (2005); *Chem. Abstr.*, **143**, 248092 (2005); (*c*) O. Knopff, Substituted cyclohexenones (to Firmenich SA), WO 2009/095804 (2009); *Chem. Abstr.*, **151**, 245405 (2009).
61. D. Helmlinger, G. Fráter, U. Müller, Methylcyclotetradec-5-en-1-ones (to Givaudan SA), EP 0 983 989 (2000); *Chem. Abstr.*, **132**, 194247 (2000).
62. (*a*) D. Munro, C. S. Sell, 14-Methyl-hexadecenolide and 14-methyl-hexadecanolide (to Quest BV), EP 0 841 333 (1998); *Chem. Abstr.*, **129**, 4590 (1998); (*b*) G. Fráter, D. Helminger, U. Müller, Macrocyclen (to Givaudan SA), EP 0 908 455 (1999); *Chem. Abstr.*, **130**, 301508 (1999); (*c*) J. Bajgrowicz, G. Fráter, P. Kraft, Thiamacrolide (to Givaudan SA), EP 0 859 003 (1998); *Chem. Abstr.*, **129**, 175655 (1998); (*d*) P. Kraft, Macrocyclic lactones as fragrances (to Givaudan SA), WO 2009/039675 (2009); *Chem. Abstr.*, **150**, 405684 (2009).
63. P. Kraft, C. Berhold, (4*E*,8*Z*)-12-Methyloxacyclotetradeca-4,8-dien-2-one and its 7a-homologue: conformationally constrained double-unsaturated macrocyclic musks by ring-closing metathesis, *Synthesis*, 543–550 (2008) and references cited therein.
64. W. Hoffmann, K. von Fraunberg, Cyclopentenderivate (to BASF AG), DE 25 13 996 (1976); *Chem. Abstr.*, **86**, 16353 (1977).
65. (*a*) W. K. Giersch, K.-H. Shulte Elte, C. Mahaim, Esters nouveaux et leur utilisation en parfumerie (to Firmenich SA), EP 0 472 966 (1992); *Chem. Abstr.*, **117**, 5713 (1992); (*b*) A. S. Williams, Odorant compounds (to Firmenich SA), EP 1 459 735 (2004); *Chem. Abstr.*, **141**, 301056 (2004).
66. A. S. Williams, Esters with musky odour and their use in perfumery (to Firmenich SA), WO 2000/014051 (2000); *Chem. Abstr.*, **132**, 194523 (2000).
67. A. S. Williams, C. Fehr, Perfuming ingredient with a floral character (to Firmenich SA), US 2004/0167059 (2004); *Chem. Abstr.*, **141**, 230344 (2004).
68. (*a*) P. Kraft, R. Cadalbert, Cycloalkanecarboxylic acid derivatives as fragrances with musk characteristics (to Givaudan SA) *EP 1 262 474* (2002); *Chem. Abstr.*, **138**, 4714 (2003); (*b*) P. Kraft, Esters and their use in perfumery (to Givaudan SA), WO 2004/050602 (2004); *Chem. Abstr.*, **141**, 38376 (2004); (*c*) P. Kraft, Alicyclic carboxylic acid oxycarbonylmethyl esters and their use as odorants (to Givaudan SA), WO 2005/108534 (2005); *Chem. Abstr.*, **143**, 465642 (2005); (*d*) P. Kraft, Aliphatic compounds as fragrances with musk characteristics (to Givaudan SA), WO 2004/050595 (2004); *Chem. Abstr.*, **141**, 38362 (2004); *(e)* J. O. Bledsoe, M. Britten-Kelly, M. A. Sprecker, R. P. Belko, M. Pawlak, M. G. Montelone, Fruity musky compositions (to International Flavors and Fragrances Inc.), EP 1 398 366 (2004); *Chem. Abstr.*, **140**, 258663 (2004);. (*f*) A. T. Levorse Jr., C. M. Smith, R. A. Weiss, Optically active ester compounds and their use in perfume compositions (to International Flavors and Fragrances Inc.), US 7,563,925 (2009); *Chem. Abstr.*, **151**, 180925 (2009).
69. (*a*) P. Kraft, W. Eichenberger, Synthesis and odor of alicyclic musks: discovery of a new class of odorants, *Eur. J. Org. Chem.*, 354–365 (2004); (*b*) M. Eh, New acyclic musks: the fourth generation of musk odorants, *Chem. Biodivers.*, **1**, 1975–1984 (2004).

70. M. Klaus, D. Helmlinger, M. Pesaro, Substituted bicyclo[4.3.0]non-1-6-en-8-yl methyl ketones and fragrance and flavour compositions (to Givaudan SA), EP 0 116 277 (1984); *Chem. Abstr.*, **102**, 61860 (1985).
71. D. Helmlinger, M. Pesaro, Novel bicyclic epoxides and compositions (to Givaudan SA), EP 0 115 274 (1984); *Chem. Abstr.*, **102**, 62066 (1985).
72. J. B. Hall, L. Kewalram Lala, Allylic oxidation of bicyclic hydrocarbons to alcohols and ketones (to International Flavors and Fragrances Inc.), US 3,847,993 (1974); *Chem. Abstr.*, **78**, 124151 (1973).
73. C. Fehr, Musk odorant compounds (to Firmenich SA), WO 2005/054220 (2005); *Chem. Abstr.*, **143**, 47784 (2005).
74. D. Munro, Novel fragrance compounds (to Quest International B.V), EP 1 036 792 (2000); *Chem. Abstr.*, **133**, 237864 (2000).
75. C. Vial, P.-A. Blanc, Use of dihydrobenzofurans as scenting ingredients (to Firmenich SA), WO 95/30667 (1995); *Chem. Abstr.*, **124**, 155724 (1996).
76. J. Kula, R. Bonikowski, M. Staniszewska, A. Krakowiak, M. W. Wieczorek, W. R. Majzner, G. D. Bujacz, Transformation of carotol into the hydroindane-derived musk odorant, *Eur. J. Org. Chem.*, 1826–1829 (2002).
77. (*a*) P. Kraft, K. Popaj, Musk or violet? Design, synthesis and odor of *seco*-derivatives of a musky carotol lead, *Tetrahedron*, **62**, 12211–12219 (2006); (*b*) P. Kraft, K. Popaj, New musk odorants: (3*E*)-4-(2′-alkyl-5,5′-dimethylcyclopent-1′-enyl)but-3-en-2-ones and (3*E*)-1-acetyl-3-alkylidene-4,4-dimethylcyclohexenones, *Eur. J. Org. Chem.*, 4806–4814 (2008).
78. (*a*) B. Lawrence, Progress in essential oils, *Perfum. Flavor.*, **34** (May), 52–56 (2009); (*b*) M. Nagesawa-Rao, S. Padmini, K. N. Ganeshaiah, R. U. Shaanker, J. R. Soneji, Material study: Indian sandalwood crisis, *Perfum. Flavor.*, **33** (October), 38–44 (2008); (*c*) P. Biggs, Raw material bulletin: sustainable Australian sandalwood, *Perfum. Flavor.*, **32** (April), 28–30 (2007). East Indian Sandalwood (*Santalum album* sp.) is included on the Convention on International Trade in Endangered Species of Wild Fauna and Flora (CITES) Red List whereby the export of timber is banned, and the species is vunerable.
79. C. Chapuis, P.-A. Blanc, Dérivés de l'aldéhyde campholénique et leur utilisation en parfumerie (to Firmenich SA), EP 0 694 520 (1995); *Chem. Abstr.*, **124**, 261418 (1996).
80. (*a*) J. A. Bajgrowicz, I. Frank, G. Fráter, M. Henning, Synthesis and structural elucidation of a new potent sandalwood-oil substitute, *Helv. Chim. Acta*, **81**, 1349–1358 (1998); (*b*) G. Brunner, L. Eberhard, J. Oetiker, F. Schröder, Tandem cyclopropanation with dibromomethane under Grignard conditions, *J. Org. Chem.*, **73**, 7543–7554 (2008); (*c*) J. A. Bajgrowicz, G. Fráter, Cyclopentanol derivatives as odorants (to Givaudan SA), EP 0 801 049 (1997); *Chem. Abstr.*, **127**, 358652 (1997); (*d*) F. Schröder, Cyclopropanation process (to Givaudan SA), WO 2009/023980 (2009); *Chem. Abstr.*, **150**, 282331 (2009).
81. J. A. Bajgrowicz, Organic compounds (to Givaudan SA), WO 2008/052379 (2008); *Chem. Abstr.*, **148**, 517464 (2008).
82. C. Chapuis, Compounds having sandalwood odors (to Firmenich SA), WO 2010/061316 (2010).
83. C. Brocke, M. Eh, A. Finke, Recent developments in the chemistry of sandalwood odorants, *Chem. Biodivers.*, **5**, 1000–1010 (2008).
84. P. A. Christenson, N. Secord, B. J. Willis, Identification of *trans*-β-santalol and *epi-cis*-β-santalol in East Indian sandalwood oil, *Phytochemistry*, **20**, 1139–1141 (1981).
85. A. Krotz, G. Helmchen, Total synthesis of sandalwood fragrances: (*Z*)- and (*E*)-β-santalol and their enantiomers, *ent*-β-santalene, *Tetrahedron: Asymmetry*, **8**, 537–540 (1990).
86. (*a*) R. L. Snowden, P. Sonnay, G. Ohloff, Stereoselective synthesis of (±)-*epi*-β-santalene and (±)-*epi*-β-santalol, *Helv. Chim. Acta*, **64**, 25–32 (1981); (*b*) G.-F. Zhong, M. Schlosser, A short and simple access to both enantiomers of *epi*-β-santalene and (*Z*)-*epi*-β-santalol, *Synlett*, 173–174 (1994); (*c*) E. J. Brunke, F. J. Hammerschmidt, H. Struwe, (+)-*epi*-β-Santalol, Isolierung aus Sandelholzöl, und Partialsynthese aus (+)-α-Santalol, *Tetrahedron Lett.*, **21**, 2405–2408 (1980); (*d*) P. A. Christenson, B. J. Willis, East Indian sandalwood oil 2. Stereoselective synthesis of (±)-*epi*-β-santalene and (±)-*epi*-β-santalol, *J. Org. Chem.*, **45**, 3068–3072 (1980).

87. (*a*) D. Solas, J. Wolinsky, Total synthesis of β-santalol, *J. Org. Chem.*, **48**, 1988–1991 (1983); (*b*) H. Monti, C. Corriol, M. Bertrand, Synthèse stéréoselective du (±)-β-santalol, *Tetrahedron Lett.*, **23**, 5539–5540 (1982); (*c*) H. C. Kretschmar, W. F. Erman, The total synthesis and geometric configuration of D,L-β-santalol, *Tetrahedron Lett.*, **1**, 41–44 (1970); (*d*) Z. Painoski, L. Rupniki, P. Cmoch, K. Stalinski, Radical cyclizations leading to the bicycle[2.2.1]heptanes framework: a new radical approach to (±)-(Z)-β-santalol, *Synlett*, 900–904 (2005), and references cited in these articles.
88. P. A. Christenson, B. J. Willis, East Indian sandalwood oil. 1. Stereoselective synthesis of (±)-β-santalene and (±)-β-santalol, *J. Org. Chem.*, **44**, 2012–2018 (1979).
89. (*a*) M. Baumann, W. Hoffmann, Synthese von *rac*-β-Santalol und *rac*-β-Dehydrosantalol, *Liebigs Ann.*, **6**, 743–950 (1979); (*b*) W. Hoffmann, K. von Fraunberg, Bicyclic aldehydes (to BASF AG), US 4,197,411 (1980); *Chem. Abstr.*, **90**, 86872 (1979).
90. C. Fehr, I. Magpantay, J. Arpagaus, X. Marquet, M. Vuagnoux, Enantioselective synthesis of (–)-β-santalol by a copper-catalyzed enyol cyclization–fragmentation reaction, *Angew. Chem.*, **121**, 7357–7359 (2009); *Angew Chem. Int. Ed.*, **48**, 7221–7223 (2009).
91. Y. Hayashi, S. Samanta, H. Gotoh, H. Ishikawa, Asymmetric Diels–Alder reactions of α,β-unsaturated aldehydes catalyzed by a diarylprolinol silyl ether salt in the presence of water, *Angew. Chem.*, **120**, 6736–6739 (2008); *Angew. Chem. Int. Ed.*, **47**, 6634–6637 (2008).
92. E. J. Corey, H. Yamamoto, Modification of the Wittig reaction to permit the stereospecific synthesis of certain trisubstituted olefins. Stereospecific synthesis of α-santalol, *J. Am. Chem. Soc.*, **92**, 226–228 (1970).
93. For representative attempts to selectively install the (*Z*)-allylic alcohol motif see: M. Schlosser, G.-F. Zhong, A shortcut to α-santalol, *Tetrahedron Lett.*, **34**, 5441–5444 (1993).
94. C. Fehr, M. Vuagnoux, Process for the preparation of β-santalol and derivatives thereof (to Firmenich SA), WO 2009/141781 (2009); *Chem. Abstr.*, **151**, 571216 (2009).
95. (*a*) C. A. Lipinski, F. Lombardo, B. W. Dominy, P. J. Feeney, Experimental and computational approaches to estimate solubility and permeability in drug discovery and developmental settings, *Adv. Drug Delivery Rev.*, **23**, 3–25 (1997); (*b*) S. Houlton, Breaking the rules, *Chem. World*, June, 58–62 (2008); (*c*) P. D. Leeson, B. Springthorpe, The influence of drug-like concepts on decision-making in medicinal chemistry, *Nat. Rev. Drug Discovery*, **6**, 881–890 (2007); (*d*) J. R. Proudfoot, The evolution of synthetic drug properties, *Bioorg. Med. Chem. Lett.*, **15**, 1087–1090 (2005); (*e*) Y. C. Martin, A bioavailability score, *J. Med. Chem.*, **48**, 3164–3170 (2005).
96. (*a*) C. M. Tice, Selecting the right compounds for screening: does Lipinski's Rule of 5 for pharmaceuticals apply to agrochemicals? *Pest Manage. Sci.*, **57** 3–16 (2001); (*b*) G. M. Ramos Tombo, D. Belluš, Chirality and crop protection, *Angew. Chem.*, **103**, 1219–1241 (1991); *Angew Chem. Int. Ed.*, **30**, 1193–1386 (1991).
97. C. Chapuis, D. Jacoby, Catalysis in the preparation of fragrances and flavours, *Appl. Catal., A*, **221**, 93–117. (2001).
98. (*a*) B. Schilling, Metabolic method to identify compounds (to Givaudan SA), WO 2006/007751 (2006); *Chem. Abstr.*, **144**, 176960 (2006); (*b*) B. Schilling, Method to identify or evaluate compounds useful in the field of fragrances and aromas (to Givaudan SA), WO 2006/007752 (2006); *Chem. Abstr.*, **144**, 134740 (2006); (*c*) B. Schilling, T. Granier, G. Fráter, A. Hanhart, Organic compounds and compositions having the ability to modulate fragrance compositions (to Givaudan SA), WO 2008/116338 (2008); *Chem. Abstr.*, **149**, 425525 (2008); (*d*) B. Schilling, W. D. Woggon, A. Chougnet, T. Granier, G. Fráter, Organic compounds (to Givaudan SA), WO 2008/116339 (2008); *Chem. Abstr.*, **149**, 397757 (2008).
99. (*a*) M. Zarzo, The sense of smell: molecular basis of odorant recognition, *Biol. Rev.*, **82**, 455–479 (2007); (*b*) M. Zarzo, The debate over rational design of odorants, *Perfum. Flavor.*, **32** (April), 20–26 (2007).
100. (*a*) J. Wang, Z. A. Luthey-Schulten, K. S. Suslick, Is the olfactory receptor a metalloprotein? *Proc. Natl. Acad. Sci. USA*, **100**, 3035–3039 (2003); (*b*) see also: L. Turin, F. Yoshii, Structure–odor relations: a modern perspective, in *Handbook of Olfaction and Gustation*, R. L. Doty (Ed.), Marcel Dekker, New York, 2003, pp. 275–295, and references quoted therein.

101. (*a*) A. Chester, S. Firestein, Neuroscience: current views on odour receptors, *Nature*, **452**, 944–945 (2008); (*b*) K. Sato, M. Pellegrino, N. Nakagawa, T. Nakagawa, L. B. Vosshall, K. Touhara, Insect olfactory receptors are heterodimeric ligand-gated ion channels, *Nature*, **452**, 1002–1006 (2008).
102. (*a*) G. Ohloff, W. K. Giersch, Stereochemistry–activity relationships in olfaction. Odorants containing a proton donor/proton acceptor unit, *Helv. Chim. Acta*, **63**, 76–94 (1980); (*b*) G. Ohloff, Chemistry of odour stimuli, *Experientia*, **42**, 271–279 (1986).
103. Y. Oka, A. Nakamura, H. Watanabe, K. Touhara, An odorant derivative as an antogonist for an olfactory receptor, *Chem. Senses*, **29**, 815–822 (2004).

8

The Biosynthesis of Volatile Sulfur Flavour Compounds

Meriel G. Jones

The School of Biological Sciences, The University of Liverpool, Liverpool, L69 7ZB, UK

8.1 Introduction: Flavours as Secondary Metabolites

Many natural biological flavours are products of secondary metabolism and pose a particular challenge in understanding both their biosynthetic pathways and roles within their native organism. The concepts of primary and secondary metabolism were adopted at the end of the nineteenth century to distinguish the pathways for the biosynthesis of the essential metabolites required in all cells (e.g. amino acids, lipids, carbohydrates, nucleotides), whether plant, animal or microbial, from the metabolites that were frequently species-specific and initially considered dispensable (e.g. isoprenoids, phenolics, steroids). These definitions have inevitably become looser as knowledge of metabolism and genetics has advanced and demonstrate both complexity and conservation in biosynthetic pathways for secondary metabolites. The biosynthesis of secondary metabolites is now seen as the result of a highly organized and controlled process, frequently requiring large numbers of enzymes and complex genetic regulation. Furthermore, many secondary metabolites are now known to be essential for survival.[1]

The early appreciation that secondary metabolites are usually restricted to a particular species has remained true. In contrast to the substantial uniformity of primary metabolism, secondary metabolism is frequently species- and indeed cell type-specific and understanding their biosynthesis requires research within the specific producer organisms. The model organism approach, which proved so powerful in advancing biological knowledge during the late twentieth century, therefore has limited value because the

The Chemistry and Biology of Volatiles Edited by Andreas Herrmann

exact biosynthetic pathway is not found elsewhere.[2] However, although the full details of the biosynthetic routes for many flavour compounds are currently incomplete, the origin of secondary metabolism through diversifying a limited number of primary biochemical pathways means that there is nevertheless considerable knowledge to guide investigators. In addition, developments from high-throughput postgenomic technologies mean that substantial amounts of new information on many species are accumulating rapidly.[3]

The notion that these compounds are in any way dispensable or 'secondary' is completely superseded through appreciation that many have important biological, especially ecological, functions.[1,4] The mere fact that plants are eaten by animals ensures that compounds that influence the level of consumption affect plant fitness (as measured by parameters including biomass and seed production) and are subject to natural selection within both the producer and the consumer. For plants that are part of human food, our tastes and agricultural practices have influenced their secondary metabolite composition. Chemical analysis has identified over 100 000 distinct secondary compounds and five times as many may exist. Although only some contribute to flavour and aroma, emerging knowledge of the diversity in sensory receptors in humans and animals adds a further complexity to this subject.[5] Flavours are appreciated following the perception of chemicals and materials within the mouth and nose, modulated by our memories.

8.2 Sulfur in Plant Biology

Sulfur is an important macronutrient in plant biology. It is an essential component of housekeeping biomolecules (the amino acids methionine and cysteine; sulfur-containing lipids; sulfur-containing extracellular polysaccharides, iron/sulfur clusters in electron transfer and redox sensing proteins; glutathione) and of compounds involved in defence against biotic and abiotic stresses in all (glutathione, elemental sulfur) or some [glucosinolates, alk(en)yl cysteine sulfoxides] plant species.[6–8] It is assimilated via the roots from sulfate ions (**1**) in the soil. Sulfate transporters driven by the cell membrane potential expel protons to take sulfate into the cytoplasm. These inorganic ions have to be assimilated into an organic form to take part in cell metabolism. This occurs mainly after the sulfate has been transported to the leaves. Scheme 8.1 shows how it occurs. The initial step is activation through reaction with the high-energy phosphate compound adenosine-5′-triphosphate (ATP) to form adenosine-5′-phosphosulfate (APS, **2**), liberating pyrophosphate. This can be converted to adenosine-3′-phosphate-5′-phosphosulfate (PAPS, **3**) by APS kinase and used for biosynthesis of some classes of sulfur compounds. One class, the glucosinolates (GS), or mustard oils, is of particular interest because these are precursors of the volatile sulfur flavour compounds in the Brassicaceae. However, most sulfur-containing metabolites are derived from cysteine (**4**), following reduction of APS to sulfite **5** and then sulfide **6**. The sulfide is added to *O*-acetylserine (**7**) to form cysteine, releasing acetate. This is the major route of sulfur into cell metabolites, including the amino acid methionine (**8**) and the important defence and redox regulation compound glutathione (γ-glutamyl cysteinyl glycine, **9**). The volatile sulfur compounds derived from alk(en)ylcysteine sulfoxides (CSOs), found mainly in the genus *Allium*, are synthesized via cysteine (**4**).

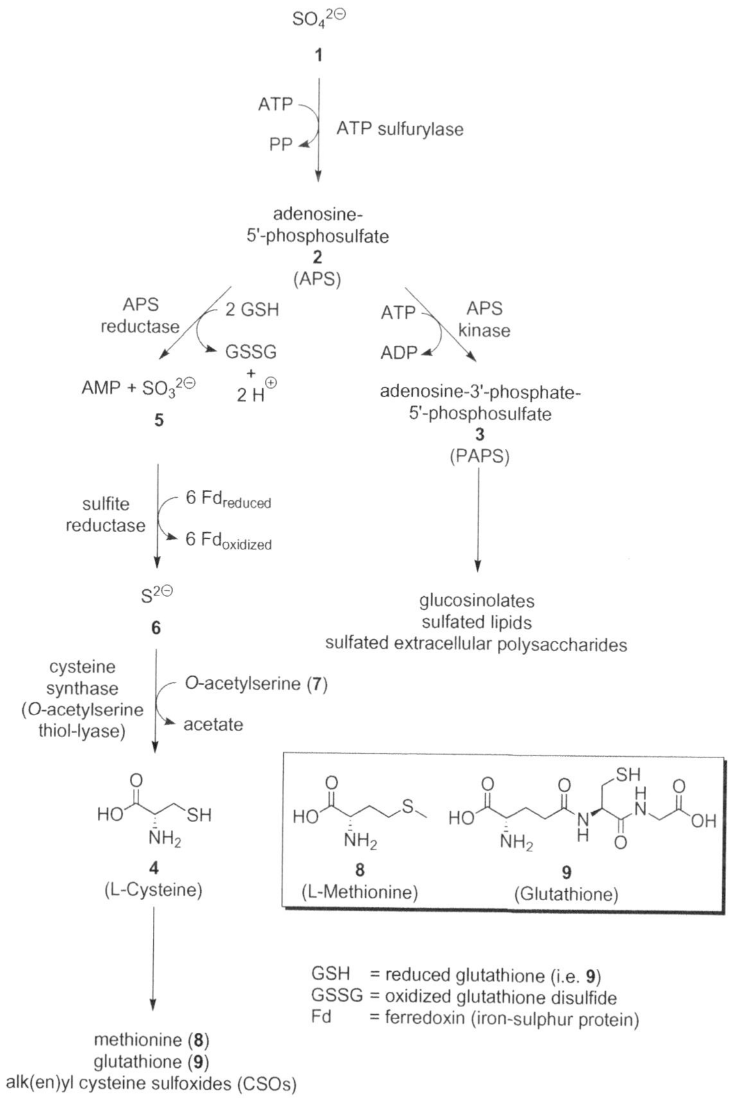

Scheme 8.1 *Sulfur assimilation in plants*

8.3 Sulfur Compounds as Flavour Volatiles

Sulfur compounds are important in flavours and odours. Many originate through chemical processes during food processing, especially cooking. This is not surprising after consideration of the reactive nature of sulfur chemistry. The importance to both attractive flavour notes and 'off' flavours means that there has been very extensive investigation of their identity through analysis and chemical synthesis. Microbial biotransformations of sulfur compounds that occur during food preservation or spoilage, and even within the mouth, have also attracted substantial attention.[9–13] A much more limited number of sulfur flavours are of plant origin, generated as the products of enzyme reactions within fresh vegetables. This chapter focuses on the biosynthesis of these natural compounds. These are best known from the onion (*Allium*) and cabbage (Brassicaceae) plant families and provide their well known flavours and odours that are important in all national cuisines. They illustrate how natural selection can generate substantial chemical diversity by utilizing sulfur chemistry, and there are excellent reviews of the products of the vapour phase chemical reactions that occur after the biological reactions.[14,15] Stable precursors are synthesized and stored within these plants for release as volatiles under appropriate conditions. The enzymatic synthesis of these precursors and the reactions that release the flavour volatiles are covered in detail here. These two groups of compounds are the best known, and most economically important, of the plant sulfur-containing flavour compounds. Other volatile sulfur compounds emitted from plants are covered briefly at the end of this chapter.

8.4 The Alk(en)yl Cysteine Sulfoxide Flavour Precursors

The most immediately striking fact about the volatiles of the *Allium* vegetables is that some, namely the onions, chives and shallots, have an acrid, tear-inducing effect, while others, notably garlic and garlic chives, have an equally distinctive, but nonlachrymatory, odour. These volatiles are derived from their differing content of four nonvolatile, odourless CSO precursors, as shown in Figure 8.1. All these compounds have two asymmetric centres, one at the C(α) in cysteine and a second at the sulfur. They are *S*-methyl-L-cysteine sulfoxide (**10**, MCSO, methiin; present in all *Allium* spp. and many Brassicaceae), *S*-2-propenyl-L-cysteine sulfoxide (**11**, ACSO, alliin, *S*-allyl-L-cysteine sulfoxide; characteristic of garlic and generating the volatiles that give its easily recognized smell), (*E*)-*S*-1-propenyl-L-cysteine sulfoxide (**12**, PeCSO, isoalliin; characteristic of onion and the precursor of a volatile tear-inducing compound, among others) and *S*-propyl-L-cysteine sulfoxide (**13**, PCSO, propiin; characteristic of onion and related species). Several additional CSOs have been identified either occasionally in trace amounts in *Allium* vegetables, in ornamental *Allium* spp. and also in a few other species with odours reminiscent of *Allium*.[16,17]

Allium spp. are used as vegetables and flavourings in food throughout the world. Around 20 species are cultivated and others are collected from the wild as part of regional cuisines. The onion (*A. cepa*) and similar species and varieties [shallots (*A. cepa aggregatum* L.), spring (bunching or Welsh) onions (*A. fistulosum* L.), chives (*A. schonopraesum*)], contain PeCSO (**12**) as the major CSO, with MCSO (**10**) and PCSO (**13**) as minor flavour precursors. The characteristic onion flavour and pungency is derived from the cleavage products of these precursors, especially PeCSO (**12**). A second group has the unmistakable odour and flavour of garlic, and includes not only garlic (*A. sativum*) but garlic chives (Chinese chives or leek, *A. tuberosum*), and leek (*A. porrum*, *A. ampelopraesum*). These species have ACSO (**11**) as the major CSO, with minor amounts of MCSO (**10**) and PCSO (**13**) and sometimes small amounts of PeCSO (**12**).[16]

Further CSOs have been identified in *Allium* spp., either restricted to a specific species or as trace amounts in several (see Figure 8.1). *S*-ethyl-L-cysteine sulfoxide (**14**) occurs in low to trace amounts in 13 *Allium* spp.[18] although it has not been consistently identified.[19] *A. siculum* contains butyl cysteine sulfoxide **15** and this has also been detected at trace levels in

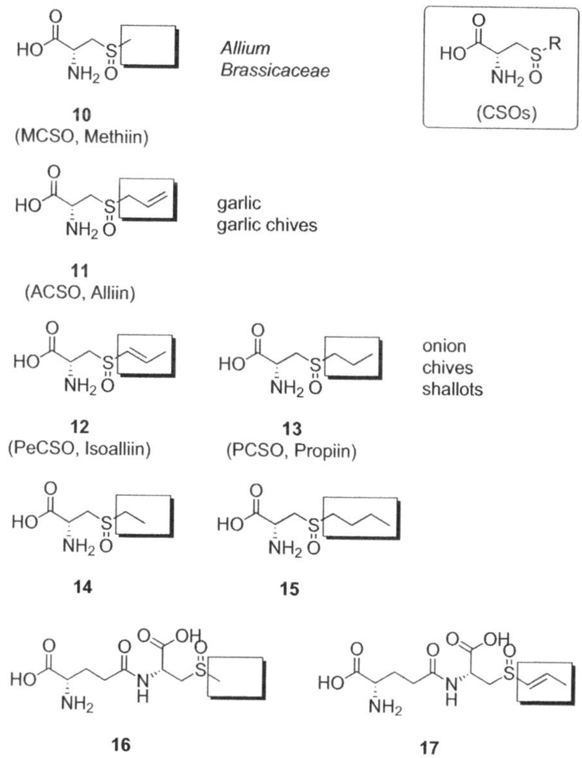

Figure 8.1 *Four widespread alk(en)yl cysteine sulfoxides (CSOs)* ***10–13*** *of Allium vegetables together with further CSOs* ***14*** *and* ***15*** *found in some wild species. Examples of γ-glutamyl peptides include γ-glutamyl-S-methylcysteine sulfoxide (**16**) and γ-glutamyl-(E)-S-1-propenyl-cysteine sulfoxide (**17**)*

some analyses of onions.[20] In addition to the CSOs, *Allium* spp. also contain significant amounts of at least 24 γ-glutamyl sulfoxide peptide derivatives (γGP; Figure 8.1, examples **16, 17**)[21] and a role for some of these in the biosynthesis of the CSOs has been suggested (see below).

8.5 Biosynthesis of the Flavour Precursors of *Allium*

Two pathways have been proposed for the biosynthesis of the CSOs and are shown in Scheme 8.2. One (Scheme 8.2a) requires γGPs as intermediates, derived from glutathione (**9**) with parallels to the pathway for its degradation while the other (Scheme 8.2b) is through direct (thio)alk(en)ylation of cysteine (**4**) or *O*-acetylserine (**7**) to *S*-methylcysteine (**18**) followed by oxidation, by analogy with the production of other secondary metabolites mediated by cysteine synthases (CSs).[17,22,23] There is experimental evidence supporting both routes and it is certainly possible that both occur, although perhaps in different tissues or times in the plant life history. The biological sources of the alk(en)yl groups remain to be

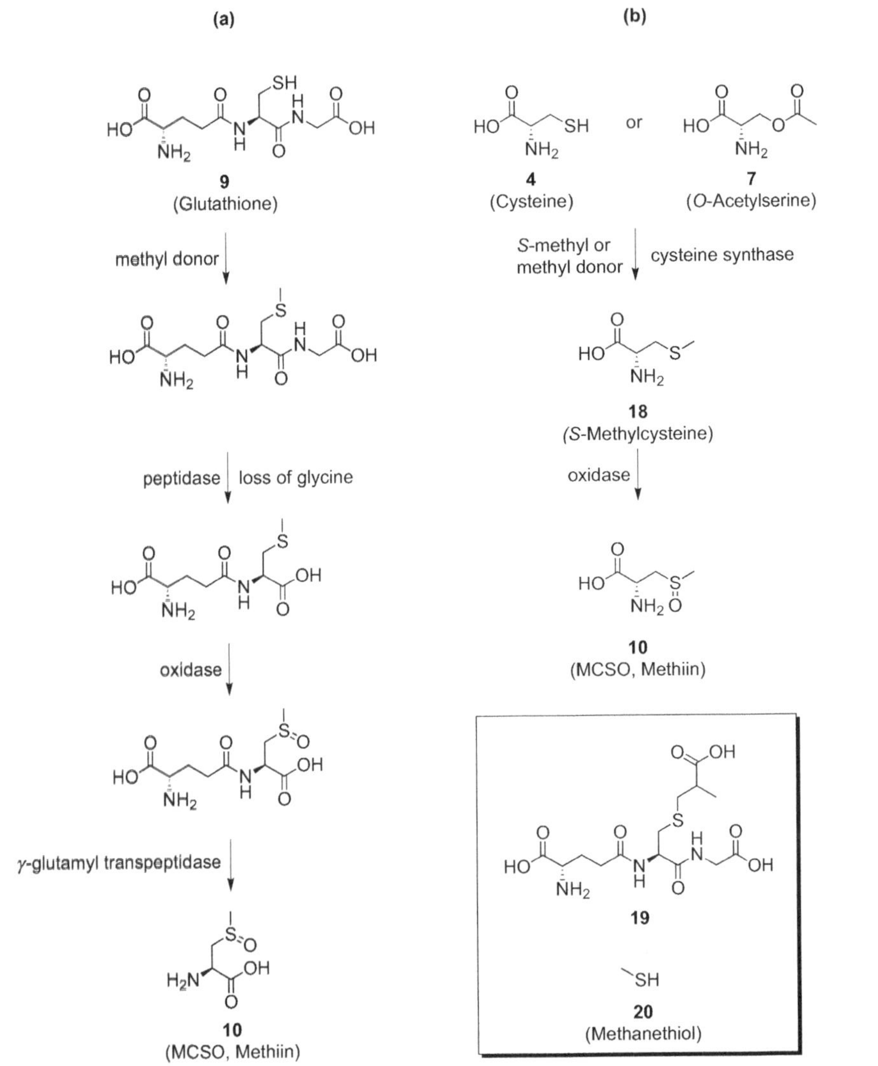

*Scheme 8.2 Proposed biosynthetic pathways for CSOs, illustrated for MCSO (**10**). (a) Via γ-glutamyl intermediates from glutathione (**9**). (b) Via cysteine (**4**) or O-acetylserine (**7**)*

identified. Similarly, although the classes of enzymes that are likely to perform some of the steps can be presumed, only a few of the proposed biosynthetic enzymes have been studied, purified or sequenced to provide detailed information.

8.5.1 The Biosynthesis of *Allium* Flavour Precursors via γ-Glutamyl Peptides

Glutathione (**9**) has a major role in redox homeostasis within all cells. Measures to prevent damage from reactive oxygen species formed during aerobic metabolism are essential in all cells and glutathione has a major role involving reversible oxidation and reduction of the sulfydryl group. It participates in the glutathione cycle to maintain enzymes and co-factors in a reduced state.[7] Any damage to *Allium* tissue initiates the release of highly oxidative sulfur volatiles that will inevitably react with glutathione. This provides a technical difficulty in working with the tissues and may allow interchange among the flavour precursors, explaining the sometimes differing findings of different research groups.

The biosynthetic pathway for CSOs via γGPs (Scheme 8.2a), proposed by Lancaster and her colleagues,[22] takes *S*-alk(en)ylation of glutathione as the starting point. This is followed by removal of the terminal glycyl residue by an enzyme with peptidase activity, oxidation at the sulfur to yield a sulfoxide by an oxidase and finally removal of the γ-glutamyl group by a transpeptidase to yield the CSO. Biosynthesis must also result in packaging the CSOs into the small intracytoplasmic storage vesicles in which they are securely isolated from alliinase, the key enzyme for volatile release.

Evidence for this pathway comes from radiolabelling studies, tissue analysis and some enzyme studies. Pulse labelling of excised onion seedling leaves with ^{35}S-sulfate showed that the label appeared first in glutathione and its derivatives, including γ-glutamyl-(*E*)-*S*-1-propenyl-L-cysteine sulfoxide **17** and only later appeared within the CSOs.[22] The appearance of the radiolabel in glutathione (**9**), following assimilation of the sulfate via cysteine biosynthesis, might be expected since the glutathione cycle is estimated to use up to one third of the cell's cysteine. Interestingly, the radiolabel appeared in *S*-2-carboxypropyl glutathione **19** in cut leaves from both onion and garlic, suggesting that this was an intermediate in the biosynthesis of ACSO (**11**), PeCSO (**12**) and PCSO (**13**).[22,23] In contrast, a later study, using intact sprouting onion sets, indicated a reverse order for the incorporation of ^{35}S, namely into CSOs with little evidence that γGPs had been labelled.[24]

Peptidase and γ-glutamyl transpeptidase activity has been identified in onion and garlic tissues. Ceci and coworkers showed that these enzyme activities were higher in rapidly growing garlic leaves than cloves.[25] One enzyme that might be involved in CSO synthesis from γGP has been partially purified from onion. It had both peptidase and γ-glutamyl transpeptidase activity and a broad specificity for glutathione and several γ-glutamyl peptides.[26] A further membrane-associated γ-glutamyl transpeptidase isolated from sprouting onion bulbs had a high substrate affinity for several glutathione conjugates although kinetic analysis suggested that γ-glutamyl-(*E*)-*S*-prop-1-enyl cysteine sulfoxide (**17**), the major γGP in onion, was unlikely to be a substrate *in vivo*.[27] These contrasting findings illustrate the difficulty of making confident assignment of specific enzyme activities to a specific pathway.

8.5.2 The Biosynthesis of *Allium* Flavour Precursors via Cysteine Synthases

There is evidence that enzymes required for the essential primary pathway for cysteine biosynthesis have evolved new substrate specificities for secondary metabolism, and potentially a route for CSO biosynthesis. Cysteine biosynthesis is at the convergence of C, N and S assimilation in plants and involves successive addition of an *O*-acetyl group and then sulfide to serine (see Scheme 8.1) catalysed sequentially by serine acetyltransferase (SAT, EC 2.3.1.30) and cysteine synthase [CS, *O*-acetylserine(thiol)lyase, EC 2.5.1.47]. Both these enzymes occur as families where different members are located within each of the cytoplasmic, mitochondrial and chloroplastic subcellular compartments.[28] The catalytic mechanism of CS has allowed new substrate specificities to evolve and several good examples are known where this enzyme is the source of secondary metabolites *in vivo* and can make use of a range of substrates *in vitro*.[29] Of relevance to CSO synthesis, *S*-methylcysteine (**18**) has been identified as a metabolite of methionine catabolism in *Arabidopsis* cell cultures provided with additional methionine.[30] Radiolabelling experiments indicated that biosynthesis of **18** occurred through reaction of methanethiol (**20**) with *O*-acetylserine (**7**), presumably catalysed by a CS. The pioneering work of Granroth in 1970 showed that garlic or onion leaves could synthesize ^{14}C-labelled alkylated cysteines when provided with ^{14}C-serine and a series of thiols, and similar findings have been made by other workers.[23,31] This led to a proposed pathway (Scheme 8.2b), mediated by a CS.

A cysteine synthase has been purified from Chinese chives and garlic that could synthesize *S*-substituted cysteines. *Allium* spp. are likely to have a higher requirement for cysteine production than most plants, and indeed cysteine levels are five- to six-fold higher in Chinese chives than in tobacco. Interestingly, the two CS purified from Chinese chive leaves were not inhibited by their substrate *O*-acetylserine (**7**).[17,32] This lack of substrate inhibition might be necessary to allow sufficient flux through the pathway.

Oxidations to form ACSO (**11**) and PeCSO (**12**) must be stereoselective to account for the structure of the products, but there is evidence of a general ability of onion leaf tissue to oxidize exogenous *S*-alk(en)yl cysteines to the corresponding sulfoxide.[23,33] The detailed nature of these enzymes remains to be discovered.

8.6 Formation of Volatiles from CSOs

The release of volatiles from CSOs involves enzymes of the carbon/sulfur (C-S) lyase family of aminotransferases (E.C. 4.4.1.*x*). All make use of pyridoxal-5′-phosphate (PLP) during the catalytic cycle and stereoselectivity around this event is the key determinant of which bond will be broken. Members of the family are involved in several plant processes including amino acid and hormone metabolism as well as the synthesis and degradation of secondary products including both CSOs and GSs. They are also present in microorganisms and can cleave further sulfur compounds such as methionine and cystathionine.[12,34] The substrate specificity of individual enzymes therefore differs, and the C-S lyase required to cleave the CSOs in *Allium* spp. has been given the name alliinase in recognition of this strong substrate specificity for sulfoxides.

8.6.1 *S*-Methyl-L-cysteine sulfoxide

S-Methyl-L-cysteine sulfoxide (MCSO, **10**) is more widely distributed than the other CSOs and therefore merits special consideration. It has been identified within some Brassicaceae as well as *Allium*, and this wider distribution suggests that its biosynthetic and breakdown pathways may differ from the other CSOs. Analysis for MCSO (**10**) has focused on species that are consumed as vegetables. Within the genus *Brassica*, the *B. oleracea* vegetables (cabbage, Brussels sprouts, broccoli, Chinese cabbage, cauliflower) have been paid particular attention as well as radish (*Raphanus sativus*). Several Brassicaceae species have been the subject of careful analysis that indicated they were devoid of MCSO, including gardencress pepperweed (*Lepidium sativum*) and horseradish (*Armoracia*

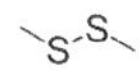

21
(Dimethyl disulfide)

Figure 8.2 Structure of dimethyl disulfide (**21**), the major thermal breakdown product from MCSO (**10**)

rusticana), although a trace could be detected in wasabi (*Eutremia wasabi*).[19,35] Analysis of the model *Brassica*/*Arabidopsis thaliana* suggests that MCSO (**10**) is absent, although a potential precursor *S*-methylcysteine (**18**) has been identified in cell cultures under specific growth conditions.[30,36] Some species within the genera *Vigna* (Family Fabaceae) and *Phaseolus* (Family Leguminosae) and the families Compositae and Umbelliferaceae also contain MCSO.[37]

The MCSO thermal breakdown products, especially the major one, dimethyl disulfide (**21**, Figure 8.2), contribute to a characteristic 'cabbage' odour.[38] This smell has also been attributed to products generated from cleavage of synthetic MCSO by purified onion alliinase,[39] confirming that MCSO is a key flavour precursor for the sensory perception of cabbage aroma. The flavour volatiles are released after tissue damage *in vivo* through the action of a cystine lyase (cystathionine-β-lyase, E.C. 4.4.1.8) on MCSO in Brassicaceae. This member of the carbon/sulfur (C-S) lyase enzyme class is related to alliinase but can be functionally distinguished through its ability to cleave L-cystine and *S*-alkylcysteines as well as CSOs.[40]

8.6.2 Release of the *Allium* CSOs

The key step in the formation of *Allium* flavour is disruption of the tissues to allow the flavour precursors to meet the lytic enzyme alliinase (alliin lyase, E.C. 4.4.1.4), as shown in Scheme 8.3. The CSOs are sequestered within vesicles in the cytoplasm while alliinase is located within the vacuole of all onion cells but only the bundle sheath cells adjacent to the phloem in garlic.[41] Alliinase was especially prominent in bundle sheath cells in green shoots of garlic chives.[42] The everyday experience that onions and garlic are essentially odourless until damaged illustrates the efficiency of the compartmentation. The fact that the volatiles are only formed within damaged tissue also offers an immediate escape route from the plant for the potentially toxic flavour volatiles.

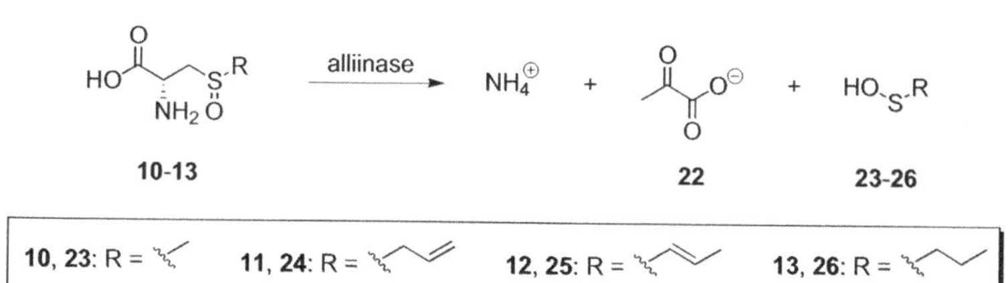

Scheme 8.3 Reaction catalysed by alliinase. A pyridoxal-5′-phosphate cofactor at the enzyme active site is crucial to the enzyme reaction that results in cleavage of the C(β)-S bond. The products are ammonia, pyruvate (**22**), and a series of unstable sulfenic acids (**23–26**). In onion, the 1-propenyl sulfenic acid (**25**) derived from PeCSO (**12**) is immediately a substrate for LFS

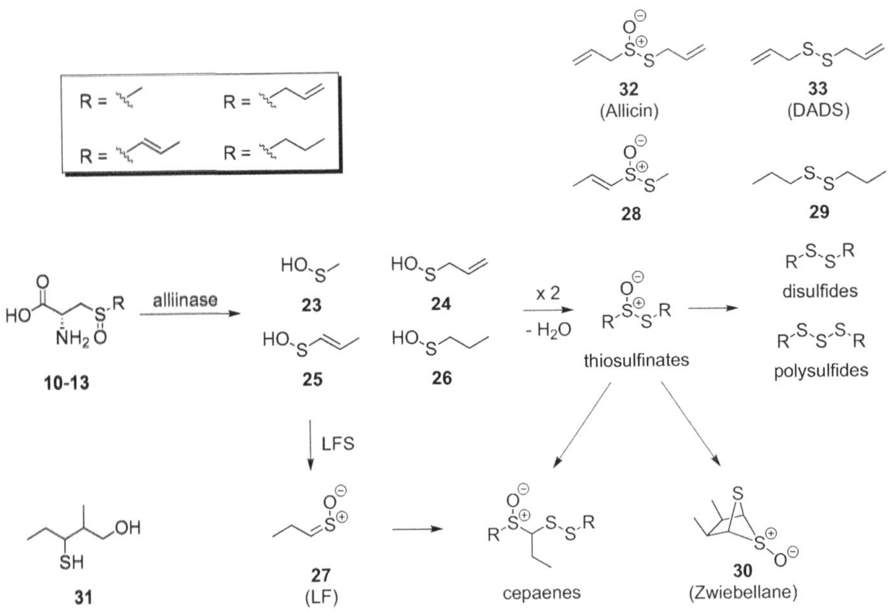

*Scheme 8.4 The flavour volatiles resulting from the action of alliinase and LFS (in onions). Unstable volatile thiosulfinates result from alliinase activity [illustrated as 1-propenyl methane thiosulfinate (**28**) and di-2-propenyl thiosulfinate (**32**)]. Formation of LF, cepaenes and some zwibellanes requires LFS. A series of disulfides [illustrated as dipropyl disulfide (**29**) and diallyl disulfide (**33**)] and polysulfides are more stable products*

Alliinase cleaves CSOs **10–14** to ammonia, pyruvate (**22**) and unstable sulfenic acids **23–26**, respectively, (Scheme 8.3) that undergo further chemical transformation resulting in a large series of sulfur compounds that change the flavour and odour over time.[14,15] The only subsequent step requiring biological catalysis is exclusive to onion and its relations and follows the release of 1-propenylsulfenic acid (**25**) from PeCSO (**12**) by alliinase, as shown in Scheme 8.4. Lachrymatory factor synthase (LFS) catalyses formation of lachrymatory factor **27** (LF, propanthial *S*-oxide) rather than thiosulfinates that would otherwise form during chemical processes.[43] LF is the source of the pungency and characteristic eye-watering response to onions and their close relatives that is absent from other *Allium* spp., like garlic.

8.7 The *Allium* Flavour Volatiles

The formation of sulfur volatiles takes an almost entirely chemical route following alliinase action on CSOs. Scheme 8.4 shows some of these products. They include well over 100 compounds that have been studied in detail and there are a number of excellent and very comprehensive accounts of this very interesting and complex sulfur chemistry.[14,15,44] The description given here has the aim of providing enough information to appreciate the effect of the presence of LFS in onion on directing the production of volatiles (see below). Pairs of sulfenic acids (e.g. **23–26**) condense within minutes of their release by alliinase to form thiosulfinates. Conversion from the essentially nonreactive sulfoxide to the strong, but thiol-specific oxidant thiosulfinate is the starting point for further reactions. Reaction with any convenient thiol will regenerate a molecule of sulfenic acid and a mixed disulfide. Depending on redox potential, the sulfenic acid can react with further thiols, generating additional mixed disulfides and polysulfides.

Alliinase cleavage of PeCSO (**12**) in onion results in a mixture of LF (**27**) with primarily 1-propenyl methane thiosulfinate (**28**) and dipropyl disulfide (**29**) as a consequence of LFS intercepting the 1-propenyl sulfenic acid (**25**) released by alliinase. However, a large series of volatiles form via further chemical reactions, including cepaenes (α-sulfinyl disulfides) and zwiebelanes (**30**) (isomers of di-1-propenyl thiosulfinate). The rich spectrum of volatiles has been well described but additional volatiles can be found using advances in analytical technology. A significant one is 3-mercapto-2-methylpentan-1-ol (**31**) first identified in cooked onion but now also found in freshly sliced tissue.[45] It has a highly distinctive meaty, sweaty and leek-like aroma. In garlic, allicin (di-2-propenyl thiosulfinate, **32**) is the major product from ACSO, decomposing primarily into diallyl disulfide (DADS, di-2-propenyl disulfide, **33**) that is immediately recognisable as garlic aroma.[14,46] Despite the lack of biological control on these vapour-phase processes, the flavours and aromas of *Allium* spp. are reproducible and instantly recognisable.

8.8 The Enzyme Alliinase

The importance of alliinase to flavour production has ensured that it has been studied in considerable detail. It has been purified from several *Allium* species. This has been facilitated by the fact that, although it is present in all tissues, it comprises a significant proportion of the protein in bulbs and cloves. Levels were estimated as up to 25% total protein in garlic cloves and 6% in onion bulbs.[47,48] Several features are consistent in alliinase from all *Allium* species. The native form of the enzyme is a homodimer of two ca. 51 kDa subunits with a carbohydrate content of around 6%. Each monomer contains one molecule of PLP that is essential for catalytic activity.[47,49] Lysine acts as the PLP-binding site and the identity of the relevant residue was confirmed by functional analysis following site-directed mutagenesis as Lys280 in Chinese chives and the corresponding Lys285 in onion. Chemical modification studies of garlic alliinase have also suggested importance for Trp182.[42,50]

The X-ray crystal structure of alliinase from garlic has been resolved.[51,52] The central and C-terminal domains have structures similar to other C-S lyases, while the N-terminal is different. This has structural motifs typically involved in protein–protein interactions and may function during transport of alliinase. Identification of a cleavage signal in onion, garlic and Chinese chive indicates alliinase is synthesized as a preproprotein with an N-terminal extension of 38–40 amino acids, providing an additional transport feature consistent with its vacuolar location.[42,47] There are also four putative glycosylation sites. Two, corresponding to Asn164 and Asn348 in garlic, have been positively identified as glycosylated *in vivo* with a branched mannose-rich hexasaccharide.[52,53]

Alliinase cleaves all available CSOs through elimination of an aminoacryl intermediate bound near the pyridoxal-5′-phosphate cofactor. Following formation of a Schiff's base and electron rearrangement, the carbon–sulfur bond is cleaved and a sulfenic acid is then released. The aminoacryl intermediate spontaneously decomposes to pyruvate and ammonia.[51,54] The enzyme has an absolute specificity for CSOs and indeed *S*-alkyl derivatives of cysteine (**4**) are inhibitors through a partially competitive mechanism, or at best poor substrates.[51,55] The compartmentation of alliinase within the vacuole minimizes exposure to any free *S*-alkyl derivatives of cysteine prior to tissue damage and the release of CSOs from cytoplasmic vesicles. There is biochemical evidence that alliinase from different *Allium* species have different substrate specificities. Several strongly suggest that onion alliinase preferentially hydrolyses PeCSO (**12**).[48,56] The substrate specificity of garlic alliinase has been measured and is highest towards the (R_cS_s) diastereoisomer of ACSO (**11**), lower towards the (R_cR_s) diastereoisomer and PeCSO (**12**) and none at all if the cysteine has the D-(S) configuration.[57]

Evidence for a small family of closely related alliinase genes, at least within garlic, supports the idea of differing substrate specificities among alliinases. The expression of these proteins differs since, for example, antibodies to garlic bulb alliinase could not detect the protein in roots despite enzymatic evidence of abundant activity.[47,49,58]

8.9 The Enzyme Lachrymatory Factor Synthase

Until 2002, alliinase was thought to be the final biological step in production of the *Allium* flavour volatiles, with all remaining volatiles generated through entirely chemical processes. The discovery of the enzyme lachrymatory factor synthase (LFS) in onion changed this view.[43] It may lead to new possibilities in manipulation of onion flavour, through breeding to separate the formation of LF (**27**) from the other flavour volatiles. A key observation was that crude alliinase from garlic could not produce LF from synthetic PeCSO (**12**). Subsequent separation of alliinase and LF formation activities within onion protein extracts led the way to identification of the gene for LFS. Recombinant LFS only produced LF from the natural isomer (*E*)-PeCSO (**12**). In the absence of LF, achieved through silencing the LFS gene,[46] onion alliinase gave higher yields of thiosulfinates formed via condensation of 1-propenyl sulfenic acid (**25**).

The consequences of silencing LFS in onions using RNA interference has provided greater insight into the importance of this enzyme for onion flavour and is also an instructive example of the consequence on aroma of manipulating one step in a biosynthetic pathway.[46] It resulted in both increased, decreased and new volatile products. Substantially higher conversion of 1-propenyl sulfenic acid (**25**) to di-1-propenyl thiosulfinate (**34**, Figure 8.3) occurred *in vivo* and this product then underwent chemical conversion to compounds that could be predicted as decomposition products but had only previously been detected in trace amounts in onion, if at all. Levels of the usual onion volatile dipropyl disulfide (**29**) decreased and 1-propenyl propyl disulfide (**35**) increased. Significant amounts of six additional compounds rarely or never encountered previously in onion were detected. Four of these could be predicted on the basis of known chemistry, namely three di-1-propenyl disulfide isomers (**36**) and a zwiebelane (**30**) while two were unexpected isomers of 2-mercapto-3,4-dimethyl-2,3-dihydrothiophenes (**37**) that might be artefacts from the analysis method but might also be produced if these onions were to be cooked. The overall

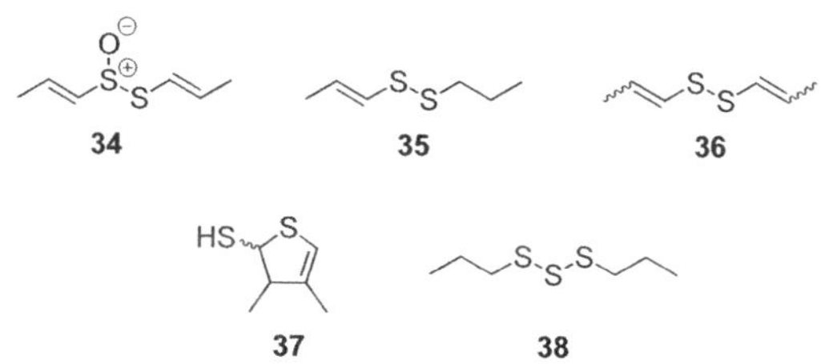

Figure 8.3 *Manipulated production of compounds **34–37** by silencing LFS. A role for dipropyl trisulfide **38** in biological responses has been proposed*

aroma from transgenic onions with LFS silenced was less pungent and sweeter than normal and lacked the eye-watering effect.

8.10 The Biological Roles of the Flavour Precursors

Biological roles as defence and storage compounds have been ascribed to the flavour precursors. The life history strategy of *Allium* spp. is that most species are biennials or perennials and are natives of arid, rocky areas. Wild *Allium* spp. generally flower in spring, passing the hot, dry summer and autumn as bulbs stocked with carbohydrates as well as flavour precursors. The CSOs therefore fulfil a need for defence to protect this vulnerable stage in the life cycle. Anecdotal and experimental evidence that high pungency onions survive storage and overwintering in adverse soil conditions better than those with a milder flavour support this view.[59] The CSOs and γGPs are also a significant reservoir of C, N and S. If they can be metabolized for rapid growth in the next season, the plant receives an additional benefit. The high level of alliinase recorded in *Allium* bulbs has also led to speculation that it may be a storage protein as well as required for release of the defence volatiles. However, the rationale for its high level may equally be that this is essential for rapid release of volatiles after tissue damage because the enzyme is comparatively inefficient and rapidly inactivated by reaction products.[49]

During vegetative growth, the volatiles provide defence, but are also attractants for specialist insect predators and fungal pathogens. *Allium* spp. are free from most species of aphids. Experiments with the generalist herbivore the green peach aphid *Myzus persicae*, demonstrates the defensive role since the aphids do not settle on material impregnated with low concentrations of garlic or onion oils. In contrast, the onion aphid *Neotoxoptera formosana* is a specialist herbivore of *Allium* spp. that sometimes forms large colonies on the leaves. It appears to use olfactory cues to locate the plants. In a behavioural response test to volatiles from *A. fistulosum* and *A. tuberosum*, the aphids were significantly attracted by sulfur volatiles identified as dipropyl trisulfide (**38**, Figure 8.3) and diallyl disulfide (**33**, Scheme 8.4), respectively.[60] Flavour volatiles are also attractants for the leek moth *Acrolepiopsis assectella* and the onion fly *Delia antiqua*.[61]

The fungus *Sclerotium cepivorum* causes white rot, one of the most important diseases of onions and garlic that can take land out of production for these crops for years, as well as causing immediate economic loss. This is because the fungus forms very long-lived resting structures (sclerotia) that only germinate when exposed to the flavour volatiles. These can be spread to fresh areas by routine agricultural practices. Strategies to remove sclerotia have included limited success from incorporating garlic powder or diallyl disulfide (**33**) into soil to trigger germination in the absence of the host plant.[62]

In addition to their value fresh, dried and as extracts in the food industry, the *Allium* flavour volatiles have attracted substantial interest because of possible benefits for human health and uses as novel microbiocides.[63]

8.11 The Glucosinolate Flavour Precursors

The glucosinolates (GS, mustard oils) are the other series of plant secondary metabolites that contain sulfur and are derived from amino acids. The sulfur-containing volatiles derived from these precursors are biologically important not only as flavours in food but through mediating both beneficial and deleterious plant–insect interactions. In addition, some may confer benefits for human health. They are found almost exclusively within the family Brassicaceae (the cabbage and mustard family). Selection by humans within a few Brassicaceae species have produced a very large number of vegetables (e.g. cabbage, broccoli, Brussel sprouts and cauliflower from *Brassica oleracea*). Glucosinolates are present throughout these plants, although levels are higher in younger tissues. Substantial progress has been made in understanding the biosynthesis of GS, aided by the fact that the model plant *Arabidopsis thaliana* is within this family.[64,65]

The GS and their lytic enzyme myrosinase (thioglucoside glucohydrolase, E. C. 3.2.3.1) form an interesting comparison with the CSOs and alliinase as a second system where stable sulfur-containing precursors are cleaved to form volatile and reactive flavours and odours. However, a major difference is that over 120 GS have been identified to date. Each plant species typically produces 30–40 GS with the exact composition varying due to environmental, developmental and genetic effects.[66,67] The reason for this large number is not known. However, the important biological roles in defence and plant–insect interactions played by the volatile GS breakdown products offer one explanation for this diversity (see below). The large diversity in precursor GS, combined with emerging knowledge of the biosynthetic pathways, provide opportunities to develop new biosynthesized S volatiles with flavour, agronomic and pharmaceutical uses. The sections below outline current knowledge of GS biosynthesis and volatile release and their biological roles.

8.12 GS and Their Biosynthetic Pathways

GS are β-thioglucoside *N*-hydroxysulfates, such as **39**–**45**, containing a side chain derived from an amino acid and a β-D-glucopyranose moiety (Figure 8.4). The GS can be divided into structural groups based on the amino acid that was the precursor to the aliphatic (**39**–**41**), indole (**42**, **43**), or aromatic (**44**, **45**) side chain in the structure.[67] About half the identified GS fall within the aliphatic group, mainly derived from methionine (**8**), with others originating from alanine, leucine, isoleucine or valine. The two further groups are the indole glucosinolates derived from tryptophan, and the aromatic glucosinolates derived from tyrosine or phenylalanine. The amino acid precursor of around one-third of GS has not yet been identified.

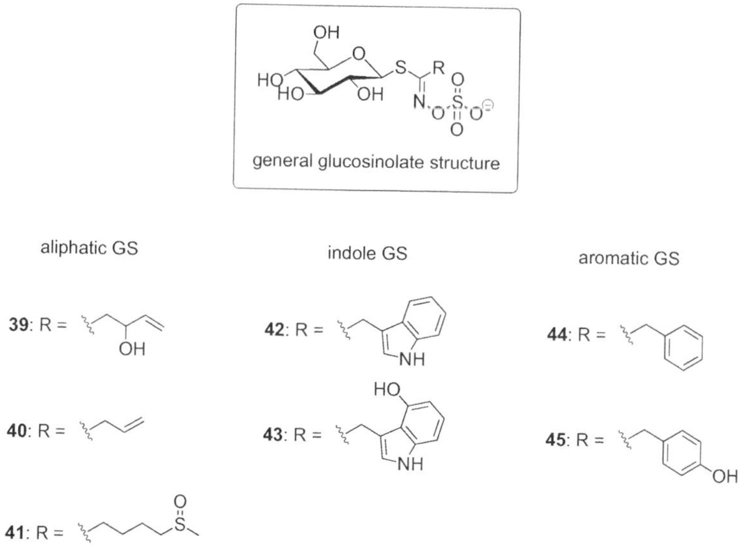

Figure 8.4 *The basic structure of glucosinolates and some representative examples. Aliphatic GS: 2-hydroxy-3-butenyl glucosinolate (**39**), allyl glucosinolate (**40**), 4-methylsulfinylbutyl glucosinolate (glucoraphanin, **41**); indole GS: indol-3-ylmethylglucosinolate (glucobrassicin, IMG, **42**), 4-hydroxyindol-3-ylmethylglucosinolate (4hIMG, **43**); aromatic GS: benzyl glucosinolate (**44**), p-hydroxybenzyl glucosinolate (sinalbin, **45**)*

The GS biosynthetic pathways have been the subject of substantial research and excellent reviews.[65,67,68] It is summarized in Scheme 8.5. The methionine and phenylalanine precursors may undergo one or more cycles of elongation of the amino acid side chain, utilizing acetyl-CoA as a donor of two-carbon units. Following deamination of the starting amino acid to a 2-oxoacid, these are added in cycles of condensation, isomerization and oxidative decarboxylation that result in loss of the original carboxylic acid functional group and its replacement with one from the acetyl group, resulting in extension of the side chain by a single carbon. After one to nine rounds of extension, transamination is required to regenerate an elongated amino acid in preparation for biosynthesis of the GS core structure. The enzymes required for these processes have started to be identified. *Arabidopsis* contains a small but highly variable family of methylthioalkylmalate synthase genes (MAM, EC 2.3.3.*x*) that determine the length of side chains in aliphatic GS.[69]

The first committed step in the GS biosynthetic pathway (Scheme 8.6) is oxidation of the amino acids to their respective aldoximes by side chain-specific cytochrome P450 monooxygenases from the CYP79 family. The enzymes involved in subsequent biosynthetic steps have much lower substrate specificity. Action by cytochrome P450 monooxygenases from the CYP83 family form *S*-alkylthiohydroximates from the aldoximes via activated, unstable nitrile oxide or *aci*-nitro intermediates. Cysteine (**4**) is probably the *S*-donor *in vivo* with the reaction mediated by a glutathione-*S*-transferase-like enzyme. Subsequent cleavage by a C-S lyase to yield a thiohydroximic acid is followed by glycosylation using an uridine-diphosphate-*S*-glucosyltransferase. A final sulfation step by a sulfotransferase modifies the desulfoglucosinolate to produce the GS core.[67]

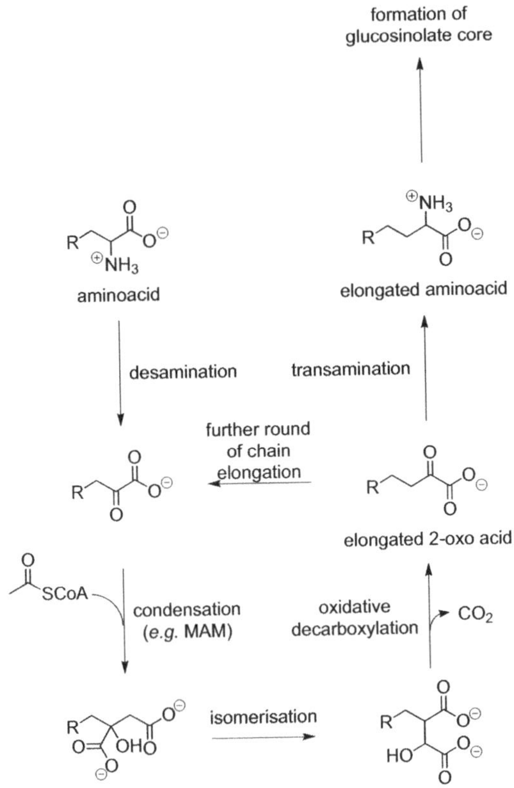

Scheme 8.5 *Chain elongation of amino acids prior to GS biosynthesis. Condensation of the 2-oxo acid with acetyl-CoA gives a 2-malate derivative that isomerises to a 3-malate derivative to give an elongated 2-oxo acid after oxidative decarboxylation. The compound is finally transaminated to provide an elongated amino acid as a precursor to formation of the GS core*

Regulation of aliphatic and indole GS biosynthesis involves tight control over a wide range of primary metabolic pathways.[70] The need for this control is apparent after considering the metabolic fates of the relevant amino acids, in addition to their essential use in protein synthesis. As a particularly important example (Figure 8.5), tryptophan (**46**) is a key intermediate for biosynthesis of the essential hormone growth regulator indole-3-acetic acid (**47**, IAA) in all plants. Within the Brassicaceae, it has a further role as a precursor for indole-GSs and is also required as a precursor for the phytoalexin camalexin (**48**), an important antibacterial compound induced at sites of infection.[70] To satisfy these competing demands for tryptophan within the plant requires sophisticated regulation.

8.13 Release of Volatile GS Hydrolysis Products

Tissue damage allows myrosinase access to GS. Interestingly, both GS and myrosinase are compartmentalized within vacuoles, although within different cell types or subcellular

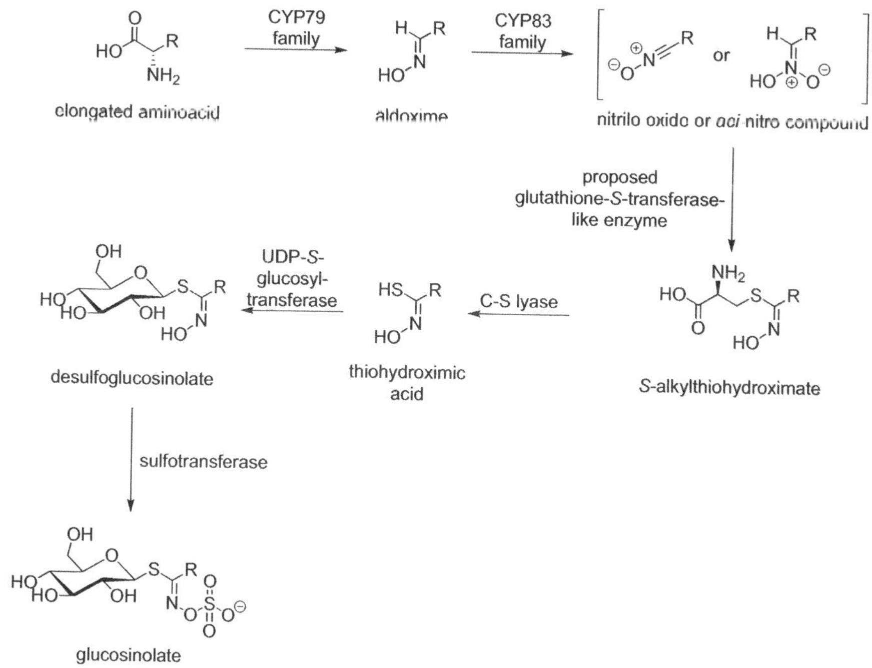

Scheme 8.6 *Biosynthesis of the GS core structure*

compartments. In *Arabidopsis*, myrosinase is only found in the myrosin cells distributed throughout the plant while GS are found in all organs but especially within sulfur-rich S cells.[71] Cleavage of the thioglucosidic bond yields glucose and an unstable aglycone that can rearrange into several possible final products (Scheme 8.7). The outcome is influenced by the structure of the side chain and also proteins such as the epithiospecifier proteins (ESPs) that occur in some *Brassica* spp. The ESPs are an interesting parallel with LFS following the CSO alliinase reaction. The GS aglycone can yield isothiocyanates (R–N=C=S) after spontaneous chemical rearrangements but an ESP promotes formation of less toxic nitriles (R–CN), sulfur-containing nitriles (such as epithionitrile **49**) and organic thiocyanates (R–SCN) at the expense of isothiocyanates.[72] The activity of ESP is regulated at the transcriptional and posttranscriptional level and may modulate the interaction between plant and insect and microbial predators (see below).

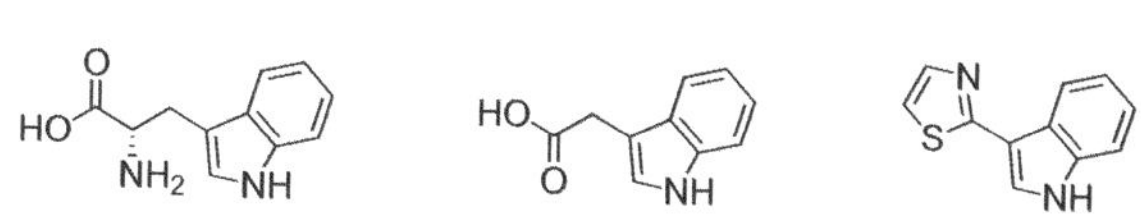

46
(L-Tryptophan)

47
(IAA)

48
(Camalexin)

Figure 8.5 *Tryptophan (**46**) is a precursor of indole glucosinolates (e.g. **42, 43**), indole-3-acetic acid (IAA; **47**) and camalexin (**48**) in addition to being required for protein synthesis*

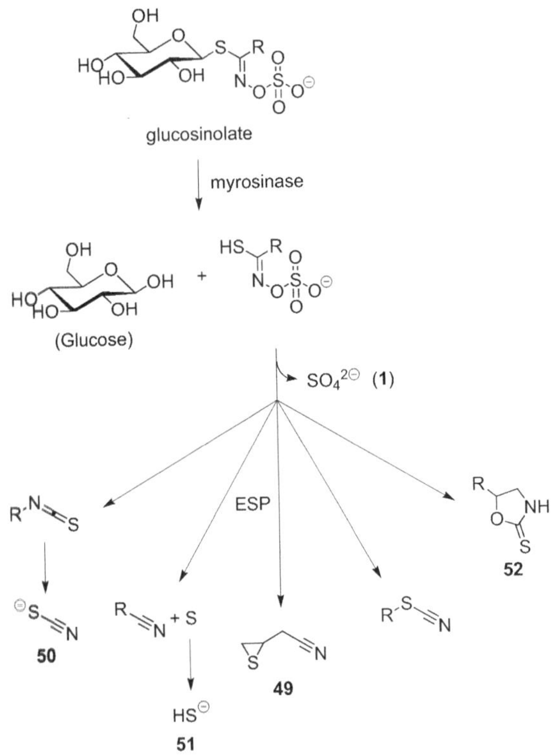

Scheme 8.7 GS hydrolysis products. The exact products depend on the structure of the specific GS and the presence of any modifying proteins such as epithiospecifier proteins (ESP) in addition to myrosinase

Methylated derivatives of the aglycone rearrangement products have been identified [methylthiocyanate, methylisothiocyanate, methanethiol, (**20**)]. These arise from methylation via thiol methyltransferases (TMTs; EC 2.1.1.9) that can utilize thiocyanate (**50**) and hydrogen sulfide ions (**51**) as well as organic sulfides as substrates, with *S*-adenosyl methionine as the methyl donor.[73] This particular family of methyltransferases may be confined to species that contain glucosinolates. *B. oleracea* has at least two genes and five isoforms of this enzyme, giving substantial potential for differentially regulated expression. Putative homologues have been identified in *Arabidopsis* (At2g43910, At2g43920, At2g43940). A role has been proposed for TMT in detoxifying the glucosinolate hydrolysis products to protect the plant and in biosynthesis of further volatile signalling molecules that may have a different spectrum of activity.

8.14 The Biological Role of Glucosinolates

The quality and quantity of GS are part of the natural defences of the Brassicaceae. It is clear that the different types of hydrolysis products have roles in defence against insects and microbial pathogens. The isothiocyanates are very reactive and toxic to microorganisms, nematodes and insects.[74] Indeed, *Brassica* spp. have long been used within agriculture as break crops within crop rotations, now termed biofumigation[75] where the whole *Brassica* plant is ploughed into the soil rather than harvested. This has the effect of suppressing soilborne pathogens and is an important component of sustainable agriculture.

The natural biological roles of the GS hydrolysis products are under very active investigation. The complexity of these roles, involving biological specificity and constant adaptation by the plants to a wide range of predators, may be a reason for the large number of GS. The GS myrosinase system, like the CSO alliinase system, is an inducible defence to the extent that tissue damage is required to bring precursor and lytic enzyme into contact. However, damage also affects GS biosynthesis with several of the transcription factors that control biosynthesis of either specific GSs or flux through the entire pathway induced by herbivory or wounding (*IQD1*, *AtDof1.1*, *TFL2*, members of the MYB family).[76] The total level of GS increases upon herbivore attack with indole GS generally increasing substantially, while aliphatic and aromatic GS increase only slightly or decline.

The flavour volatiles repel generalist herbivores but are attractants for some specialized insects that have evolved detoxifying mechanisms and indeed may require GS as feeding stimulants.[77] For example, the small white butterfly (*Pieris rapae*) produces a nitrile-specifier protein to direct the hydrolysis products to nitriles rather than toxic isothiocyanates.[78] Furthermore, the volatiles are attractants not only for these specialist herbivores but also for carnivorous insects that either parasitize or consume the herbivores and are therefore a further layer in the plant's defences.[79] Modulation of the volatiles via ESPs is an important component of these signaled interactions.[80] Other evidence for subtle and dynamic effects on GS biosynthesis includes identification that the genetic locus for GS side chain elongation enzymes in the plant *Arabidopsis* is a measure of its resistance to insect populations.[79] There is also evidence that GS influence microbial populations. Isothiocyanates can inhibit bacterial and fungal growth *in vitro* and *in vivo* and GS affects the bacterial population composition on *Brassica* leaves.[81]

Finally, GS affect nutritional quality for animals and humans as well as flavour. Adverse effects of high GS consumption have been recorded in animals that include low thyroid function (goitre) caused by oxazolidine-2-thiones (**52**), cyclized products of GSs that have a hydroxyl group at C(2), such as 2-hydroxy-3-butenyl glucosinolate (**39**). Animals may also only gain weight slowly on a diet high in some GSs because of poor palatability due to the many bitter flavour volatiles.[82] Plant breeding has reduced levels of GS in many crops for this reason. A comparison of wild *B. oleracea* populations with modern crop varieties, where the former have been under natural and the latter under human selection for over 1000 years, indicated that the GS composition differed both qualitatively and quantitatively.[83] Plants from the wild populations had up to three times as much GS as cultivated varieties and responded to insect damage by increasing total GS levels further, something that did not occur in the cultivated varieties.

However, antioxidant and cancer protection qualities have been identified for several isothiocyanate volatiles, leading to suggestions that their consumption could provide health benefits. Epidemiological studies have indicated that *Brassica* vegetables within the diet may lower the overall cancer risk, and interest has focused on the isothiocyanate sulforaphane [4-methylsulfinylbutyl isothiocyanate (**53**), Figure 8.6] because of its effects in cell culture and animal model studies, although other isothiocyanates have similar effects. It is

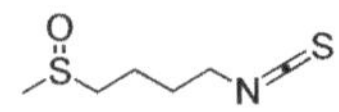

53
(Sulforaphane)

Figure 8.6 *GS hydrolysis product 4-methysulfinylbutyl isothiocyanate (**53**) is implicated in anticancer benefits*

derived from 4-methylsulfinylbutyl glucosinolate (**41**, glucoraphanin) which is particularly high in broccoli florets and seedlings (broccoli sprouts).[84] However, proposals to increase the health benefits of *Brassica* vegetables through enhancing the level of GSs requires consideration of the flavour consequences. It is clear that volatiles from several GS contribute to flavour, although the contribution of each one is unclear. Studies that have measured both GS composition in leaves or florets and the response of a consumer taste panel indicated that bitterness was related to total GS level and not the presence of a specific GS.[85] Since agronomic performance is also an important factor in crops, this offers considerable scope for plant breeders to develop new plant varieties with a combination of acceptable taste, higher GS levels and good agronomic traits to meet demand for health benefits.

8.15 Application of Transgenic Technology to Applied Aspects of GS Biosynthesis

Transgenic technology has had an important place in increasing fundamental knowledge of GS biosynthesis. It may also be valuable in developing plants with improved disease resistance and conferring increased human health benefits through engineering GS and flavour volatile production. These ideas will require very substantial development before reaching products acceptable to the public, at least within Europe,[86] but the following examples indicate possible developments. Addition of single CYP97 genes from evolutionarily distant species into *Arabidopsis* resulted in the production of novel GS and plants with reduced susceptibility to bacterial and fungal disease through both direct effects of GS and modulation of jasmonate- and salicylic acid-mediated defence pathways. Engineering novel GS biosynthesis into a plant species has been demonstrated by adding the final three steps of the benzylglucosinolate pathway from *Arabidopsis* (comprising a C-S lyase, glycosyltransferase, sulfotransferase) into tobacco. This gave low, but detectable levels of GS within this usually GS-free species.[87] This opens the possibility of adding plant or human health benefits to a variety of economically important plant species.

8.16 Volatile Sulfur Compounds from Other Plants

8.16.1 Complex Organic Sulfur Volatiles

Other plants, in addition to the *Allium* and *Brassica* species, have flavours or aromas that originate from volatile sulfur compounds. In many cases, C-S lyase enzymes are implicated in the release of the volatile but the biosynthetic pathway for the flavour

54 55 56 57 58 59

Figure 8.7 Further volatile sulfur compounds

precursor is generally unknown.[10,88] These include 1,3-dithiabutane (**54**, Figure 8.7), found in cooked petai beans (*Parkia speciosa*) and 1,2,4-trithiolane (**55**), identified in palm species of the genus *Geonoma* as well as petai. The strongly lachrymatory compound thiobenzaldehyde-*S*-oxide (**56**) is produced immediately upon damage of *Petiveria alliacea*.[89] These volatiles are products of the plants, but others may come from microbial action during either food preparation or its consumption. Sulfur volatiles have been identified as significant source of aroma in plant products that have undergone microbial fermentations. These include white wine, where 3-sulfanylhexan-1-ol (**57**) and 4-methyl-4-sulfanylpentan-2-one (**58**) are important flavour components. These are produced by the yeast *Saccharomyces cerevisiae* during the alcoholic fermentation of the grape juice through the action of Irc7p, a putative cystathionine β-lyase, on stable cysteine-*S*-conjugates.[90] However, these volatiles may also be released by the oral microflora.[12] Sulfur compounds are generated in cysteine and methionine catabolism by anaerobic bacteria that inhabit the mouth and are a prominent feature of oral malodour. Hydrogen sulfide (H_2S), along with methanethiol (**20**), dimethylsulfide (**59**) and dimethyldisulfide (**21**) are the most important constituents. However, the bacterial β-C-S lyases that cleave cysteine into hydrogen sulfide, pyruvate and ammonia[91] can make use of other substrates to form more interesting flavour compounds, such as **57**.[9,13]

8.16.2 Simple Sulfur Volatiles

Many plant species release reduced sulfur compounds into the atmosphere including hydrogen sulfide (H_2S), carbonyl sulfide (O=C=S), carbon disulfide (S=C=S), dimethyl sulfide (**59**) and methanethiol (**20**).[92] These may be emitted continuously during growth or in association with attack by a pathogen, although their importance as defence compounds is uncertain. Although not, perhaps, flavour compounds *sensu stricto*, they are perceived as strong odours and also shed interesting light on plant sulfur metabolism. The biochemical sources of several of these compounds are still speculative but current ideas on the origin of hydrogen sulfide and methanethiol are outlined below.

8.16.3 Hydrogen Sulfide

Hydrogen sulfide (H_2S) is the major reduced sulfur compound released by plants[93] but its origin is still uncertain. Candidates include L- and D-cysteine desulfhydrases (E.C. 4.4.1.1) that degrade cysteine into H_2S and alanine, or H_2S, ammonia and pyruvate respectively.[8,93] Several enzymes involved in iron/sulfur centre biosynthesis or catalysing sulfur transfers show cysteine desulfhydrase activity *in vitro*. Another candidate is CS that consumes sulfide during cysteine biosynthesis but may produce H_2S in a partial reverse reaction. One technical difficulty in determining the origin of H_2S is the difficulty of measuring its concentration accurately *in vivo*.[94]

8.16.4 Methanethiol

It has been known since at least the mid1980s that some plants and cell cultures can continuously emit these volatiles: methanethiol (**20**, Scheme 8.2), its oxidation product dimethyl disulfide (**21**, Figure 8.2) and dimethyl sulfide (**59**, Figure 8.7).[30,95] In particular, this occurs in plant tissues provided with high levels of the amino acid methionine (**8**, Scheme 8.1), suggesting a catabolic product of methionine.[95] Methionine is an essential amino acid within the human diet and its catabolism has thus attracted considerable attention. A reverse-transulfurylation pathway route has therefore been well characterized in animals and fungi but key enzymes are absent from plants.

An alternative catabolic pathway known from bacteria and protozoa requires methionine-γ-lyase (MGL, EC 4.4.1.11), a member of the γ-family of PLP dependent lyases. This cleaves methionine via α,γ-elimination to yield α-ketobutyrate, methanethiol and ammonia. Identification of a gene for this cytosolic enzyme within the *Arabidopsis* genome (At1g64660) followed by biochemical and functional characterization of the recombinant protein,[95] provided a route for methanethiol biosynthesis and also insight into control of plant sulfur metabolism. In addition to release into the atmosphere, the methanethiol could be converted into cysteine through a new, and still unknown biochemical pathway. The gene for MGL is widely distributed among plant species, suggesting that most, if not all, can synthesize methanethiol. Efficient assimilation into cysteine in most species and environments may be the reason why the smell has only been noted from a few. Interest in the mechanism of MGL has been increased by the observation that many cancer cells have an increased requirement for methionine, with MGL thus having antitumour activity. The crystal structure of a bacterial MGL has been solved, implicating cysteine and tyrosine residues on the protein in the catalytic mechanism.[96]

8.17 Conclusion

Sulfur flavour and odour volatiles bring together some of the most interesting areas of chemistry and biology. The chemical reactivity of the compounds makes analysis challenging while the biochemical strategies that have evolved to utilize this natural potential are both intriguing and only partly understood. Ranging from complex biosynthetic and developmental pathways to olfactory receptors that initiate specific animal behaviour, volatile sulfur compounds have a profound influence on biology and on the research of many biologists, chemists and molecular biologists.

References

1. F. Hadacek, Secondary metabolites as plant traits: current assessment and future perspectives, *Crit. Rev. Plant Sci.*, **21**, 273–322 (2002).
2. W. Schwab, R. Davidovich-Rikanati, E. Lewinsohn, Biosynthesis of plant-derived flavor compounds, *Plant J.*, **54**, 712–732 (2008).
3. For example see: The Plant Genomes Database, http://www.plantgdb.org, or KEGG Plant, http://www.genome.jp/kegg/plant/, and contrast with The Arabidopsis Information resource, http://www.arabidopsis.org, to see potential for development.
4. N. Dudareva, F. Negre, D. A. Nagegowda, I. Orlova, Plant volatiles: recent advances and future perspectives, *Crit. Rev. Plant Sci.*, **25**, 417–440 (2006).

5. Reviewed in: A. A. Bachmanov, G. K. Beauchamp, Taste receptor genes, *Annu. Rev. Nutr.*, **27**, 389–414 (2007).
6. (*a*) T. Leustek, M. N. Martin, J.-A. Bick, J. P. Davies, Pathways and regulation of sulfur metabolism revealed through molecular and genetic studies, *Annu. Rev. Plant Physiol. Mol. Biol.*, **51**, 141–165 (2000); (*b*) M. Y. Hirai, K. Saito, Analysis of systemic sulfur metabolism in plants using integrated '-omics' strategies, *Mol. Biosyst.*, **4**, 967–973 (2008).
7. A. J. Meyer, R. Hell, Glutathione homeostasis and redox regulation by sulfhydryl groups, *Photosynth. Res.*, **86**, 435–457 (2005).
8. C. Kruse, R. Jost, M. Lipschis, B. Kopp, M. Hartmann, R. Hell, Sulfur-enhanced defence: effects of sulfur metabolism, nitrogen supply and pathogen lifestyle, *Plant Biol.*, **9**, 608–619 (2007).
9. I. Blank, Sensory relevance of volatile organic sulfur compounds in food, in *Heteroatomic Aroma Compounds*, ACS Symp. Ser., Vol. 826, G. A. Reineccius, T. A. Reineccius (Eds), American Chemical Society, Washington, D. C., 2002, pp. 25–53.
10. C. Starkenmann, M. Troccaz, K. Howell, The role of cysteine and cysteine-*S*-conjugates as odour precursors in the flavour and fragrance industry, *Flavour Fragr. J.*, **23**, 369–381 (2008).
11. C. Jacob, A. Anwar, The chemistry behind redox regulation with a focus on sulphur redox systems, *Physiol. Plant.*, **113**, 469–480 (2008).
12. H. E. Spinnler, N. Martin, P. Bonnarme, Generation of sulphur flavour compounds by microbial pathways, in *Heteroatomic Aroma Compounds*, ACS Symposium Series Vol. 826, G. A. Reineccius, T. A. Reineccius (Eds), American Chemical Society, Washington, D. C., 2002, pp. 54–72.
13. C. Starkenmann, B. Le Calve, Y. Niclass, I. Cayeux, S. Beccucci, M. Troccaz, Olfactory perception of cysteine-*S*-conjugates from fruits and vegetables, *J. Agric. Food Chem.*, **56**, 9575–9580 (2008).
14. E. Block, The organosulfur chemistry of the genus *Allium* – implications for the organic chemistry of sulfur, *Angew. Chem.*, **104**, 1158–1203 (1992); *Angew. Chem. Int. Ed.*, **31**, 1135–1178 (1992).
15. V. Lanzotti, The analysis of onion and garlic, *J. Chromatogr. A*, **1112**, 3–22 (2006).
16. R. M. Fritsch, M. Keusgen, Occurrence and taxonomic significance of cysteine sulphoxides in the genus *Allium* L. (Alliaceae), *Phytochemistry*, **67**, 1127–1135 (2006).
17. Reviewed by: M. G. Jones, J. Hughes, A. Tregova, J. Milne, A. B. Tomsett, H. A. Collin, Biosynthesis of the flavour precursors of onion and garlic, *J. Exp. Bot.*, **55**, 1903–1918 (2004).
18. R. Kubec, M. Svobodova, J. Velíšek, Distribution of *S*-alk(en)ylcysteine sulphoxides in some *Allium* species. Identification of a new flavor precursor: *S*-ethylcysteine sulphoxide (ethiin), *J. Agric. Food Chem.*, **48**, 428–433 (2000).
19. R. Kubec, E. Dadakova, Quantitative determination of *S*-alk(en)ylcysteine-*S*-oxides by micellar electrokinetic capillary chromatography, *J. Chromatogr. A*, **1212**, 154–157 (2008).
20. R. Kubec, S. Kim, D. M. McKeon, R. A. Musah, Isolation of *S*-*n*-butyl cysteine sulfoxide and six *n*-butyl-containing thiosulfinates from *Allium siculum*, *J. Nat. Prod.*, **65**, 960–964 (2002).
21. J. R. Whitaker, Development of flavor, odor and pungency in onion and garlic, *Adv. Food Res.*, **22**, 73–133 (1976).
22. J. E. Lancaster, M. L. Shaw, γ-Glutamyl peptides in the biosynthesis of *S*-alk(en)yl-L-cysteine sulphoxides (flavour precursors) in *Allium*, *Phytochemistry*, **28**, 455–460 (1989).
23. B. Granroth, Biosynthesis and decomposition of cysteine derivatives in onion and other *Allium* species, *Ann. Acad. Sci. Fenn., Ser. A*, **154**, 1–71 (1970).
24. S. J. Edwards, G. Britton, H. A. Collin, The biosynthetic pathway of the *S*-alk(en)yl-L-cysteine sulphoxides (flavour precursors) in species of *Allium*, *Plant Cell, Tissue Organ Cult.*, **38**, 181–188 (1994).
25. L. N. Ceci, O. A. Curzio, A. B. Pomilio, γ-Glutamyl transpeptidase/γ-glutamyl peptidase in sprouted *Allium sativum*, *Phytochemistry*, **31**, 441–444 (1992).
26. J. E. Lancaster, M. L. Shaw, Characterisation of purified γ-glutamyl transpeptidase in onions: evidence for *in vivo* role as a peptidase, *Phytochemistry*, **36**, 1351–1358 (1994).
27. M. Shaw, M. D. Pither-Joyce, J. A. McCallum, Purification and cloning of a gamma-glutamyl transpeptidase from onion (*Allium cepa*), *Phytochemistry*, **66**, 515–522 (2005).
28. F. H. Haas, C. Heeg, R. Querioz, A. Bauer, M. Wirtz, R. Hell, Mitochondrial serine acetyltransferase as a pacemaker of cysteine synthesis in plant cells, *Plant Physiol.*, **148**, 1055–1067 (2008).

29. F. Ikegami, I. Murakoshi, Enzymic synthesis of non-protein β-substituted alanines and some higher homologues in plants, *Phytochemistry*, **35**, 1089–1104 (1994).
30. F. Rébeillé, S. Jabrin, R. Bligny, K. Loizeau, B. Gambonnet, V. van Wilder, R. Douce, S. Ravanel, Methionine catabolism in *Arabidopsis* cells is initiated by a γ-cleavage process and leads to *S*-methylcysteine and isoleucine syntheses, *Proc. Natl. Acad. Sci. USA*, **103**, 15687–15692 (2006).
31. (*a*) A. Turnbull, I. J. Galpin, J. L. Smith, H. A. Collin, Comparison of the onion plant (*Allium cepa*) and onion tissue culture. IV. Effect of shoot and root morphogenesis on flavour precursor synthesis in onion tissue culture, *New Phytol.*, **87**, 257–268 (1981); (*b*) C. L. Prince, M. L. Shuler, Y. Yamada, Altering flavor profiles in onion (*Allium cepa* L.) root cultures through directed biosynthesis, *Biotechnol. Prog.*, **13**, 506–510 (1997).
32. (*a*) F. Ikegami, S. Itagaki, I. Murakoshi, Purification and characterization of two forms of cysteine synthase from *Allium tuberosum*, *Phytochemistry*, **32**, 31–34 (1993); (*b*) Y. Urano, T. Manabe, M. Noji, K. Saito, Molecular cloning and functional characterization of cDNAs encoding cysteine synthase and serine acetyltransferase that may be responsible for high cellular cysteine content in *Allium tuberosum*, *Gene*, **257**, 269–277 (2000).
33. C. Ohsumi, T. Hayashi, K. Sano, Formation of alliin in the culture tissues of *Allium sativum*. Oxidation of *S*-allyl-L-cysteine, *Phytochemistry*, **33**, 107–111 (1993).
34. G. A. Kiddle, R. N. Bennett, A. J. Hick, R. M. Wallsgrove, C-S lyase activities in leaves of crucifers and non-crucifers, and the characterization of three classes of C-S lyase activities from oilseed rape (*Brassica napus* L.), *Plant, Cell Environ.*, **22**, 433–445 (1999).
35. R. Kubec, M. Svobodova, J. Velíšek, Gas-chromatographic determination of *S*-methylcysteine sulphoxide in cruciferous vegetables, *Eur. Food Res. Technol.*, **213**, 386–388 (2001).
36. K. Saito, M. Y. Hirai, personal communication.
37. N. J. Benevegna, G. L. Case, R. D. Stele, Occurrence and metabolism of *S*-methyl-L-cysteine and *S*-methyl-L-cysteine sulfoxide in plants and their metabolism and toxicity in animals, in *Toxicants of Plant Origin*, Vol. III, P. R. Cheeke (Ed.), CRC Press, Boca Raton, 1989, pp. 203–228.
38. R. Kubec, V. Drhová, J. Velíšek, Thermal degradation of *S*-methylcysteine and its sulfoxide–important flavor precursors of *Brassica* and *Allium* vegetables, *J. Agric. Food. Chem.*, **46**, 4334–4340 (1998).
39. D. J. Thomas, K. L. Parkin, Immobilisation and characterization of C-S lyase from onion (*Allium cepa*) bulbs, *Food Biotechnol.*, **5**, 139–159 (1991).
40. A. Hamamoto, M. Mazelis, The C-S lyases of higher plants: isolation and properties of homogeneous cystine lyase from broccoli (*Brassica oleracea* var *botrytis*) buds, *Plant Physiol.*, **80**, 702–706 (1986).
41. (*a*) J. E. Lancaster, H. A. Collin, Presence of alliinase in isolated vacuoles and of alkyl cysteine sulphoxides in the cytoplasm of bulbs of onion (*Allium cepa*), *Plant Sci. Lett.*, **22**, 169–176 (1981); (*b*) J. E. Lancaster, P. H. S. Reynolds, M. L. Shaw, E. M. Dommisse, J. Munro, Intracellular localization of the biosynthetic pathway to flavour precursors in onion, *Phytochemistry*, **28**, 461–464 (1989); (*c*) G. S. Ellmore, R. S. Feldberg, Alliin lyase localization in bundle sheaths of the garlic clove (*Allium sativum*), *Am. J. Bot.*, **81**, 89–94 (1994).
42. T. Manabe, A. Hasumi, M. Sugiyama, M. Yamazaki, K. Saito, Alliinase [*S*-alk(en)yl-L-cysteine sulfoxide lyase] from *Allium tuberosum* (Chinese chive), *Eur. J. Biochem.*, **257**, 21–30 (1998).
43. S. Imai, N. Tsuge, M. Tomotake, Y. Nagatome, H. Sawada, T. Nagata, H. Kumagai, An onion enzyme that makes the eyes water, *Nature*, **419**, 685 (2002).
44. (*a*) E. Block, M. Thiruvazhi, P. J. Toscano, T. Bayer, S. Grisoni, S.-H. Zhao, *Allium* chemistry: structure, synthesis, natural occurrence in onion (*Allium cepa*), and reactions of 2,3-dimethyl-5,6-dithiabicyclo[2.1.1]hexane *S*-oxides, *J. Am. Chem. Soc.*, **118**, 2790–2798 (1996); (*b*) P. Rose, M. Whiteman, P. K. Moore, Y. Z. Zhu, Bioactive *S*-alk(en)yl cysteine sulfoxide metabolites in the genus *Allium*: the chemistry of potential therapeutic agents, *Nat. Prod. Rep.*, **22**, 351–368 (2005).
45. S. Widder, C. S. Luntzel, T. Dittner, W. Pickenhagen, 3-Mercapto-2-methyl-pentan-1-ol, a powerful new aroma compound, *J. Agric. Food Chem.*, **48**, 418–423 (2000).
46. C. C. Eady, T. Kamoi, M. Kato, N. G. Porter, S. Davis, M. Shaw, A. Kamoi, S. Imai, Silencing onion lachrymatory factor synthase causes a significant change in the sulfur secondary metabolite profile, *Plant Physiol.*, **147**, 2096–2106 (2008).

47. E. M. J. van Damme, K. Smeets, S. Torrekens, F. Vanleuven, W. J. Peumans, Isolation and characterisation of alliinase cDNA clones from garlic (*Allium sativum* L.) and related species, *Eur. J. Biochem.*, **209**, 751–757 (1992).
48. L. P. Nock, M. Mazelis, The C S lyases of higher plants: direct comparison of the physical properties of homogeneous alliin lyase of garlic (*Allium sativum*) and onion (*Allium cepa*), *Plant Physiol.*, **85**, 1079–1083 (1987).
49. A. Rabinkov, X. Z. Zhu, G. Grafi, G. Galili, D. Mirelman, Alliin lyase (Alliinase) from garlic (*Allium sativum*)–biochemical characterization and cDNA cloning, *Appl. Biochem. Biotechnol.*, **48**, 149–171 (1994).
50. (*a*) N. Kitamura, N. Shimomura, J. Iseki, M. Honma, S. Chiba, S. Tahara, J. Mizutani, Cysteine-*S*-conjugate beta-lyase activity and pyridoxal phosphate binding site of onion alliin lyase, *Biosci. Biotechnol. Biochem.*, **61**, 1327–1330 (1997); (*b*) Y. N. Jin, Y. H. Choi, C. H. Yang, Identification of an essential tryptophan residue in alliinase from garlic (*Allium sativum*) by chemical modification, *Bull. Korean Chem. Soc.*, **22**, 68–76 (2001).
51. E. B. Kuettner, R. Hilgenfled, M. S. Weiss, The active principle of garlic at atomic resolution, *J. Biol. Chem.*, **277**, 46402–46407 (2002).
52. L. J. W. Shimon, A. Rabinkov, I. Shin, T. Miron, D. Mirelman, M. Wilchek, F. Frolow, Two structures of alliinase from *Allium sativum* L.: *apo* form and ternary complex with aminoacrylate reaction intermediate covalently bound to the PLP cofactor, *J. Mol. Biol.*, **366**, 611–625 (2007).
53. A. Rabinkov, M. Wilchek, D. Mirelman, Alliinase (alliin lyase) from garlic (*Allium sativum*) is glycosylated at ASN^{146} and forms a complex with a garlic mannose-specific lectin, *Glycoconjugate J.*, **12**, 690–698 (1995).
54. The mechanism is animated at http://www.biologie.uni-hamburg.de/lehre/bza/taste/1lk9/1lk9m.htm
55. S. Schwimmer, C. A. Ryan, F. F. Wong, Specificity of L-cysteine sulfoxide lyase and partially competitive inhibition by *S*-alkyl-L-cysteines, *J. Biol. Chem.*, **239**, 777–782 (1964).
56. J. E. Lancaster, M. L. Shaw, W. M. Randle, Differential hydrolysis of alk(en)yl cysteine sulphoxides by alliinase in onion macerates: flavour implications, *J. Sci. Food Agric.*, **78**, 367–372 (1998).
57. Quoted in: R. Bentley, Role of sulphur chirality in the chemical processes of biology, *Chem. Soc. Rev.*, **34**, 609–624 (2005).
58. M. G. Jones, H. A. Collin, A. Tregova, L. Truman, L. Brown, R. Cosstick, J. Hughes, J. Milne, M. C. Wilkinson, A. B. Tomsett, B. Thomas, The biochemical and physiological genesis of alliin in garlic, *Med. Arom. Plant Sci. Biotechnol.*, **1**, 21–24 (2007).
59. (*a*) B. Smith, personal communication; (*b*) G. A. Chope, L. A. Terry, P. J. White, The effect of the transition between controlled atmosphere and regular atmosphere storage on bulbs of onion cultivars SS1, Carlos and Renate, *Postharvest Biol. Technol.*, **44**, 228–239 (2007).
60. (*a*) M. Hori, Settling inhibition and insecticidal activity of garlic and onion oils against *Myzus persicae* (Sulzer) (Homoptera: Aphididae), *Appl. Entomol. Zool.*, **31**, 605–612 (1996); (*b*) M. Hori, Onion aphid (*Neotoxoptera formosana*) attractants, in the headspace of *Allium fistulosum* and *A. tuberosum* leaves, *J. Appl. Entomol.*, **131**, 8–12 (2007).
61. J. Romeis, D. Ebbinghaus, J. Scherkenbeck, Factors accounting for the variability in the behavioural response of the onion fly (*Delia antique*) to *n*-dipropyl disulphide, *J. Chem. Ecol.*, **29**, 2131–2142 (2003).
62. (*a*) J. R. Coley-Smith, Interactions between *Sclerotium cepivorum* and cultivars of onion, leek, garlic and *Allium fistulosum*, *Plant Pathol.*, **35**, 362–369 (1986); (*b*) R. M. Davis, J. J. Hao, M. K. Romberg, J. J. Nunez, R. F. Smith, Efficacy of germination stimulants of sclerotia of *Sclerotium cepivorum* for management of white rot of garlic, *Plant Dis.*, **91**, 204–208 (2007).
63. (*a*) G. Griffiths, L. Trueman, T. Crowther, B. Thomas, B. Smith, Onions–a global benefit to health, *Phytother. Res.*, **16**, 603–615 (2002); (*b*) C. Jacob, A scent of therapy: pharmacological implications of natural products containing redox-active sulfur atoms, *Nat. Prod. Rep.*, **23**, 851–863 (2006); (*c*) G. A. Benavides, G. L. Squadrito, R. W. Mills, H. D. Patel, T. S. Isbell, R. P. Patel, V. M. Darley-Usmar, J. E. Doeller, D. W. Kraus, Hydrogen sulfide mediates the vasoactivity of garlic, *Proc. Natl. Acad. Sci. USA*, **104**, 17977–17982 (2007); (*d*) M. Corzo-Martínez, N.

Corzo, M. Villamiel, Biological properties of onions and garlic, *Trends Food Sci. Technol.*, **18**, 609–625 (2007).
64. M. D. Mikkelsen, B. L. Petersen, C. E. Olsen, B. A. Halkier, Biosynthesis and metabolic engineering of glucosinolates, *Amino Acids*, **22**, 279–295 (2002).
65. C. D. Grubb, S. Abel, Glucosinolate metabolism and its control, *Trends Plant Sci.*, **11**, 89–100 (2006).
66. J. W. Fahey, A. T. Zalcmann, P. Talalay, The chemical diversity and distribution of glucosinolates and isothiocyanates among plants, *Phytochemistry*, **56**, 5–51 (2001).
67. B. A. Halkier, J. Gershenzon, Biology and biochemistry of glucosinolates, *Annu. Rev. Plant Biol.*, **57**, 303–333 (2006).
68. X. F. Yan, S. X. Chen, Regulation of plant glucosinolate metabolism, *Planta*, **226**, 1343–1352 (2007).
69. J. Kroymann, S. Donnerhacke, D. Schnabelrauch, T. Mitchell-Olds, Evolutionary dynamics of an *Arabidopsis* insect resistance quantitative trait locus, *Proc. Natl. Acad. Sci. USA*, **100**, 14587–14592 (2003).
70. S. Malitsky, E. Blum, H. Less, I. Venger, M. Elbaz, S. Morin, Y. Eshed, A. Aharoni, The transcript and metabolite networks affected by the two clades of *Arabidopsis* glucosinolate biosynthesis regulators, *Plant Physiol.*, **148**, 2021–2049 (2008).
71. H. Ueda, C. Nishiyama, T. Shimada, Y. Koumoto, Y. Hayashi, M. Kondo, T. Takahashi, I. Ohtomo, M. Nishimura, I. Hara-Nishimura, AtVAM3 is required for normal specification of idioblasts, myrosin cells, *Plant Cell Physiol.*, **47**, 164–175 (2006).
72. (*a*) M. Burow, M. Rice, B. Hause, J. Gershenzon, U. Wittstock, Cell- and tissue-specific localization and regulation of the epithiospecifier protein in *Arabidopsis thaliana*, *Plant Mol. Biol.*, **64**, 173–185 (2007); (*b*) U. Wittstock, M. Burow, Tipping the scales – specifier proteins in glucosinolate hydrolysis, *IUBMB Life*, **59**, 744–751 (2007).
73. (*a*) J. Attieh, K. F. Kleppinger-Sparace, C. Nunes, S. A. Sparace, H. S. Saini, Evidence implicating a novel thiol methyltransferase in the detoxification of glucosinolate hydrolysis products in *Brassica oleracea* L., *Plant, Cell Environ.*, **23**, 165–174 (2000); (*b*) J. Attieh, R. Djiana, P. Koonjul, C. Étienne, S. A. Sparace, H. S. Saini, Cloning and functional expression of two plant thiol methyltransferases: a new class of enzymes involved in the biosynthesis of sulfur volatiles, *Plant Mol. Biol.*, **50**, 511–521 (2002).
74. For a review, see: U. Wittstock, D. J. Kliebenstein, V. M. Lambrix, M. Reichelt, J. Gershenzon, Glucosinolate hydrolysis and its impact on generalist and specialist insect herbivores, in *Integrative Phytochemistry: from Ethnobotany to Molecular Ecology*, J. T. Romero (Ed.), Elsevier, Amsterdam, 2003, pp. 101–125.
75. B. J. Smith, J. A. Kirkegaard, G. N. Howe, Impacts of *Brassica* break crops on soil biology and yield of following wheat crops *Aust. J. Agric. Res.* **55**, 1–11 (2004).
76. S. Textor, J. Gershenzon, Herbivore induction of the glucosinolate-myrosinase defense system: major trends, biochemical bases and ecological significance *Phytochem. Rev.* **8**, 149–170 (2009).
77. L. M. Schoonhoven, J. J. A. van Loon, M. Dicke, *Insect–Plant Biology*, 2nd Ed., Oxford University Press, Oxford, 2005.
78. U. Wittstock, N. Agerbirk, E. J. Stauber, C. E. Olsen, M. Hippler, T. Mitchell-Olds, J. Gershenzon, H. Vogel, Successful herbivore attack due to metabolic diversion of a plant defence, *Proc. Natl. Acad. Sci. USA*, **101**, 4859–4864 (2004).
79. R. Mumm, M. Burow, G. Bukovinszkine'Kiss, E. Kazantzidou, U. Wittstock, M. Dicke, J. Gershenzon, Formation of simple nitriles upon glucosinolate hydrolysis affects direct and indirect defense against the specialist herbivore, *Pieris rapae*, *J. Chem. Ecol.*, **34**, 1311–1321 (2008).
80. R. J. Hopkins, N. M. van Dam, J. J. A. van Loon, Role of glucosinolates in insect–plant relationships and multitrophic interactions, *Annu. Rev. Entomol.*, **54**, 57–83 (2009).
81. (*a*) K. F. M. J. Tierens, B. P. H. Thomma, M. Brouwer, J. Schmidt, K. Kistner, A. Porzel, B. Mauch-Mani, B. P. A. Cammue, W. F. Broekaert, Study of the role of antimicrobial glucosinolate-derived isothiocyanates in resistance of *Arabidopsis* to microbial pathogens, *Plant Physiol.*, **125**, 1688–1699 (2001); (*b*) A. Sellam, B. Iacomi-Vasilescu, P. Hudhomme, P. Simoneau, *In vitro* antifungal activity of brassinin, camalexin and two isothiocyanates against the crucifer pathogens

Alternaria brassicicola and *Alternaria brassicae*, *Plant Pathol.*, **56**, 296–301 (2007); (*c*) S. Ruppel, A. Krumbein, M. Schreiner, Composition of the phyllospheric microbial populations on vegetable plants with different glucosinolate and carotenoid compositions, *Microb. Ecol.*, **56**, 364 372 (2008).

82. K. R. Anilakumar, F. Khanum, A. S. Bawa, Dietary role of glucosinolate derivatives: a review, *J. Food Sci. Technol.*, **43**, 8–17 (2006).
83. R. Gols, T. Bukovinszky, N. M. van Dam, M. Dicke, J. M. Bullock, J. A. Harvey, Performance of generalist and specialist herbivores and their endoparasitoids differ on cultivated and wild *Brassica* species, *J. Chem. Ecol.*, **34**, 132–143 (2008).
84. For reviews, see: (*a*) J. D. Clarke, R. H. Dashwood, E. Ho, Multi-targeted prevention of cancer by sulforaphane, *Cancer Lett.*, **269**, 291–304 (2008); (*b*) J. D. Hayes, M. O. Kelleher, I. M. Eggleston, The cancer chemopreventive actions of phytochemicals derived from glucosinolates. *Eur. J. Nutr.*, **47**, 73–88 (2008); (*c*) R. Mithen, K. Faulkner, R. Magrath, P. Rose, G. Williamson, J. Marquez, Development of isothiocyanate-enriched broccoli and its enhanced ability to induce phase 2 detoxification enzymes in mammalian cells, *Theor. Appl. Gen.*, **106**, 727–734 (2003).
85. (*a*) H. Y. Baik, J. Juvik, E. H. Jeffery, M. A. Wallig, M. Kushad, B. P. Klein, Relating glucosinolate content and flavor of broccoli cultivars, *J. Food Sci.*, **68**, 1043–1050 (2003); (*b*) G. Padilla, M. E. Cartea, P. Velasco, A. de Haro, A. Ordás, Variation of glucosinolates in vegetable crops of *Brassica rapa*, *Phytochemistry*, **68**, 536–545 (2007).
86. I. Lassen, Commonplaces and social uncertainty: negotiating public opinion, *J. Risk Res.*, **11**, 1025–1045 (2008).
87. F. Geu-Flores, C. E. Olsen, B. A. Halkier, Towards engineering glucosinolates into non-cruciferous plants, *Planta*, **229**, 261–270 (2009).
88. C. Vermeulen, L. Gijs, S. Collin, Sensorial contribution and formation pathways of thiols in foods: a review, *Food Rev. Int.*, **21**, 69–137 (2005).
89. (*a*) E. Frérot, A. Velluz, A. Bagnoud, E. Delort, Analysis of the volatile constituents of cooked petai beans (*Parkia speciosa*) using high-resolution GC/ToF-MS, *Flavour Fragr. J.*, **23**, 434–440 (2008); (*b*) R. Kubec, S. Kim, R. A. Musah, The lachrymatory principle of *Petiveria alliacea*, *Phytochemistry*, **63**, 37–40 (2003).
90. C. Thibon, P. Marullo, O. Claisse, C. Cullin, D. Dubourdieu, T. Tominaga, Nitrogen catabolic repression controls the release of volatile thiols by *Saccharomyces cerevisiae* during wine fermentation, *FEMS Yeast Res.*, **8**, 1076–1086 (2008).
91. (*a*) Y. Yoshida, M. Negishi, A. Amano, T. Oho, Y. Nakano, Differences in the βC-S lyase activities of viridans group streptococci, *Biochim. Biophys. Res. Commun.*, **300**, 55–60 (2003); (*b*) M. Fernández, M. Zúñiga, Aminoacid catabolic pathways of lactic acid bacteria, *Crit. Rev. Microbiol.*, **32**, 155–183 (2006).
92. P. Schröder, Plants as sources of atmospheric sulfur, in *Sulfur Nutrition and Assimilation in Higher Plants: Regulatory, Agricultural and Environmental Aspects*, L. J. De Kok, I. Stulen, H. Rennenberg, C. Brunold, W. E. Rauser, (Eds), SPB Academic Publishers, Den Haag, 1993, pp. 253–270.
93. J. Papenbrock, A. Reimenschneider, A. Kamp, H. N. Schulz-Vogt, A. Schmidt, Characterisation of cysteine-degrading and H_2S releasing enzymes of higher plants–from the field to the test-tube and back, *Plant Biol.*, **9**, 582–588 (2007).
94. (*a*) J. D. Schwenn, U. Schriek, H. H. Kiltz, Dissimilation of methionine in cell suspension cultures from *Catharanthus roseus* L., *Planta*, **158**, 540–549 (1983); (*b*) A. Schmidt, H. Rennenberg, L. G. Wilson, P. Filner, Formation of methanethiol from methionine by leaf tissue, *Phytochemistry*, **24**, 1181–1185 (1985).
95. A. Goyer, E. Collakova, Y. Shachar-Hill, A. D. Hanson, Functional characterization of a methionine gamma-lyase in Arabidopsis and its implication in an alternative to the reverse *trans*-sulfuration pathway, *Plant Cell Physiol.*, **48**, 232–242 (2007).
96. D. Kudou, S. Misaki, M. Yamashita, T. Tamura, N. Esaki, K. Inagaki, The role of cysteine 116 in the active site of the antitumor enzyme L-methionine γ-lyase from *Pseudomonas putida*, *Biosci. Biotechnol. Biochem.*, **72**, 1722–1730 (2008).

9

Thermal Generation of Aroma-Active Volatiles in Food

Christoph Cerny

Firmenich SA, Division Recherche et Développement,
7 Rue de la Bergère, B. P. 148, 1217 Meyrin 2, Switzerland

9.1 Introduction

The cooking and heating of food is closely related to the emergence of civilization and probably started not very long after the discovery and control of fire half a million years ago. Most likely by accident our early ancestors discovered that meat and other foods were more enjoyable and easier to chew after being heated. These days, three-star chefs such as Ferra, Blumenthal and Keller have brought the modern culinary art to a high level of sophistication, but the benefits of cooking still remain much the same as before the ice age. Cooked foods taste better, are more palatable and easier to digest, offer a larger variety of possible dishes and are microbially safer than raw ones.

Volatiles are generated during the cooking of food and their number and concentration increase during the course of it. As an example, Table 9.1 compares the compound classes and the number of volatiles of raw and cooked beef.[1] The complexity in cooked beef is extraordinary and the number of identified volatiles increases tenfold compared to the raw state. While just one sulfur compound is found in raw beef, in cooked beef sulfur volatiles (124 compounds) are the most numerous chemicals. Many of the aroma compounds that give meat its delicious typical flavour contain a sulfur atom.

The number of thermal food processes is large, including the roasting of coffee, cocoa, and nuts, boiling, braising, frying, and grilling of meat, baking of bread, sterilizing of preserved food, pasteurizing of milk and microwaving of ready to eat dishes. During heating, the compounds present in the food react, and constituents such as proteins, peptides

The Chemistry and Biology of Volatiles Edited by Andreas Herrmann

Table 9.1 *Type and number of volatile compounds in raw and cooked beef*[1]

Compound class	Raw beef	Cooked beef
Hydrocarbons	19	73
Alcohols	11	32
Carbonyl compounds, aldehydes	10	57
Carbonyl compounds, ketones	8	49
Acids		9
Esters and lactones		11
Bases		91
Sulfur compounds	1	124
Ethers		5
Halogenated compounds		6
Nitriles and amides		1
Phenols		3
Furans		22
Oxazol(in)es		5
Total	49	488

and amino acids, together with reducing sugars, form nonvolatile dark-coloured compounds, mainly through non-enzymatic browning (the Maillard reaction). The same reaction also accounts for part of the concurrent generation of flavour compounds. There is no generally recognized definition of what a 'flavour compound' is. In the context of the present chapter, flavour compounds are classified into the following three categories:

- Volatile aroma compounds (odorants).
- Taste compounds that evoke sweet, bitter, umami, salty and/or sour sensations on the tongue. In principle they can be either volatile or nonvolatile.
- Trigeminal compounds (cool, hot, pungent, tingling, sharp), either volatile or nonvolatile.

This chapter covers only volatile flavour compounds and specifically those which are perceivable by our sense of smell, the aroma compounds. However, some taste or taste-enhancing products that are thermally generated and have been discovered in food in recent years are not included in this book, because they are nonvolatile.[2] In order to be classified as an aroma compound, a chemical must not just be volatile, but must also have an odour and must be present in concentrations above its odour threshold; otherwise its contribution to the aroma is questionable. As an example, Table 9.2 compares the number of volatiles identified in roasted coffee with the number of aroma compounds. Volatiles are considered aroma compounds if they are sensorily detected by gas chromatography–olfactometry (GC-O) or when their concentration exceeds the odour threshold of the compound. GC-O is a bioassay, which combines sensory and instrumental analysis. During GC-O, part of the eluent at the end of the GC capillary exiting the column oven has its odour evaluated by a panellist at a sniffing port. The remainder is detected by a classical detector, such as a flame ionization detector (FID) or mass spectrometry (MS) detector. From Table 9.2, it is obvious that only a small fraction (5%) of the volatiles plays a significant role in the aroma of roasted coffee.[1,3] Certain chemical classes like sulfur and carbonyl compounds are more relevant and comprise a larger number of odour-active species, while others, such as hydrocarbons, play no role at all. Certainly, it cannot be

Table 9.2 Type and number of volatile and aroma compounds in roasted coffee

Compound class	Volatile compounds[a]	Aroma compounds[b]
Hydrocarbons	80	0
Alcohols	24	1
Carbonyl compounds, aldehydes	37	8
Carbonyl compounds, ketones	85	4
Acids	28	2
Esters and lactones	36	0
Bases	215	6
Sulfur compounds	100	11
Acetals	1	0
Ethers	2	0
Halogenated compounds	1	0
Nitriles and amides	4	0
Phenols	49	3
Furans	127	4
Pyrans	6	0
Oxazol(in)es	35	0
Anhydrides	4	0
Total	834	39

Note: [a]See ref. 1.
[b]As detected by aroma extract dilution analysis (AEDA) and by calculation of odour activity values (OAV), according to ref. 3.

excluded that future research will reveal more aroma compounds in roasted coffee; however, the picture that only relatively few aroma compounds have an impact on the odour is unlikely to change.

This chapter shows several examples of chemical reactions that have great importance for the thermal generation of aroma-active substances in food, as well as giving an overview of the odorants generated. Fat oxidation, which also plays a key part in the thermal generation of aroma-active volatiles, is only briefly covered because the relevant odorants and mechanisms are very similar to the ones that occur in nonthermal oxidation and biogeneration in food.

9.2 The Maillard Reaction

The Maillard reaction is named after the French chemist Maillard (1878–1936), who investigated the reaction between glucose and glycine and postulated the formation of a Schiff base as the first step in the formation of the resulting brown-coloured product.[4] In general terms, the Maillard reaction is considered to be the reaction between a reducing sugar and an amino acid, peptide, or protein, and is often also called non-enzymatic browning. Maillard reaction products contribute to the colour, aroma and taste generation in heated foods (e.g. baked bread, fried meat, roasted cocoa). In contrast, under harsh cooking conditions of high temperature and low water activity, unhealthy mutagenic Maillard products, such as heterocyclic aromatic amines or acrylamide, can also be formed.[5] Furthermore, the reaction between essential amino acids and sugars can lead to nutritional depletion of proteins because the availability of lysine and other essential amino acids decreases, for example during the spray-drying of milk.[6] However, recent research shows that also health-promoting agents, with antioxidant or chemopreventive properties, are formed in the Maillard reaction, for example in coffee.[7] The reaction proceeds not only at high temperatures during cooking, but also under the moderate thermal conditions inside our body, *in vivo*, where reducing sugars such as glucose and proteins are present in tissues and blood, although the reaction is much slower. The *in vivo* Maillard reaction and the resulting advanced glycation end products are claimed to contribute substantially to the physiological effects of ageing, long-term consequences of diabetes, eye cataracts and other diseases.[8]

The thermal formation of aroma-active volatiles from the Maillard reaction plays a role in the production and preparation of a wide range of foods, such as:

- Savoury (boiled, roasted, fried, grilled, sauté): meat, fish, seafood, potato, vegetables, eggs.
- Sweet and bakery (roasted, toasted, grilled, baked): bread, cake, cocoa, nut, vanilla, popcorn.
- Beverages (malted, roasted, dried): coffee, beer, whisky, tea, chocolate drinks.

Chemically, the Maillard reaction can be divided into early stages (Amadori and Heyns rearrangement, formation of deoxyosones, fragmentation) and late stages with further reactions of intermediates and the generation of odorants and other compounds. An excellent review on the Maillard reaction is available from Nursten,[9] as well as several proceedings of Maillard symposia, for example the one edited by Schleicher *et al.*[10]

9.2.1 The Amadori Rearrangement

In the very first stage of the Maillard reaction, an amino compound adds via a nucleophilic attack to the carbonyl group of a reducing sugar, which leads to the formation of Schiff base **1**. In Scheme 9.1, for example, the reaction is depicted with an amino acid and an aldohexose

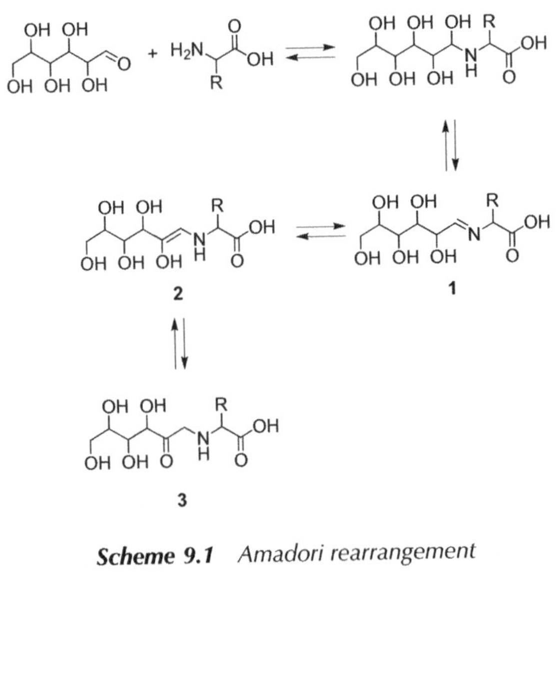

Scheme 9.1 *Amadori rearrangement*

as reaction partners. It should be noted that the sugar structures are shown in the open-chain form to facilitate comprehension, while in aqueous solution they occur mainly in the cyclic hemi-acetal form. Imine **1** can further react and rearrange to the intermediate enaminol **2** and finally through keto-enol tautomerization produce aminoketose **3**, the so-called 'Amadori compound'.

The Amadori rearrangement is named after the Italian chemist Amadori (1886–1941), who studied the condensation products between glucose and *p*-phenetidine and other amino compounds.[11] Similar rearrangement products are obtained with ketoses (e.g. fructose) instead of aldoses. In this case, the intermediate compounds are not aminoketoses, but aminoaldoses (Heyns compounds). Amadori compounds are found in many heat-treated foods, such as dried vegetables, malt and cocoa. Concentrations of up to 3.8% have been found in dried tomato powder for the Amadori compound from glucose and glutamic acid.[12]

9.2.2 Deoxyosones

Degradation of Amadori compounds produces a variety of α-dicarbonyl compounds with an intact sugar carbon chain, as illustrated in Scheme 9.2, using as example the reactions of the hexose-derived Amadori compound **3**. 1,2-Enolization leads to enaminol **2** which, upon the loss of one molecule of water and hydrolysis of the resulting Schiff base **4**, forms 3-deoxy-1,2-hexodiulose (**5**), also called '3-deoxyosone'.[13] 2,3-Enolization of **3**, in contrast, leads to enediol **6** which can proceed through different reaction sequences. Abstraction of the amino acid produces 1-deoxy-2,3-hexodiulose (**7**), the so-called '1-deoxyosone'. If instead, the hydroxyl group is lost in **6** at C(4) of the sugar moiety, the 4-deoxyosone derivative **8** results. Amadori compounds were recently shown to possess great carbonyl mobility[14] with formation of 1-amino-1,4-dideoxy-5,6-hexodiulose (**10**) after a shift of the carbonyl group down to the end of intermediate **9**. In the case of deoxyosones **8** and **10**, the amino acid remains part of the molecule. 1,4-Dideoxy-2,3-hexodiulose (1,4-dideoxyosone, **11**) can be formed from **7** by a reduction step and loss of water.[15] The formation of **11** is particularly important in the Maillard reaction of di- and oligosaccharides under near anhydrous conditions. In the case of Amadori compounds from di- and oligosaccharides (e.g. from maltose, **12**), the glucose moiety is 'peeled off' from the C(4) carbon to form **11** during the course of the reaction.[16]

9.2.3 Retro-Aldolization

Compounds with a β-hydroxy carbonyl structure can easily undergo retro-aldol reactions, thus leading to smaller carbon fragments from sugars or deoxyosones. Scheme 9.3 shows the formation of fragments from deoxyosones **5**, **7**, and **11**. Oxopropanal (**13**) and glyceraldehyde (**14**), which are quite reactive intermediates, can be formed from 3-deoxyosone (**5**). Aldehyde **14** can subsequently undergo retro-aldolization to yield formaldehyde (**15**) and glycolaldehyde (**16**), which react further. Alternatively, **16** can arise from the degradation of both 1-deoxyosone (**7**) and 1,4-dideoxyosone (**11**) by retro-aldol reaction, forming **17** or **18** as co-products, respectively. The retro-aldol product 2,3-butanedione (**18**) is an intermediate and can undergo further reactions. As such, **18** plays a role in the aroma of coffee, fish, dairy products and many other food products. Apart from the Maillard reaction, there is a nonthermal, microbial formation pathway.

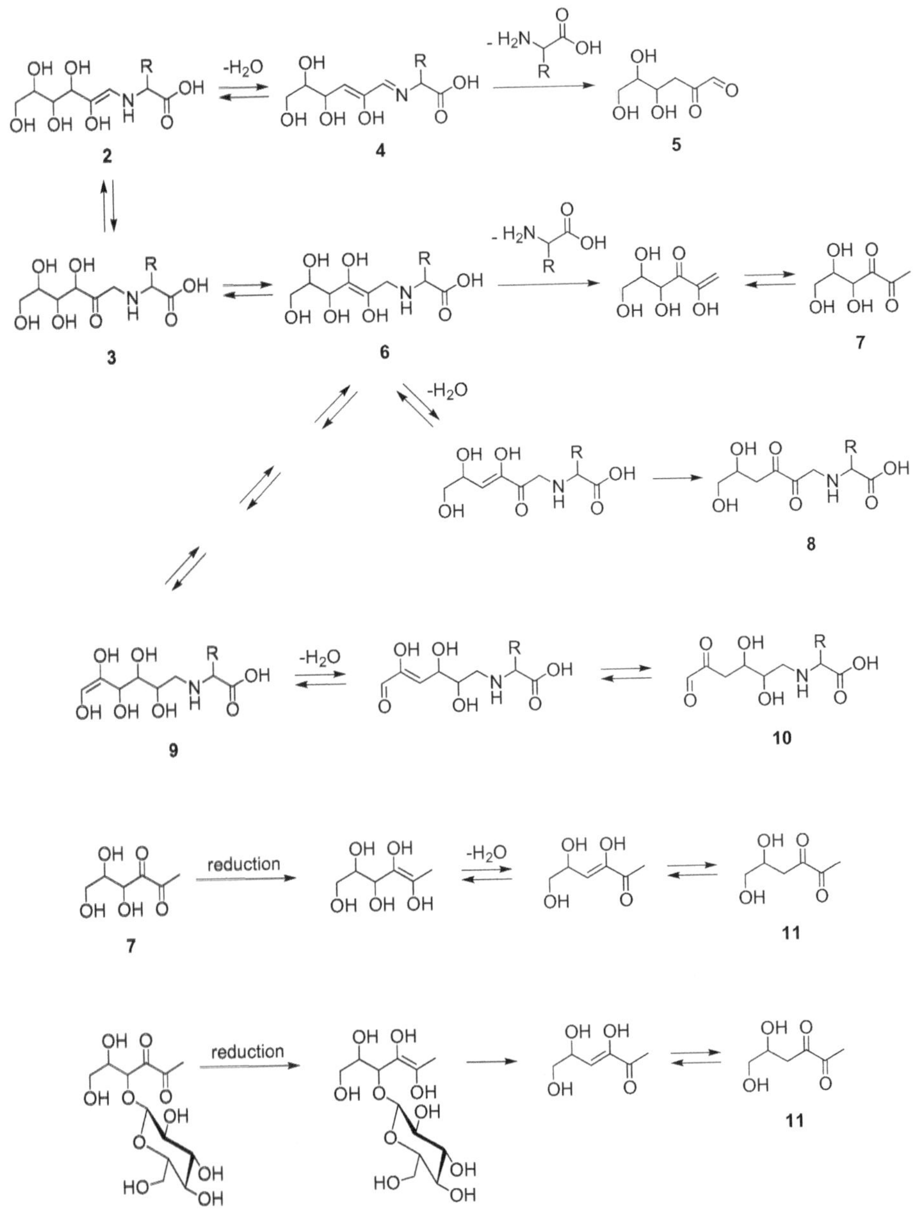

Scheme 9.2 *Formation of deoxyosones from Amadori compounds*

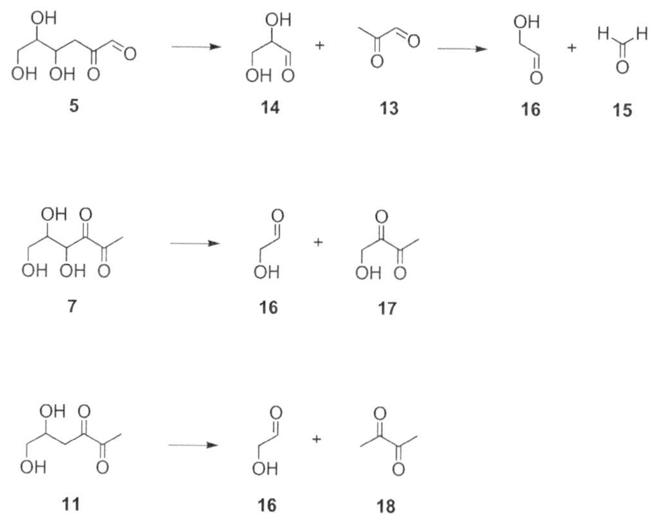

Scheme 9.3 Retro-aldolization of deoxyosones

9.3 Formation of Aroma Compounds in the Later Stages of the Maillard Reaction

9.3.1 2-Furfurylthiol

The formation of 2-furfurylthiol (**19**) is important in a variety of heated foods (see Table 9.3),[17,18] but it is certainly a key odorant in coffee. The compound has an extremely low odour threshold in water of 0.036 $\mu g\ l^{-1}$ and dilute solutions have a typical coffee-like, sulfurous odour.[17] Studies have revealed furfural (**20**) as an important intermediate in its formation.[19] The formation pathway, proposed in Scheme 9.4, starts from pentose sugar **21** and proceeds to the intermediate 3-deoxyosone **22** which, through cyclization and loss of water, gives rise to **20**. Reaction with hydrogen sulfide then forms **19**. In coffee, the relevant sugar precursor is most likely arabinose, stemming from the degradation of arabinogalactan present in the bean, and hydrogen sulfide originates from the protein fraction in coffee.

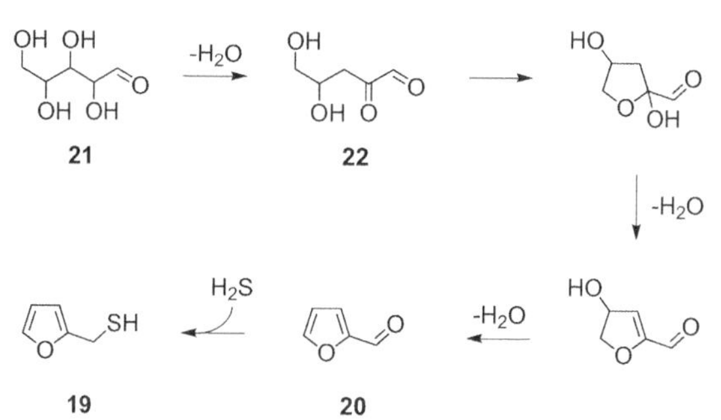

Scheme 9.4 Formation of 2-furfurylthiol (**19**) from pentoses

Table 9.3 *Odours and odour thresholds of selected aroma compounds formed in the Maillard reaction*

No.	Compound	Odour	Precursors	Food (examples)	Odour threshold[a] [$\mu g\,l^{-1}$]
19	2-Furfurylthiol	Coffee-like	Pentoses, H_2S	Coffee, roasted and stewed meat, roasted sesame, bread	0.036[b]
23	4-Hydroxy-2,5-dimethyl-3(2*H*)-furanone (Furaneol)	Caramel-like	Hexoses	Strawberry, pineapple, cheese, soy sauce, boiled, roasted and braised meat, bread, malt, beer, coffee, tea, cocoa	40[b]
27	2,3,5-Trimethylpyrazine	Roasted	Reducing sugars	Potatoes, grilled, roasted and fried meat, bread, fish, roasted nuts, beer, soy sauce, shrimps, roasted sesame, cheese, cocoa, coffee	71–400[c]
28	3-Ethyl-2,5-dimethylpyrazine	Roasted, earthy	Reducing sugars, alanine	Potatoes, grilled, roasted and fried meat, bread, fish, roasted nuts, beer, soy sauce, shrimps, roasted sesame, cocoa, coffee	0.04–9[c]
29	2-Ethenyl-3,5-dimethylpyrazine	Earthy	Reducing sugars	Coffee, beef, nuts	
34	2-Acetyl-1-pyrroline	Roasted, popcorn	Reducing sugars, proline, ornithine	Bread, popcorn, rice, cooked meat, fish and seafood, yeast, cocoa	0.053[b]

Note: [a]Odour detection threshold (orthonasal) in $\mu g\,l^{-1}$ water.
[b]See ref. 17.
[c]See ref. 18.

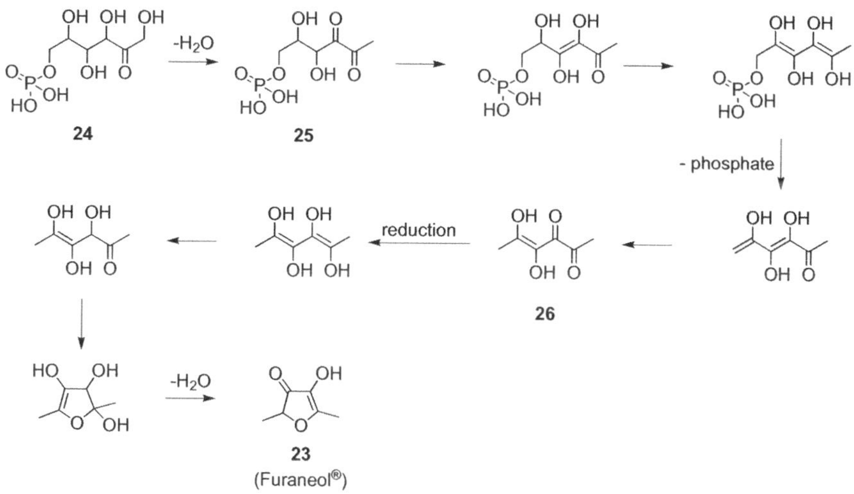

Scheme 9.5 *Formation of 4-hydroxy-2,5-dimethyl-3(2H)-furanone (Furaneol, **23**) from fructose-6-phosphate (adapted from ref. 21)*

9.3.2 4-Hydroxy-2,5-dimethyl-3(2*H*)-furanone

The formation pathways for 4-hydroxy-2,5-dimethyl-3(2*H*)-furanone (Furaneol, **23**), which smells like caramel, are broadly diversified. In strawberry, pineapple and other fruits, it is generated through plant biochemical routes, but there exist also microbial pathways which are responsible for its formation (e.g. in cheese and soy sauce) and there are various mechanisms for its generation through the Maillard reaction in heated foods like meat, bread and cocoa (see Table 9.3). Important precursors for the thermal formation of **23** in food are rhamnose, glucose, fructose and their respective phosphate esters. For example, raw beef contains glucose ($1.1\,g\,kg^{-1}$), fructose ($0.3\,g\,kg^{-1}$), glucose-6-phosphate ($1.8\,g\,kg^{-1}$) and fructose-6-phosphate (**24**, $0.5\,g\,kg^{-1}$) as the main hexose sugars.[20] Under aqueous thermal conditions, **24** is a much more efficient precursor than the other three hexoses,[21] and therefore it can be regarded as the key precursor of **23** in beef and other meats. The concentration of **23** in cooked meat increases with longer cooking times, from $0.1\,mg\,kg^{-1}$ after roasting beef for 1 min, to $2.0\,mg\,kg^{-1}$ after 15 min and $12.0\,mg\,kg^{-1}$ after stewing for 4 h.[20] The reaction pathway from **24** leads via 1-deoxyosone-6-phosphate **25** and loss of phosphate to open-chain molecule **26**, followed by a reduction step, cyclization and further loss of water to **23** (Scheme 9.5).

9.3.3 Alkyl and Alkenylpyrazines

Alkyl and alkenylpyrazines contribute mainly roasted aroma notes to heated foods, especially when they have been prepared at higher process temperatures, as in deep-frying, oven and pan roasting, or barbecuing. They are virtually not formed below 100 °C and their concentration increases with increasing temperatures and heating time, with a reasonably high stability towards thermal degradation.

Aroma threshold and odour impact strongly depend on the alkyl substitution pattern.[22] Alkylpyrazines with relatively low thresholds, like 2,3,5-trimethylpyrazine (**27**, odour threshold 50 ng l^{-1} in air; 71–400 μg l^{-1} water), 3-ethyl-2,5-dimethylpyrazine (**28**, 3.6 ng l^{-1} in air) or 2-ethenyl-3,5-dimethylpyrazine (**29**, 0.012 ng l^{-1} in air), contribute to the aroma of foods like coffee, cocoa, bread, roasted nuts, French fries and beef steak (Table 9.3). Their formation is strongly linked to the degradation of reducing sugars in the Maillard reaction. The vicinal dicarbonyl compounds, like the deoxyosones (e.g. **30**) or fragmentation products, such as **13** and **18** (Scheme 9.3), are important intermediates which can react with an amino acid via a Strecker reaction to give rise to α-aminocarbonyl compounds (Scheme 9.6), such as for example, 1-aminoacetone (**31**), 3-amino-2-butanone (**32**), or 2-amino-4,5-dihydroxy-3-pentanone (**33**), as well as affording aldehydes, depending on the amino acid, such as acetaldehyde in case of alanine as reaction partner. Two molecules of **31**

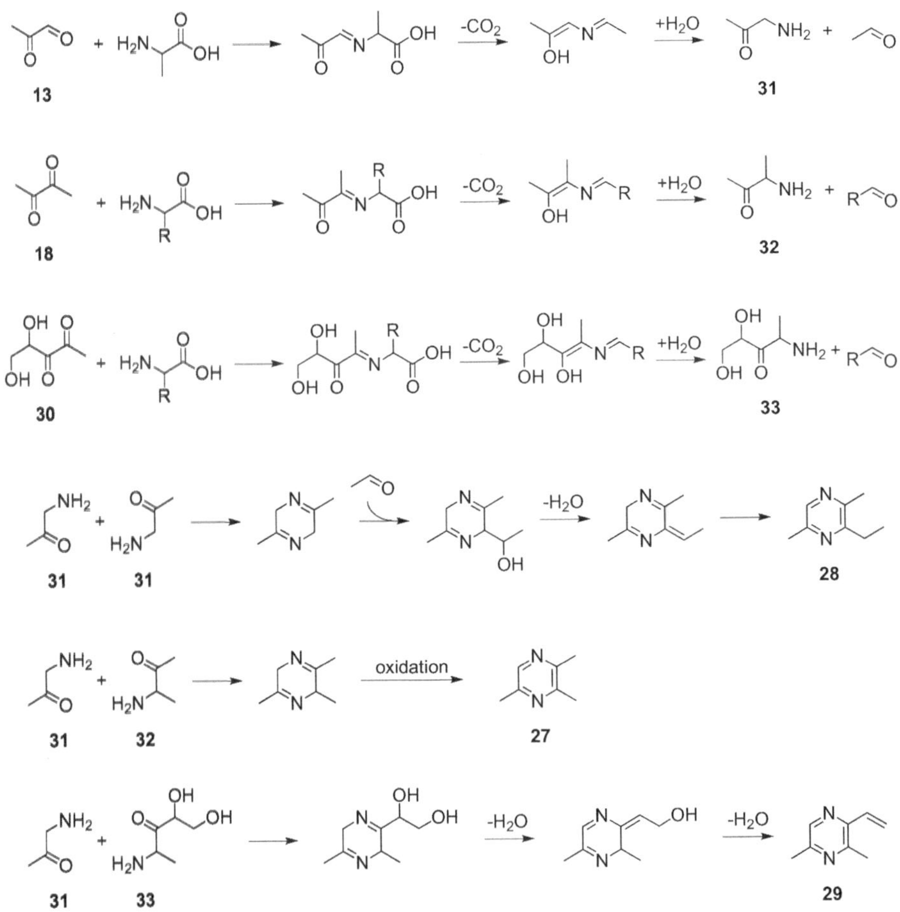

Scheme 9.6 *Formation of alkyl and alkenylpyrazines* ***27–29***

can then condense to form 2,5-dimethyl-3,6-dihydropyrazine, which can subsequently react with acetaldehyde and dehydrate to yield **28**. The formation of trimethylpyrazine **27** can be explained by condensation of the two aminocarbonyl compounds **31** and **32** and oxidation of the intermediate dihydropyrazine. When the α-aminocarbonyl compound **33** combines with **31** (Scheme 9.6), the resulting dihydropyrazine, after the loss of two water molecules, finally forms alkenylpyrazine **29**.

9.3.4 2-Acetyl-1-pyrroline

2-Acetyl-1-pyrroline (**34**), which smells like popcorn, has an extremely low odour threshold of 0.053 $\mu g\, l^{-1}$ in water, was identified for the first time in boiled rice,[23] to which it adds the characteristic basmati rice odour. While the formation pathway in rice still remains vague,[24] it can also be formed microbially, for example in fermented sausages and cheese. However, in many food products (Table 9.3) the Maillard reaction remains the key pathway, as in different kinds of bread, cooked meat, fish and seafood, as well as in cocoa and popcorn. For its formation, two intermediates play a significant role. One of them is 2-oxopropanal (**13**), which can be formed through the degradation of reducing sugars, followed by the fragmentation of intermediate deoxyosones such as **5** (Scheme 9.3). For the other intermediate, 1-pyrroline (**35**), two main precursors exist: the amino acid proline, which is more important in popcorn, tortilla chips and most food of animal origin, and the amino acid ornithine (**36**), which does not occur in proteins, but has a biochemical role in urea metabolism. The latter occurs in fairly high amounts in yeast, and consequently plays a key role as precursor for **34** in bread aroma and for the odour of other yeast-derived foods.

The formation of **34** from **36** (Scheme 9.7) starts with a Strecker degradation (see next section) leading to 4-aminobutanal (**37**) and then to 1-pyrroline (**35**) upon cyclization.[25] One possible next step according to Hofmann and Schieberle[26] is the condensation of the unstable intermediate **35** with the hydrated form of **13** to generate 2-(1,2-dioxopropyl)-pyrrolidine which, after oxidation by air to 2-(1,2-dioxopropyl)-1-pyrroline (**38**), followed by formation of the hydrate, rearrangement and decarboxylation, yields 2-acetylpyrrolidine (**39**). Compound **39** is easily oxidized by air and yields the target compound 2-acetyl-1-pyrroline (**34**). This reaction pathway is complex and comprises multiple reaction steps. Although experiments with $^{13}C(6)$-labelled glucose[26] could elucidate alternative mechanisms to explain the formation of **34**, these are equally complex. In consequence, the efficiency of all these formation pathways is minute and in fact the yields and amounts of **34** formed in food are incredibly low (around 10 $\mu g\, kg^{-1}$ in bread). However, because this odorant is so powerful due to its low odour threshold, even low concentrations in bread are sufficient to provoke the characteristic freshly baked aroma.

9.4 The Strecker Degradation

When the German chemist Strecker (1822–1871) heated a solution of alloxan [2,4,5,6-(1*H*,3*H*)-pyrimidinetetraone] and the amino acid alanine, he observed that the colour turned purple and carbon dioxide and acetaldehyde evolved, the pungent odour of the latter being perceptible throughout the experiment.[27] The essential feature in alloxan for the generation

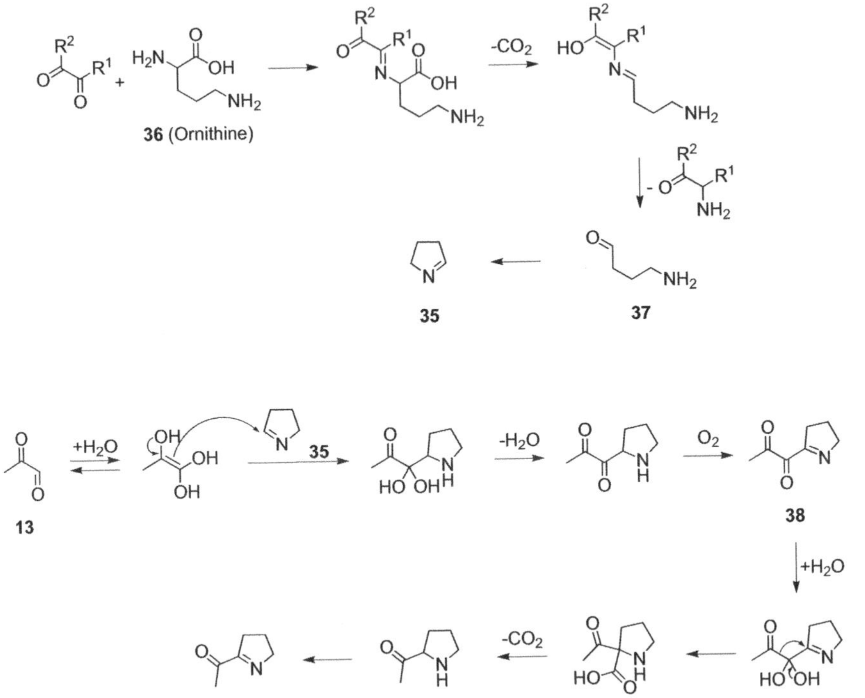

Scheme 9.7 Formation of 2-acetyl-1-pyrroline (**34**) from ornithine (**36**) (adapted from refs. 25, 26)

of carbon dioxide and acetaldehyde is its α-dicarbonyl structure, and that is why the reaction also works with other α-dicarbonyl compounds. Later the chemical reaction was named the 'Strecker degradation' and the corresponding aldehyde a 'Strecker aldehyde'. The Strecker degradation and the Maillard reaction are closely linked, because the Maillard reaction produces a large variety of α-dicarbonyl compounds which can undergo Strecker degradation. The Strecker degradation can of course also occur with α-dicarbonyls which have been formed microbially, as in fermented foods, and even fat oxidation products, such as 4,5-epoxy-(2*E*)-decenal, can produce the corresponding Strecker aldehydes.[28] Scheme 9.8 explains the individual steps of the reaction, starting with the condensation of α-dicarbonyl (**40**) and amino acid **41** to form the Schiff base **42**. Then, in the subsequent decarboxylation, favoured by the electron-withdrawing character of the second carbonyl group, α-aminocarbonyl compound **43** is formed. Due to the labile imino structure, **43** readily hydrolyses, releasing **44** and the Strecker aldehyde **45**.

The structures of aroma-relevant Strecker aldehydes **46–51** which are found in food are illustrated in Figure 9.1. They occur in many different classes of food that have gone through a thermal process, such as vegetables, rice, bread, malt, meat, fish, seafood, coffee and cocoa (see Table 9.4).[17,18] Aldehydes from proteinaceous amino acids other than those listed in Table 9.4 play only a subordinate aroma role due to their lack of only

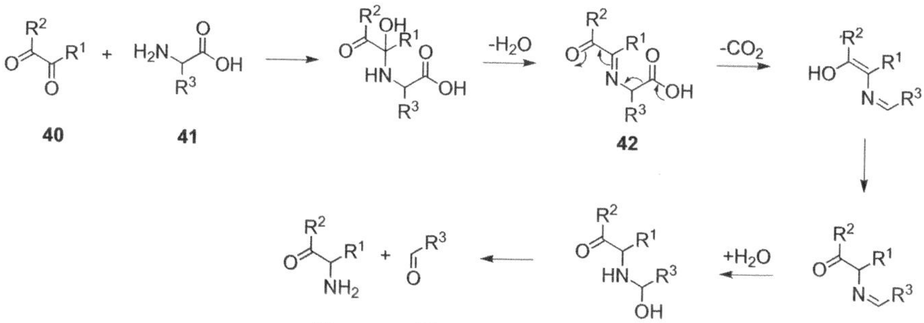

Scheme 9.8 Strecker degradation

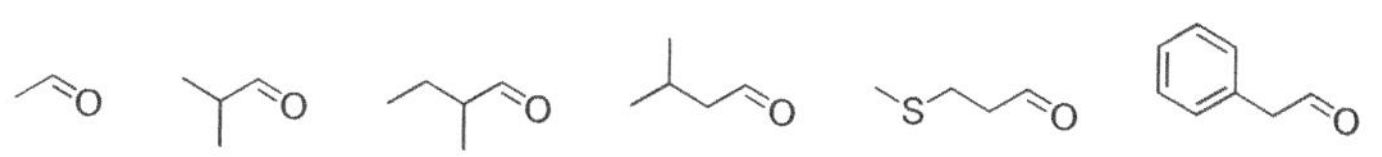

Figure 9.1 Odour-active Strecker aldehydes

weak odour activities. Acetaldehyde (**46**) and the branched alkyl aldehydes, 2-methylpropanal (**47**), 2-methylbutanal (**48**) and 3-methylbutanal (**49**), are derived from the amino acids alanine, valine, isoleucine and leucine, respectively. The sulfur-containing amino acid, methionine, produces methional (**50**) which has a potato-like odour and a very low odour threshold of around $0.43\,\mu g\,l^{-1}$ in water. Methional (**50**) is not very stable and decomposes to propenal and methanethiol (**52**), which has a cabbage-like odour and which can oxidize to dimethyl disulfide (**53**), with a sulfurous, vegetable-like odour, as shown in Scheme 9.9.

Dimethyl sulfide (**54**) can be formed analogously from the amino acid *S*-methylmethionine (**55**),[29] which is not present in proteins, but is found for example in tomato, malt and asparagus. Scheme 9.10 proposes its formation via a Strecker degradation.

Table 9.4 Odours and odour thresholds of aroma-active Strecker aldehydes

No.	Compound	Odour	Precursor	Odour threshold[a] [$\mu g\,l^{-1}$]
46	Acetaldehyde	Pungent	Alanine	25[b]
47	2-Methylpropanal	Green	Valine	0.49[b]
48	2-Methylbutanal	Chocolate, malty	Isoleucine	1.5[b]
49	3-Methylbutanal	Malty	Leucine	0.50[b]
50	Methional	Potato-like	Methionine	0.43[b]
51	Phenylacetaldehyde	Honey, flowery	Phenylalanine	4[c]

Note: [a]Odour detection threshold (orthonasal) in $\mu g\,l^{-1}$ water.
[b]See ref. 17.
[c]See ref. 18.

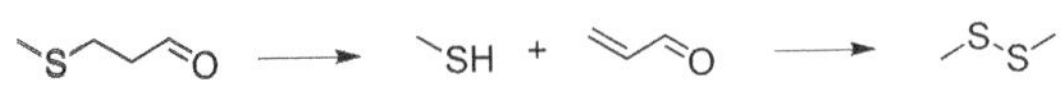

50 52 53

Scheme 9.9 Formation of methanethiol (**52**) and dimethyl disulfide (**53**) from methional (**50**)

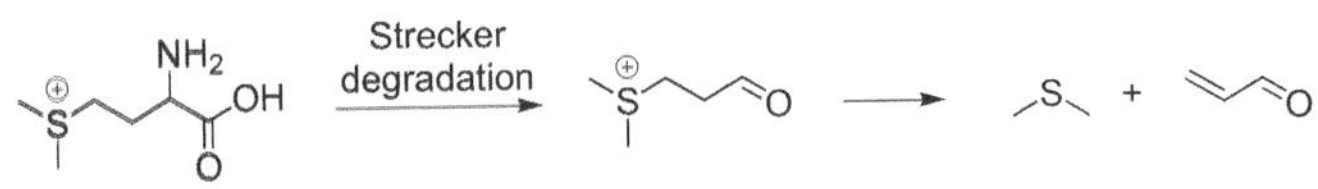

55 54

Scheme 9.10 Formation of dimethyl sulfide (**54**) from S-methylmethionine (**55**)

9.5 Caramelization

Caramelization is, after the Maillard reaction, the second most important chemical reaction which provokes non-enzymatic browning of food, the formation of brown colour and caramel-like, sweet, nutty odours. Similar to the Maillard reaction, sugar degradation is key to caramelization, but, in contrast, amino acids do not participate or catalyse the degradation. More drastic reaction conditions are required and, unlike the Maillard reaction (which occurs even at room temperature, e.g. during the storage of food, or under physiological conditions in live organisms, e.g. in the blood stream), caramelization takes place only at higher temperatures, typically in a baking/roasting environment above 160 °C and generally at low water activity.[30] Caramelization under food-relevant conditions, as in the production of sugar candy and confectionery or when glazing vegetables during cooking, should be clearly distinguished from the production of caramel colorants, which are food additives. In the latter case, extreme pH conditions, are employed, either a very acidic or a very alkaline environment, often with ammonia.

The following relevant chemical reactions are likely to occur during caramelization:

- Cleavage of sucrose into glucose and fructose,
- Dehydration,
- Fragmentation (dicarbonyl cleavage, retro-aldolization),
- Enolization and isomerization.

The structures of some odour-active compounds in caramelized food are displayed in Figure 9.2.

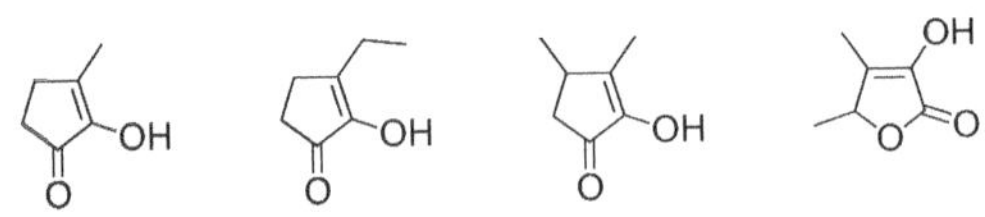

56 57 58 61

Figure 9.2 Aroma-active volatiles from caramelization

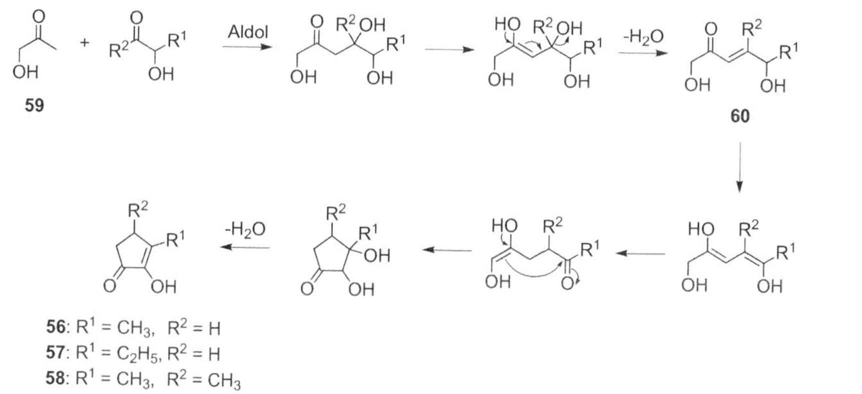

Scheme 9.11 *Formation of 2-hydroxy-2-cyclopenten-1-one derivatives* ***56–58*** *(adapted from ref. 31)*

The formation of the 2-hydroxy-2-cyclopenten-1-one derivatives **56**, **57** and **58** starts with C_3 and C_4 hydroxy carbonyl compounds from sugar degradation and fragmentation during caramelization (Scheme 9.11).[31] For example, 2-hydroxypropanal and 1-hydroxypropanone (**59**) aldolize ($R^1 = CH_3$; $R^2 = H$), and the resulting intermediate then undergoes isomerization and dehydration to dihydroxycarbonyl compound **60**, which after an enolization step then cyclizes and dehydrates to 2-hydroxy-3-methyl-2-cyclopenten-1-one (**56**). Hydroxy ketone **56** has a maple-like, burnt smell.[32] When 1-hydroxypropanone (**59**) reacts with 2-hydroxybutanal ($R^1 = C_2H_3$; $R^2 = H$), 2-hydroxy-3-ethyl-2-cyclopenten-1-one (**57**) results, having a sweet, sugary odour. Finally 2-hydroxy-3,4-dimethyl-2-cyclopenten-1-one (**58**), with a sweet, burnt aroma, forms when 3-hydroxy-2-butanone ($R^1 = CH_3$; $R^2 = CH_3$) is the reaction partner of **59**.

The highly potent odorant lactone 3-hydroxy-4,5-dimethyl-2(5*H*)-furanone (**61**, Figure 9.2) has a extremely low nasal odour threshold of 0.001 $\mu g\,l^{-1}$ in water,[18] possessing a characteristic savoury, spicy, fenugreek-like aroma. It is noteworthy that, like many other odorants which arise in the course of caramelization reactions, **61** can also be formed via a Maillard pathway. Additionally there also exist biochemical pathways, both in plants and in microorganisms. Scheme 9.12 illustrates its (thermal) formation from the sugar breakdown fragments, 2,3-butanedione (**18**) and hydroxyacetaldehyde (**16**), which add to each other in an aldol-type reaction, giving **62**. Loss of water and isomerization, followed by cyclization, produces 2-hydroxy-4,5-dimethyl-3(2*H*)-furanone (**63**), which rearranges into **61**.

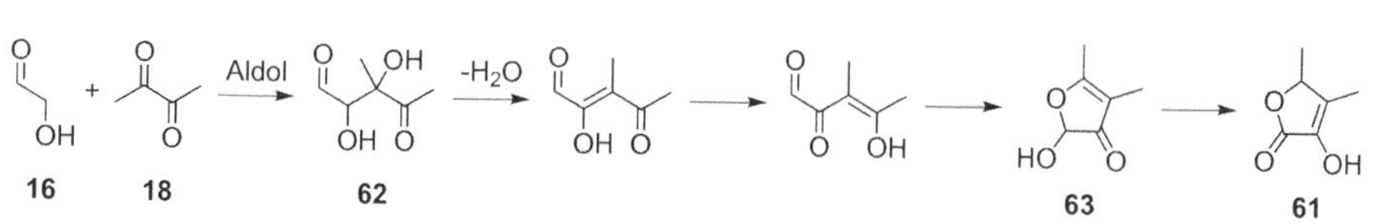

Scheme 9.12 *Formation of 3-hydroxy-4,5-dimethyl-2(5H)-furanone (**61**) (adapted from ref. 31)*

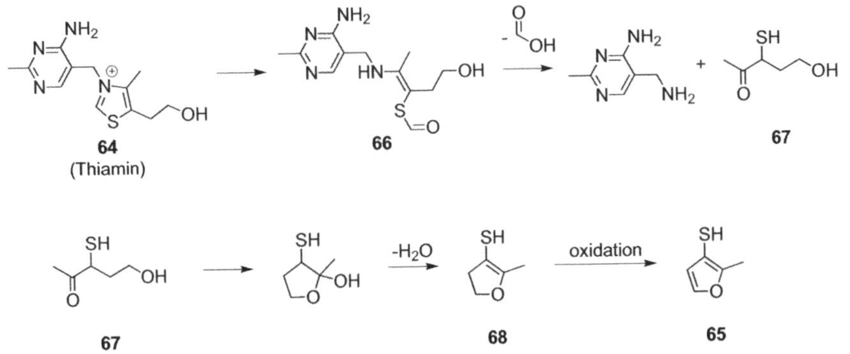

Scheme 9.13 Formation of 2-methyl-3-furanthiol (**65**) from thiamin (**64**)

Compound **61** is found in a large variety of foods and plays an important organoleptic role, for example in coffee and roasted meat.

9.6 Thiamin Degradation

Thiamin (Vitamin B_1, **64**) has an essential biochemical role as co-enzyme in the metabolism of carbohydrates and branched amino acids. Furthermore, it also has an important role as an aroma precursor in meat, because thiamin degradation is the main formation pathway for 2-methyl-3-furanthiol (**65**), a key odorant for the sulfurous meaty odour, for example of a beef bouillon.[20] During the boiling of beef approximately 2 $\mu g\,kg^{-1}$ of **65** are formed. At a thiamin content around 1–2 $mg\,kg^{-1}$ in beef this represents a quite poor molar yield for **65**. Nevertheless, due to the very low odour threshold (0.0004 $\mu g\,l^{-1}$), this low concentration is still more than 2000 times above the odour threshold and hence this compound has a huge impact on the aroma.

In the first phase of the degradation, the thiazole ring of **64** is hydrolysed (Scheme 9.13), and the resulting intermediate **66** loses formic acid and 2-methyl-4-amino-5-(aminomethyl) pyrimidine to generate the key intermediate, 5-hydroxy-3-mercapto-2-pentanone (**67**). Upon cyclization and loss of water, 4,5-dihydro-2-methyl-3-furanthiol (**68**) is formed which then oxidizes to **65**.

9.7 Ferulic Acid Degradation

Phenols are main contributors to the characteristic odour of smoke and smoked food products. Ferulic acid (**69**) is an important constituent of wood lignocellulose, but is also widespread in plants and foods like cereals, coffee, nuts and vanilla. Ferulic acid is not aroma-active itself, but is a flavour precursor which is converted into various odorants upon thermal treatment.[33]

Scheme 9.14 illustrates a range of compounds which can be generated from ferulic acid in this way. Thermal decomposition of ferulic acid generates 4-vinylguaiacol (**70**), a volatile with a spicy, clove-like odour (Table 9.5),[17] which upon oxidation in the presence of oxygen

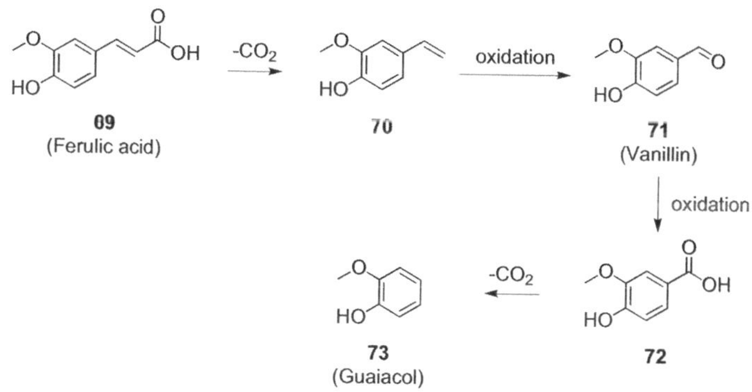

Scheme 9.14 *Ferulic acid degradation*

Table 9.5 *Odours and odour thresholds of compounds formed from ferulic acid*

No.	Compound	Odour	Odour threshold[a] [$\mu g\, l^{-1}$]
69	Ferulic acid	No odour	
70	4-Vinylguaiacol	Spicy, clove-like	5.1[b]
71	Vanillin	Vanilla-like	53[b]
72	Vanillic acid	Vanilla-like	
73	Guaiacol	Burnt, smoky	0.84[b]

Note: [a]Odour detection threshold (orthonasal) in $\mu g\, l^{-1}$ water.
[b]See ref. 17.

forms vanillin (**71**). Vanillin (**71**) itself can oxidize, yielding the corresponding carboxylic acid **72** and finally, through the loss of carbon dioxide, producing guaiacol (**73**), which is a typical product in smoked, roasted and grilled food.

9.8 Fat Oxidation

Fats and lipids are among the main constituents of our diet. Fat, carbohydrates and proteins essentially form the energy content of our food, fat providing 39 kJ g^{-1}. Some foodstuffs, like egg yolk, butter and bacon, are particularly rich in fat. In contrast, vegetables and fruits generally contain little fat. However, even the minuscule amounts of lipids in some fruits and vegetables can play a role in the development of the respective characteristic aroma (e.g. melon, cucumber). The main lipids in food are triglycerides; others include free fatty acids, phospholipids, lipoproteins and waxes, all of which can undergo oxidation.

We can classify lipid oxidation into several classes:

- Lipid photo-oxidation: the oxidation of lipids in the presence of singlet oxygen ($^1O_2$), an oxygen species which can be formed by activation via light radiation.
- Enzymatic lipid oxidation: catalysed by lipoxygenases, enzymes which occur in many plants, but also in animal tissues and promote the oxygenation of fatty acids, such as linoleic and linolenic acid. The resulting hydroperoxides can be further degraded by other

enzymes, such as hydroperoxide lyases. The resulting decomposition products are often aroma-active aldehydes. In cucumber, for example, (2*E*,6*Z*)-nonadienal is an enzymatic degradation product of linolenic acid and, once the cucumber is cut and the tissue disrupted, the enzymes come into contact with the cucumber lipid, generating the very characteristic odour,

- Lipid autoxidation: reaction with oxygen by a radical mechanism. This can occur at low temperature (e.g. room temperature) but is accelerated at higher temperature during cooking (e.g. during deep-frying). The variety and amounts formed are dependent on the temperature employed.

In general, unsaturated fatty acids are increasingly susceptible to fat oxidation in the following order:

linolenic acid (18 : 3) > linoleic acid (18 : 2) > oleic acid (18 : 1) > stearic acid (18 : 0).

Because of this, many of the volatiles formed play a role in flavour generation during cooking. In particular, the odour of deep-fried foods is largely dominated by aldehydes, such as (2*E*,4*E*)-decadienal (**74**), which has a fried, fatty smell and stems from the lipid oxidation of linoleic acid (**75**, R = H; Scheme 9.15). The type of aldehyde formed largely depends on the precursor fatty acid which is present, either as a free fatty acid or ester-bound, as in mono-, di- and triglycerides, phospho- and glycolipids and waxes. Grosch has published a detailed review of the different steps which are involved in lipid autoxidation.[34]

Scheme 9.15 describes the oxidation of **75** and the formation of the corresponding aroma-relevant aldehydes. The first step in this radical mechanism is the formation of hydroperoxyoctadienoate (HPOD) derivatives **76**–**79**. Initially, a hydrogen atom is abstracted from a carbon which is neighbouring a double bond, thus generating a radical. In the case of **75**, either one of the carbon atoms C(8), C(11) or C(14) can lose hydrogen. The position of one of the double bonds then shifts, followed by the addition of oxygen to form a peroxy radical. The peroxy radical can abstract a hydrogen atom (e.g. from another molecule of fatty acid) and form a hydroperoxide. The principal peroxides from **75** are depicted in Scheme 9.15 (**76**–**79**). In the next step, the β-cleavage of peroxides, a bond on either side of the hydroperoxide carbon can break. In the case of 9-hydroperoxy-10,12-octadecadienoic acid (**76**, R = H, 9-HPOD), either two C_9 fragments or a C_{10} and a C_8 fragment result. Fragments containing a carboxy group are not aroma-active, while the other fragments can form aldehydes, which are very often aroma-active.

The odorant aldehydes that are generated from **76** are (3*Z*)-nonenal (**80**), which can rearrange into (2*E*)-nonenal (**81**), and (2*E*,4*Z*)-decadienal (**82**), which can isomerize into (2*E*,4*E*)-decadienal (**74**). Table 9.6 lists the compounds together with their odour characteristics. In analogy to **76**, the other hydroperoxides **77**–**79** undergo β-cleavage, forming the carbonyl compounds **83**–**87** displayed in Scheme 9.15. As an exception, 4,5-epoxy-(2*E*)-decenal (**88**) is formed from 13-hydroperoxy-9,11-octadienoate **79** (13-HPOD) via a second hydroperoxide intermediate (**89**), followed by β-cleavage. Compound **88** has a fatty, metallic smell, a very low odour threshold of 1 $\mu g\, l^{-1}$ in oil and 0.038 $\mu g\, l^{-1}$ in water, and is often characteristic of the flavour of fatty food.

Fat oxidation greatly contributes to the well appreciated flavour of numerous foods. Especially in cooked items like potato chips and crisps, boiled and roasted meat, fish and many dishes which use fat and oil in cooking, it plays a key role. The fatty acid composition

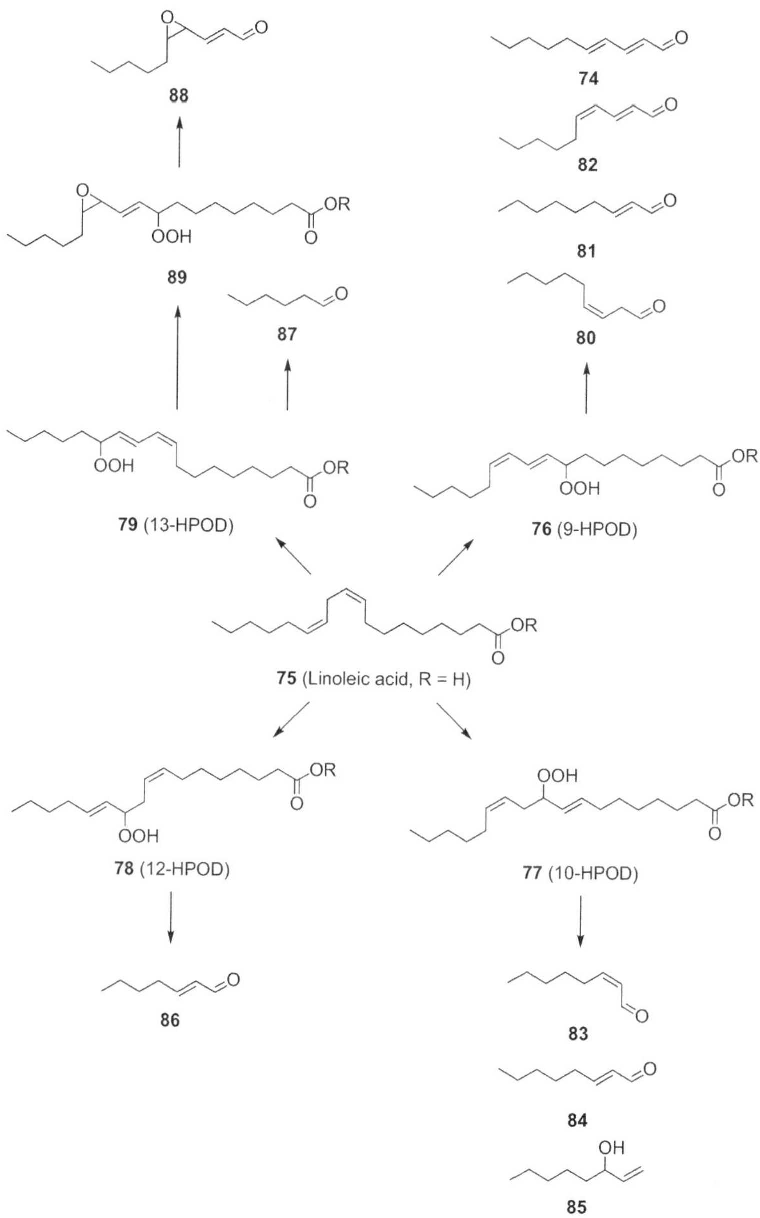

Scheme 9.15 *Linoleic acid degradation*

and the resulting odorants from fat oxidation are of paramount importance for the animal-specific note in meat aroma. The reverse side of the coin, as far as aroma is concerned, is off flavour formation, which is also frequently associated with lipid oxidation. One example is the warmed-over flavour (WOF) of meat. This can develop when cooked meat is refrigerated and later warmed again for consumption, resulting in cardboard-like, rancid aroma notes

Table 9.6 *Odours and odour thresholds of compounds resulting from linoleic acid degradation*

No.	Compound	Odour	Odour threshold[a] [$\mu g\, l^{-1}$]
74	(2*E*,4*E*)-Decadienal	Fatty, deep-fried	0.027[b]
80	(3*Z*)-Nonenal	Cucumber-like	
81	(2*E*)-Nonenal	Tallowy, cardboard-like	0.19[b]
82	(2*E*,4*Z*)-Decadienal	Fatty, deep-fried	
83	(2*Z*)-Octenal	Nutty	
84	(2*E*)-Octenal	Fatty, nutty	4[c]
85	1-Octen-3-ol	Mushroom-like	0.1[c]
86	(2*E*)-Heptenal	Fatty	13–80[b]
87	Hexanal	Green, leaf-like	2.4[b]
88	4,5-Epoxy-(2*E*)-decenal	Fatty, metallic	0.038[b]

Note: [a]Odour detection threshold (orthonasal) in $\mu g\, l^{-1}$ water.
[b]See ref. 17.
[c]See ref. 18.

(commonly found with in-flight meals). Fish and chicken are most at risk, followed by pork and beef. WOF in boiled chicken is mainly caused by an increase of **87** and a decrease in the amount of **74** and furfurylthiol (**19**).[35]

9.9 Conclusion

The thermal generation of volatiles during cooking and processing is one of the key mechanisms for the formation of aroma in food. The chemical reactions which are mainly responsible include the Maillard reaction, the Strecker degradation, caramelization, the degradation of thiamin and ferulic acid and lipid autoxidation. The resulting aroma compounds belong to various chemical classes and are often heterocyclic products (furans, pyrazines, pyrrolines), aldehydes, or are sulfur-containing molecules. Knowledge about key odorants in food, their precursors and their formation pathways helps to favour pleasant aromas and avoids the formation of off flavours.

References

1. Boelens Aroma Chemical Information Service, VCF 00 Database of volatile compounds in food, BACIS, Huizen, The Netherlands, 1999.
2. (*a*) K. Shima, N. Yamada, E. Suzuki, T. Harada, Novel brothy taste modifier isolated from beef broth, *J. Agric. Food Chem.*, **46**, 1465–1468 (1998); (*b*) O. Frank, H. Ottinger, T. Hofmann, Characterization of an intense bitter-tasting 1*H*,4*H*-quinolizinium-7-olate by application of the taste dilution analysis, a novel bioassay for the screening and identification of taste-active compounds in food, *J. Agric. Food Chem.*, **49**, 231–238 (2001); (*c*) H. Ottinger, T. Hofmann, Identification of the taste enhancer alapyridaine in beef broth and evaluation of its sensory impact by taste reconstitution experiments, *J. Agric. Food Chem.*, **51**, 6791–6796 (2003); (*d*) H. Schlichtherle-Cerny, M. Affolter, C. Cerny, Taste-active glycoconjugates of glutamate: new umami compounds, in *Challenges in Taste Chemistry and Biology*, T. Hofmann, C. T. Ho, W. Pickenhagen (Eds), American Chemical Society, Washington D. C., 2004, pp. 210–222.

3. (*a*) I. Blank, A. Sen, W. Grosch, Potent odorants of the roasted powder and brew of Arabica coffee, *Z. Lebensm. Unters. Forsch.*, **195**, 239–245 (1992); (*b*) P. Semmelroch, G. Laskawy, I. Blank, W. Grosch, Determination of potent odourants in roasted coffee by stable isotope dilution assays, *Flavour Fragr. J.*, **10**, 1–7 (1995); (*c*) M. Czerny, F. Mayer, W. Grosch, Sensory study on the character impact odorants of roasted Arabica coffee, *J. Agric. Food Chem.*, **47**, 695–699 (1999); (*d*) W. Grosch, Sensory studies on the key odorants of roasted coffee, in *Caffeinated Beverages: Health Benefits, Physiological Effects, and Chemistry*, T. H. Parliment, C. T. Ho, P. Schieberle (Eds), American Chemical Society, Washington D. C., 2000, pp. 202–209.
4. L. C. Maillard, Action des acides aminés sur les sucres; formation des mélanoïdines par voie méthodique, *C. R. Hebd. Séances Acad. Sci.*, **154**, 66–68 (1912).
5. M. Friedman, Biological effects of Maillard browning products that may affect acrylamide safety in food: biological effects of Maillard products, *Adv. Exp. Med. Biol.*, **561**, 135–156 (2005).
6. P. A. Finot, Historical perspective of the Maillard reaction in food science, *Ann. N. Y. Acad. Sci.*, **1043**, 1–8 (2005).
7. V. Somoza, Five years of research on health risks and benefits of Maillard reaction products: an update, *Mol. Nutr. Food Res.*, **49**, 663–672 (2005).
8. A. Cerami, The role of the Maillard reaction *in vivo*, in *Maillard Reactions in Chemistry, Food, and Health*, T. P. Labuza, G. A. Reineccius, V. M. Monnier, J. O'Brien, J. W. Baynes (Eds), The Royal Society of Chemistry, Cambridge, 1994, pp. 1–10.
9. H. Nursten, *The Maillard Reaction – Chemistry, Biochemistry, and Implications*, The Royal Society of Chemistry, Cambridge, 2005.
10. E. Schleicher, V. Somoza, P. Schieberle (Eds), *The Maillard Reaction – Recent Advances in Food and Biomedical Sciences* (Ninth International Symposium on the Maillard Reaction), *Ann. N. Y. Acad. Sci.*, **1126**, 1–340 (2008).
11. M. Amadori, Prodotti di condensazione tra glucosio e *p*-fenetidina. Nota I, *Atti. Reale Accad. Naz. Lin.*, **2**, 337–342 (1925).
12. K. Eichner, M. Reutter, R. Wittmann, Detection of Amadori compounds in heated foods, in *Thermally Generated Flavors: Maillard, Microwave, and Extrusion Processes*, T. H. Parliament, M. J. Morello, R. J. McGorrin (Eds), American Chemical Society, Washington D. C., 1994, pp. 42–54.
13. F. Ledl, E. Schleicher, New aspects of the Maillard reaction in foods and in the human body, *Angew. Chem.*, **102**, 597–626 (1990); *Angew. Chem. Int. Ed.*, **29**, 565–594 (1990).
14. K. M. Biemel, J. Conrad, M. O. Lederer, Unexpected carbonyl mobility in aminoketose: the key to major Maillard crosslinks, *Angew. Chem.*, **114**, 826–828 (2002); *Angew. Chem. Int. Ed.*, **41**, 801–804 (2002).
15. W. Nedvidek, F. Ledl, P. Fischer, Detection of 5-hydroxymethyl-2-methyl-3(2*H*)-furanone and of α-dicarbonyl compounds in reaction mixtures of hexoses and pentoses with different amines, *Z. Lebensm. Unters. Forsch.*, **194**, 322–328 (1992).
16. A. Hollnagl, L. W. Kroh, Degradation of oligosaccharides in nonenzymatic browning by formation of α-dicarbonyl compounds via a 'peeling off' mechanism, *J. Agric. Food Chem.*, **48**, 6219–6226 (2000).
17. M. Czerny, M. Christlbauer, M. Christlbauer, A. Fischer, M. Granvogl, M. Hammer, C. Hartl, N. Moran Hernandez, P. Schieberle, Re-investigation on odour thresholds of key food aroma compounds and development of an aroma language based on odour qualities of defined aqueous odorant solutions, *Eur. Food Res. Technol.*, **228**, 265–273 (2008).
18. M. Rychlik, P. Schieberle, W. Grosch, *Compilation of Odor Thresholds, Odor Qualities and Retention Indices of Key Food Odorants*, Deutsche Forschungsanstalt für Lebensmittelchemie, Garching, 1998.
19. (*a*) T. Shibamoto, Formation of sulfur- and nitrogen-containing compounds from the reaction of furfural with hydrogen sulfide and ammonia, *J. Agric. Food Chem.*, **25**, 206–208 (1977); (*b*) L. J. Farmer, D. S. Mottram, F. B. Whitfield, Volatile compounds produced in Maillard reactions involving cysteine, ribose and phospholipid, *J. Sci. Food Agric.*, **49**, 347–368 (1989).
20. W. Grosch, G. Zeiler-Hilgart, C. Cerny, H. Guth, Studies on the formation of odorants contributing to meat flavors, in *Progress in Flavors Precursors Studies*, P. Schreier, P. Winterhalter (Eds), Allured Publishing, Carol Stream, 1993, pp. 329–342.

21. P. Schieberle, Formation of furaneol in heat-processed foods, in *Flavor Precursors*, R. Teranishi, G. Takeoka, M. Güntert (Eds), American Chemical Society, Washington D. C., 1992, pp. 164–174.
22. R. Wagner, M. Czerny, J. Bielohradsky, W. Grosch, Structure–odour-activity relationships of alkylpyrazines, *Z. Lebensm. Unters. Forsch.*, **208**, 308–316 (1999).
23. R. G. Buttery, L. C. Ling, B. O. Juliano, J. G. Turnbaugh, Cooked rice aroma and 2-acetyl-1-pyrroline, *J. Agric. Food Chem.*, **31**, 823–826 (1983).
24. A. Adams, N. de Kimpe, Chemistry of 2-acetyl-1-pyrroline, 6-acetyl-1,2,3,4-tetrahydropyridine, 2-acetyl-2-thiazoline, and 5-acetyl-2,3-dihydro-4*H*-thiazine: extraordinary Maillard flavor compounds, *Chem. Rev.*, **106**, 2299–2319 (2006).
25. P. Schieberle, The role of free amino acids present in yeast as precursors of the odorants 2-acetyl-1-pyrroline and 2-acetyltetrahydropyridine in wheat bread crust, *Z. Lebensm. Unters. Forsch.*, **191**, 206–209 (1990).
26. T. Hofmann, P. Schieberle, 2-Oxopropanal, hydroxy-2-propanone, and 1-pyrroline – important intermediates in the generation of the roast-smelling food flavor compounds 2-acetyl-1-pyrroline and 2-acetyltetrahydropyridine, *J. Agric. Food Chem.*, **46**, 2270–2277 (1998).
27. A. Strecker, Notiz ueber eine eigenthuemliche Oxydation durch Alloxan, *Liebigs Ann. Chem.*, **123**, 363–365 (1862).
28. R. Zamora, E. Gallardo, F. J. Hidalgo, Strecker degradation of phenylalanine initiated by 2,4-decadienal or methyl 13-oxooctadeca-9,11-dienoate in model systems, *J. Agric. Food Chem.*, **55**, 1308–1314 (2007).
29. R. Tressl, D. Bahri, M. Holzer, T. Kossa, Formation of flavor components in asparagus. 2. Formation of flavor components in cooked asparagus, *J. Agric. Food Chem.*, **25**, 459–463 (1977).
30. L. W. Kroh, Caramelization in food and beverages, *Food Chem.*, **51**, 373–379 (1994).
31. H. D. Belitz, W. Grosch, P. Schieberle, *Food Chemistry*, Springer, Berlin, 2004.
32. O. Nishimura, S. Mihara, Investigation of 2-hydroxy-2-cyclopenten-1-ones in roasted coffee, *J. Agric. Food Chem.*, **38**, 1038–1041 (1990).
33. W. Fiddler, W. E. Parker, A. E. Wasserman, R. C. Doerr, Thermal decomposition of ferulic acid, *J. Agric. Food Chem.*, **15**, 757–761 (1967).
34. W. Grosch, Reactions of hydroperoxides – products of low molecular weight, in *Autoxidation of Unsaturated Lipids*, H. W. Chan (Ed.), Academic Press, London, 1987, pp. 95–137.
35. J. Kerler, W. Grosch, Character impact odorants of boiled chicken: changes during refrigerated storage and re-heating, *Z. Lebensm. Unters. Forsch.*, **205**, 232–238 (1997).

10

Human Olfactory Perception

Alan Gelperin

Monell Chemical Senses Center, 3500 Market Street, Philadelphia, PA 19104, USA; and Princeton Neuroscience Institute, Department of Molecular Biology, Princeton University, Princeton, NJ 08544, USA

10.1 Introduction

The scientific study of human olfactory perception, as for any sensory modality, perforce involves the use of rigorous psychophysical and psychometric methods, that is administering tightly controlled physical stimuli to human subjects while obtaining verbal or motor responses that indicate the subject's evaluation of the stimuli. The special challenges inherent in applying controlled and quantitative odorant stimuli often require sophisticated instrumentation for generating known concentrations of selected odorants[1] and collaborations between analytical chemists and psychophysicists to directly measure and control the chemical properties of the stimuli used to elicit subject responses adequate to the task of defining the relevant psychometric functions.[2–6] Assessments of olfactory thresholds require careful attention to methodology and careful comparison of the several different methods currently employed,[7] as do methods for determining odour quality.[8]

A central problem in the study of olfactory information processing and the olfactory percepts that result from neural processing of olfactory stimuli is the lack of clear stimulus dimensions in olfaction, analogous to wavelength in vision and frequency in audition.[9] The relationship between molecular properties of an odorant molecule and its sensory properties is not yet understood.[10] To overcome this limitation, a large empirical literature has been developed to relate human olfactory perception to chemical structures grouped into classes of related compounds.[11] Using this prior psychophysical information on responses of human subjects to olfactory stimuli of varied chemical structure,[12] one can begin to make predictions of perceptual responses to new compounds or mixtures of compounds, albeit

The Chemistry and Biology of Volatiles Edited by Andreas Herrmann

with mixed results.[6] The large body of empirical data on human olfactory responses is extremely useful to practitioners of basic and applied olfaction but the central fact remains that studies of olfactory perception and its neural basis are trying to relate two domains of measurement – neural activity patterns and reports of olfactory percepts – with no natural dimensions. No simple scheme relating odorant chemical structure or olfactory receptor activation to perception is likely to be of broad explanatory value without incorporating prior empirical knowledge of psychophysical responses to the odorant ligands being considered.[13,14] These issues have recently been reviewed.[15]

Human olfactory ability is often viewed as of minor importance relative to other distal sensory modalities such as vision or audition. This facile view underestimates the actual extent of human olfactory sensitivity[16,17] and neglects the role of olfaction in many aspects of human interactions, including social communication,[18,19] mate choice,[20,21] scent tracking and odour localization,[22–24] flavour evaluation and retronasal olfaction,[25] subcortical effects on response selection,[26,27] olfactory effects on food selection,[28] olfactory alterations in pregnancy,[29,30] epigenetic transmission of odour preferences[31,32] and protective alarm responses,[33] among others. Even though a variety of genetic mechanisms have limited the repertoire of expressed human olfactory receptor genes,[34] the several hundred human olfactory receptor genes that are expressed represent the largest gene superfamily in the human genome[35] and provide ample combinatorial and computational complexity for molecular recognition and pattern analysis far beyond any technical device yet known.[36] In short, human olfactory abilities and the role of olfaction in human interactions are greatly underrated.[16]

10.2 Historical Perspective on Olfactory Perception

One of the most insightful students of the mammalian olfactory system, Adrian (First Baron Adrian, 1889–1977), grappled in the later phases of his distinguished research career with the relationship between electrical activity in the olfactory system and the observed phenomena of olfactory perception as known at that time.[37,38]

Adrian used newly developed methods for recording brain surface potentials to study human electroencephalogram (EEG) rhythms in epileptic patients, and hence it was a natural progression to use the same EEG recording methods to study rhythmic surface potentials in the olfactory bulb (OB – the first processing centre for olfactory sensory input in the brain) and their modification by odorant applications.[39] It was also natural for someone with medical training and experience with brain injured patients from the First World War to seek a relationship between human perceptual phenomena and brain electrical activity. Adrian was one of the first to draw attention to the prominent gamma band (25–80 Hz) oscillations in surface potentials recorded from the OB.[38] The computational significance of oscillatory local field potentials later became a focus of work in olfaction[40] and in other sensory systems, particularly vision, as a possible mechanism of feature binding,[41] a mechanism to unite the elements of a percept into a perceptual unit.

Another notable figure in the history of olfactory perception and cortical electrophysiology is Freeman, whose pioneering work using both single unit recording and arrays of field potential electrodes in rabbit olfactory bulb shows clearly that learning about an odorant changes its representation in the OB of an awake animal.[42] Freeman's focus on the dynamics of network activity during olfactory perception and the trajectory of bulbar network activity in high-dimensional phase space is reminiscent of very recent work on the dynamics of network activity viewed as trajectories in phase space during perception of odours in animal olfactory systems.[43]

10.3 Human Olfactory Pathway

Volatile molecules inhaled from the external world or exhaled after liberation from ingested substances interact with receptors in ciliary membranes[44] of olfactory receptor neurons lining the upper nasal passageways.[45] Computational fluid dynamics studies of the human nasal cavity are beginning to clarify how inhaled or exhaled volatiles interact with the olfactory receptor sheet, which has direct consequences for the probability of uptake of volatile odorants into the mucus layer bathing the cilia of the olfactory receptor neurons.[46] In the vertebrate lineage, olfactory receptor genes comprise the largest multigene family.[47] Humans express several hundred olfactory receptors[48] with each olfactory receptor neuron expressing one or a very small number[49] of receptor genes. Each odorant molecule interacts with a subset of the population of expressed olfactory receptors with varying binding constants and consequent levels of G protein activation of the sensory neurons.[50] Thus even a monomolecular odorant elicits a pattern of activity in a population of receptors. In addition, each olfactory receptor neuron is activated by a set of olfactory ligands, so a complex natural odorant stimulus with dozens or hundreds of molecular constituents (such as the scent of rose)[51] elicits a complex pattern of sensory activity to be decoded by central processing. The ability to reliably identify a sensory percept such as rose odour in the face of variation of odorant composition and concentration is remarkable. Odorants can also inhibit the responses of receptors to their cognate ligands, thereby adding further information to the sensory pattern.[52] Progress has recently been made in identifying the central neural regions involved in this process.[53]

The primary olfactory receptors in the olfactory epithelium project their axons to the OB, where they synapse with the apical dendrites of the output cells of the OB, the mitral and tufted cells, in spherical regions of neuropil called glomeruli.[54] There are multiple layers and mechanisms for lateral interactions between input fibres and mitral/tufted cells at the glomerular layer and in deeper layers of the OB (e.g. in the granule cell layer). In addition to the intrinsic synaptic interactions in the OB from cellular elements in the bulb, several types of feedback and modulatory interactions occur in the OB, including cholinergic,[55] adrenergic,[56] serotonergic, dopaminergic and peptidergic inputs from more central regions.[57,58] The OB is a site of several forms of behaviourally related synaptic plasticity and may well play a role in odour perception through its extensive feedforward and feedback connections with other regions.[59]

The OB projects to the primary olfactory cortex, which is comprised of the anterior olfactory cortex, the olfactory tubercle, the piriform cortex, parts of the amygdala and the entorhinal cortex (Scheme 10.1). The piriform cortex has two zones, the anterior piriform cortex (APC) and the posterior piriform cortex (PPC), with distinctively different anatomical organizations and computational functions.[60] Odour-specific adaptation, necessary for discrimination of target odorants from background, is likely to be a property of both odour-

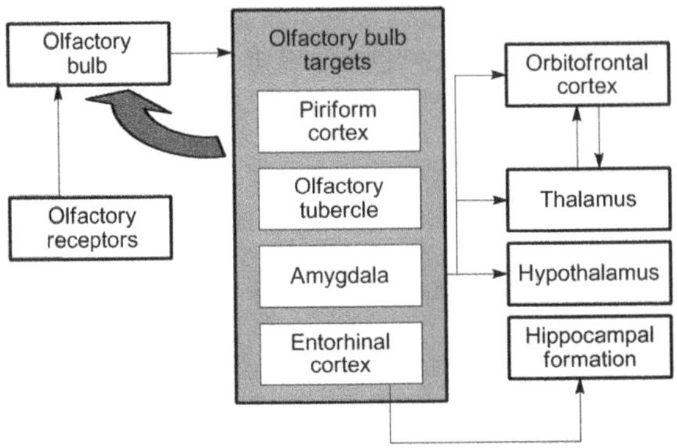

Scheme 10.1 *Diagram of the human olfactory pathway. The extensive feedback to the olfactory bulb from the piriform cortex is emphasized by the grey arrow. Modified with permission from Fig. 15.1.C in Purves et al., Neuroscience, 4th Ed., Sinauer Associates, Sunderland, USA, 2008*

specific adaptation in piriform cortical neurons and synaptic adaptation of inputs to piriform cortex from bulbar output neurons.[61] Animal studies buttress the view that the human piriform cortex performs pattern completion so that a complex odorant mixture is still classified as rose, even if one component is missing.[62] From the piriform cortex, inputs are sent to another series of structures including the caudal orbitofrontal cortex (OFC), the agranular insula, the hippocampus, the dorsomedial nucleus of the thalamus, the medial and lateral hypothalamus and the ventral striatum and pallidum. The region that receives the major corticocortical projections from the piriform cortex is the caudal OFC, at least in primates,[63] and as such has traditionally been considered to constitute the secondary olfactory cortex. In addition to this direct link, the OFC also receives indirect projections from several areas of the primary olfactory cortex through a relay in the dorsomedial nucleus of the thalamus.[64] This caudal subregion of the OFC projects to other parts of the OFC extensively.[65] Several lines of evidence suggest that orbitofrontal cortex is likely to be a site of multisensory integration of olfactory and non-olfactory stimuli influencing olfactory perception.[66]

10.4 Functional Studies in Human Subjects

The ability to study cellular and network aspects of neural activity directly in human subjects is limited to recording opportunities presented by neurosurgical patients with electrode arrays implanted to localize a region of brain pathology prior to its removal.[67] To date these types of studies have rarely involved olfactory processing structures. Alternative strategies for recording odorant-elicited neural population activity involve surface electrode recording of local odorant-evoked potentials.[68] Chemosensory event-related potentials[69] are useful for the clinical assessment of olfactory function[70] but are less useful as measures of central odour percept formation. Measurements of facial expressions using electromyography and physiological reactions using electrocardiogram and electrodermal activity have recently aided the analysis of responses of human subjects processing odour cues as both novel/familiar and pleasant/unpleasant.[71] A critical aspect of studies using physiological measures during odorant-elicited activity is validation using subject self reports of olfactory percepts.[72]

One early study of odour perception in neurosurgical patients presents electrode recordings from the human OB while patients respond to the presentation of various monomolecular and complex odorants, including cigarette smoke.[73] The most prominent features of the electrophysiological recordings from the OB are the gamma oscillations (25–80 Hz) elicited by the presentation of the odorant stimuli. Odorant-evoked gamma oscillations have been shown to be a prominent feature of olfactory network responses in a variety of mammalian species.[40] Based on animal work, these gamma oscillations in the OB are likely to be coherent with oscillations in the piriform cortex, which is synaptically linked with the OB by extensive feedforward and feedback connections. These connections and their oscillatory response may be essential to odour object recognition and feature extraction from background, as measured behaviourally.[74]

The first study of central processing of olfactory stimuli in the intact human brain was done using positron emission tomography (PET) imaging.[75] The advent of functional magnetic resonance imaging (fMRI), discovered by Ogawa at Bell Laboratories,[76] allows the identification of brain regions activated by a repeated regime of sensory stimulation causing regional neural activations that increase regional cerebral blood flow (CBF) and concomitantly increase venous blood oxygenation. Soon after the appearance of the blood oxygenation level-dependent (BOLD) contrast fMRI method, techniques were developed to apply controlled and repetitive olfactory stimuli to human subjects during BOLD fMRI imaging.[77] These methods have been extended to study many aspects of human brain activation by odours, for example to show that the right orbitofrontal cortex, right fusiform cortex and right hypothalamus respond to volatile components of natural human sexual sweat,[78] among many others.[79] fMRI studies have also clarified how pleasant and unpleasant components of odorant mixtures are represented by activity in the medial orbitofrontal cortex, the dorsal anterior cingulate and mid-orbitofrontal cortex.[80] The orbitofrontal cortex also appears to code the reward value for a variety of sensory inputs, including olfactory inputs.[81]

A recent study by Gottfried and coworkers[53] provides compelling evidence that distributed patterns of neural activation in the human posterior piriform cortex are crucial to the formation of odour percepts, independent of the precise chemical nature of the odorant stimulus triggering the percept.[14] Use of multivariate analysis of the fMRI images allowed detection of activation patterns at a finer spatial scale than the traditional univariate analysis, which was critical to identifying the distributed patterns of posterior piriform activation representing odour percepts. Subjects were exposed to odours selected from three sets of odours, each set comprised of three odours. Each odour set represented a different category of odorant (i.e. minty, woody, citrus). Independent psychophysical tests verified that the members of each odour set were perceived as belonging to the same set by the subjects and that the three sets of odorants were classified as distinct categories. The analysis of the fMRI image series obtained during odorant sampling clearly showed that only distributed activity in the posterior piriform cortex predicted the category of the odorant sampled by the subject, independent of the particular chemical structure of the odorant representing one of the three categories of odorants.[53]

Concentration-invariant odorant feature extraction in the posterior piriform cortex very likely involves integration of information from non-olfactory inputs and both contextual and background stimuli.[82] The posterior piriform cortex receives reciprocal connections from high-order association areas of the cerebral cortex such as orbitofrontal cortex, the insula and infralimbic regions of the prefrontal cortex, the amygdala, and the perirhinal and entorhinal cortices.[83] These connections are likely to provide the contextual and non-olfactory information needed to allow concentration-invariant odorant feature extraction.

A longstanding debate has been how well the human working memory functions for odours, with some even questioning whether humans actually possess a working memory for odours.[84] Working memory is a component of short-term memory during which stored information is actively monitored or manipulated. Although behavioural studies have demonstrated that our working memory for odours is inferior to that of the visual working memory,[85] initial imaging studies demonstrated that olfactory working memory is processed by modality-independent neuronal networks[86] as well as sensory-specific activation in the orbitofrontal cortex.[86,87] However, an fMRI study of olfactory working memory recently found a double dissociation, whereby remembering nameable odorants was reflected in sustained activity in the prefrontal language areas, while remembering unnameable odorants was reflected in sustained activity in the primary olfactory cortex. These findings suggest a novel dedicated mechanism in the primary olfactory cortex, where odour information is maintained in temporary storage to subserve ongoing tasks.[88]

Two-photon laser scanning microscopy (TPLSM) has proven to be an efficient tool to study CBF and cellular activity in depth in the rodent brain, which has relevance to understanding the mechanisms coupling neural activity and changes in local blood flow underlying the signals measured with fMRI. The OB is a good model to study neurovascular coupling using TPLSM. By combining intra- and extracellular recordings, TPLSM of CBF in individual capillaries and local application of drugs, it was shown that odour application triggers odorant-specific and concentration-dependent increases in CBF in glomeruli. It was also demonstrated that the activation of neurons is required to trigger blood flow responses.[89]

fMRI studies have recently been reported using human sweat collected under two conditions, stress and nonstress, as the olfactory stimulus.[90] The stimuli were perceived with a low intensity, so that under the stimulus conditions used only about half of the odour presentations were detected by the participants. The fMRI results (event-related design) show that chemosensory anxiety signals activate brain areas involved in the processing of social emotional stimuli (fusiform gyrus) and in the regulation of empathic feelings (insula, precuneus, cingulate cortex). In addition, neuronal activity within attentional (thalamus, dorsomedial prefrontal cortex) and emotional (cerebellum, vermis) control systems was observed. Although subjects could not differentiate the two types of chemosensory stimuli, they produced distinctively different patterns of brain activation, presumably reflecting different processing or storage of the olfactory information that did not rise to the level of consciousness.

Perceptual rivalry in the olfactory system has been demonstrated during the presentation of two different odours, each odorant delivered to a different nostril. Subjects alternate odour percepts between the two different odorants. Binaral rivalry involves adaptations at both the peripheral sensory neurons and in the cortex.[91]

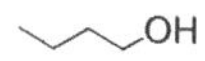

1

Figure 10.1 Structure of n-butyl alcohol (**1**)

10.5 Functional Studies in Brain-Damaged Subjects

Studies of olfactory discrimination and detection in patients who have undergone surgical removal of a brain region in the central olfactory pathway can provide critical insights into the relative roles of central neural structures in human olfactory perception. One such study involved measurements of olfactory discrimination and detection in 106 patients with a variety of central neural deficits, including unilateral removal of the right or left temporal lobe, right or left frontal lobe, left parietal lobe, or right frontal and temporal lobes, along with measurements on 20 normal control subjects.[92] Primary sensory loss was determined by measuring detection thresholds for *n*-butyl alcohol (**1**, Figure 10.1), measured separately in each nostril, and did not differ across subject groups or across nostrils. The discrimination task involved monorhinal presentation of pairs of unfamiliar odorants, which the subjects judged as the same or different in quality. The results showed a significant deficit in discrimination confined to the nostril ipsilateral to the lesion in patients with temporal lobe removals. Patients with frontal lobe excisions were also impaired and, for patients with right frontal lesions including the orbital cortex, the impairment was found in both nostrils. No significant deficits were found in patients with left parietal lesions. Normal subjects showed consistently better performance in the right than in the left nostril. The results clearly demonstrate the importance of the orbitofrontal cortex in olfactory discrimination, as confirmed by subsequent studies using functional imaging.[87] Temporal lobe lesions may degrade olfactory performance by disrupting the input to the orbitofrontal cortex. The role of the posterior piriform cortex in odour quality assessments and the role of plasticity in olfactory cortex are active topics of research,[93] fundamental to an understanding of human olfactory perception. The nostril difference in the normal subjects, together with the birhinal impairment in patients with right orbitofrontal damage, suggest a relative advantage of the right orbital region in olfactory processing.[92]

10.6 Single Odorants, Binary Mixtures and Complex Odour Objects

Studies of the chemical determinants of the olfactory potency of vapours by humans often produce unexpected results as demonstrated clearly in studies of the concentration detection functions for the odours of the homologous *n*-alkylbenzenes toluene (**2**), ethylbenzene (**3**), butylbenzene (**4**), hexylbenzene (**5**) and octylbenzene (**6**, Figure 10.2).[4] Olfactory detection thresholds showed a U-shaped trend as a function of alkyl chain length, indicating a loss of odour potency beyond a certain molecular size. Measured concentration detection functions for a series of homologous ketones [propanone (**7**), 2-pentanone (**8**), 2-heptanone (**9**), 2-nonanone(**10**)] showed a linear function.[3]

The first step in moving from perception of simple monomolecular odorants to complex odour objects is to compare perceptual responses and brain activation patterns produced by

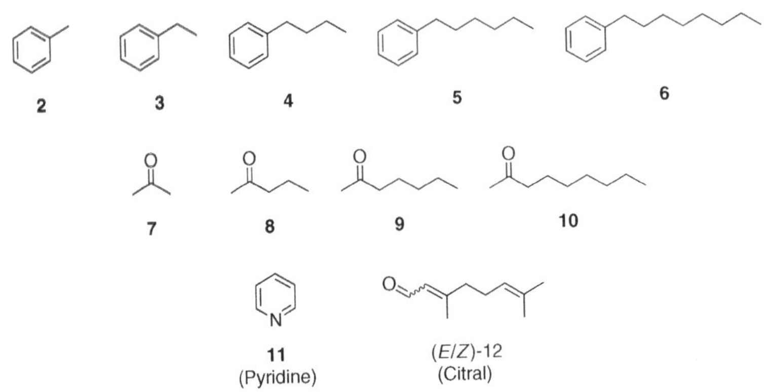

Figure 10.2 *Structures of a homologous series of n-alkylbenzenes* ***2-6*** *and ketones* ***7-10****, pyridine (****11****) and citral (****12****)*

presentation of pairs of odorants given either as monomolecular stimuli or as a binary mixture. Boyle *et al.* asked whether single and binary odour mixtures activate different regions in the human brain.[94] They obtained data from PET scans using pyridine (**11**), citral (**12**) and five mixtures of pyridine and citral in proportions varying from 10 : 90 to 90 : 10, including a 50 : 50 mixture. Comparing mixtures with single odorants showed activation in the left cingulate, right parietal and superior frontal cortices and bilateral activation in the anterior and lateral orbitofrontal cortices. They also found that brain activity in the anterior orbitofrontal cortex responded specifically to binary odour mixtures irrespective of the proportions of the components in the mixture, with marked deactivation of the same area for the single components. These findings show that binary odour mixtures, and their individual components, are processed differently by the human brain. Two separate, yet interconnected, mechanisms seem to regulate human mixture perception. A certain portion of the human brain (anterior orbitofrontal cortex) detects the mere presence of an odour mixture, acting like an on/off detector of an odour mixture, whereas other areas (lateral orbitofrontal cortex) seem to process the mixture ratio. This organization of one detection mechanism and subsequent more fine-tuned and specialized mechanisms seems to be ubiquitous over all neuronal sensory systems. A common example is the visual system with basic processing in primary visual cortex and subsequent specialized processing in the higher areas.

In psychophysical studies of the ability of human subjects to discriminate between binary odour mixtures and the monomolecular components of the mixtures, measurements of the latencies of subject responses are found to be very informative and provide significant predictive ability about the subject's decision-making ability. There is a significant relationship between the latency for making a same/different discrimination between odour pairs and the accuracy of discriminations of odour quality between binary mixtures and their components.[95] Difficult discriminations, for example those between 50 : 50 mixtures and the components of the mixture, require more decision-making time than less difficult discriminations, for example, discriminations between the pure monomolecular components of a binary mixture. This is an example of a general phenomenon in sensory psychophysics known as speed–accuracy trade off.[96]

Interactions between components of binary odour mixtures are seen clearly when stimuli are presented at perithreshold concentrations, for example, in measurements of perithreshold mixture interactions between maple lactone (ML, **13**, Figure 10.3) and selected carboxylic acids.[5] This study was the first clear demonstration of synergy in odour detection by humans from an experiment that combined precise stimulus control, vapour phase calibration of stimuli and a clear statistical definition of synergy. Sub-threshold odour stimulation with carboxylic acids can also have a synergistic effect on the rated intensity of supra-threshold stimuli.[97]

The nonlinear interactions between components of a mixture, even a simple two-component (binary) mixture, when the components are sampled simultaneously, have limited the ability to predict sensory properties of a mixture from measurements of the sensory properties of the individual components. This has recently been shown with regard to predicting the rated pleasantness of a binary mixture from the rated pleasantness of the components of the mixture.[6] Study of the perceived pleasantness of five different binary mixtures (0 : 100, 25 : 75, 50 : 50, 75 : 25, 100 : 0) constructed from a set of six odorants, three odorants with positive pleasantness values [L-carvone (**14**), linalool (**15**), phenylethyl alcohol (**16**, Figure 10.3)], and three with negative pleasantness values [valeric acid (**17**), isovaleric acid (**18**), butanoic acid (**19**)],[11] showed that the pleasantness of the mixture fell between the pleasantness values of its separate constituents and that the pleasantness rating of the mixture was strongly influenced by the relative intensities of the components in the mixture. Attempts to develop a predictive relationship between the pleasantness of components of a binary mixture and the pleasantness of a mixture of those components met with limited success, in that the predictive model needed to incorporate substantial

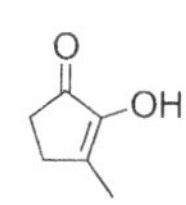

13
(Maple lactone)

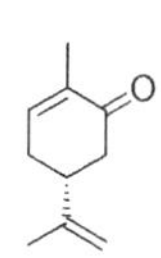

(*R*)-**14**
(L-Carvone)

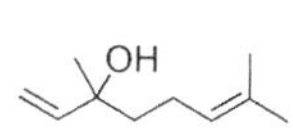

(±)-**15**
(Linalool)

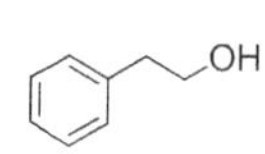

16
(Phenethyl alcohol)

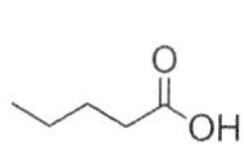

17
(Valeric acid)

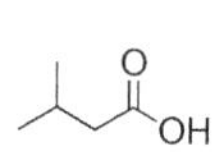

18
(Isovaleric acid)

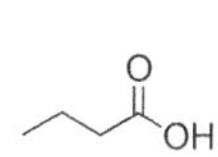

19
(Butanoic acid)

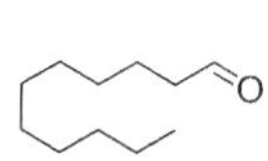

20
(Undecanal)

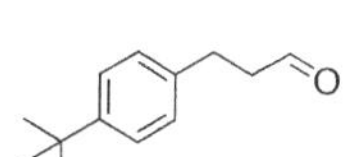

21
(Bourgeonal)

Figure 10.3 *Structures of compounds investigated as binary mixtures*

empirical data on the pleasantness of the mixture components and could only explain the behaviour of mixtures of intermediate pleasantness.[6]

Interactions in binary mixtures of odorants sometimes reveal clear signs of antagonism between the components. Given that undecanal (**20**) has been indicated as an antagonist for bourgeonal-sensitive receptors in the human olfactory epithelium, experiments compared the perception of mixtures of isointense concentrations of bourgeonal (**21**) and undecanal (**20**) with perception of mixtures of isointense concentrations of bourgeonal and *n*-butanol (**1**).[98] Results showed that the bourgeonal–undecanal mixture was dominated by the antagonist's quality, but only when mixed at higher concentration, consistent with effects produced by low-affinity receptor antagonism.

10.7 Olfactory Versus Trigeminal Odorant Identification

The human oral and nasal cavities are not only the gateways for odorant delivery to the olfactory sensory epithelium but are also richly supplied with sensory endings of the trigeminal nerve.[99] Thus the perceptual effects of a given odorant or odorant mixture must be parsed between the olfactory pathway and the various elements of the trigeminal pathway.[100] The ability to lateralize an odorant stimulus delivered to a single nostril is taken as evidence for trigeminal activation by that stimulus.[23,101] Intranasal carbon dioxide is a stimulus often used to activate the trigeminal pathway and map the central neural structures responding to trigeminal stimuli in humans.[102] These considerations lead to the identification of a few odorants as purely olfactory and many other odorants as mixed stimulants for both the olfactory and trigeminal pathways, at least at moderate to high concentrations. The issue of directional localization of odour sources by human subjects, introduced in the modern literature by von Békésy,[103] and the central neural structures activated during an odour localization task, have recently been revisited.[104]

One approach to the differentiation of odorants as purely olfactory or mixed olfactory–trigeminal is to determine a subject's ability to identify the odorants when exposure is restricted to the oral cavity, which possesses only trigeminal endings. The results of oral cavity only (OCO) exposure can be compared with both retronasal and orthonasal presentation of the same stimuli. For example, the purely olfactory odorants phenylethyl alcohol (**16**), coumarin (**22**), octanoic acid (**23**) and vanillin (**24**, Figure 10.4) are consistently identified accurately when presented retronasally but cannot be identified when presented OCO. To further evaluate OCO processing of purely olfactory odorants, subjects were asked to discriminate vapour phase phenylethyl alcohol, coumarin, octanoic acid and vanillin and, as a control, the trigeminal stimulus peppermint extract, from their glycerin

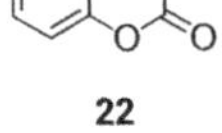

22
(Coumarin)

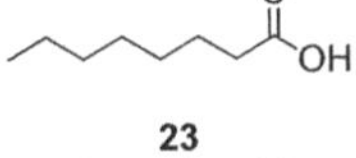

23
(Octanoic acid)

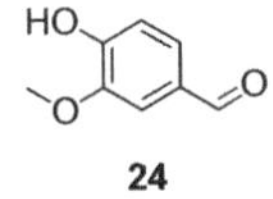

24
(Vanillin)

*Figure 10.4 Structures of pure olfactory stimulants coumarin (**22**), octanoic acid (**23**) and vanillin (**24**)*

solvent, all presented OCO. None of the purely olfactory odorants could be discriminated when presented OCO, but, as expected, peppermint extract was consistently discriminated when presented OCO. These results show that the oral cavity trigeminal system is unresponsive to the tested odorants in vapour phase. The results also confirm that, as expected, the oral cavity trigeminal system provides differential response information for vapour phase stimuli that engage trigeminal sensory endings.[105] Note that a compound would have to be tested in the nose as well as tested OCO to be fully confident that the compound was a pure olfactory stimulant with no concomitant trigeminal activation.

OCO identifications of air phase trigeminal stimuli (i.e. pure chemicals that are discriminated from their solvents by anosmics and are usually lateralized without sniffing by normosmics) were compared with retronasal identifications. Results suggest that the oral cavity trigeminal system may be different from and generally provides less differential information than the nasal cavity trigeminal system.[106] The oral and nasal trigeminal systems may have different patterns and densities of innervation and receptor expression, in addition to differences in the structure of the mucosa. Intranasal trigeminal sensation may be modified in patients who experience olfactory loss.[107]

10.8 Orthonasal Versus Retronasal Odour Perception

Results from psychophysical, electrophysiological, and imaging studies suggest that there are clear differences in the perception of odorant stimuli reaching the olfactory receptor epithelium via an orthonasal versus a retronasal pathway.[108] The differences between ortho- and retronasal perception of odours are thought to be, at least partly, due to differential absorption of odours into the mucus coating of the olfactory epithelium, which appears to differ in relation to the direction of the airflow across the olfactory epithelium.[109]

Perceptual interactions of components of a binary mixture are dramatically influenced by the route of access (orthonasal versus retronasal) taken by the volatile compounds to the olfactory receptor epithelium. For example, perceptual interactions in woody/fruity mixtures were compared during ortho- versus retronasal stimulation, based on both perceived odour quality and event-related potentials measured with surface electrodes.[110] Synergy or masking of the fruity component by the woody component was observed, depending on the level of the woody component. Synergy was reflected by larger N1 amplitude of the event-related potential.

The distinction between orthonasal and retronasal perception is seen clearly when cysteine-*S* conjugates in fruits and vegetables are acted on by mouth bacteria to liberate volatile thiols.[111] The poorer sensitivity to odorants sampled via the retronasal route may be due to the greater absorption of odorants by the nasopharyngeal mucus compared to the nasal mucus during retronasal sampling, thereby reducing the peak concentration of odorants reaching the nasal sensory epithelium and perhaps also slowing the buildup of odorants in the nasal mucosa. There is also the issue of airflow for orthonasal and retronasal odour sampling. When smelling orthonasally, people naturally sniff, effectively forcing odour molecules into the nose. There is no equivalent retronasal sniff.

Interestingly, the solubility of an odorant in mucus, rather than its solubility in water, predicts which component of a binary odour mixture is perceived first during retronasal perception.[112]

25 26

Figure 10.5 *Structures of butyl (**25**) and ethyl mercaptan (**26**) subject to specific anosmia*

10.9 Specific Anosmias

Widespread phenotypic diversity in human olfaction is, in part, attributable to prevalent genetic variation in genes for olfactory receptors (ORs), owing to copy number variation, deletion alleles and deleterious single nucleotide polymorphisms.[35] Chemosensory deficits restricted to a single odorant, called specific anosmias, are widespread in the general population, suggesting that a specific element in the olfactory system, mostly likely an OR type, is missing. For example, about one person in 1000 is insensitive to butyl mercaptan (**25**, Figure 10.5), the foul-smelling odorant added to natural gas to help detect gas leaks.[113] More serious anosmias are the inability to detect hydrogen cyanide (HCN; one in ten people), or ethyl mercaptan (**26**). By contrast to specific anosmias, the term anosmia is used to represent the total absence of a functional sense of smell.

A strong association was observed between the single nucleotide polymorphism variants in OR11H7P and sensitivity to the odorant isovaleric acid (**18**). This association was largely due to the low frequency of homozygous pseudogenized genotypes in individuals with specific hyposmia (reduced olfactory sensitivity) to this odorant, implying a possible functional role of OR11H7P in isovaleric acid detection. This predicted receptor–ligand functional relationship was further verified using the *Xenopus* oocyte expression system, whereby the intact allele of OR11H7P coded for olfactory receptors that exhibited a response to isovaleric acid applied to the surface of the oocyte expressing the OR11H7P gene.[114]

Recent comparative work in fruit flies and humans has tried to bridge the gap between activation of specific ORs and odour perception, including the effects of single gene deletions.[115] Human OR1G1 recognizes a group of odorants that share both 3D structural and perceptual qualities, and may contribute to the coding of waxy, fatty, and rose odours in humans.[116] A deletion or inactivation of OR1G1 could result in specific anosmias to these odour categories. Studies of the response of the human olfactory receptor OR1D2 to a broad array of odorants found that there is no simple, direct correlation between a molecule's ability to activate this receptor and the odour percept elicited in the brain. In a parallel study on specific anosmia, no evidence was found for odorant-specific anosmia to either musk or amber, but rather to specific molecules within these categories.[13]

Genotypic variation in OR7D4 accounts for a significant proportion of the valence (pleasantness or unpleasantness) and intensity variance in perception of steroidal odours. Studies of the OR7D4 receptor and androstenone (**27**, Figure 10.6) perception demonstrated a link between the function of a human odorant receptor *in vitro* and odour perception.[117] Further coordinated studies of human OR repertoires and odour perception will attempt to rationalize the observed population differences in the human functional olfactory repertoire.[118] Loss of functionality of human OR genes may reflect the changing role of olfaction in our evolutionary lineage.[119] The diversity of odour perception phenotypes is clearly shown in the responses of a large group of human subjects to a family of musk-like odours.[120]

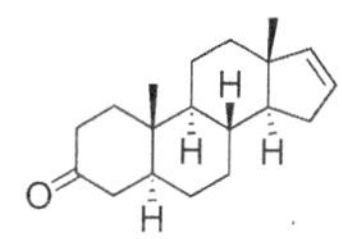

27
(Androstenone)

Figure 10.6 *Structure of androstenone (**27**)*

The clear demonstration of human specific anosmias depends on the use of odorants that do not stimulate the trigeminal system. Androstenone (**27**), one of the most widely studied odorants used to characterize human specific anosmias, is an odorant that can produce a concentration-dependent degree of trigeminal stimulation.[121] This potential trigeminal component of androstenone stimulation may explain the diversity of the reported prevalence of specific anosmia for androstenone.

A genome-wide screen for genetic loci linked to hyposmia in a genetically restricted population of Hutterites identified a 45 centiMorgan region on chromosome 4q linked to individuals with severe hyposmia.[122] Genetic linkage studies such as these, performed on either genetically isolated populations or via large numbers of twins,[123] have proved extremely useful in identifying candidate genes involved in both olfaction[115,124] and taste.[125]

10.10 MHC-Correlated Odour Preferences in Human Subjects

Both mice and humans have been shown to prefer the body odours of potential mates that have dissimilar major histocompatibility complex (MHC) genotypes, which would result in offspring with increased heterozygosity. The human MHC is called HLA (human leukocyte antigen). This result suggests that people who share, for example, HLA-A2, have a similar preference for a variety of odours, including perfume ingredients. Recent results suggest that perfumes are selected for self application to amplify in some way body odours that relate to a person's immunogenetic status.[126]

Extremely high variability in genes of the MHC in vertebrates is assumed to be a consequence of frequency-dependent parasite-driven selection and mate preferences based on promotion of offspring heterozygosity at the MHC loci, or potentially, genome-wide inbreeding avoidance. Where effects have been found, mate choice studies on rodents and other species usually find preferences for MHC dissimilarity in potential partners. Rodent studies also show that individual odour types result from an interaction between background genes and MHC genes.[127] Studies on MHC-associated mate choice in humans support the view that olfactory and visual channels may work in a complementary way (i.e. odour preference for MHC dissimilarity, visual preference for MHC similarity) to achieve an optimal level of genetic variability.[128]

Does the number of MHC alleles usually present in individuals represent an optimal balance between the advantages of presenting an increased range of MHC-coded peptides versus the disadvantages of an increased loss of T cells? An adaptive immune response is usually initiated only if an MHC molecule presents pathogen-derived peptides to T cells.

Every MHC molecule can present only peptides that match its peptide-binding groove. Thus, it seems advantageous for an individual to express many different MHC molecules to be able to resist many different pathogens. However, although MHC genes are the most polymorphic genes of vertebrates, each individual has only a very small subset of the diversity present at the population level. This is an evolutionary paradox. A recent review of data on infection studies and mate choice experiments concludes that overall the evidence suggests that intermediate intra-individual MHC diversity is optimal.[129]

Studies in wild-derived house mice support the idea that females are more attracted to outbred males, and suggest that such preferences may be stronger when expressed by inbred than outbred females, which is consistent with the 'good genes as heterozygosity' hypothesis.[130] In several species, including rodents and fish, it has been shown that the MHC influences mating preferences and, in some cases, that this may be mediated by preferences based on body odour. A recent study supports the hypothesis that genes in the MHC (HLA in humans) locus influence mate choice in some human populations.[131]

Humans show MHC-dependent mating preferences, based on studies of mating patterns in isolated religious enclaves, where couples in this genetically restricted population were much less likely to share MHC haplotypes than expected by chance.[132] Human mate choices are influenced by paternally, but not maternally, inherited MHC alleles communicated by body odours.[20] These effects may explain why human couples who mate with their third or fourth cousins apparently optimize their reproductive success.[21,133]

10.11 Odour Deprivation and Odour Perception

A large body of animal work indicates that significant periods of odour deprivation lead to substantial loss in odour sensitivity and discrimination ability measured after termination of the period of odour deprivation[134] and numerous cellular changes in the olfactory pathway.[135] Work in mice provides evidence for an odour-elicited activity-based mechanism that enhances synaptogenesis of adult-born periglomerular neurons during their initial phases of development.[136] The reduction in odour-elicited sensory inputs due to deprivation could influence the dynamics and survival of synaptic connection at multiple stages in the olfactory pathway.

The first preliminary report of the effects of odour deprivation on odour perception and odour-induced neural activation in human subjects has recently appeared.[137] Perceptual and imaging data were analysed from one female subject who was admitted to the hospital for seven days of bilateral odour deprivation. Following deprivation, the subject showed modest improvements in both odour detection thresholds and identification scores, whereas her ability to assess odour quality similarity between perceptually related odorants became more variable. Multivariate (pattern-based) fMRI analysis revealed that in posterior piriform cortex, odour deprivation increased the spatial correlation between qualitatively similar odorants, reflecting the loss of quality-coding specificity within this region. These initial results clearly require replication with a larger subject population, but are consistent with the idea that sustained olfactory exposure is critical for maintaining the ability to make neural and hence perceptual discriminations between odour objects, particularly closely related odour objects.[136] The replication and extension of these results will provide further tests of the causal link between patterns of activation in posterior piriform cortex and variations in odour perception in humans.

An animal model of early odour deprivation suggests that changes in the transcription of brain-derived neurotrophic factor may underlie some of the longterm consequences of the early stress of maternal separation and consequent odour deprivation.[138]

10.12 Age-Related Decline in Olfactory Perception

Odour recognition memory in rodents may provide a valuable model of cognitive aging in humans. Signal detection analyses were used to distinguish odour recognition based on recollection versus odour recognition based on familiarity. Aged rats were selectively impaired in odour recollection, with relative sparing of odour familiarity, and the deficits in odour recollection were correlated with spatial memory impairments. These results complement electrophysiological findings indicating age-associated deficits in the ability of hippocampal neurons to differentiate contextual information, and this information processing impairment may underlie the common age-associated decline in olfactory and spatial memory.[139]

In another test of the age-dependence of olfactory discrimination, young, middle-aged and senior human subjects performed tasks designed to examine whether odour quality discrimination varies independently of sensitivity. One task entailed detection of 2-heptanone (**9**, Figure 10.2). A second task (match to sample) employed sets of 2-heptanone and homologues of 2-heptanone and nonketones. Subjects were asked to discriminate odorants presented either at intensity-matched concentrations far above threshold, but fixed across subjects, or at levels adjusted to neutralize differences in sensitivity. The young and middle-aged groups showed the same absolute sensitivity, but the senior group showed poorer sensitivity. Performance in quality discrimination, however, declined progressively with age. Performance was not associated with absolute sensitivity, no matter how examined. These data suggest largely independent processing of odour quality and intensity.[140]

The National Geographic Smell Survey provided a rich source of information on aging and odour processing.[141] Since much of human food choice is based on olfactory input, the following question was included with the scratch and sniff samples in the smell survey: Would you eat something that smelled like this? Two of the odours in the survey were food-related and two were fragrance-related. Another question asked was: Would you apply something that smelled like this to your body? Answers were affected in part by the age and gender of the respondent and by the perceived pleasantness and intensity of the odour.

Work on olfactory decline in the elderly is often approached by the study of food choices, given that olfaction is a major component of flavour. The elderly, especially those with poor olfaction, are more willing to accept novel foods than are younger adults, and elderly subjects are more willing to accept foods with unpleasant odours than are young subjects.[142] The increased willingness to try novel foods among elderly subjects with poor olfaction was found to be due to decreased rejection of foods with unpleasant odours and not due to decreased food neophobia per se. The loss of gustatory and olfactory ability with normal aging has been the subject of several reviews.[143]

10.13 New Neurons in Adult Brains

Rodent and human olfactory systems,[144] among others, are in a constant state of sensory cell turnover in the periphery and a constant state of cell addition in the olfactory bulb.[145,146] Based on animal work, these cellular neurogenic events are likely to be altered by the subject's history of odorant exposure, both as regards its composition and its novelty.[147] Another factor that can influence the dynamics of central neurogenesis is sleep deprivation.[146] Circulating signals like the acidic derivative of vitamin A, retinoic acid (RA), may regulate neurogenetic activity in resident stem cells in the adult nervous system, particularly in the olfactory pathway.[148] Identification of adult neural stem cells and their role in the continuous production of neurons throughout life in the dentate gyrus and the subventricular zone of the lateral ventricles may allow the development of novel therapeutic strategies to induce regeneration in the damaged brain.[149]

10.14 Epidemiological Studies of Human Olfaction

Olfactory perception is important for environmental and nutritional safety. Population-based epidemiological studies of olfaction aim to understand the magnitude of the health burden, identify modifiable risk factors, and develop and test prevention and treatment strategies for olfactory impairment. However, measuring olfaction in large studies is challenging, requiring repeatable, efficient methods that can measure changes in olfactory function over time. Two large cohort studies (the epidemiology of hearing loss study, EHLS; and the beaver dam offspring study, BOSS) included olfactory testing. In both studies, the San Diego odor identification test (SDOIT)[150] was used to measure olfactory function. Subjects were asked to identify eight common household odours (such as coffee and chocolate). Olfactory impairment was defined as correctly identifying fewer than six out of eight odorants after two trials. The EHLS participants were 53–95 years of age at the time of the first measurement (1998–2000), and participants in the BOSS were 21–84 years of age. The prevalence of olfactory impairment in the EHLS was 25% overall, more common in men than in women and increased with age. Five years later olfaction was measured a second time and the majority (84%) of the EHLS participants were classified the same. Among those with impairment at the base line nearly one-third (31%) improved to unimpaired. This heterogeneity in olfactory impairment has unique implications for data analyses and predicting outcomes and associations. Preliminary data from the BOSS study suggest the prevalence of olfactory impairment may be lower in younger generations. All these factors point to a continuing need for epidemiological studies of olfaction.[151]

When studied in a large population of older adults with normal cognition at baseline, odour identification ability is associated with the five-year incidence of cognitive impairment, and olfactory impairment contributes to the prediction of cognitive decline.[152] A common source of cognitive decline is Parkinson's disease (PD). Not only is olfactory dysfunction commonly associated with PD, but in addition impaired olfaction can predate clinical PD in men by at least four years and may be a useful screening tool to detect those at high risk for development of PD in later life.[153]

Although smell loss has several potential etiologies (e.g. head trauma, allergic rhinitis, enlarged adenoids) that are common among children, studies evaluating the prevalence of olfactory dysfunction in the pediatric population are rare. Several challenges confront the clinician or researcher hoping to evaluate odour identification ability in young children. Children are likely to be unfamiliar with many of the odour stimuli used in adult tests and have limited ability to read and identify labels to select from alternative choices, which is the typical adult response option. Consequently, specialized forms of olfactory tests must be developed for this population. Based on the format of the SDOIT[154] and the delivery system of the brief smell identification test,[155] a shortform odour identification test utilizing standardized odour stimuli (in which participants matched six odorants to pictures of the odour source) was developed and delivered via scratch and sniff plates. The pilot version of this test was administered to children between the ages of 3–17 as part of the pre-surgical intake evaluation at the A. I. duPont Hospital for Children and as part of basic research studies at the Monell Center. The hospital study population is broad and includes children undergoing ear, nose and throat surgery, as well as control subjects (children undergoing general surgery), with approximately 50 children per week eligible for evaluation. To improve correct interpretation of the results, stimulus familiarity was evaluated by having the child's parent/guardian also complete the test and answer a short questionnaire about the child's experience with the various odour stimuli. The authors suggest that the test may be able to classify children with olfactory dysfunctions as such, and a larger trial is presently ongoing.[156] In addition to the classic scratch and sniff tests, a new clinical test for children has recently been developed. This test uses pictures to describe the odour objects and seems suitable for use on children above the age of four years, although it has not yet been validated as a clinical tool.[157]

10.15 Active Sampling and Olfactory Perception

Olfaction, like vision and audition, incorporates a mechanism for active acquisition of the stimulus. Normal nasal breathing results in transport of odorant-laden air over the nasal mucosa, where absorption of odorant molecules into the nasal mucosa leads to interactions between the molecules and the binding sites of olfactory receptor proteins embedded in the ciliary membranes of OR neurons (ORNs).[158] The flow rate and flow path resulting from the vigour of each inhalation and their repetition rate interact to directly impact transport of odorant molecules from the vapour phase to the fluid phase of the mucus bathing the ORN dendrites.[45] In rodents, odorant deposition is highly dependent on solubility and correlates with the locations of different types of receptors for highly water-soluble versus highly fat-soluble odorants. The way in which the water or nasal mucus solubility of odorants influences the uptake patterns at the olfactory epithelium, and consequent receptor activation expressed in patterns of glomerular activation, is still a topic of active research.[15]

The pattern of active odorant sampling (sniffing) is a critical determinant of both odorant deposition into the nasal mucus and the temporal pattern of activation of the ORNs. The ability of an ORN to follow the temporal pattern of activation during active sniffing is controlled in part by Ca^{2+}/calmodulin-mediated negative feedback to the cyclic nucleotide-gated channel in the ciliary membrane of the ORN.[158] Surprisingly, if this channel is rendered resistant to fast desensitization by genetic modification in mice, the Ca^{2+}/calmodulin-mediated negative feedback is found to function primarily to control ORN response termination. The importance of sniff frequency for the dynamics of glomerular activation has also been highlighted in recent optical measurements of calcium dynamics in presynaptic terminals of rat ORNs during sniffing.[159] These studies have shown that OR sensory endings are not able to follow volleys of activation at normal sniffing frequencies (5–10 Hz). These results in animal systems have direct implications for the ability or lack thereof of humans to identify and discriminate odours sampled by rapid sniffing. Sniffing may also play a role in enhancing the ability to localize an odour source.[22]

The cortical and subcortical networks controlling sniffing are referred to as the olfactomotor system.[160] Using methods for sniff measurement concurrent with olfactory tasks in humans,[161] it was found that the olfactomotor system generates sniffs that are sufficiently vigorous to insure maximal coverage of the sensory epithelium by the inspired air stream and inversely proportional to odorant concentration, as measured by average sniff airflow velocity, maximum airflow velocity, volume and duration of sniff. The olfactomotor system includes the cerebellum as clear sniff and odorant-induced activation of the cerebellum has been measured by fMRI. Olfactory perceptual impairments in patients with unilateral cerebellar lesions are found to be selective to nasal inputs on the side opposite the cerebellar lesion.[162] The results provide further evidence for an olfacto-cerebellar pathway and suggest that this pathway connects each nostril primarily to the contralateral cerebellum. The ability of a sniff to adjust odour sampling depending on odorant concentration and familiarity is thus seen to be an integral part of the olfactory percept.[163] In fact, humans can use sniff sampling to follow scent trails while crawling on natural surfaces when deprived of visual and auditory cues.[24] The subjects improved their performance with practice and performed better with input from both nostrils as opposed to input from a single nostril, thus directly demonstrating the utility of computing an olfactory percept based on inputs from the two nostrils. In general bilateral odorant thresholds are usually equivalent to those obtained unilaterally from the more sensitive nostril (if there is a difference), and people with unilateral olfactory loss are often completely unaware of it.

Event-related fMRI has been successfully used to study the complete collection of neural loci activated by sniffing. Results show that sniffing activates a bilateral cortical and subcortical sensorimotor network. The activations are localized within the primary sensorimotor cortex, lateral premotor cortex, supplementary motor area, anterior cingulate, insula, basal ganglia, thalami, mesencephalon, upper pons, cerebellar vermis, piriform cortex, entorhinal cortex and parahippocampal gyrus.[164]

10.16 Human Olfactory Imagery

Human subjects can be instructed to engage in olfactory imagery or can be presented with a real olfactory stimulus, while parameters of respiration are monitored. A significant increase in respiratory volume between olfactory perception and the unstimulated baseline condition is found, and a statistically comparable increase in respiratory volume between olfactory imaging and the baseline condition is also observed. Thus olfactory perception and olfactory imagery have similar effects on respiratory profiles, perhaps based on a common underlying mechanism.[165] Sniffing-specific activation occurs in the hippocampus and piriform cortex during voluntary sniffing,[166] which may contribute to the olfactory percept.

In other senses, the mere act of imagining the visual or auditory object initiates activity within the primary sensory cortical areas processing the input from the sense we are imagining. The same is true for olfaction. When subjects were asked to imagine odours, the primary olfactory cortex (piriform cortex) was activated in the same manner as when they actually smelled an odour, albeit to a lesser extent. Interestingly, the degree to which the olfactory cortex was activated corresponded with subject reports of how successful they were at imagining the odour in question.[167] A more recent fMRI study measured region-specific brain activations in subjects who alternated between smelling and imagining pleasant and unpleasant odours. Activity induced by imagining the odours mimicked that commonly induced by odours in a hedonic-specific pattern. For both real and imagined odours, unpleasant stimuli induced greater activity than pleasant stimuli in the left frontal portion of piriform cortex and left insula. These findings show clearly that in olfaction, as in other sensory modalities, primary sensory cortical structures are activated during mental imagery of sensory events.[168] They also suggest that the large inter-individual variability in odour imaging ability is reflected in variability of regional brain activation during odour imagery.

Olfactory hallucinations (phantosmias), characterized as odour perception in the absence of an odorant, can occur in subjects in the absence of sensory or neurological disturbances[169] but can also be associated with nonmotor manifestations of Parkinson's disease.[170] Two possible mechanisms have been suggested for phantosmia: either an altered inhibitory input from the primary olfactory neurons, or the presence of peripheral olfactory or trigeminal signals that 'trigger' a central perception. Olfactory perceptual disturbances can also be associated with temporal lobe epilepsy.[171]

10.17 Top-Down Influences on Olfactory Perception

The mapping of neural activity to olfactory percepts is also dramatically influenced by contextual cues that provide top-down influences not only to the primary olfactory cortices but also to the first central olfactory processing centre, the OB.[56,57,172] The sources of these top-down influences are presumably the feedback and modulatory pathways responsible for the striking effects of cognitive factors such as odour names,[173] emotional tone[174] and concurrent visual input,[175] selected to elicit positive or negative emotional responses,[176] on odour processing and evaluation.[177] The classic demonstration of contextual cues influencing hedonic aspects of odour perception is the presentation of an identical olfactory stimulus, isovaleric acid (**18**, Figure 10.3), in containers labelled either 'food' or 'body'. The stimulus is rated as significantly more unpleasant when sampled from a container labelled 'body' than when sampled from a container labelled 'food'.[178] Cognitive and affective states in humans are clearly also modified by the global olfactory surround.[179] The mind's nose can have powerful effects on memory systems, particularly autobiographical memory,[180] even in the absence of overt odour stimulation.[181]

The global brain state, exemplified by states of attentiveness during wakefulness, sleep stages and anesthesia, has a dramatic effect on the processing of olfactory information.[182] For example, recent evidence indicates that olfactory information processing occurs during sleep and that the emotional tone of dreams can be influenced significantly depending upon the hedonic characteristic of the olfactory stimulus applied during sleep.[183] A more complete treatment of these issues is available.[182]

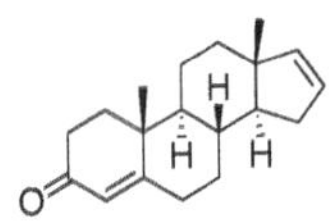

28
(Androstadienone)

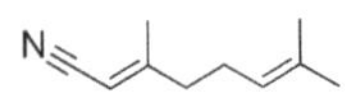

29
(Citralva)

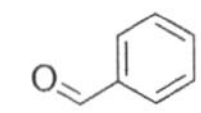

30
(Benzaldehyde)

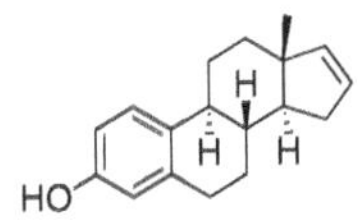

31

Figure 10.7 *Structures of androstadienone (**28**), citralva (geranyl nitrile, **29**), benzaldehyde (**30**) and estra-1,3,5(10),16-tetraen-3-ol (**31**)*

10.18 Reproductive State and Olfactory Sensitivity

The gender and reproductive status of humans can have marked effects on the perceptual processing of odours.[184] Gender differences in general indicate that women are more sensitive to odours than men,[30] whereas the effects of hormone levels are more variable. Olfactory event-related potential measurements revealed that in spite of gender-independent perceptual responses some odours produce gender-specific patterns of cortical activation.[69] Early work by McClintock and colleagues indicated that unidentified components of male body odours could influence perceptual processing by female subjects.[18,26] More recent work indicates that smelling a single component of human sweat, pure androstadienone (4,16-androstadien-3-one, **28**, Figure 10.7), leads to alterations in human cortisol levels in female subjects.[185] Allocation of attentional resources to emotionally significant aspects of stimuli was enhanced by subliminal exposure to **28** in both male and female subjects.[27] A critical assessment of this work has recently appeared.[186]

The induction of enhanced olfactory sensitivity by odour exposure in humans was first demonstrated when subjects, both men and women, who initially could not smell androstenone (**27**) developed the ability to smell this compound after repeated brief exposures.[187] Olfactory sensitivity to androstenone also was found to vary during adolescence.[188] This phenomenon has now been extended to several common odorants but remarkably only female subjects of reproductive age showed increased threshold sensitivity to odorants such as citralva (**29**), benzaldehyde (**30**) and lemon–orange smell, sometimes by six orders of magnitude.[189] The exposure-induced enhancement of responses to androstadienone (**28**) has been replicated with both psychophysical methods and odour-evoked potentials to show a gender-specific effect of repeated exposure on late components of the odour-evoked potential response.[190] The genetic and environmental factors contributing to the perceived pleasantness of androstenone (**27**) have recently been clarified in a study involving 917 twin subjects.[191] The data suggest that both intensity and pleasantness of androstenone are moderately influenced by genetic factors (21–28% heritability) and that the traits are modified by an overlapping set of genes.

In some cases the gender preferences rather than the actual gender of the subjects have a strong bearing on the perceptual processing of certain classes of social chemosignals: for example, lesbian women process the progesterone derivative **28** and the estrogenlike steroid estra-1,3,5(10),16-tetraen-3-ol (**31**, Figure 10.7) more like heterosexual men than heterosexual women.[192] Preferences for body odour samples obtained from (*a*) heterosexual males and gay males, (*b*) heterosexual males and heterosexual females, (*c*) heterosexual females and lesbians and (*d*) gay males and lesbians indicate that differences in body odour are detected and responded to, at least in part, on the basis of an individual's gender and sexual orientation.[193] Subsequent work showed that heterosexual women had a hedonic rating of body odour samples from heterosexual and homosexual men that was biased to favour the body odours of heterosexual men.[194] PET studies show that human body odours are processed in specialized neural circuits distinct from the circuits processing nonsocial odours.[195] Romantic attachment appears to deflect attention from odour cues of nonmates rather than increase sensitivity to odour cues from the current mate.[196] The identity of a variety of components of human skin odours has recently been elucidated,[197] opening the way to identification of the changes in body odour or odour type correlated with sexual preference.

Olfactory sensitivity may be dramatically altered during pregnancy, as nearly two-thirds of pregnant women rate their olfactory sensitivity to be enhanced during pregnancy.[29] However, these self reports of increased olfactory sensitivity do not always correlate with the results of measurements of olfactory sensitivity using standardized tests of olfactory function, such as the University of Pennsylvania smell identification test.[198] The general issue of sex differences and gonadal hormone influences on human olfactory function has recently been reviewed.[30]

In general women are reported to be more sensitive to odorant stimuli than men but the precise aspects of olfactory information processing that may underlie this difference have not yet been clearly defined. To approach this issue, a large cohort of subjects was recently tested for odour sensitivity and hedonic evaluation and the results were evaluated for gender differences.[199] It was found that women evaluate the pleasantness of perceived odours in a more extreme manner than men without significant differences in hedonic polarity (pleasantness/unpleasantness).

Gender-specific effects have also been found recently in studies of cross-adaptation of stress-related odours (SROs).[200] Subjects provided hedonic and intensity ratings of the SRO and of each of the potential cross-adapting agents prior to 2.5 min of induced olfactory adaptation to each agent. During adaptation, possible cross-adaptation was evaluated by intermittent ratings of the perceived intensity of the SRO. Some potential cross-adapting agents did reduce the impact of the SRO; however, the same chemicals were not necessarily effective for male and female SROs. The effective chemicals depended upon the gender of the donor of the SRO and the gender of the subject, suggesting a gender-specific response to both the SRO stimuli used and the fragrance chemicals used to cross-adapt them.[200] Gender-specific effects have also been documented in the importance of smell for human mate selection, in that women, but not men, ranked body odour as more important for attraction than 'looks' or any social factor except 'pleasantness'.[201]

10.19 Olfaction, Hunger and Satiety

In conditioning studies, pairings of odours with sweet, sour and bitter tastes often impact ratings of sensory quality for the odour presented alone. Pleasantness ratings for odours paired with sucrose are increased for participants who like sweet tastes, and conversely decreased pleasantness is reported for quinine paired odours. However, when hungry subjects experience an odour paired with sucrose, liking for the odour is not increased when subjects are tested while sated, suggesting that expression of acquired liking for odours depends on the subject's current motivational state. Overall, the literature suggests that, once an odour is experienced in a food-related context, that specific odour acquires the ability to modify both the preparatory and the satiety-related components of ingestion.[202]

Interestingly, paired presentations of odorants and tastes can result in odorants influencing taste properties.[203] In addition, an odorant paired with sweet or sour taste can modify the taste of other solutions.[204] fMRI studies confirm the activation of gustatory cortex when such odorants are presented on their own.[205] It will be interesting to see if odour–taste synesthesia is affected by hunger state.

Olfactory perception in humans is heavily influenced by emotional reactions that reflect previous experiences and associations with odorant stimuli. A recent review summarizes the various factors that affect the hedonic responses of humans to odours. Factors that influence individual variations in odour perception include gender, reproductive status and changes occurring throughout the life span. Finally, relevant factors affecting odour perception also arise from the experimental context in which the smell is perceived.[206] It will be a challenge to unravel the effects of hunger and satiety on human olfactory perception in the face of the many other confounding factors outlined above.

10.20 Odour Perception Bias by Odour Names

Will an odour presented with a positive, neutral, or negative name be perceived differently depending on the odour name? Recent results provide a clear affirmative answer to this question.[173] A group of subjects was asked to rate 15 odours for their pleasantness, intensity and arousal. In a second experiment, participants passively smelled ten odours while their skin conductance (SC), heart rate (HR) and sniffing were recorded. Significant overall effects of odour names on perceived pleasantness, intensity and arousal were found. Pleasantness showed the most robust effect of odour names: the same odours were perceived as more pleasant when presented with positive rather than with neutral or negative names and when presented with neutral rather than with negative names. In addition, odorants were rated as more intense when presented with negative rather than with neutral or positive names and as more arousing when presented with positive rather than with neutral names. Furthermore, SC and sniff volumes, but not HR, were modified by odour names. The SC changes could not be accounted for by sniffing changes. In control experiments, odour names presented with odourless water did not produce any effect on SC and sniff volumes, ruling out the possibility that the name-related findings were triggered by an emotional reaction to odour names. These results show clearly the critical importance of odour labels for olfactory perception.

The emotional associations of an odour name can have an influence on odour perception, given the demonstration that showing subjects pictures with strong emotional content can alter olfactory sensitivity, pleasantness and intensity ratings.[176] Olfactory performance was assessed in subjects after presentation of affective pictures. Olfactory sensitivity was significantly reduced following presentation of an unpleasant picture for all subjects and following presentation of a pleasant picture for males only. A recent study applying odours with different hedonic values while recording chemosensory event-related potentials with multisite EEG showed an effect of odour hedonic value on the latencies of N1 (first negative peak) and P2 (second positive peak) components of the chemosensory evoked potentials with current source densities most prominent in the frontal lobe in participants receiving odours with negative hedonic ratings.[177]

10.21 Olfaction and Disease States

A growing body of evidence suggests that decline in olfactory function may herald the onset of the clinical signs of Parkinson's disease (PD)[207] and Alzheimer's disease.[208] Moreover, neuroanatomical studies of brains from asymptomatic older individuals have identified pathological hallmarks of these diseases in regions involved in processing olfactory input, suggesting that these brain regions may be targeted early in the disease process. Two small prospective studies and estimates from functional imaging support the possibility that olfactory impairment begins two to seven years before patients meet the diagnostic criteria for PD.[209] Recent work suggests that the olfactory deficit in PD is due to changes in central neural circuits rather than changes in the olfactory epithelium.[210] A large multicentre study of olfactory deficits in PD patients showed that 96.7% of PD patients present with significant olfactory loss when compared to young normosmic subjects. This figure falls to 74.5%, however, when adjusted to age-related norms. Thus, olfactory dysfunction can be considered as a reliable marker of PD.[211]

Cognitive deficits associated with more general psychiatric conditions, such as schizophrenia and depression, can degrade olfactory perception and lead to deficits in odour identification, recognition and discrimination with no accompanying reliable change in odour sensitivity.[212] First-degree relatives of schizophrenia patients exhibit specific impairments in odorant-elicited olfactory evoked potentials.[213] Recordings of the electro-olfactogram from the olfactory epithelium of schizophrenic patients suggest a role for altered sensory processing in the olfactory deficits of these patients.[214] Depressive symptoms in general can be related to a reduced olfactory sensitivity.[215] There also may be an association among autoimmunity, psychiatric disorders and smell impairment.[216] Human immunodeficiency virus (HIV) infection is associated with a decrease in olfactory ability[217] that is likely due, at least in part, to rhinosinusitis commonly associated with HIV infection.[218]

Olfactory dysfunction is clearly associated with schizophrenia,[219] as evidenced by studies of both schizophrenic patients and their first-degree relatives.[213] A component of altered olfactory function in schizophrenic patients is due to altered function of olfactory receptors, as assessed by both *in situ* odour-evoked electro-olfactogram recordings,[214] olfactory event-related potentials[220] and studies of fresh biopsy samples of the olfactory epithelium.[221] Recent psychophysical evidence implicates a deficit in intracellular cyclic adenosine monophosphate (cAMP) signalling in schizophrenic patients as a causative factor in olfactory dysfunction.[222]

Olfactory function was studied in pairs of monozygotic twins where one twin was diagnosed with schizophrenia and the other twin had no such diagnosis. Olfaction of affected monozygotic twins was globally impaired. Partial olfactory impairment of their unaffected co-twins may point to a genetic cause of olfactory impairment in schizophrenia. The influence of genetic factors was most evident for olfactory acuity and least evident for smell identification.[223]

Although blindness is not a disease, the literature on olfactory sensitivity of blind subjects is interesting. An early study showed that blind subjects had poorer absolute olfactory sensitivity but outperformed the sighted at odorant identification.[224] Investigations of olfactory discrimination and identification abilities in early blind subjects and age-matched sighted controls indicate that early blind subjects have better access to the semantic information from odour perception. This accounts for part of the improved olfactory performances in odour identification in the blind.[225]

The mechanisms of a broad class of human diseases affecting olfactory ability, termed ciliopathies, a class of genetic disorders affecting the structure and function of cilia,[226] have recently highlighted the relationship of ciliopathies to the pathogenesis of human sensory perception.[227] The cilia at the apical ends of olfactory receptor neurons bear the olfactory receptor proteins essential for odorant binding, so it is not surprising that a general ciliary dysfunction has a dramatic effect of olfactory function. More generally, formal olfactory recognition testing can help differentiate between several categories of human pathologies, including illnesses and movement disorders.[210] The clinical assessment of smell disorders has recently been reviewed.[228]

10.22 Prenatal and Postnatal Influences on Infant Odour/Flavour Preferences

A complete understanding of human odour perception must include the factors that shape adult odour and flavour perception, as a result of exposures and experiences during infancy, including exposure to potential odorants *in utero*. It has been shown that prenatal and early postnatal exposure to a flavour (carrot) enhanced the infants' enjoyment of that flavour in solid foods during weaning.[229] Volatile compounds in mother's milk and the influence of the maternal diet on the volatile compounds in mother's milk have recently been identified.[230]

The development of flavour perception, including the olfactory component of flavour, represents an interaction of innate responses and learning in infants and children. Understanding of various factors that determine flavour choice and hence food ingestion, particularly an understanding of the early factors, is important in designing strategies to enhance the health of the infant, child and adult.[32] Clarification of prenatal and postnatal influences on flavour preferences is also critical for optimizing the sensory properties of oral medications for children.[231]

In a study assessing the influence of school grade and gender on odour sensitivity and identification and the role of verbal capacity on odour identification in children (4–12 years), it was found that odour sensitivity and identification scores increased with grade and that identification performance was better in girls than in boys. However, when performance was controlled for verbal ability, the gender differences disappeared but the grade differences remained.[232] The role of olfactory cues in the everyday life of children in grades two, four and five has been described, based on the children's responses to the children's olfactory behaviors in everyday life (COBEL) questionnaire.[233]

Maternal alcohol consumption, particularly the emotional context in which it occurs, also influences the responses of children to the odours of alcoholic beverages.[234]

A brief window of time immediately after delivery may be a particularly sensitive period for olfactory learning by human neonates.[235] Fifty-five vaginally delivered newborns were exposed to an odorant for 30 min, beginning 4–37 min after birth (early exposure) or 12 h postpartum (late exposure). Several days later, measurements of head orientation responses by the infants to the exposure odour versus an unfamiliar odour or an odourless control stimulus were tested. Infants in the early exposure group, but not in the late exposure group, became familiar with the exposure odour and retained a memory trace of it during the test trials, as indicated by their robust head orientation responses to the odour.

An olfactory test kit suitable for children, based on retronasal smelling of aromas combined with a sweet taste, has recently been developed.[236] This may facilitate further studies of pediatric populations, which are critical to early identification and remediation of early olfactory dysfunction.

10.23 Future Directions

Advances in genetic techniques,[124] optical imaging of neural activity,[237] optical activation of neural activity,[238] fMRI image analysis methods[239] and new methods for the analysis of very large sets of neural data offer real hope for fundamental advances in our understanding of animal and human chemosensory processing, particularly olfactory information processing.[33,240] The bridge from animal to human work on olfaction has recently been strengthened by the demonstration that mice and humans show some similarities in olfactory preferences.[241] The advent of large-scale functional screens of expressed ORs, combined with genetic association studies, have the potential to link variations in OR expression to human chemosensory phenotypes.[35] Humans are indeed a very advantageous system for the analysis of olfactory information processing[242] particularly in light of the increased appreciation for the role of olfaction in human communication and the availability of new analytical tools appropriate for use in intact human subjects.

Acknowledgements

I thank my colleagues at the Monell Chemical Senses Center for their generous and insightful comments, particularly Gary Beauchamp, Beverly Cowart, Pamela Dalton, Johan Lundström, Julie Mennella, Marcia Pelchat, George Preti, Paul Wise and Charles Wysocki. Support for the preparation of this material was provided by a grant from the Army Research Office.

Note added in proof. A useful review of the functions of human olfaction in ingestive behaviour, detecting environmental hazards and social communication has just appeared.[243]

References

1. B. N. Johnson, N. Sobel, Methods for building an olfactometer with known concentration outcomes, *J. Neurosci. Methods*, **160**, 231–245 (2007).
2. (*a*) R. L. Doty, D. G. Laing, Psychophysical measurement of human olfactory function, including odorant mixture assessment, in *Handbook of Olfaction and Gustation*, R. L. Doty (Ed.), Marcel Dekker, New York, 2003, pp. 338–381; (*b*) P. M. Wise, T. Miyazawa, M. Gallagher, G. Preti, Human odor detection of homologous carboxylic acids and their binary mixtures, *Chem. Senses*, **32**, 475–482 (2007).
3. J. E. Cometto-Muñiz, M. H. Abraham, Olfactory psychometric functions for homologous 2-ketones, *Behav. Brain Res.*, **201**, 207–215 (2009).
4. J. E. Cometto-Muñiz, M. H. Abraham, Olfactory detectability of homologous *n*-alkylbenzenes as reflected by concentration–detection functions in humans, *Neuroscience*, **161**, 236–248 (2009).
5. T. Miyazawa, M. Gallagher, G. Preti, P. M. Wise, Synergistic mixture interactions in detection of perithreshold odors by humans, *Chem. Senses*, **33**, 363–369 (2008).
6. H. Lapid, D. Harel, N. Sobel, Prediction models for the pleasantness of binary mixtures in olfaction, *Chem. Senses*, **33**, 599–609 (2008).
7. P. M. Wise, N. Bien, C. J. Wysocki, Two rapid odor threshold methods compared to a modified method of constant stimuli, *Chemosens. Percept.*, **1**, 16–23 (2008).
8. P. M. Wise, M. J. Olsson, W. S. Cain, Quantification of odor quality, *Chem. Senses*, **25**, 429–443 (2000).
9. (*a*) R. Haddad, H. Lapid, D. Harel, N. Sobel, Measuring smells, *Curr. Opin. Neurobiol.*, **18**, 438–444 (2008); (*b*) R. Haddad, R. Khan, Y. K. Takahashi, K. Mori, D. Harel, N. Sobel, A metric for odorant comparison, *Nat. Methods*, **5**, 425–429 (2008).
10. (*a*) C. Margot, D. P. Simmons, D. Skuy, Amber-woody scent: alcohols with divergent structure present common olfactory characteristics and sharp enantiomer differentiation, *Helv. Chim. Acta*, **87**, 2662–2684 (2004); (*b*) C. S. Sell, On the unpredictability of odor, *Angew. Chem.*, **118**, 6402–6410 (2006); *Angew. Chem. Int. Ed.*, **45**, 6254–6261 (2006).
11. R. M. Khan, C. H. Luk, A. Flinker, A. Aggarwal, H. Lapid, R. Haddad, N. Sobel, Predicting odor pleasantness from odorant structure: pleasantness as a reflection of the physical world, *J. Neurosci.*, **27**, 10015–10023 (2007).
12. M. Zarzo, D. T. Stanton, Understanding the underlying dimensions in perfumers' odor perception space as a basis for developing meaningful odor maps, *Atten. Percept. Psychophys.*, **71**, 225–247 (2009).
13. A. Triller, E. A. Boulden, A. Churchill, H. Hatt, J. Englund, M. Spehr, C. S. Sell, Odorant–receptor interactions and odor percept: a chemical perspective, *Chem. Biodivers.*, **5**, 862–886 (2008).
14. C. Margot, A noseful of objects, *Nat. Neurosci.*, **12**, 813–814 (2009).
15. M. Leon, B. A. Johnson, Is there a space-time continuum in olfaction? *Cell Mol. Life Sci.*, **66**, 2135–2150 (2009).
16. G. M. Shepherd, The human sense of smell: are we better than we think? *PLoS Biol.*, **2**, 572–575 (2004).
17. G. M. Shepherd, Outline of a theory of olfactory processing and its relevance to humans, *Chem. Senses*, **30** (Suppl. 1), i3–i5 (2005).
18. M. K. McClintock, S. Jacob, B. Zelano, D. J. Hayreh, Pheromones and vasanas: the functions of social chemosignals, *Nebr. Symp. Motiv*, **47**, 75–112 (2001).
19. (*a*) C. J. Wysocki, G. Preti, Facts, fallacies, fears, and frustrations with human pheromones, *Anat. Rec.*, **281A**, 1201–1211 (2004); (*b*) T. K. Saxton, A. Lyndon, A. C. Little, S. C. Roberts, Evidence that androstadienone, a putative human chemosignal, modulates women's attributions of men's attractiveness, *Horm. Behav.*, **54**, 597–601 (2008).
20. S. Jacob, M. K. McClintock, B. Zelano, C. Ober, Paternally inherited HLA alleles are associated with women's choice of male odor, *Nat. Genet.*, **30**, 175–179 (2002).

21. J. N. Lundstrom, C. J. Wysocki, M. J. Olsson, G. Preti, K. Yamazaki, Biological basis of the third-cousin crush, *Science*, **320**, 1160–1161 (2008).
22. J. Frasnelli, G. Charbonneau, O. Collignon, F. Lepore, Odor localization and sniffing, *Chem. Senses*, **34**, 139–144 (2009).
23. A. M. Kleemann, J. Albrecht, V. Schopf, K. Haegler, R. Kopietz, J. M. Hempel, J. Linn, V. L. Flanagin, G. Fesl, M. Wiesmann, Trigeminal perception is necessary to localize odors, *Physiol. Behav.*, **97**, 401–405 (2009).
24. J. Porter, B. Craven, R. M. Khan, S. J. Chang, I. Kang, B. Judkewitz, J. Volpe, G. Settles, N. Sobel, Mechanisms of scent-tracking in humans, *Nat. Neurosci.*, **10**, 27–29 (2007).
25. (*a*) H. Breer, The sense of smell: reception of flavors, *Ann. N. Y. Acad. Sci.*, **1126**, 1–6 (2008); (*b*) J. Frasnelli, M. Ungermann, T. Hummel, Ortho- and retronasal presentation of olfactory simuli modulates odor percepts, *Chemosens. Percept.*, **1**, 9–15 (2008).
26. S. Jacob, L. H. Kinnunen, J. Metz, M. Cooper, M. K. McClintock, Sustained human chemosignal unconsciously alters brain function, *Neuroreport*, **12**, 2391–2394 (2001).
27. T. A. Hummer, M. K. McClintock, Putative human pheromone androstadienone attunes the mind specifically to emotional information, *Horm. Behav.*, **55**, 548–559 (2009).
28. K. Asahina, V. Pavlenkovich, L. B. Vosshall, The survival advantage of olfaction in a competitive environment, *Curr. Biol.*, **18**, 1153–1155 (2008).
29. E. L. Cameron, Measures of human olfactory perception during pregnancy, *Chem. Senses*, **32**, 775–782 (2007).
30. R. L. Doty, E. L. Cameron, Sex differences and reproductive hormone influences on human odor perception, *Physiol. Behav.*, **97**, 213–228 (2009).
31. J. A. Mennella, C. E. Griffin, G. K. Beauchamp, Flavor programming during infancy, *Pediatrics*, **113**, 840–845 (2004).
32. G. K. Beauchamp, J. A. Mennella, Early flavor learning and its impact on later feeding behavior, *J. Pediatr. Gastroenterol. Nutr.*, **48** (Suppl 1), S25–S30 (2009).
33. A. Gelperin, Neural computations with mammalian infochemicals, *J. Chem. Ecol.*, **34**, 928–942 (2008).
34. Y. Hasin, T. Olender, M. Khen, C. Gonzaga-Jauregui, P. M. Kim, A. E. Urban, M. Snyder, M. B. Gerstein, D. Lancet, J. O. Korbel, High-resolution copy-number variation map reflects human olfactory receptor diversity and evolution, *PLoS Genet.*, **4**, e1000249 (2008).
35. Y. Hasin-Brumshtein, D. Lancet, T. Olender, Human olfaction: from genomic variation to phenotypic diversity, *Trends Genet.*, **25**, 178–184 (2009).
36. A. D. Wilson, M. Baietto, Applications and advances in electronic-nose technologies, *Sensors*, **9**, 5099–5148 (2009).
37. E. D. Adrian, The action of the mammalian olfactory organ, *J. Laryngol. Otol.*, **70**, 1–14 (1956).
38. E. D. Adrian, Electrical oscillations from the olfactory organ, *J. Physiol.*, **136** (Suppl.), 29P–30P (1957).
39. E. D. Adrian, Olfactory reactions in the brain of the hedgehog, *J. Physiol.*, **100**, 459–473 (1942).
40. A. Gelperin, Olfactory computations and network oscillations, *J. Neurosci.*, **26**, 1663–1668 (2006).
41. S. Yu, D. Huang, W. Singer, D. Nikolic, A small world of neuronal synchrony, *Cereb. Cortex*, **18**, 2891–2901 (2008).
42. W. J. Freeman, The physiology of preception, *Sci. Am.*, **264**, 78–85 (1991).
43. M. Rabinovich, R. Huerta, G. Laurent, Transient dynamics for neural processing, *Science*, **321**, 48–50 (2008).
44. D. P. McEwen, P. M. Jenkins, J. R. Martens, Olfactory cilia: our direct neuronal connection to the external world, *Curr. Top. Dev. Biol.*, **85**, 333–370 (2008).
45. K. Zhao, P. Dalton, G. C. Yang, P. W. Scherer, Numerical modeling of turbulent and laminar airflow and odorant transport during sniffing in the human and rat nose, *Chem. Senses*, **31**, 107–118 (2006).
46. (*a*) K. Zhao, P. Dalton, The way the wind blows: implications of modeling nasal airflow, *Curr. Allergy Asthma Rep.*, **7**, 117–125 (2007); (*b*) Y. Liu, M. R. Johnson, E. A. Matida, S. Kherani, J. Marsan, Creation of a standardized geometry of the human nasal cavity, *J. Appl. Physiol.*, **106**,

784–795 (2009); (*c*) S. Ishikawa, T. Nakayama, M. Watanabe, T. Matsuzawa, Flow mechanisms in the human olfactory groove: numerical simulation of nasal physiological respiration during inspiration, expiration, and sniffing, *Arch. Otolaryngol., Head Neck Surg.*, **135**, 156–162 (2009).

47. A. Gottlieb, T. Olender, D. Lancet, D. Horn, Common peptides shed light on evolution of olfactory receptors, *BMC Evol. Biol.*, **9**, article no. 91 (2009).
48. D. Krautwurst, Human olfactory receptor families and their odorants, *Chem. Biodivers.*, **5**, 842–852 (2008).
49. N. E. Rawson, G. Gomez, Cell and molecular biology of human olfaction, *Microsc. Res. Tech.*, **58**, 142–151 (2002).
50. M. Spehr, S. D. Munger, Olfactory receptors: G-protein coupled receptors and beyond, *J. Neurochem.*, **109**, 1570–1583 (2009).
51. (*a*) I. Flament, C. Debonneville, A. Furrer, Volatile constituents of rose, in *Bioactive Voltiles from Plants*, ACS Symp. Ser., Vol. 525, R. Teranishi, R. G. Buttery, H. Sugisawa (Eds), American Chemical Society, Washington D. C., 2003, pp. 269–281; (*b*) J. T. Knudsen, R. Eriksson, J. Gershenzon, B. Stahl, Diversity and distribution of floral scent, *Bot. Rev.*, **72**, 1–120 (2006).
52. (*a*) R. C. Araneda, Z. Peterlin, X. Zhang, A. Chesler, S. Firestein, A pharmacological profile of the aldehyde receptor repertoire in rat olfactory epithelium, *J. Physiol.*, **555**, 743–756 (2004); (*b*) S. D. Munger, T. Leinders-Zufall, F. Zufall, Subsystem organization of the mammalian sense of smell, *Annu. Rev. Physiol.*, **71**, 115–140 (2009).
53. J. D. Howard, J. Plailly, M. Grueschow, J. D. Haynes, J. A. Gottfried, Odor quality coding and categorization in human posterior piriform cortex, *Nat. Neurosci.*, **12**, 932–938 (2009).
54. D. J. Zou, A. Chesler, S. Firestein, How the olfactory bulb got its glomeruli: a just so story? *Nat. Rev. Neurosci.*, **10**, 611–618 (2009).
55. A. S. Ghatpande, A. Gelperin, Presynaptic muscarinic receptors enhance glutamate release at the mitral/tufted to granule cell dendrodendritic synapse in the rat main olfactory bulb, *J. Neurophysiol.*, **101**, 2052–2061 (2009).
56. W. Doucette, J. Milder, D. Restrepo, Adrenergic modulation of olfactory bulb circuitry affects odor discrimination, *Learn. Mem.*, **14**, 539–547 (2007).
57. (*a*) M. T. Shipley, M. Ennis, Functional organization of olfactory system, *J. Neurobiol.*, **30**, 123–176 (1996); (*b*) G. M. Shepherd, W. R. Chen, C. A. Greer, Olfactory bulb, in *The Synaptic Organization of the Brain*, G. M. Shepherd (Ed.), Oxford University Press, New York, 2004, pp. 133–169.
58. S. Matsutani, N. Yamamoto, Centrifugal innervation of the mammalian olfactory bulb, *Anat. Sci. Int.*, **83**, 218–227 (2008).
59. N. Mandairon, C. Linster, Odor perception and olfactory bulb plasticity in adult mammals, *J. Neurophysiol.*, **101**, 2204–2209 (2009).
60. L. B. Haberly, Olfactory cortex, in *The Synaptic Organization of the Brain*, G. M. Shepherd (Ed.), Oxford University Press, New York, 1990, pp. 317–345.
61. C. Linster, A. V. Menon, C. Y. Singh, D. A. Wilson, Odor-specific habituation arises from interaction of afferent synaptic adaptation and intrinsic synaptic potentiation in olfactory cortex, *Learn. Mem.*, **16**, 452–459 (2009).
62. D. C. Barnes, R. D. Hofacer, A. R. Zaman, R. L. Rennaker, D. A. Wilson, Olfactory perceptual stability and discrimination, *Nat. Neurosci.*, **11**, 1378–1380 (2008).
63. (*a*) S. T. Carmichael, M. C. Clugnet, J. L. Price, Central olfactory connections in the macaque monkey, *J. Comp. Neurol.*, **346**, 403–434 (1994); (*b*) E. T. Rolls, H. D. Critchley, A. Treves, Representation of olfactory information in the primate orbitofrontal cortex, *J. Neurophysiol.*, **75**, 1982–1296 (1996).
64. (*a*) T. P. Powell, W. M. Cowan, G. Raisman, The central olfactory connexions, *J. Anat.*, **99**, 791–813 (1965); (*b*) L. B. Buck, The molecular architecture of odor and pheromone sensing in mammals, *Cell*, **100**, 611–618 (2000).
65. S. T. Carmichael, J. L. Price, Architectonic subdivision of the orbital and medial prefrontal cortex in the macaque monkey, *J. Comp. Neurol.*, **346**, 366–402 (1994).
66. G. M. Shepherd, Perspectives on olfactory processing, conscious perception, and orbitofrontal cortex, *Ann. N. Y. Acad. Sci.*, **1121**, 87–101 (2007).

67. R. Q. Quiroga, L. Reddy, G. Kreiman, C. Koch, I. Fried, Invariant visual representation by single neurons in the human brain, *Nature*, **435**, 1102–1107 (2005).
68. T. S. Lorig, The application of electroencephalographic techniques to the study of human olfaction: a review and tutorial, *Int. J. Psychophysiol.*, **36**, 91–104 (2000).
69. J. N. Lundstrom, S. Seven, M. J. Olsson, B. Schaal, T. Hummel, Olfactory event-related potentials reflect individual differences in odor valence perception, *Chem. Senses*, **31**, 705–711 (2006).
70. P. Rombaux, A. Mouraux, S. Collet, P. Eloy, B. Bertrand, Usefulness and feasibility of psychophysical and electrophysiological olfactory testing in the rhinology clinic, *Rhinology*, **47**, 28–35 (2009).
71. S. Delplanque, D. Grandjean, C. Chrea, G. Coppin, L. Aymard, I. Cayeux, C. Margot, M. I. Velazco, D. Sander, K. R. Scherer, Sequential unfolding of novelty and pleasantness appraisals of odors: evidence from facial electromyography and autonomic reactions, *Emotion*, **9**, 316–328 (2009).
72. K. Katata, N. Sakai, K. Doi, H. Kawamitsu, M. Fujii, K. Sugimura, K. Nibu, Functional MRI of regional brain responses to 'pleasant' and 'unpleasant' odors, *Acta Oto-Laryngol.*, **129**, 85–90 (2009).
73. J. R. Hughes, D. E. Hendrix, N. S. Wetzel, J. W. Johnston Jr., Correlations between electrophysiological activity from the human olfactory bulb and the subjective response to odoriferous stimuli, in *Olfaction and Taste III*, C. Pfaffman (Ed.), Rockefeller Press, New York, 1969, pp. 172–191.
74. L. M. Kay, J. Beshel, J. Brea, C. Martin, D. Rojas-Libano, N. Kopell, Olfactory oscillations: the what, how and what for, *Trends Neurosci.*, **32**, 207–214 (2009).
75. R. J. Zatorre, M. Jones-Gotman, A. C. Evans, E. Meyer, Functional localization and lateralization of human olfactory cortex, *Nature*, **360**, 339–340 (1992).
76. S. Ogawa, D. W. Tank, R. Menon, J. M. Ellermann, S. G. Kim, H. Merkle, K. Ugurbil, Intrinsic signal changes accompanying sensory stimulation: functional brain mapping with magnetic resonance imaging, *Proc. Natl. Acad. Sci. USA*, **89**, 5951–5955 (1992).
77. T. S. Lorig, D. G. Elmes, D. H. Zald, J. V. Pardo, A computer-controlled olfactometer for fMRI and electrophysiological studies of olfaction, *Behav. Res. Methods Instrum. Comput.*, **31**, 370–375 (1999).
78. W. Zhou, D. Chen, Encoding human sexual chemosensory cues in the orbitofrontal and fusiform cortices, *J. Neurosci.*, **28**, 14416–14421 (2008).
79. (*a*) J. N. Lundstrom, M. J. Olsson, B. Schaal, T. Hummel, A putative social chemosignal elicits faster cortical responses than perceptually similar odorants, *NeuroImage*, **30**, 1340–1346 (2006); (*b*) J. Plailly, A. J. Radnovich, M. Sabri, J. P. Royet, D. A. Kareken, Involvement of the left anterior insula and frontopolar gyrus in odor discrimination, *Hum. Brain Mapp.* **28**, 363–372 (2007); (*c*) J. Plailly, J. D. Howard, D. R. Gitelman, J. A. Gottfried, Attention to odor modulates thalamocortical connectivity in the human brain, *J. Neurosci.*, **28**, 5257–5267 (2008); (*d*) C. Zelano, J. Montag, B. Johnson, R. Khan, N. Sobel, Dissociated representations of irritation and valence in human primary olfactory cortex, *J. Neurophysiol.*, **97**, 1969–1976 (2007); (*e*) S. Lombion, A. Comte, L. Tatu, G. Brand, T. Moulin, J. L. Millot, Patterns of cerebral activation during olfactory and trigeminal stimulations, *Hum. Brain Mapp.*, **30**, 821–828 (2009).
80. F. Grabenhorst, E. T. Rolls, C. Margot, M. A. da Silva, M. I. Velazco, How pleasant and unpleasant stimuli combine in different brain regions: odor mixtures, *J. Neurosci.*, **27**, 13532–13540 (2007).
81. (*a*) E. T. Rolls, F. Grabenhorst, The orbitofrontal cortex and beyond: from affect to decision-making, *Prog. Neurobiol.*, **86**, 216–244 (2008); (*b*) E. T. Rolls, Functions of the orbitofrontal and pregenual cingulate cortex in taste, olfaction, appetite and emotion, *Acta Physiol. Hung.*, **95**, 131–164 (2008); (*c*) Z. F. Mainen, A. Kepecs, Neural representation of behavioral outcomes in the orbitofrontal cortex, *Curr. Opin. Neurobiol.*, **19**, 84–91 (2009).
82. R. J. Stevenson, D. A. Wilson, Odour perception: an object-recognition approach, *Perception*, **36**, 1821–1833 (2007).
83. (*a*) L. B. Haberly, Parallel-distributed processing in olfactory cortex: new insights from morphological and physiological analysis of neuronal circuitry, *Chem. Senses*, **26**, 551–576

(2001); (*b*) D. J. Calu, M. R. Roesch, T. A. Stalnaker, G. Schoenbaum, Associative encoding in posterior piriform cortex during odor discrimination and reversal learning, *Cereb. Cortex*, **17**, 1342–1349 (2007).

84. T. L. White, Olfactory memory: the long and short of it, *Chem. Senses*, **23**, 433–441 (1998).
85. T. Engen, The sense of smell, *Annu. Rev. Psychol.*, **24**, 187–206 (1973).
86. L. A. Dade, R. J. Zatorre, A. C. Evans, M. Jones-Gotman, Working memory in another dimension: functional imaging of human olfactory working memory, *NeuroImage*, **14**, 650–660 (2001).
87. I. Savic, B. Gulyas, M. Larsson, P. Roland, Olfactory functions are mediated by parallel and hierarchical processing, *Neuron*, **26**, 735–745 (2000).
88. C. Zelano, J. Montag, R. Khan, N. Sobel, A specialized odor memory buffer in primary olfactory cortex, *PLoS One*, **4**, e4965 (2009).
89. P. Tiret, E. Chaigneau, J. Lecoq, S. Charpak, Two-photon imaging of capillary blood flow in olfactory bulb glomeruli, *Methods Mol. Biol.*, **489**, 81–91 (2009).
90. A. Prehn-Kristensen, C. Wiesner, T. O. Bergmann, S. Wolff, O. Jansen, H. M. Mehdorn, R. Ferstl, B. M. Pause, Induction of empathy by the smell of anxiety, *PLoS One*, **4**, e5987 (2009).
91. W. Zhou, D. Chen, Binaral rivalry between the nostrils and in the cortex, *Curr. Biol.*, 1561–1565, (2009).
92. R. J. Zatorre, M. Jones-Gotman, Human olfactory discrimination after unilateral frontal or temporal lobectomy, *Brain*, **114** (Pt 1A), 71–84 (1991).
93. (*a*) J. P. Royet, J. Hudry, D. H. Zald, D. Godinot, M. C. Gregoire, F. Lavenne, N. Costes, A. Holley, Functional neuroanatomy of different olfactory judgments, *NeuroImage*, **13**, 506–519 (2001); (*b*) W. Li, J. D. Howard, T. B. Parrish, J. A. Gottfried Aversive learning enhances perceptual and cortical discrimination of indiscriminable odor cues, *Science*, **319**, 1842–1845 (2008).
94. J. A. Boyle, J. Djordjevic, M. J. Olsson, J. N. Lundstrom, M. Jones-Gotman, The human brain distinguishes between single odorants and binary mixtures, *Cereb. Cortex*, **19**, 66–71 (2009).
95. P. M. Wise, W. S. Cain, Latency and accuracy of discriminations of odor quality between binary mixtures and their components, *Chem. Senses*, **25**, 247–265 (2000).
96. D. Rinberg, A. Koulakov, A. Gelperin, Speed–accuracy tradeoff in olfaction, *Neuron*, **51**, 351–358 (2006).
97. T. Miyazawa, M. Gallagher, G. Preti, P. M. Wise, The impact of subthreshold carboxylic acids on the odor intensity of suprathreshold flavor components, *Chemosens. Percept.*, **1**, 163–167 (2008).
98. M. Brodin, M. Laska, M. J. Olsson, Odor interaction between bourgeonal and its antagonist undecanal, *Chem. Senses*, **34**, 625–630 (2009).
99. P. M. Wise, K. Zhao, C. J. Wysocki, Dynamics of nasal chemesthesis, *Ann. N. Y. Acad. Sci.*, **1170**, 206–214 (2009).
100. W. L. Silver, T. E. Finger, The anatomical and electrophysiologial basis of peripheral nasal trigeminal chemoreception, *Ann. N. Y. Acad. Sci.*, **1170**, 202–205 (2009).
101. (*a*) P. M. Wise, S. E. Toczydlowski, K. Zhao, C. J. Wysocki, Temporal integration in nasal lateralization of homologous propionates, *Inhal. Toxicol.*, **21**, 819–827 (2009); (*b*) P. M. Wise, S. E. Toczydlowski, C. J. Wysocki, Temporal integration in nasal lateralization of homologous alcohols, *Toxicol. Sci.*, **99**, 254–259 (2007); (*c*) P. M. Wise, T. M. Canty, C. J. Wysocki, Temporal integration in nasal lateralization of ethanol, *Chem. Senses*, **31**, 227–235 (2006); (*d*) P. M. Wise, T. Radil, C. J. Wysocki, Temporal integration in nasal lateralization and nasal detection of carbon dioxide, *Chem. Senses*, **29**, 137–142 (2004); (*e*) C. J. Wysocki, B. J. Cowart, T. Radil, Nasal trigeminal chemosensitivity across the adult life span, *Percept. Psychophys.*, **65**, 115–122 (2003).
102. (*a*) E. Iannilli, C. Del Gratta, J. C. Gerber, G. L. Romani, T. Hummel, Trigeminal activation using chemical, electrical, and mechanical stimuli, *Pain*, **139**, 376–388 (2008); (*b*) M. Bensafi, E. Iannilli, J. Gerber, T. Hummel, Neural coding of stimulus concentration in the human olfactory and intranasal trigeminal systems, *Neuroscience*, **154**, 832–838 (2008).
103. G. von Békésy, Olfactory analogue to directional hearing, *J. Appl. Physiol.*, **19**, 369–373 (1964).

104. J. Porter, T. Anand, B. Johnson, R. M. Khan, N. Sobel, Brain mechanisms for extracting spatial information from smell, *Neuron*, **47**, 581–592 (2005).
105. D. Stephenson, B. P. Halpern, No oral-cavity-only discrimination of purely olfactory odorants, *Chem. Senses*, **34**, 121–126 (2009).
106. V. Parikh, A. P. Lee-Lim, B. P. Halpern, Retronasal and oral-cavity-only identifications of air-phase trigeminal stimuli, *Chemosens. Percept.*, **2**, 9–24 (2009).
107. (*a*) J. Frasnelli, B. Schuster, T. Hummel, Interactions between olfaction and the trigeminal system: what can be learned from olfactory loss, *Cereb. Cortex*, **17**, 2268–2275 (2007); (*b*) P. Rombaux, A. Mouraux, T. Keller, T. Hummel, Trigeminal event-related potentials in patients with olfactory dysfunction, *Rhinology*, **46**, 170–174 (2008).
108. (*a*) D. M. Small, J. C. Gerber, Y. E. Mak, T. Hummel, Differential neural responses evoked by orthonasal versus retronasal odorant perception in humans, *Neuron*, **47**, 593–605 (2005); (*b*) G. Bender, T. Hummel, S. Negoias, D. M. Small, Separate signals for orthonasal versus retronasal perception of food but not nonfood odors, *Behav. Neurosci.*, **123**, 481–489 (2009).
109. (*a*) D. E. Hornung, M. M. Mozell, Factors influencing the differential sorption of odorant molecules across the olfactory mucosa, *J. Gen. Physiol.*, **69**, 343–361 (1977); (*b*) T. Hummel, Retronasal perception of odors, *Chem. Biodivers.*, **5**, 853–861 (2008).
110. A. Ishii, N. Roudnitzky, N. Beno, M. Bensafi, T. Hummel, C. Rouby, T. Thomas-Danguin, Synergy and masking in odor mixtures: an electrophysiological study of orthonasal versus retronasal perception, *Chem. Senses*, **33**, 553–561 (2008).
111. C. Starkenmann, B. Le Calve, Y. Niclass, I. Cayeux, S. Beccucci, M. Troccaz, Olfactory perception of cysteine-*S*-conjugates from fruits and vegetables, *J. Agric. Food Chem.*, **56**, 9575–9580 (2008).
112. F. J. Wilkes, D. G. Laing, I. Hutchinson, A. L. Jinks, E. Monteleone, Temporal processing of olfactory stimuli during retronasal perception, *Behav. Brain Res.*, **200**, 68–75 (2009).
113. P. M. Patterson, B. A. Lauder, The incidence and probable inheritance of smell blindness to normal butyl mercaptan, *J. Hered.*, **39**, 295–297 (1948).
114. I. Menashe, T. Abaffy, Y. Hasin, S. Goshen, V. Yahalom, C. W. Luetje, D. Lancet, Genetic elucidation of human hyperosmia to isovaleric acid, *PLoS Biol.*, **5**, e284 (2007).
115. A. Keller, L. B. Vosshall, Influence of odorant receptor repertoire on odor perception in humans and fruit flies, *Proc. Natl. Acad. Sci. USA*, **104**, 5614–5619 (2007).
116. G. Sanz, T. Thomas-Danguin, E. H. Hamdani, C. Le Poupon, L. Briand, J.-C. Pernollet, E. Guichard, A. Tromelin, Relationships between molecular structure and perceived odor quality of ligands for a human olfactory receptor, *Chem. Senses*, **33**, 639–653 (2008).
117. A. Keller, H. Zhuang, Q. Chi, L. B. Vosshall, H. Matsunami, Genetic variation in a human odorant receptor alters odour perception, *Nature*, **449**, 468–472 (2007).
118. (*a*) Y. Gilad, O. Man, S. Paabo, D. Lancet, Human specific loss of olfactory receptor genes, *Proc. Natl. Acad. Sci. USA*, **100**, 3324–7332 (2003); (*b*) Y. Gilad, D. Lancet, Population differences in the human functional olfactory repertoire, *Mol. Biol. Evol.*, **20**, 307–314 (2003).
119. (*a*) E. R. Liman, Use it or lose it: molecular evolution of sensory signaling in primates, *Pflugers Arch.*, **453**, 125–131 (2006); (*b*) M. Nei, Y. Niimura, M. Nozawa, The evolution of animal chemosensory receptor gene repertoires: roles of chance and necessity, *Nat. Rev. Genet.*, **9**, 951–963 (2008).
120. A. N. Gilbert, S. E. Kemp, Odor perception phenotypes: multiple, specific hyperosmias to musks, *Chem. Senses*, **21**, 411–416 (1996).
121. J. A. Boyle, J. N. Lundstrom, M. Knecht, M. Jones-Gotman, B. Schaal, T. Hummel, On the trigeminal percept of androstenone and its implications on the rate of specific anosmia, *J. Neurobiol.*, **66**, 1501–1510 (2006).
122. J. M. Pinto, S. Thanaviratananich, M. G. Hayes, R. M. Naclerio, C. Ober, A genome-wide screen for hyposmia susceptibility loci, *Chem. Senses*, **33**, 319–329 (2008).
123. (*a*) P. M. Wise, J. L. Hansen, D. R. Reed, P. A. Breslin, Twin study of the heritability of recognition thresholds for sour and salty taste, *Chem. Senses*, **32**, 749–754 (2007); (*b*) A. Knaapila, H. Tuorila, K. Silventoinen, M. J. Wright, K. O. Kyvik, K. Keskitalo, J. Hansen,

J. Kaprio, M. Perola, Environmental effects exceed genetic effects on perceived intensity and pleasantness of several odors: a three-population twin study, *Behav. Genet.*, **38**, 484–492 (2008).
124. A. Keller, L. B. Vosshall, Better smelling through genetics: mammalian odor perception, *Curr. Opin. Neurobiol.*, **18**, 364–369 (2008).
125. (*a*) A. A. Bachmanov, D. R. Reed, X. Li, G. K. Beauchamp, Genetics of sweet taste preferences, *Pure Appl. Chem.*, **74**, 1135–1140 (2002); (*b*) M. G. Tordoff, H. Shao, L. K. Alarcon, R. F. Margolskee, B. Mosinger, A. A. Bachmanov, D. R. Reed, S. McCaughey, Involvement of T1R3 in calcium–magnesium taste, *Physiol. Genomics*, **34**, 338–348 (2008).
126. M. Milinski, C. Wedekind, Evidence for MHC-correlated perfume preferences in humans, *Behav. Ecol.*, **12**, 140–149 (2001).
127. A. Willse, J. Kwak, K. Yamazaki, G. Preti, J. H. Wahl, G. K. Beauchamp, Individual odortypes: interaction of MHC and background genes, *Immunogenetics*, **58**, 967–982 (2006).
128. J. Havlicek, S. C. Roberts, MHC-correlated mate choice in humans: a review, *Psychoneuroendocrinology*, **34**, 497–512 (2009).
129. B. Woelfing, A. Traulsen, M. Milinski, T. Boehm, Does intra-individual major histocompatibility complex diversity keep a golden mean? *Philos. Trans R. Soc. London, B*, **364**, 117–128 (2009).
130. P. Ilmonen, G. Stundner, M. Thoss, D. J. Penn, Females prefer the scent of outbred males: good-genes-as-heterozygosity? *BMC Evol. Biol.*, **9** Article No. 104 (2009).
131. R. Chaix, C. Cao, P. Donnelly, Is mate choice in humans MHC-dependent? *PLoS Genet.*, **4**, e1000184 (2008).
132. C. Ober, Studies of HLA, fertility and mate choice in a human isolate, *Hum. Reprod. Update*, **5**, 103–107 (1999).
133. A. Helgason, S. Palsson, D. F. Gudbjartsson, T. Kristjansson, K. Stefansson, An association between the kinship and fertility of human couples, *Science*, **319**, 813–816 (2008).
134. P. C. Brunjes, Unilateral naris closure and olfactory system development, *Brain Res. Rev.*, **19**, 146–160 (1994).
135. K. C. Biju, T. G. Mast, D. A. Fadool, Olfactory sensory deprivation increases the number of proBDNF-immunoreactive mitral cells in the olfactory bulb of mice, *Neurosci. Lett.*, **447**, 42–47 (2008).
136. Y. Livneh, N. Feinstein, M. Klein, A. Mizrahi, Sensory input enhances synaptogenesis of adult-born neurons, *J. Neurosci.*, **29**, 86–97 (2009).
137. J. A. Gottfried, K. N. Wu, Perceptual and neural pliability of odor objects, *Ann. N. Y. Acad. Sci.*, **1170**, 324–332 (2009).
138. B. Zimmerberg, H. E. Foote, T. A. Van Kempen, Olfactory association learning and brain-derived neurotrophic factor in an animal model of early deprivation, *Dev. Psychobiol.*, **51**, 333–344 (2009).
139. H. Eichenbaum, R. J. Robitsek, Olfactory memory: a bridge between humans and animals in models of cognitive aging, *Ann. N. Y. Acad. Sci.*, **1170**, 658–663 (2009).
140. W. S. Cain, R. A. de Wijk, S. Nordin, M. Nordin, Independence of odor quality and absolute sensitivity in a study of aging, *Chemosens. Percept.*, **1**, 24–33 (2008).
141. C. J. Wysocki, M. L. Pelchat, The effects of aging on the human sense of smell and its relationship to food choice, *Crit. Rev. Food Sci. Nutr.*, **33**, 63–82 (1993).
142. M. L. Pelchat, You can teach an old dog new tricks: olfaction and responses to novel foods by the elderly, *Appetite*, **35**, 153–160 (2000).
143. (*a*) R. L. Doty, P. Shaman, S. L. Applebaum, R. Giberson, L. Siksorski, L. Rosenberg, Smell identification ability: changes with age, *Science*, **226**, 1441–1443 (1984); (*b*) B. J. Cowart, Relationships between taste and smell across the adult life span, *Ann. N. Y. Acad. Sci.*, **561**, 39–55 (1989); (*c*) S. S. Schiffman, Taste and smell losses in normal aging and disease, *J. Am. Med. Assoc.*, **278**, 1357–1362 (1997); (*d*) C. Murphy, C. R. Schubert, K. J. Cruickshanks, B. E. Klein, R. Klein, D. M. Nondahl, Prevalence of olfactory impairment in older adults, *J. Am. Med. Assoc.*, **288**, 2307–2312 (2002); (*e*) C. Murphy, The chemical senses and nutrition in older adults, *J. Nutr. Elder*, **27**, 247–265 (2008).
144. M. A. Curtis, M. Kam, U. Nannmark, M. F. Anderson, M. Z. Axell, C. Wikkelso, S. Holtas, W. M. van Roon-Mom, T. Bjork-Eriksson, C. Nordborg, J. Frisen, M. Dragunow, R. L. Faull, P. S.

Eriksson, Human neuroblasts migrate to the olfactory bulb via a lateral ventricular extension, *Science*, **315**, 1243–1249 (2007).

145. (*a*) E. Gould, How widespread is adult neurogenesis in mammals? *Nat. Rev. Neurosci.*, **8**, 481–488 (2007); (*b*) A. Nissant, C. Bardy, H. Katagiri, K. Murray, P.-M. Lledo, Adult neurogenesis promotes synaptic plasticity in the olfactory bulb, *Nat. Neurosci.*, **12**, 728–730 (2009); (*c*) C. G. Gross, Three before their time: neuroscientists whose ideas were ignored by their contemporaries, *Exp. Brain Res.*, **192**, 321–334 (2009).
146. P. Meerlo, R. E. Mistlberger, B. L. Jacobs, H. C. Heller, D. McGinty, New neurons in the adult brain: the role of sleep and consequences of sleep loss, *Sleep Med. Rev.*, **13**, 187–194 (2009).
147. (*a*) C. Rochefort, G. Gheusi, J. D. Vincent, P.-M. Lledo, Enriched odor exposure increases the number of newborn neurons in the adult olfactory bulb and improves odor memory, *J. Neurosci.*, **22**, 2679–2689 (2002); (*b*) A. Veyrac, J. Sacquet, V. Nguyen, M. Marien, F. Jourdan, A. Didier, Novelty determines the effects of olfactory enrichment on memory and neurogenesis through noradrenergic mechanisms, *Neuropsychopharmacology*, **34**, 786–795 (2009).
148. N. E. Rawson, A. S. LaMantia, A speculative essay on retinoic acid regulation of neural stem cells in the developing and aging olfactory system, *Exp. Gerontol.*, **42**, 46–53 (2007).
149. R. Vandenbosch, L. Borgs, P. Beukelaers, S. Belachew, G. Moonen, L. Nguyen, B. Malgrange, Adult neurogenesis and the diseased brain, *Curr. Med. Chem.*, **16**, 652–666 (2009).
150. E. M. Krantz, C. R. Schubert, D. S. Dalton, W. Zhong, G. H. Huang, B. E. Klein, R. Klein, F. J. Nieto, K. J. Cruickshanks, Test–retest reliability of the San Diego odor identification test and comparison with the brief smell identification test, *Chem. Senses*, **34**, 435–440 (2009).
151. C. R. Schubert, K. J. Cruickshanks, C. Murphy, G.-H. Huang, B. E. Klein, R. Klein, F. J. Nieto, J. S. Pankow, T. S. Tweed, Olfactory impairment in adults, *Ann. N. Y. Acad. Sci.*, **1170**, 531–536 (2009).
152. C. R. Schubert, L. L. Carmichael, C. Murphy, B. E. Klein, R. Klein, K. J. Cruickshanks, Olfaction and the 5-year incidence of cognitive impairment in an epidemiological study of older adults, *J. Am. Geriatr. Soc.*, **56**, 1517–1521 (2008).
153. G. W. Ross, H. Petrovitch, R. D. Abbott, C. M. Tanner, J. Popper, K. Masaki, L. Launer, L. R. White, Association of olfactory dysfunction with risk for future Parkinson's disease, *Ann. Neurol.*, **63**, 167–173 (2008).
154. C. Murphy, J. A. Anderson, S. Markison, Psychophysical assessment of chemosensory disorders in clinical populations, in *Olfaction and Taste XI*, K. S. N. Kunihara, N. Suzuki, H. Ogawa (Eds), Springer-Verlag, Tokyo, 1994, pp. 609–613.
155. R. L. Doty, A. Marcus, W. W. Lee, Development of the 12-item cross-cultural smell identification test (CC-SIT), *Laryngoscope*, **106**, 353–356 (1996).
156. P. Dalton, J. A. Mennella, B. J. Cowart, C. Maute, E. A. Pribitkin, J. S. Reilly, Evaluating the prevalence of olfactory dysfunction in a pediatric population, *Ann. N. Y. Acad. Sci.*, **1170**, 537–542 (2009).
157. D. G. Laing, C. Segovia, T. Fark, O. N. Laing, A. L. Jinks, J. Nikolaus, T. Hummel, Tests for screening olfactory and gustatory function in school-age children, *Otolaryngol. – Head Neck Surg.*, **139**, 74–82 (2008).
158. J. Reisert, D. Restrepo, Molecular tuning of odorant receptors and its implication for odor signal processing, *Chem. Senses*, **34**, 535–545 (2009).
159. (*a*) D. W. Wesson, R. M. Carey, J. V. Verhagen, M. Wachowiak, Rapid encoding and perception of novel odors in the rat, *PLoS Biol.*, **6**, e82 (2008); (*b*) M. Wachowiak, D. W. Wesson, N. Pirez, J. V. Verhagen, R. M. Carey, Low-level mechanisms for processing odor information in the behaving animal, *Ann. N. Y. Acad. Sci.*, **1170**, 286–292 (2009).
160. B. N. Johnson, J. D. Mainland, N. Sobel, Rapid olfactory processing implicates subcortical control of an olfactomotor system, *J. Neurophysiol.*, **90**, 1084–1094 (2003).
161. B. N. Johnson, C. Russell, R. M. Khan, N. Sobel, A comparison of methods for sniff measurement concurrent with olfactory tasks in humans, *Chem. Senses*, **31**, 795–806 (2006).
162. J. D. Mainland, B. N. Johnson, R. Khan, R. B. Ivry, N. Sobel, Olfactory impairments in patients with unilateral cerebellar lesions are selective to inputs from the contralesional nostril, *J. Neurosci.*, **25**, 6362–6371 (2005).

163. J. Mainland, N. Sobel, The sniff is part of the olfactory percept, *Chem. Senses*, **31**, 181–196 (2006).
164. B. Koritnik, S. Azam, C. M. Andrew, P. N. Leigh, S. C. Williams, Imaging the brain during sniffing: a pilot fMRI study, *Pulm. Pharmacol. Ther.*, **22**, 97–101 (2009).
165. A. M. Kleemann, R. Kopietz, J. Albrecht, V. Schopf, O. Pollatos, T. Schreder, J. May, J. Linn, H. Bruckmann, M. Wiesmann, Investigation of breathing parameters during odor perception and olfactory imagery, *Chem. Senses*, **34**, 1–9 (2009).
166. K. Simonyan, Z. S. Saad, T. M. Loucks, C. J. Poletto, C. L. Ludlow, Functional neuroanatomy of human voluntary cough and sniff production, *NeuroImage*, **37**, 401–409 (2007).
167. J. Djordjevic, R. J. Zatorre, M. Petrides, J. A. Boyle, M. Jones-Gotman, Functional neuroimaging of odor imagery, *NeuroImage*, **24**, 791–801 (2005).
168. M. Bensafi, N. Sobel, R. M. Khan, Hedonic-specific activity in piriform cortex during odor imagery mimics that during odor perception, *J. Neurophysiol.*, **98**, 3254–3262 (2007).
169. D. Leopold, Distortion of olfactory perception: diagnosis and treatment, *Chem. Senses*, **27**, 611–615 (2002).
170. B. N. Landis, P. R. Burkhard, Phantosmias and Parkinson disease, *Arch. Neurol.*, **65**, 1237–1239 (2008).
171. S. Jacek, R. J. Stevenson, L. A. Miller, Olfactory dysfunction in temporal lobe epilepsy: a case of ictus-related parosmia, *Epilepsy Behav.*, **11**, 466–470 (2007).
172. S. J. Sara, The locus coeruleus and noradrenergic modulation of cognition, *Nat. Rev. Neurosci.*, **10**, 211–223 (2009).
173. J. Djordjevic, J. N. Lundstrom, F. Clement, J. A. Boyle, S. Pouliot, M. Jones-Gotman, A rose by any other name: would it smell as sweet? *J. Neurophysiol.*, **99**, 386–393 (2008).
174. (*a*) D. Chen, J. Haviland-Jones, Human olfactory communication of emotion, *Percept. Mot. Skills*, **91**, 771–781 (2000); (*b*) D. Chen, P. Dalton, The effect of emotion and personality on olfactory perception, *Chem. Senses*, **30**, 345–351 (2005).
175. (*a*) J. A. Gottfried, R. J. Dolan, The nose smells what the eye sees: crossmodal visual facilitation of human olfactory perception, *Neuron*, **39**, 375–386 (2003); (*b*) M. L. Dematte, D. Sanabria, C. Spence, Olfactory discrimination: when vision matters? *Chem. Senses*, **34** 103–109 (2009).
176. O. Pollatos, R. Kopietz, J. Linn, J. Albrecht, V. Sakar, A. Anzinger, R. Schandry, M. Wiesmann, Emotional stimulation alters olfactory sensitivity and odor judgment, *Chem. Senses*, **32**, 583–589 (2007).
177. J. H. Laudien, S. Wencker, R. Ferstl, B. M. Pause, Context effects on odor processing: an event-related potential study, *NeuroImage*, **41**, 1426–1436 (2008).
178. I. E. de Araujo, E. T. Rolls, M. I. Velazco, C. Margot, I. Cayeux, Cognitive modulation of olfactory processing, *Neuron*, **46**, 671–679 (2005).
179. (*a*) P. Walla, Olfaction and its dynamic influence on word and face processing: cross-modal integration, *Prog. Neurobiol.*, **84**, 192–209 (2008); (*b*) S. T. Weber, E. Heuberger, The impact of natural odors on affective states in humans, *Chem. Senses*, **33**, 441–447 (2008); (*c*) R. S. Herz, Aromatherapy facts and fictions: a scientific analysis of olfactory effects on mood, physiology and behavior, *Int. J. Neurosci.*, **119**, 263–290 (2009).
180. (*a*) R. S. Herz, J. Eliassen, S. Beland, T. Souza, Neuroimaging evidence for the emotional potency of odor-evoked memory, *Neuropsychologia*, **42**, 371–378 (2004); (*b*) J. S. Jellinek, Proust remembered: has Proust's account of odor-cued autobiographical memory recall really been investigated? *Chem. Senses*, **29** 455–458; and author reply 459–461 (2004).
181. J. Willander, M. C. Larsson, The mind's nose and autobiographical odor memory, *Chemosens. Percept.*, **1**, 210–215 (2008).
182. A. Gelperin, Brain states and olfaction, in *Encyclopedia of Neuroscience*, M. D. Binder, N. Hirokawa, U. Windhorst (Eds), Springer, Berlin, 2009, pp. 490–492.
183. M. Schredl, D. Atanasova, K. Hormann, J. T. Maurer, T. Hummel, B. A. Stuck, Information processing during sleep: the effect of olfactory stimuli on dream content and dream emotions, *J. Sleep Res.*, **18**, 285–290 (2009).
184. (*a*) F. Tubaldi, C. Ansuini, R. Tirindelli, U. Castiello, Odours grab his hand but not hers, *Perception*, **37**, 1886–1889 (2008); (*b*) A. Chopra, A. Baur, T. Hummel, Thresholds and

chemosensory event-related potentials to malodors before, during, and after puberty: differences related to sex and age, *NeuroImage*, **40**, 1257–1263 (2008).
185. C. Wyart, W. W. Webster, J. H. Chen, S. R. Wilson, A. McClary, R. M. Khan, N. Sobel, Smelling a single component of male sweat alters levels of cortisol in women, *J. Neurosci.*, **27**, 1261–1265 (2007).
186. C. J. Wysocki, G. Preti, *Human Pheromones: What's Purported, What's Supported*, Sense of Smell Institute, http://www.senseofsmell.org, 2009, pp. 1–21.
187. C. J. Wysocki, K. M. Dorries, G. K. Beauchamp, Ability to perceive androstenone can be acquired by ostensibly anosmic people, *Proc. Natl. Acad. Sci. USA*, **86**, 7976–7978 (1989).
188. K. M. Dorries, H. J. Schmidt, G. K. Beauchamp, C. J. Wysocki, Changes in sensitivity to the odor of androstenone during adolescence, *Dev. Psychobiol.*, **22**, 423–435 (1989).
189. P. Dalton, N. Doolittle, P. A. Breslin, Gender-specific induction of enhanced sensitivity to odors, *Nat. Neurosci.*, **5**, 199–200 (2002).
190. N. Boulkroune, L. Wang, A. March, N. Walker, T. J. Jacob, Repetitive olfactory exposure to the biologically significant steroid androstadienone causes a hedonic shift and gender dimorphic changes in olfactory-evoked potentials, *Neuropsychopharmacology*, **32**, 1822–1829 (2007).
191. A. Knaapila, H. Turila, K. Silventoinen, M. J. Wright, K. O. Kyvik, L. F. Cherkas, K. Keskitalo, J. Hansen, N. G. Martin, T. D. Spector, J. Kaprio, M. Perola, Genetic and environmental contributions to perceived intensity and pleasantness of androstenone odor: an international twin study, *Chemosens. Percept.*, **1**, 34–42 (2008).
192. H. Berglund, P. Lindstrom, I. Savic, Brain response to putative pheromones in lesbian women, *Proc. Natl. Acad. Sci. USA*, **103**, 8269–8274 (2006).
193. Y. Martins, G. Preti, C. R. Crabtree, T. Runyan, A. A. Vainius, C. J. Wysocki, Preference for human body odors is influenced by gender and sexual orientation, *Psychol. Sci.*, **16**, 694–701 (2005).
194. M. J. Sergeant, T. E. Dickins, M. N. Davies, M. D. Griffiths, Women's hedonic ratings of body odor of heterosexual and homosexual men, *Arch. Sex Behav.*, **36**, 395–401 (2007).
195. J. N. Lundstrom, J. A. Boyle, R. J. Zatorre, M. Jones-Gotman, Functional neuronal processing of body odors differs from that of similar common odors, *Cereb. Cortex*, **18**, 1466–1474 (2008).
196. J. N. Lundstrom, M. Jones-Gotman, Romantic love modulates women's identification of men's body odors, *Horm. Behav.*, **55**, 280–284 (2009).
197. M. Gallagher, C. J. Wysocki, J. J. Leyden, A. I. Spielman, X. Sun, G. Preti, Analyses of volatile organic compounds from human skin, *Br. J. Dermatol.*, **159**, 780–791 (2008).
198. (*a*) R. L. Doty, P. Shaman, M. Dann, Development of the University of Pennsylvania smell identification test: a standardized microencapsulated test of olfactory function, *Physiol. Behav.*, **32**, 489–502 (1984); (*b*) J. H. Cho, Y. S. Jeong, Y. J. Lee, S. C. Hong, J. H. Yoon, J. K. Kim, The Korean version of the sniffin' stick (KVSS) test and its validity in comparison with the cross-cultural smell identification test (CC-SIT), *Auris, Nasus, Larynx*, **36**, 280–286 (2009).
199. N. Thuerauf, U. Reulbach, J. Lunkenheimer, B. Lunkenheimer, R. Spannenberger, A. Gossler, C. Maihofner, S. Bleich, J. Kornhuber, K. Markovic, Emotional reactivity to odors: olfactory sensitivity and the span of emotional evaluation separate the genders, *Neurosci. Lett.*, **456**, 74–79 (2009).
200. C. J. Wysocki, J. Louie, J. J. Leyden, D. Blank, M. Gill, L. Smith, K. McDermott, G. Preti, Cross-adaptation of a model human stress-related odour with fragrance chemicals and ethyl esters of axillary odorants: gender-specific effects, *Flavour Fragr. J.*, **24**, 209–218 (2009).
201. R. S. Herz, M. Inzlicht, Sex diferences in response to physical and social factors involved in human mate selection – the importance of smell for women, *Evol. Hum. Behav.*, **23**, 359–364 (2002).
202. M. R. Yeomans, Olfactory influences on appetite and satiety in humans, *Physiol. Behav.*, **89**, 10–14 (2006).
203. D. Labbe, N. Martin, Impact of novel olfactory stimuli at supra and subthreshold concentrations on the perceived sweetness of sucrose after associative learning, *Chem. Senses*, **34**, 645–651 (2009).

204. R. J. Stevenson, R. A. Boakes, J. Prescott, Changes in odor sweetness from implicit learning of a simultaneous odor–sweetness association: an example of learned synesthesia, *Learn. Motiv.*, **29**, 113–132 (1998).
205. G. M. Shepherd, Smell images and the flavour system in the human brain, *Nature*, **444**, 316–321 (2006).
206. Y. Yeshurun, N. Sobel, An odor is not worth a thousand words: from multidimensional odors to unidimensional odor objects, *Annu. Rev. Psych.*, **61**, 219–241 (2010).
207. R. L. Doty, D. A. Deems, S. Stellar, Olfactory dysfunction in parkinsonism: a general deficit unrelated to neurologic signs, disease stage, or disease duration, *Neurology*, **38**, 1237–1244 (1988).
208. M. W. Albers, M. H. Tabert, D. P. Devanand, Olfactory dysfunction as a predictor of neurodegenerative disease, *Curr. Neurol. Neurosci. Rep.*, **6**, 379–386 (2006).
209. M. W. Albers, K. S. Marder, A tale of two etiologies: loss and recovery of olfactory function, *Ann. Neurol.*, **63**, 132–134 (2008).
210. M. Witt, K. Bormann, V. Gudziol, K. Pehlke, K. Barth, A. Minovi, A. Hahner, H. Reichmann, T. Hummel, Biopsies of olfactory epithelium in patients with Parkinson's disease, *Mov. Disord.*, **24**, 906–914 (2009).
211. A. Haehner, S. Boesveldt, H. W. Berendse, A. Mackay-Sim, J. Fleischmann, P. A. Silburn, A. N. Johnston, G. D. Mellick, B. Herting, H. Reichmann, T. Hummel, Prevalence of smell loss in Parkinson's disease – a multicenter study, *Parkinsonism Relat. Disord.*, **15**, 490–494 (2009).
212. B. Atanasova, J. Graux, W. El Hage, C. Hommet, V. Camus, C. Belzung, Olfaction: a potential cognitive marker of psychiatric disorders, *Neurosci. Biobehav. Rev.*, **32**, 1315–1325 (2008).
213. B. I. Turetsky, C. G. Kohler, R. E. Gur, P. J. Moberg, Olfactory physiological impairment in first-degree relatives of schizophrenia patients, *Schizophr. Res.*, **102**, 220–229 (2008).
214. B. I. Turetsky, C. G. Hahn, S. E. Arnold, P. J. Moberg, Olfactory receptor neuron dysfunction in schizophrenia, *Neuropsychopharmacology*, **34**, 767–774 (2009).
215. O. Pollatos, J. Albrecht, R. Kopietz, J. Linn, V. Schoepf, A. M. Kleemann, T. Schreder, R. Schandry, M. Wiesmann, Reduced olfactory sensitivity in subjects with depressive symptoms, *J. Affect. Disord.*, **102**, 101–108 (2007).
216. (*a*) S. Kivity, O. D. Ortega-Hernandez, Y. Shoenfeld, Olfaction – a window to the mind, *Isr. Med. Assoc. J.*, **11**, 238–243 (2009); (*b*) O. D. Ortega-Hernandez, S. Kivity, Y. Shoenfeld, Olfaction, psychiatric disorders and autoimmunity: is there a common genetic association? *Autoimmunity*, **42**, 80–88 (2009).
217. G. M. Zucco, G. Ingegneri, Olfactory deficits in HIV-infected patients with and without AIDS dementia complex, *Physiol. Behav.*, **80**, 669–674 (2004).
218. H. K. Prasad, K. M. Bhojwani, V. Shenoy, S. C. Prasad, HIV manifestations in otolaryngology, *Am. J. Otolaryngol.*, **27**, 179–185 (2006).
219. B. I. Turetsky, C. G. Hahn, K. Borgmann-Winter, P. J. Moberg, Scents and nonsense: olfactory dysfunction in schizophrenia, *Schizophr. Bull.*, **35**, 1117–1131 (2009).
220. B. I. Turetsky, P. J. Moberg, K. Owzar, S. C. Johnson, R. L. Doty, R. E. Gur, Physiologic impairment of olfactory stimulus processing in schizophrenia, *Biol. Psychiatry*, **53**, 403–411 (2003).
221. C. G. Hahn, G. Gomez, D. Restrepo, E. Friedman, R. Josiassen, E. A. Pribitkin, L. D. Lowry, R. J. Gallop, N. E. Rawson, Aberrant intracellular calcium signaling in olfactory neurons from patients with bipolar disorder, *Am. J. Psychiatry*, **162**, 616–618 (2005).
222. B. I. Turetsky, P. J. Moberg, An odor-specific threshold deficit implicates abnormal intracellular cyclic AMP signaling in schizophrenia, *Am. J. Psychiatry*, **166**, 226–233 (2009).
223. T. Ugur, M. Weisbrod, E. Franzek, U. Pfuller, H. Sauer, Olfactory impairment in monozygotic twins discordant for schizophrenia, *Eur. Arch. Psychiatry Clin. Neurosci.*, **255**, 94–98 (2005).
224. C. Murphy, W. S. Cain, Odor identification: the blind are better, *Physiol. Behav.*, **37**, 177–180 (1986).
225. I. Cuevas, P. Plaza, P. Rombaux, A. G. De Volder, L. Renier, Odour discrimination and identification are improved in early blindness, *Neuropsychologia*, **47**, 3079–3083 (2009).

226. M. Adams, U. M. Smith, C. V. Logan, C. A. Johnson, Recent advances in the molecular pathology, cell biology and genetics of ciliopathies, *J. Med. Genet.*, **45**, 257–267 (2008).
227. P. M. Jenkins, D. P. McEwen, J. R. Martens, Olfactory cilia: linking sensory cilia function and human discasc, *Chem. Senses*, **34**, 451–464 (2009).
228. B. J. Cowart, I. M. Young, R. S. Feldman, L. D. Lowry, Clinical disorders of smell and taste, *Occup. Med.* **12**, 465–483 (1997).
229. (*a*) J. A. Mennella, C. P. Jagnow, G. K. Beauchamp, Prenatal and postnatal flavor learning by human infants, *Pediatrics*, **107**, E88 (2001); (*b*) J. A. Mennella, Flavour programming during breast-feeding, *Adv. Exp. Med. Biol.*, **639**, 113–120 (2009).
230. H. Hausner, M. Philipsen, T. H. Skov, M. A. Petersen, W. L. P. Bredie, Characterization of the volatile composition and variations between infant formulas and mother's milk, *Chemosens. Percept.*, **2**, 79–93 (2009).
231. J. A. Mennella, G. K. Beauchamp, Optimizing oral medications for children, *Clin. Ther.*, **30**, 2120–2132 (2008).
232. S. Monnery-Patris, C. Rouby, S. Nicklaus, S. Issanchou, Development of olfactory ability in children: sensitivity and identification, *Dev. Psychobiol.*, **51**, 268–276 (2009).
233. C. Ferdenzi, S. Mustonen, H. Tuorila, B. Schaal, Children's awareness and uses of odor cues in everyday life: a Finland–France comparison, *Chemosens. Percept.*, **1**, 190–198 (2008).
234. J. A. Mennella, C. A. Forestell, Children's hedonic responses to the odors of alcoholic beverages: a window to emotions, *Alcohol*, **42**, 249–260 (2008).
235. O. Romantshik, R. H. Porter, V. Tillmann, H. Varendi, Preliminary evidence of a sensitive period for olfactory learning by human newborns, *Acta Paediatr.*, **96**, 372–376 (2007).
236. B. Renner, C. A. Mueller, J. Dreier, S. Faulhaber, W. Rascher, G. Kobal, The candy smell test: a new test for retronasal olfactory performance, *Laryngoscope*, **119**, 487–495 (2009).
237. B. Kuhn, W. Denk, R. M. Bruno, *In vivo* two-photon voltage-sensitive dye imaging reveals top-down control of cortical layers 1 and 2 during wakefulness, *Proc. Natl. Acad. Sci. USA*, **105**, 7588–7593 (2008).
238. (*a*) J. Y. Lin, M. Z. Lin, P. Steinbach, R. Y. Tsien, Characterization of engineered channelrhodopsin variants with improved properties and kinetics, *Biophys. J.*, **96**, 1803–1814 (2009); (*b*) O. G. Ayling, T. C. Harrison, J. D. Boyd, A. Goroshkov, T. H. Murphy, Automated light-based mapping of motor cortex by photoactivation of channelrhodopsin-2 transgenic mice, *Nat. Methods*, **6**, 219–224 (2009).
239. (*a*) E. T. Rolls, F. Grabenhorst, L. Franco, Prediction of subjective affective state from brain activations, *J. Neurophysiol.*, **101**, 1294–1308 (2009); (*b*) P. A. Bandettini, What's new in neuroimaging methods? *Ann. N.Y. Acad. Sci.*, **1156**, 260–293 (2009); (*c*) K. Tabelow, V. Piech, J. Polzehl, H. U. Voss, High-resolution fMRI: overcoming the signal-to-noise problem, *J. Neurosci. Methods*, **178**, 357–365 (2009).
240. D. B. Katz, H. Matsunami, D. Rinberg, K. Scott, M. Wachowiak, R. I. Wilson, Receptors, circuits, and behaviors: new directions in chemical senses, *J. Neurosci.*, **28**, 11802–11805 (2008).
241. N. Mandairon, J. Poncelet, M. Bensafi, A. Didier, Humans and mice express similar olfactory preferences, *PLoS One*, **4**, e4209 (2009).
242. C. Zelano, N. Sobel, Humans as an animal model for systems-level organization of olfaction, *Neuron*, **48**, 431–454 (2005).
243. R. J. Stevenson, An initial evaluation of the functions of human olfaction, *Chem. Senses*, **35**, 3–20 (2010).

11

Perfumery – The Wizardry of Volatile Molecules

Christophe Laudamiel

DreamAir LLC, 210 Eleventh Ave, Suite 1002, New York, NY 10001, USA

11.1 The Big Picture

'The multifarious nuances of the sense of smell embody the archetypal longing for the lower forms of existence, for direct unification with circumambient nature [...].'[1]

Volatile molecules are to perfumers what music notes are to music composers and colours to painters.[2] Everyone is quite familiar with how vision works; photons (tiny packs of energy of different wavelengths, meaning in plain language of different colours) travel from a light source, are reflected by objects and enter the eyes where they are transformed into nerve signals transporting the information to the brain. In the case of audition, air vibrations are created by the vocal cords (or by some loudspeakers) and reach the ears where they are transformed into nerve signals also travelling to the brain. And for olfaction, molecules are emitted by a scent source (a flower, the sea, your skin) and reach your nose where they create nerve signals, also relaying the information to the brain (Table 11.1). The brain is a huge data processor which tries to make sense of all the stimuli coming from the senses and draws conclusions, stores memories and decides on actions or inactions. Just like the eyes are made to receive photons of light, whether from natural or manmade sources, ears are made to hear vibrations, natural or manmade. The nose is made to smell molecules, natural or manmade. One should not be fearful of smelling. Numerous similarities exist with the other senses, such as true figurative perceptions of the world or conceptual interpretations, illusions or 'trompe nez' effects, inspirations, artistic and commercial applications and all other things associated with the senses, their biology, underlying science and outspoken arts.

The Chemistry and Biology of Volatiles Edited by Andreas Herrmann

Table 11.1 Correspondences between music, visual arts and perfumery

Body part (places of biological reception)	Physico-chemical vehicle	Informative main units	By-units or other important declinations	Disciplines
Ears	Vibrations	Frequencies/notes sounds/phonemes	Harmonics/timbre, muffler	Music/language
Eyes	Photons	Wavelengths/colours	Black, white, shades	Visual arts
Nose	Molecules	Osmonemes/simple odours/notes	Impurities, isomers, diluents[a]	Perfumery

Note: [a]One notices how the vocabulary for perfumery sounds often negative versus other disciplines. This has to be changed. We often talk about 'impurities'. Instead, sommeliers talk about 'terroir', painters about 'shades', others about spiking: all much more poetic. A diluent in perfumery can be used to make something less strong or more round (like white in paint or a mute on a trumpet). A diluent is usually an odourless solvent or a fragrance molecule of very low odour such as benzyl salicylate (Figure 11.1). It is true that diluents often suffer from bad press because they have been used to cheapen compositions unknowingly to consumers, but charlatanism should not overshadow other noble uses of these ingredients.

The debate about natural essences versus manmade molecules often sounds primitive and sterile. It is driven by a lack of basic scientific knowledge from a public that has rarely had the opportunity to get academically educated about the sense of smell. Nature has been creating and selecting, just like a chemist and then a perfumer, over several million years. In nature, this process is called 'mutation' – in perfumery, it is called 'trial and error'. The debate should elevate, however, just as when one understands and appreciates the differences between a real flower, a photograph of it, a 3D rendition of it in an Imax theatre and a painting of it. In this chapter, 'molecule' is the term retained as the actual objective description of what the nose 'smells', to encompass both molecules and mixtures thereof, including essences extracted from nature.

And just like a lighting designer knows how to use dimmed lights, spot lights, lasers, stroboscopes or pyrotechnics, with precise rules and intensity levels, to achieve an effect but remain safe, perfumers (since the 1980s) have had to respect strict rules on maximum levels for ingredients, gathered in complex regulatory tables, for safe usages per usage level and final applications (fine fragrances for skin, shampoos, candles, baby products, etc.). Furthermore, unlike eyes or ears, which do not get renewed, the olfactory neurons in the nose are renewed regularly[3] (every couple of weeks). This is a very rare feat among neurons – a little like the teeth of a shark or the limbs of a salamander.

11.2 Wizardry No. 1: Full Holograms Create Real Emotions

If the grand principles of all the senses are very similar, each sense has then its own specificities. An interesting difference between olfaction and vision (or audition) lies in the fact that molecules are matter. One can touch a group of them, they have a weight, they cannot be transported via the internet in a pdf file and cannot be burnt on a compact disc or shared on a server. This implies concretely some intense logistics and philosophically the unsettling fact that, each time you 'smell something', one actually consumes a tiny part of that thing. That 'thing' has to be kept fresh and has to be replenished regularly, or one has to catch the thing before it withers. Imagine, if each time a visitor looked at La Joconde, a tiny piece of it was consumed and disappeared forever. Imagine the logistics involved and ask yourself whether La Joconde would be as popular if this painting had to be shipped to anyone who cannot go to Paris, instead of having a look at it in a book or on the internet.

The upside to this is that a molecule reaching your nose, whether flying from a jasmine flower, or from a laboratory bottle containing that molecule, is exactly the same molecule. The nose smells, but the nose does not see. The nose does not know if that volatile molecule travelling through the air comes from real jasmine or from a reconstitution of it in your bottle. Of course, the reconstitution must be good! This means it is easier to make an airport tunnel smell like real jasmine than to use pictures, a video, or a trompe l'oeil to make people believe that they are walking through a real jasmine plantation in that tunnel. In most cases, one can always tell that it is a photograph and not the real object. I have never seen anyone trying to walk through a mural. There is the problem of translating a 3D format into a 2D format. The pigments used by a photo laboratory, or a printer, are not the same molecules as the ones used by nature. Hence, differences in hues that the eyes notice very well. Similarly, one can usually tell if a symphony orchestra or a compact disk is playing next door. Recording truncates some dimensions. A bad perfumery reconstitution also truncates some dimensions or uses other molecules (rather than the original ones) usually in a cheaper way. In several instances, some key natural molecules are still not available to the perfumer because the chemists have not found a way to reproduce them and their impurities. However, in many instances, a good reconstitution can fool you, and your smell, big time!

In perfumery, an unsettling wizardry occurs. It is amazing to realize how true to life the things one smells seem to be. Since our sense of smell is three-dimensional (around you and inside you, so there are no 3D–2D problems like in photography) and because one can actually smell the exact same components from a bottle that one can from the real object, the brain believes that you are there. A rose naturally contains 300 components, most of them at very low levels. The brain does not smell those 300 components individually, and you do not have to have 300 components to reconstitute a true to life rose. Given that the nose, even an expert nose, can detect distinctively only six to eight facets at any given point in time, and that natural extracts may contain several hundreds of molecules, if the perfumer has captured the key combination allowing recognition of the natural pattern, you are transported! Make sure that, when you do an odour evaluation, the real rose in front of you does not fool you on your appreciation of its scent versus smelling it from a jar. The same is true with a peony.

We should call this effect the 'full hologram effect'. 'Full' because, unlike visual holograms, the scent hologram is not hollow. It comes as a full volume – the matter is actually there. The real molecules are indeed present in the air, yet the source – the real flower or your loved one – is nowhere to be seen. There is only one real Joconde but there are millions of real Shalimar (Guerlain) bottles – so many chances of meeting one's grandmother again if the background odour of the person does not disturb the scent too much and if the regulations of the International Fragrance Association (IFRA) and the European Union (EU) have not done ravages to the original scent formula in the meantime.

That was Perfumery 101. Now for the more advanced readers. Here is how wicked the sense of smell is. One can also create the same full hologram by using different molecules than the ones that emanate from the original! Imagine creating a hologram made of reflections of totally different objects but each having some patches of colours in common with the original object. The wizard physicist would carefully choose these unrelated objects and assemble them in such a way that those common patches would show on the outside of the hologram to recreate the vision of the original object. He would hide inside the hologram all the other different objects, giving the illusion that the original object has been reconstituted. A perfumer does that regularly. A formula is worked to have the desired rendition perceptively on the outside and, inside the fragrance, unwanted odour facets of the same molecules cancel each other out or are hidden. To make another analogy, one could construct a Rubik's cube by either exactly copying the whole object, or by retaining only the outside facets of each small cube and changing all the inside faces of those mini cubes. One could even change the internal mechanism of the cube. No one would see the difference, except upon using it. This, too, has an interpretation in perfumery. Those mini cubes each represent a fragrance molecule with its different odour facets. A formula can be created in such a way that its outside facets (i.e. its overall perceived odour characters or 'what it smells like') smell like the real object from nature or like the exact emotion the perfumer wants. However, if one scratches the surface (by removing one ingredient at a time), one will realize that the formula is not built in the expected way. This does not mean it is better or worse, it all depends. The brain does not see if the hologram is hollow or not, or if your newly constructed Rubik's cube is identical to the original on the inside. The brain gets excited or relaxed just the same. If, however, you learn by some other means that it is not the real object (e.g. if you pass your hand through a visual hologram, if someone tells you about the 'scent hologram' or, if poorly done, you might smell it), you might be disappointed. Alternatively, you might prefer the re-engineered version because it is cleaner, or more robust, or longer-lasting. For instance, it is usual for most people to prefer lavenders re-engineered with vanillin (**1**, Figure 11.1) and other sweet and floral notes, to the real pure and natural lavender essence. In the case of a good quality product, one might actually appreciate the work involved all the more, depending on who did the work, how clever it appears to be, how original it is, etc. This is not to say that a beautiful natural is not still a beautiful natural.

This next paragraph is to be applied to interpretations of not only odours (lavender, a sea breeze, an old attic, etc.) but also to emotions. A perfumer creates an object that smells like certain concrete things or that expresses, or provides, a certain emotion. There are several ways to reach one particular emotion (e.g. happiness, softness, sexyness) and because one ingredient usually smells of several odour facets and possesses different behaviours depending on what else is contained in the formula, this approach applies. A Granny Smith apple can be fruity, green, watery and sweet. These are four odour facets. If one takes ethyl 2-methyl butyrate (**2**, Figure 11.1), it also smells apple-like, fruity (on the ripe side, especially if one uses a lot of it), green, sweet, springy, etc. The odour facets of a volatile, expressing themselves in the nose, depend on the nature of that volatile and its concentration. A

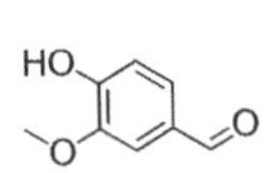

1
(Vanillin)

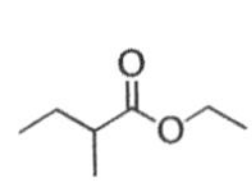

2

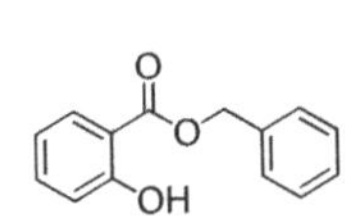

(Benzyl salicylate)

Figure 11.1 *Structures of vanillin (**1**), ethyl 2-methylbutyrate (**2**) and benzyl salicylate*

perfumer has to navigate swiftly among hundreds of ingredients (sometimes more than 2000) to carefully pick the appropriate one to reach the effect he or she wants. He or she then decides on the concentration of that ingredient in the formula – sometimes with a lot of trial and error. Concentrations can vary from 0.0001% to 100%, going through 0.0002%, 0.0003% and so on, meaning through all digits and ranges, not just in multiples of ten. For instance, in a fine fragrance, ethyl 2-methyl butyrate (**2**) smells watery, green and springy at concentrations around 0.01% and smells sticky fruity at concentrations around 0.2%. In a detergent formulation, one may have to multiply these concentrations by a factor of five or ten. A lily of the valley ingredient might be light, watery and not really floral at concentrations around 0.5–1.0% and might be strongly floral at concentrations around 2–10%. Furthermore, the by-odours of each ingredient must also be taken care of: concealed, wrapped, etc. For example some lily of the valley (or peony) notes can smell very plastic, as well as floral. These by-odours are also present in nature. Think about it next time you smell a peony, or a Casablanca lily, and you will see them (like shoe polish for peonies, pepper for freesias or burnt rubber for the large white lilies).

Another example: vanillin (**1**) is a very common and pleasant ingredient. It shows a comforting effect from little to very sweet. At times, depending on what else is inside the fragrance, it also shows a dry-woody and rubbery note. Usually we dislike these latter effects and the perfumer has three choices if vanillin is absolutely needed and its effect is otherwise liked for this fragrance: (*a*) to decrease the amount of vanillin in the formula, (*b*) to remove it entirely, or (*c*) to determine what ingredient(s) clashes with vanillin and either decrease or remove that ingredient.

Given that there are usually several odour patches or several emotions in a perfume, those building blocks are multiplied by two, three, four or even ten. For instance, one may combine a happy top with a mysterious background and touch it up with a lindenblossom note. Lindenblossom, like honeysuckle, lilac or lilly of the valley, does not have natural extracts available and never had in perfumery. So lindenblossom also has to be created by the perfumer like a full hologram. The happy top, the mysterious background and the lindenblossom touch are not to be understood here as top, middle and base notes that would disappear or develop over several hours. Usually, we describe a top note as the part of the scent which evaporates rather quickly (1–30 min), the middle note less quickly (1–2 h) and the base note lasts several hours (or more). Described here are the spatial dimensions that one smells together at each point in time: the intricate volumes that one smells together at the same second. Upon each smelling, the brain sees a foreground and a background in the scent, sometime a hole in, or a hat to, that volume and, sometimes, everything is just flat. A scent does not have to be constructed in a top, middle and base pattern. However, in every case and at every moment, one usually 'sees' volumes in his or her mind, something is in the front of the scent, something in the back, a peak, maybe again a hat or a hole, or an overall feeling of a volume shaped like a round hill.

Only then can one add the time dimension (when one uses a scent over several hours). The scent gets stripped by evaporation, like when one peels an onion. The scent should not collapse or show imperfections during that process (the hologram has to be full, not hollow; the inside of the Rubik's cube has to be clean and robust and not rusty nor collapse when one starts using it). A detergent fragrance must not collapse after being washed by the washing machine or dried by the tumble drier. A candle fragrance cannot start burning; a skin fragrance usually should not collapse, or disappear, after 1 h on skin.

11.3 Volatiles Need a Language Wizard

The general public has a hard time understanding the magic of odour or perfume creation.[4,5] One reason, beside a lack of general education from elementary school onwards, resides in the fact that many of the concepts or know-how associated with it still have no names. It is well known in philosophy that naming a thing, or a phenomenon, helps us to better understand its concept, its existence or its mechanism. A name provides a grasp for the brain. Linguists need to be put at work to help perfumery. We have a name for *viewers* and *spectators*, for *listeners* and *audiences*, but not for '*sniffers*' and *public smelling*: a performance of volatiles in a theatre or scent installations in a gallery or simply fragrances in a store.

This delay in the development of a common vocabulary and of new formulation techniques, as compared to other disciplines, results from the fact that we have not yet discovered the elementary primary notes that would allow the composition of all perceptible scents. Musical notes are very objectively related to specific sound frequencies and we name them precisely. Colours can be decomposed into the so-called three primary colours and we have a very discreet way of naming many of them. One could well imagine a set of 340 or more primary molecules, or molecular pieces or other singular entities, which would form the basis of anything the human nose can smell. Let us call them 'osmonemes', like phonemes in linguistics (but yet to be discovered, confirmed or proven wrong). One possibility is that each of them would excite one, and only one, receptor of the 340 or so different active receptors located in the human nose. Such a discovery would allow us to reach one of the holy grails of perfumery and we would suddenly need to create some 340 new words to describe precisely the entire smelling space.

We would still need more than these 340 molecules, or osmonemes, to create odours and perfumes, because different molecules of similar smells are perceived at different places in a perfume. For instance, different fruity molecules are necessary if a fruity note should stay on the skin for hours. When one wants incense to strike on top, the same ingredient is not used as when the incense needs to strike in the back only. A certain green note harmonizes or clashes with vanilla notes in a way that is different from another green note of the same green type. Different molecules of the same scent offer variations in intensities, which are not always achieved by just increasing or decreasing the scent: one can either dim a bright light using a dimmer or one can shield oneself from it, like at the dentist's. These are two different effects. Furthermore, different molecules of a similar odour are also necessary in the perfumer's catalogue because they usually behave differently depending on the end use: on skin, hair or fabric, in a candle or in a detergent.

Unlike the words 'red' and 'yellow', which were coined specifically for vision, the odour families, which are used to categorize volatile molecules, are named using other disciplines:

- Colour names: the green family, the white flower family (although not all white flowers belong to the so-called 'white flower' family).
- Botanical names: spicy (cardamom, saffron), citrus (lemon, mandarin) or floral families (with main subfamilies: rose, jasmine, lily of the valley and subfamilies of those subfamilies: lilac, honeysuckle, orange flower).
- Geological or geographical places: amber (a fossil), Chypre (an island), Oriental (a region).

- Temperature associations: a cold scent or a warm scent, hot spices (clove, cinnamon) or cold spices (cardamom, ginger).
- Chemical names: families of alcohols, aldehydes (although all aldehydes do not fit into the aldehyde family, odourwise!), acids, acetates and other linear esters, cyclic esters or lactones, pyrazines, ozone, oximes. The chemical classification of odours is also specified with subfamilies such as saturated, hydrogenated, unsaturated, linear, cyclic, etc.

The chemical way of classifying is probably the most consistent of all classifications because it brings us closer to the elementary building blocks of olfaction. The olfactory description of a volatile, an odour or a fragrance is very consistent among people with training, which means perfumers. This is a direct result of the vocabulary used for the chemical units. Perfumers refer to the same common standards when they describe a scent. However, we still have not discovered the odour units themselves, the so-called 'osmonemes'.

Beyond the perfumers, even within fragrance houses, scent descriptions are less secure. This vocabulary – one-third conceptual but vague (e.g. 'amber'), one-third figurative (e.g. 'green') and one-third molecular (e.g. 'acetate') – is obscure and deterrent to the public. Try objectively describing a symphony without knowing the instrument names, without having ever heard each instrument on its own, or without knowing major keys and minor keys. In other disciplines, the raw data (e.g. wavelength of x nm) has been translated into poetic words by the public (red, blue, yellow). In perfumery, we are missing the knowledge of the hardcore elementary units (if they ever existed in that form) and the words used for them.

So, none of the current empirical methods for describing odours is ideal and this brings numerous imperfections, exceptions and overlap within the families. No satisfactory discreet classification has been found to date.

Figure 11.2 shows the 12 main odour families proposed to categorize all volatiles by their odours. We could say that each of these 12 families can then be divided into 11 subfamilies (which are all the other families except itself: fruity floral, green sweet, etc.) then each subfamily can be divided again into subfamilies: fruity floral green, green sweet

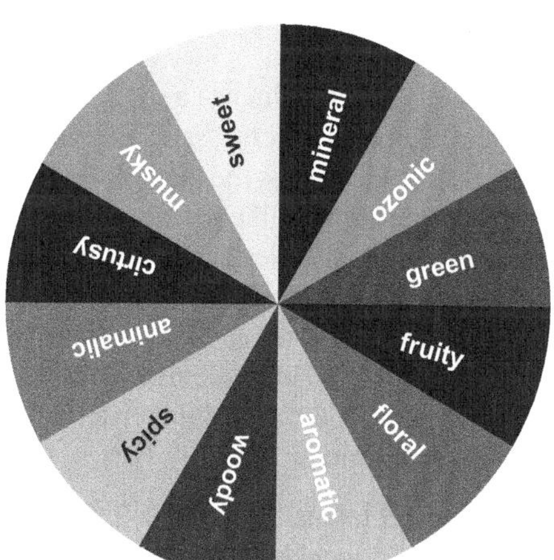

Figure 11.2 *Twelve main odour families used to categorize volatiles*

woody, etc. However, this classification quickly gets very complicated and very imprecise. It is also possible to introduce figurative names, such as rose, jasmine and muguet for the subfamilies of the floral notes, for instance, or cut grass, leafy, artichoke or fir for the green family, yet everything still remains vague unless perfumers talk among themselves.

11.4 Wizardry No. 2: The Perfumer in the Jungle of Volatiles to Create Emotions

A perfumer is someone who masters the behaviour and the perception of volatile molecules by the nose and the brain. For instance, besides scented molecules, surrounding colours or geographies can be important factors for the final perception of a scent. The term 'volatile molecules' applies here in the true general sense: molecules that are able to 'escape' from a substrate and 'fly' (*volare* in latin) into the nose via air currents or winds, breathing or other conscious nose aspirations.

There are four categories of volatiles that need to be considered by the perfumer:

- The ones that smell and are safe. Some may be safe only below a certain limit, as with many vitamins and minerals. As mentioned at the beginning of this chapter, perfumers have tables of levels not to surpass.
- The ones that: (*a*) smell but are not safe even at low concentration, or (*b*) smell but their safe usage level is so low that the human nose would no longer perceive them. Perfumers do not use any of these.
- The ones that smell and could be used but have become extinct for whatever reason (biotope destruction, lack of labour for harvest, loss of traditional agriculture, excessive legislation and rules, bankruptcy of a laboratory, budget plans). Unfortunately, perfumers cannot use these any longer but they always hope to.
- The ones that do not smell: either because they are poorly volatile or because they just do not smell. As strange as it sounds, some of these are important. For instance, either to dilute very potent molecules or to attract, retain (like a glue) or slowly release some other volatile molecules.

In summary, anything that smells or can influence a smell, that is safe and sustainable, is, can be or should be used by the perfumer. The term 'safe' has been the subject of a very large debate over the past 30 years. Before the 1970s, the pendulum swung to one extreme, when almost anything could be used in perfumery, like lead in paint (causing the still persisting belief that today's perfumes create so many allergies). Now we are at the opposite extreme, when natural ingredients and molecules are scrutinized with excess zeal by the authorities, leading to an impoverishment of the world odour and perfume heritage and qualities.

The perfumer nowadays has around 2000 ingredients available on the palette: molecules, naturals and different qualities of each. There are 'eight' criteria that a perfumer uses to pick a volatile molecule and add it to a formula:

- The way it smells in character and intensity.
- The way it smells in character and intensity.
- The way it smells in character and intensity.
- The way it smells in character and intensity.

- The way it smells in the presence of the other perfume ingredients in the formula.
- The way it helps to hide the smell of something else (e.g. to hide an unpleasant odour in the product, a malodour in the environment, or very often an unwanted odour facet from the different facets of another ingredient in the formula). As mentioned earlier, most ingredients, even single molecules, have several odour facets. The perfumer usually needs only one of these facets for a given fragrance, such as the floral note in a peony, and the chemical shoe-polish-like note found naturally in peonies must be concealed. A key natural molecule in jasmine is very fruity and floral but also nail-polish-like. The perfumer interweaves that molecule with other ingredients to mask that side effect.
- The way it is long-lasting upon use (for instance on skin or on laundry).
- The way it ages or does not age: some volatiles are very fragile and do not age well unless they are kept protected from oxygen, light and higher temperatures (e.g. citrus oils), some age well by remaining themselves all along: they are extremely stable and do not change with time, just like an antique pottery buried in dirt (e.g. certain musky or woody molecules); some age very well like a wine (e.g. patchouli oil or rose absolute).

Many factors directly influence these 'eight' criteria:

- Molecular configurations (molecular shape and/or vibrations and maybe other factors yet to be discovered). This includes size, branches, saturations and several kinds of isomerism. The human nose can smell a lot, to everyone's surprise or despair!
- Molecular functions (alcohols, aldehydes, acids, esters, nitriles, oximes, amines, etc.).
- Geographical origins for naturals (including climate, soils, light, varietals, etc.).
- Extraction methods for the naturals: alembic for steam distillation to obtain oils, fat or solvent extractions to obtain absolutes, liquid CO_2 and hydrofluorocarbon extractions, tinctures (like cold macerations) or infusions (like making a hot tea in alcohol), etc.
- The part of a plant which is extracted (roots, leaves, petals, etc.).
- The synthesis routes used in different laboratories: different routes to create the same molecule can result in the presence of different impurities (like different shades) which smell and behave differently.
- The way those molecules are presented: either pure, diluted in a solvent or trapped in a release system such as Schiff bases, Michael adducts or special esters (which act like a glue), cyclodextrins or zeolites (which act like a buoy or a sponge), encapsulates (which act like invisible mini egg shells ready to be broken to release a scent).

One of the principles governing our universe is called the second principle of thermodynamics. It says that the universe keeps on changing to move to its most disorderly state, which means exactly towards its most stable state. One drop of milk in coffee, before mixing, is order but it is not stable. After mixing, the café au lait is stable and will never return back to a cup of coffee with a drop of milk in it, but is in disorder. Everything is homogenous and all the molecules are mixed up, and the properties of the café au lait are different than those of the coffee alone or the milk alone. If one is familiar with entropy, another way to put it is to say that the entropy of the universe keeps on increasing. Perfumery is no exception to the laws of the universe: when adding matter (an ingredient) to a material disorderly system (the rest of the fragrance mixture), the state of that ingredient, including its volatility, depends on the natures, levels and energy states of all the other ingredients

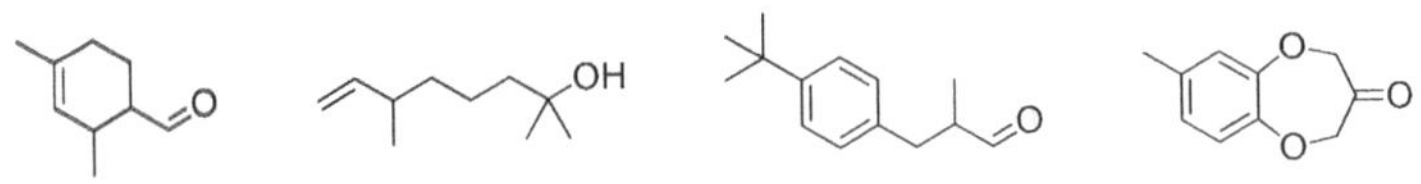

3 (Triplal®) 4 (Dihydromyrcenol) 5 (Lilial®) 6 (Calone®)

Figure 11.3 *Structures of Triplal (**3**), dihydromyrcenol (**4**), Lilial (**5**) and Calone (**6**)*

already in the mixture. Calculating real chemical potentials in real life perfumery is a nightmare (if even possible today) versus calculating ideal potentials, in the case of idealized modulated systems not representing the reality of perfumery as we can regularly smell. This influences the immediate rendition and the evolution over time of the fragrance (on skin or left alone in a bottle). I suspect that an interdependence, either between fragrance molecules, or between fragrance molecules and receptors, or between receptors, also affects the perception of different scents in the nose and in ways not yet explained. As a net result one can state that each perfume is different and it is impossible for a perfumer to predict with precision what the addition of a certain ingredient, at a certain dosage, is going to create or influence in the perfume. There are still no computer programs available to guide this process and surprises are happening every day. It is a trial and error process. Of course, the expertise of the perfumer helps the end result to be reached faster and influences the quality of that end result.

As an illustration of this process (and of the eternal despair of the perfumer), some common perfumery ingredients – patchouli oil, vanillin (**1**), Triplal (**3**), dihydromyrcenol (**4**), Lilial (**5**) and Calone (**6**; Figure 11.3) – were mixed in a 1 : 1 ratio (by weight) in a vial and the resulting scents are described in Table 11.2. As smelling the pure ingredients is usually too strong for a fair evaluation (like a spotlight shining into your eyes when you are not used to it), the ingredients were first diluted in alcohol (ethanol) in the weight percentages indicated (1%, 0.1% or 0.01%, depending on the strength of the ingredient). Evaluations were made by dipping the tips of smelling strips (made of blotter paper) into the vials, letting the alcohol evaporate for a minute and smelling gently by breathing normally. This was done by mixing only two ingredients at a time. A finished fragrance may contain anywhere from 15 to 80 ingredients, some being single molecules, others being naturals which are themselves no more and no less than a combination of 50–300 molecules. So the complexity can be appreciated here. Note that one would obtain drastically different results if, instead of blotter paper, one was to evaluate on skin, in a detergent or in a candle. This means that for each category a different table must be created.

One can flip through a book or a picture collection on a computer within seconds and see hundreds of photos – this is impossible with scents. The retention time of volatiles in the nose lasts from half a second (often green and citrus notes of terpene structures) to several minutes (certain woody or musky notes) before the molecules are evacuated again by respiration or degraded by enzymes in the nose. The intake is, in any case, limited by one's own respiration, which is also counted in several seconds.

This is extremely long compared to the intake of images and sounds. The physical perception of one scent by your nose and your brain depends on what other scent(s) was smelt in the minutes before and which is still lingering inside the nose. The musical

Table 11.2 *Olfactive profiles of 1 : 1 mixtures (by weight) of selected perfumery ingredients*

Mixtures	Patchouli oil (Indonesian iron-free)	Patchouli oil (Indonesian iron free)	Triplal (**3**)	Triplal (**3**)	dihydromyrcenol (**4**)	dihydromyrcenol (**4**)
50 : 50 (by weight)	1.0%	0.1%	0.1%	0.01%	1.0%	0.1%
Vanillin (**1**) ex. lignin 0.1%	Patchouli softened but muffled	Chalky vanilla and dry, a little patchouli	Green chalky dry	Very green springy, nice sweet, Dior Addict/Alexander McQueen type	Functional sweetness, technical	Very nice fine fragrance-like, but after 30 min is no longer nice
Vanillin (**1**) ex. lignin 0.1%	Nice patchouli-, opium-like character, more hay as it dries down	Dry dusty patchouli, not pleasant, somewhat cocoa in the back	Crunchy green but a little oily	Very green with dry chalky woody bad vanilla in the back, after 20 min only dry vanilla is left	Soft woody dihydromyrcenol, still bitter functional though	Soft woody dihydromyrcenol, still functional, but more friendly
Lilial (**5**) 1%	Dry functional dihydromyrcenol-like/terpenic functional citrus effect with woody undertones	Like a poor very diluted Gucci Rush, no bloom, no personality	Floral nice muguet, a bit withered though, and muffled, flat	Floral nice muguet, wet but too flat and oily	Functional lemon, slightly floral in the back	Functional rough woody, scratchy
Lilial (**5**) 0.1%	Green floral patchouli, slightly functional citrus terpenic	Nice elegant floral fresh on top with clean patchouli in the back	Nice floral fresh, slightly oily, like lilies	Fresh but too oily	Floral fresh, but functional on top and flat muffled overall	Functional floor cleaner
Calone (**6**) 1%	Nice wet and floral springy patchouli	Wet, sea-like, ozonic, clean patchouli, dry blond wood effect in the back	Withered flowers with oily olive-oil effect	Nice watery sea breeze and leafy springy	Functional dry, turpentine-like	Slightly algae, marine very wet, slightly functional
Calone (**6**) 1%	Very woody and dark green	Nice wet patchouli	Green and wet but does not bloom	Nice watery green, wet leaves, springy hyacinth-type	Very wet but flat, muffled	Citrus-leaf note, a little artificial but breathes nicely

equivalent is playing a C on the piano and, depending on the note you played before, the C could sound like a C#, a B or even an A at times! Table 11.2 was completed over several days to try to isolate and evaluate each scent on its own; you would get slightly different evaluations depending on which order you navigate through the table. Furthermore, the personal interpretation for each scent, personal impressions or emotional reactions, depends on the context of the previous exposures in your own personal past (pleasant or unpleasant), general education and memories (starting from four months in your mother's womb when your nose first opens, up to today) and, of course, training and experience.

A perfumer is someone who knows how to juggle with all these factors, aspects and data of appreciations. Getting a comfortable grasp of these ingredients and the parameters influencing them, with a working feeling of their numerous and precise dosages depending on the sensations you want to create, requires five to ten years of experience to feel at ease. No-one can sing them for you, no-one can describe for you what they smell like and how they behave. You have to smell them for yourself, and in combinations, in order for your brain to register the conclusions.[4]

The perfumer also has to learn how to detach himself from the name of the ingredients to create emotions. Often, the emotions do not have much to do with the actual ingredient because we do not have neutral names for ingredients. To describe the softness in common in coconut, gardenia or cream, the technical word 'lactonic' might be used. Yet, lactones can smell either very soft creamy milky but also very peachy or very musky. Red can mean something from neutral to joyful, masculine, feminine or dangerous. People usually do not roll their eyes *a priori* upon hearing 'red'. Perfumers do not have neutral names such as *yellow*, *warm yellow* or *light bright yellow*. They are often only allowed to say *lemon*, *freesia*, *primrose*, *banana* or *pineapple*, all yellow, but all biased. If one says 'lemon', or 'coconut', most people in perfumery will roll their eyes, as these scents are very common, almost not luxurious enough. If one says 'patchouli', half of the people will be transported some place, sometimes where they do not want to go – to the 1970s! A real perfumer will say: 'wait, it depends on which kind and how it is used'. People do not know that around 20% of the Paris 1738 stench created for the book and movie '*Perfume, the story of a murderer*' is based on a jasmine reconstitution.[5] How many perfumers will tell you that a big reason why they do not like to open up formulas to customers is that, once people see actual ingredients, they do not like the fragrance any longer? Not because the ingredients are of lesser quality than they thought, but because they realize their fragrance has some ingredients they hate. But they hate them because of their names, or their smells as pure ingredients, which has nothing to do with what they accomplish or how they appear in the fragrance. Patchouli is a key ingredient in perfumery. Who thinks patchouli when wearing Coco Mademoiselle or Narcisso Rodriguez? For that matter Blue Lenor or Green Downy? L'Eau D'Issey without its woods and clove is nothing. Who thinks a rose note is the key for the smell and the intensity of CkOne? If you were to use the term 'rose', chances are that it would kill the creation in question because the name (not the smell which can be very useful, modern, strong and not even recognisable under many different forms) brings with it a lot of baggage, tradition or déjà vu. A perfumer concentrates on the emotions and the ingredients equally, but has learnt how to dissociate one from the other. It is a very complicated gymnastic: smelling without looking or without listening unconsciously to the name.[6]

11.5 Wizardry No. 3: End Results Are Music to the Nose

All these volatile ingredients are trapped in many individual bottles surrounding the perfumer's compounding desk. The setting looks like a perfume organ or a perfume piano with a keyboard of more than 1000 keys. Each bottle of a raw ingredient is already a world in itself, each like a genie trapped in an Aladdin's lamp: each ingredient has always several facets, it is much more than a pot of one colour. No-one has ever been able to pack and carry around in a thorough manner this perfume orchestra. Reducing it to 200–300 ingredients is very limiting. Thus, a perfumer who is creating for a lot of different projects can never travel with his work instrument in its entirety. With time, excellent perfumers have developed (in order to survive and to do all the things they need to do) the ability 'to create at a distance'. This is not taught at school but should be and it is a very unique juggling exercise. Before leaving the office, the perfumer remembers the last version of a fragrance in its construction phase. While on a trip, the perfumer is given some feedback on the phone. He or she then has to modify the formula in his or her mind, trying to predict how it is going to smell, although, as explained earlier, this is a perilous exercise. From the comments of his or her colleagues at work smelling the new trials or from the comments of the client smelling the new work, the perfumer steers the modification of the fragrance at a distance without smelling it. This can go on for several weeks before the perfumer is back in the office and smells where the scent stands, or before a sample reaches him or her on the trip. The keys to this exercise are: (*a*) know your fragrance ingredients like the back of your hand, even the very advanced ones that are rarely used or very sensitive and (*b*) have a strong ability to listen and interpret what the smelling person is trying to tell you. This is a real skill when you remember that there is no standard vocabulary among clients or the public. Hopefully, one day, perfumers will have 'creative and recording studios' available at many places around the world.

In the fascinating book '*Musicophilia*',[7] – Sacks explains how '*music, which has no necessary relation to the world, [...] has great power [...]. This propensity to music – this "musicophilia" – shows itself in infancy, is manifest and central in every culture, and probably goes back to the very beginning of our species. It may be developed or shaped by the cultures we live in, by the circumstances of life or by the particular gifts or weaknesses we have as individuals – but it lies so deep in human nature that one is tempted to think of it as innate.*'

No-one today questions the 'usefulness' of music, its safety or its greatness. It is amazing to see how this could be translated word for word to the world of perfumery. Hopefully, one day, someone will be able to transpose the word 'music' with 'perfumery' and with all the corresponding specificities – the academic advancements of olfactory science and perfume art will be equivalent to those of the music world.

And he goes on by citing the musician Pinker: '*As far as biological cause and effect are concerned, music is useless... many of the arts [...] may be byproducts of two other traits: motivational systems that give us pleasure when we experience signals that correlate with adaptive outcomes (safety, sex, esteem, information-rich environments), and the technological know-how to create purified and concentrated doses for these signals.*' How incredible it is to see the equivalence in music of concentrated natural essences and pure molecules? The concentrated forms, the extreme pure forms and intensities (of a musical note or of a molecule) have allowed artists and designers to reach new grounds.

The hormonal rush felt while listening to music, or smelling a scent, does not just bring a certain kind of relaxation or excitement, albeit all beneficial biologically, physically or mentally. Smelling, the magic of volatiles, also constitutes good aerobics for the brain.[8] One can feel the brain at work when one trains to become a perfumer, or simply when one performs wine-tasting on a regular basis. Smelling most likely creates new neurological connections in the brain, like when you learn a new foreign language. I believe that 'fragrançophilia' exists too, and originated even earlier than music. On our brand new (almost silent) planet, the first tiny bacteria communicated with volatile molecules, for food and to procreate, to find each other or to repulse enemies. Just as we do not question whether music originated from mountain streams and arousing birdsongs, I am looking forward to the times when perfumery and its magic will be a given and simply enjoyed. Smelling sessions, games or salons – perfumery for perfumery's sake – existed in ancient Rome, in ancient India,[9] in traditional Japan and in the Maya kingdom[10] even though the logistic, the ingredients and the extraction techniques they had at the time were quite rudimentary. We are at the beginning of a grand resurrection.

New volatile notes are being developed every year. Hopefully, fewer and fewer will disappear due to negligence or laxness, thanks to newfound public recognition for cultural and human heritage purposes. In parallel, besides the usual alcohol and candles, we need more and more scent instruments to be designed, the scent piano by de Cupere (the 'Olfactiano'), or the scent organ by Fläkt-Woods (the Yamaha or Steinbeck of this world in scent pianos), to be able to play all these notes and ultimately these emotions. Some very special logistics have to be put in place from production to supply, creative studios, dispatch and International Air Transport Association (IATA) rules, in order to allow all this to flourish.

Nature opened the way, as Adorno and Horkhiemer were saying,[1] circumambient molecules have always imposed us, we have to let go, enjoy and elevate the whole discipline to a fully fledged art form *now*.

References

1. T. W. Adorno, M. Horkheimer, *Dialektik der Aufklärung*, Querido Verlag, Amsterdam, 1947.
2. (*a*) E. Roudnistka, *Le Parfum, Que-sais-je?*, Presses Universitaires de France, Paris, 1987; (*b*) E. Roudnistka E. Souriau, *L'Esthetique en Question*, Presses Universitaires de France, Paris, 1977; (*c*) A. Gilbert, *What the Nose Knows*, Random House, 2008; (*d*) R. Herz, *The Scent of Desire: Discovering Our Enigmatic Sense of Smell*, William Morrow/HarperCollins Publishers, New York, 2008.
3. See for example: (*a*) J. A. Gogos, J. Osborne, A. Nemes, M. Mendelsohn, R. Axel, Genetic ablation and restoration of the olfactory topographic map, *Cell*, **103**, 609–620 (2000); (*b*) L. Astic, D. Saucier, Neuronal plasticity and regeneration in the olfactory system of mammals: morphological and functional recovery following olfactory bulb differentiation, *Cell. Mol. Life Sci.*, **58**, 538–545 (2001).
4. For further description of the creative processes leading to a finished fragrance, see: (*a*) C. Laudamiel, Creative processes in perfumery, in *Stakes & Professions in Perfumery*, Editions d'Assalit, Toulouse, 2007, pp. 97–103; (*b*) C. Laudamiel, Haute perfumery and haute cuisine, in *The Big Fat Duck Cookbook*, H. Blumenthal, Bloomsbury Publishing, London, 2008, pp. 478–482.

5. C. Laudamiel, C. Hornetz, B. D. Mookherjee, S. Patel, Form virgin education to real education, *Chem. Biodivers.*, **5**, 1159–1169 (2008).
6. J. Lehrer, *Proust Was a Neuroscientist*, Houghton Mifflin Books, New York, 2007.
7. O. Sacks, *Musicophilia: Tales of Music and the Brain*, A. A. Knopf Inc., Toronto, 2007.
8. L. C. Katz, M. Rubin, *Keep Your Brain Alive: 83 Neurobic Exercises to Help Prevent Memory Loss and Increase Mental Fitness*, Workman Publishing Company, New York, 1999.
9. J. McHugh, *Sandalwood and Carrion*, PhD thesis, Department of Sanskrit and Indian Studies, Harvard University, 2008.
10. A. Craig Copetas, *Davos Aromas Deodorize Subprime Stench, Charm Dimon, Kissinger*, Bloomberg News, New York, 17 January 2008.

12

Microencapsulation Techniques for Food Flavour

Youngjae Byun,[a] Young Teck Kim,[a] Kashappa Goud H. Desai[b] and Hyun Jin Park[*b]*

[a]Department of Packaging Science, Clemson University, Clemson, SC 29634-0320, USA
[b]Department of Food Technology, 306 School of Life Science and Biotechnology, Korea University, Seoul, 136-701, Korea

12.1 Demands

Food flavour is a significant factor for the acceptability and quality of food by taste and smell. There is increasing interest in the flavour stability of foods, but it is not easy to control. The most delicate flavours are volatile substances. Manufacturing, packaging, or storage can cause losses or modifications in overall flavours by evaporation, oxidation, exposure to light, or ingredient interactions. Therefore, there is increasing demand for the encapsulation of food flavours prior to use.[1–4]

12.2 Microencapsulation in the Food Industry

Microencapsulation is defined as the technology of packaging active materials within another material. This technique is well developed and accepted within the agricultural, pharmaceutical, chemical, cosmetic, oil, print and food industries.[5–8] Microencapsulation techniques in the food industry are more focused on the protection of flavours and spices.[9–14] Microencapsulation of food flavours results in a longer shelf life of products because of an increased resistance to oxidation and volatilization.

The microencapsulation of flavours has been widely used in the food industry (e.g. soups, beverages, cookies, chewing gums, fish coatings, breath fresheners) for several

*Note: corresponding author.

The Chemistry and Biology of Volatiles Edited by Andreas Herrmann

reasons: (*a*) it protects the core material from the outside environment, (*b*) it decreases or slows evaporation of the flavour into the outside environment, (*c*) it modifies the physical characteristics of the original material to make it easier to handle, (*d*) it allows a modulated perception resulting from a controlled release of the flavour and (*e*) it separates components within a mixture that would otherwise react with one another.[10–13]

12.3 Techniques and Materials for Flavour Microencapsulation

Various techniques are employed to form flavour microcapsules, including spray drying, formation of inclusion complexes, extrusion, coacervation, co-crystallization, fluidized bed coating, spray chilling, spray cooling and supercritical fluids (Tables 12.1 and 12.2).[1,3,15–53] Among these techniques, the main commercial microencapsulation techniques used in the food industry are spray drying and extrusion.

The processes for selecting an appropriate microencapsulation method and suitable coating materials are interdependent.[3,4,11] Based on the coating material or method applied, the appropriate method or coating material is selected. Selection of a coating material depends on the active core material and the desired properties of the microcapsules. Cost considerations in the food industry are much more stringent than in the pharmaceutical or cosmetic industries.

The composition of the coating material is the main determinant of the functional properties of the microcapsule. Generally, flavours are encapsulated within a solid matrix such as food-approved carbohydrate, protein, sugar, or fat.[1,2,14,54] An ideal coating material for flavour encapsulation should exhibit the following characteristics:[11] it should (*a*) disperse or emulsify the active material and stabilize the emulsion produced, (*b*) be nonreactive with the material to be encapsulated, (*c*) hold the active material within its structure, (*d*) provide maximum protection to the active material against environmental conditions, (*e*) be soluble in solvents acceptable in the food industry (e.g. water, ethanol) and (*f*) be inexpensive and have food grade status.

Because no single coating material can meet all the criteria listed above, in practice either coating materials are employed in combinations, or modifiers such as oxygen scavengers, antioxidants, chelating agents, or surfactants are added. Some commonly used biocompatible and food grade coating materials are listed in Table 12.3.[2,35,54–65] Chemical modifications of the existing coating materials to manipulate their properties are also being considered. These modified coating materials exhibit better physical and mechanical properties when compared to individual coating materials.[11]

The common encapsulation techniques for flavour compounds and various wall materials for each technique are described below and summarized in Tables 12.1–12.3.

12.3.1 Spray Drying

Spray drying encapsulation is the oldest and most widely used microencapsulation technique in the food industry; and it protects flavour compounds against degradation and oxidation.[6,66,67] The processing costs of spray drying are lower than that of most other microencapsulation techniques. For instance, the cost of spray drying is 30–50 times cheaper than freeze drying.[68,69] In addition, this technique has several other advantages

Table 12.1 *Flavour microencapsulation techniques and the processes involved in each*

Microencapsulation technique	Major steps in microencapsulation	References
Spray drying	1. Preparation of the emulsion 2. Homogenization of the emulsion 3. Atomization of the emulsion 4. Dehydration of the atomized particles	Chattopadhyays *et al.*,[15] Krishnan *et al.*,[16] Varavinit *et al.*,[17] Murúa-Pagola *et al.*[18]
Extrusion	1. Preparation of a molten coating solution 2. Adding of the flavour compound into the molten polymer 3. Cooling or passing of the mixture through a dehydration liquid	Miller and Mutka,[19] Porzio and Popplewell,[20] Reichert *et al.*,[21] Saleeb and Pickup[22]
Cyclodextrin inclusion complexes	1. Preparation of cyclodextrin solution, paste or slurry 2. Adding of flavor compound into the solution, paste or slurry	Gornas *et al.*,[23] Li *et al.*,[24] Liu *et al.*,[25] Reineccius *et al.*,[26] Yoshi *et al.*[27,28]
Helical inclusion complexes	1. Preparation of starch solution, paste or slurry 2. Adding of flavor compound into the solution, paste or slurry	Godet *et al.*,[29] Heinemann *et al.*,[30,31] Itthisoponkul *et al.*,[32] Nuessli *et al.*[33,34]
Coacervation	1. Formation of immiscible chemical phase 2. Deposition of the coating 3. Solidification of the coating	Li *et al.*,[24] Daniels and Mittermaier,[35] Dong *et al.*,[36] Leclercq *et al.*,[37] Palmieri *et al.*[38]
Emulsion with freeze drying	1. Preparation of the emulsion 2. Homogenization of the emulsion 3. Dehydration of the emulsion by freeze drying	Cho and Park,[39] Choi *et al.*,[40] Flink and Gejl-Hansen,[41] Kaushik and Roos[42]
Co-crystallization	1. Preparation of supersaturated sucrose solution 2. Adding of flavour compound into the solution 3. Emission of substantial heat after the solution reaches the sucrose crystallization temperature	Beristain *et al.*,[43] Bhandari *et al.*,[44] Chen *et al.*,[45,46] Miller and Graham[47]
Spray chilling or spray cooling	1. Preparation of the emulsion 2. Homogenization of the emulsion 3. Atomization of the emulsion	Madene *et al.*,[1] Taylor,[3] Blenford,[48] Lamb[49]
Rapid expansion of supercritical solutions (RESS)	1. Preparation of supercritical fluids 2. Flavour compound is dissolved in the supercritical fluid 3. Depressurizing the solution	Jung and Perrut,[50] Pillips and Stella,[51] Shariati and Peters[52]
Gas (or supercritical fluids) anti-solvent (GAS)	1. Flavour compound is dissolved in organic solvent 2. Supercritical fluid is introduced in the vessel 3. The solution is drained 4. Wash the precipitated particles	Jung and Perrut,[50] Shariati and Peters,[52] Cocero *et al.*[53]
Particles from gas-saturated solutions (PGSS)	1. Flavour compound is suspended in a liquid carrier 2. Mixing of supercritical solvent in the solution 3. Depressurizing the solution	Jung and Perrut,[50] Shariati and Peters,[52] Cocero *et al.*[53]

Table 12.2 *Flavour microencapsulation techniques: advantages and drawbacks*

Microencapsulation technique	Cost	Load [%]	Particle size [μm]	Storage stability	Facility to scale-up
Spray drying	Low	Medium	10–100	High	High
Extrusion	Medium	Medium	200–500	High	High
Cyclodextrin inclusion complexes	High	Low	5–50	Low	Low
Helical inclusion complexes	Low	Low	1–40	Low	Low
Fluidized bed coating	Medium	Low	5–100	Medium	High
Coacervation	High	Medium	50–100	Low	Low
Emulsion with freeze drying	High	High	1–300	Medium	Low
Co-crystallization	Low	High	200–300	Medium	Medium
Spray chilling or spray cooling	Low	Low	20–200	Medium	High
Supercritical fluids	High	Low	100	Medium	High

such as higher retention of flavour compounds, good stability of microcapsules, and large scale production in a continuous process.[1,2] There is, however, one disadvantage of using spray drying for flavour microencapsulation; it needs additional agglomeration processes.

Shahidi and Han suggested that microencapsulation by spray drying involves four stages: preparation of the dispersion or emulsion, homogenization of the dispersion, atomization of the emulsion, and dehydration of the atomized particles.[13] In the detailed method, the material for encapsulation is homogenized with the carrier material during the first stage. The mixture is then fed into a spray dryer and atomized with a nozzle. Water is evaporated by the hot air contacting the atomized material. The microcapsules are collected in the bottom of the drier.[3]

It is important to prevent flavour loss during spray drying by hot air. Generally, the retention of flavour in microcapsules is manipulated by varying the spray drying conditions and the composition of the wall material. The main operation conditions in spray drying are feed temperature and air inlet temperature.[69,70] Feed temperature affects the viscosity of the solution. When the feed temperature is increased, the viscosity of the mixture decreases, resulting in a smaller final product size. A feed temperature that is too high, however, can result in the degradation of flavours. Air inlet temperature affects the drying rate and thus

Table 12.3 *Various wall materials used in microencapsulation techniques*

Category	Wall materials	References
Carbohydrates	Starch, modified starch, maltodextrin, dextran, cyclodextrin, corn syrup solids, gum arabic or acacia gum, carboxymethylcellulose, modified cellulose	Reineccius,[2] Godshall,[54] Bhandari *et al.*,[55] Greener and Fennema,[56] Rosenberg *et al.*,[57] Shiga *et al.*,[58] Yoshii *et al.*[59]
Proteins	Wey protein, soy protein, gluten, sodium caseinate, casein, gelatin, albumin	Baranauskiene *et al.*,[60] Ono,[61] Rosenberg and Sheu[62]
Lipids	Wax, paraffin, beeswax, diacylglycerols, oils, fats	Daniels and Mittermaier,[35] Kamper and Fennema,[63] Kim and Bainau,[64] Popplewell and Porzio[65]

controls the final water content of the microcapsules. When the air inlet temperature is decreased, the drying rate is decreased, resulting in a final product with a high water content. An air inlet temperature that is too high, however, can cause intense evaporation and result in the cracking of microcapsules and a degradation or loss of flavour.[69,71]

Generally, gum arabic, maltodextrins, modified starch, whey proteins, soy proteins, alginate, carboxymethylcellulose, guar gum and mixtures thereof are used as a wall material for spray drying (Table 12.4).[15–18,57–60,62,72–77] Among those wall materials, gum arabic has been widely used in spray drying over the past few decades. It produces stable emulsions with most flavours; for instance, spray drying using gum arabic as the wall material provides very good retention of citral (**1**) and linalyl acetate (**2**, Figure 12.1).[57] Krishnan *et al.* reported that gum arabic is a better wall material for the encapsulation of cardamom oleoresin as compared to maltodextrin and octenyl-succinylated waxy maize starch.[16] The shortage of gum arabic in the 1970s, however, caused its price to increase. For that reason, there is a lot of research aimed at developing alternative and inexpensive polymers that could replace gum arabic.[78]

A potential substitute for gum arabic that can be used as an encapsulating wall material is mesquite gum.[72,73] A blend of 60% of gum arabic and of 40% mesquite gum resulted in the encapsulation of orange peel oil with 93.5% retention.[79] Another mixture consisting of 40% of mesquite gum and 60% of maltodextrins was capable of retaining 84.6% of the starting oil.[74] Cardamom-based oil microcapsules were successfully produced using mesquite gum and showed 83.6% retention.[75] These results confirm that mesquite gum can encapsulate flavour well and should be considered as an alternative wall material in spray drying.

Proteins have also been reported as a potential substitute for gum arabic in spray drying. Rosenberg and Sheu demonstrated the use of whey protein isolate and a mixture of whey protein isolate and lactose as wall materials for the encapsulation of ethyl butyrate (**3**) and ethyl caprylate (**4**, Figure 12.1).[62] Ester retention in a mixture of whey protein isolate and

Table 12.4 *Various wall materials are used in the microencapsulation of food flavours by spray drying*

Wall materials	References
Acetylated and phophorlyated waxy maize starches	Murúa-Pagola *et al.*[18]
Gellan gum	Cho and Park[76]
Gum Arabic	Chattopadhyaya *et al.*,[15] Krishnan *et al.*,[16] Rosenberg *et al.*,[57] Baranauskiene *et al.*,[60] Cho and Park[76]
N-Lok, a modified food starch	Murúa-Pagola *et al.*,[18] Cho and Park,[76] Finney *et al.*[77]
Maltodextrin	Krishnan *et al.*,[16] Shiga *et al.*,[58] Yoshii *et al.*,[59] Beristain *et al.*,[74] Cho and Park[76]
Mesquite gum	Goycoolea *et al.*,[72] Beristain and Vernon-Carter,[73] Beristain *et al.*,[74] Beristain *et al.*[75]
Octenyl succinylated waxy maize starch	Krishnan *et al.*,[16] Murúa-Pagola *et al.*[18]
Oxidized starch	Chattopadhyaya *et al.*[15]
Sago and tapioca starch stearates	Varavinit *et al.*[17]
Skimmed milk powder	Baranauskiene *et al.*[60]
Whey protein isolate	Baranauskiene *et al.*,[60] Rosenberg and Sheu[62]

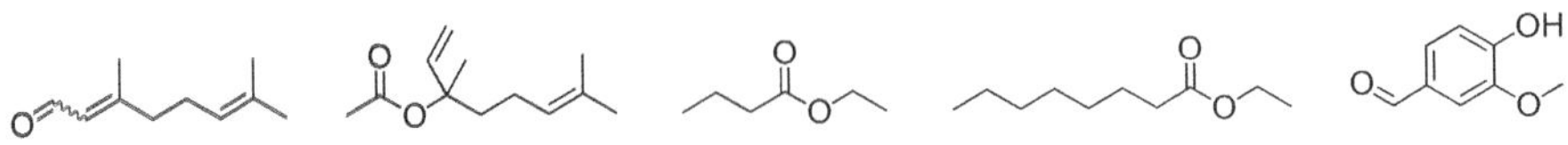

1 (Citral) 2 (Linalyl acetate) 3 (Ethyl butyrate) 4 (Ethyl caprylate) 5 (Vanillin)

Figure 12.1 *Chemical structures of compounds encapsulated by spray drying*

lactose was higher than in whey protein isolate alone. Baranauskiene *et al.* compared encapsulating properties of skimmed milk powder (SMP) and whey protein isolate concentrate (WPC) for the encapsulation of oregano essential oil (EO), marjoram and citronella aroma extracts (AE).[60] SMP retained from 80.6% (citronella AE) to 84.7% (majoram AE) of total flavour content, while WPC retained from 73.2% (oregano EO) to 81.0% (citronella AE) of total flavour content.

Modified starch has been reported to be the most interesting alternative to gum arabic. The native starch has little affinity for hydrophobic flavour compounds due to its hydrophilic properties; however, when starch molecules are modified appropriately, they contain both hydrophilic and hydrophobic properties. Chattopadhyaya *et al.* demonstrated the use of oxidized starch as a substitute for gum arabic in the encapsulation of vanillin (**5**, Figure 12.1).[15] Encapsulation efficiencies using gum arabic, oxidized corn starch and oxidized amaranth starch were 57.8, 60.9 and 58.6%, respectively. Varavinit *et al.* demonstrated the use of sago and tapioca starch stearates for flavour encapsulation.[17] A higher degree of stearic acid substitution resulted in a higher encapsulation efficiency. Murúa-Pagola *et al.* noted that the modified starches prepared using reactive extrusion showed good potential for use as wall materials for encapsulation by spray drying.[18] They prepared acetylated, *n*-octenylsuccinylated and phophorylated waxy maize starches, using reactive extrusion in a single-screw extruder. The *n*-octenylsuccinylated starch exhibited good emulsifying capabilities and total oil retention of 94.8%. Starch phosphate and acetates, respectively, had total oil retention rates of 55.8 and 61.3%.

12.3.2 Extrusion

Extrusion is the second largest production method of flavour microcapsules in the food and flavour industries. It provides a very long shelf life for encapsulated flavours because the core material is completely enclosed within molten carbohydrate. The microencapsulation of flavour compounds by extrusion is depicted in Scheme 12.1. In general for this technique the molten carbohydrate is prepared by heating and then the flavour compounds, along with an emulsifier, are added to a sealed reactor. The mixture is extruded into cold isopropanol to form an encapsulating matrix. Next, the extruded filaments are dried and sized for desired extent.[19] Swisher first introduced the basic extrusion technique.[80] After his introduction, many patents have been obtained on flavour encapsulation by extrusion.[19,20,65,81] Double capillary,[82] multi-orifice centrifugal[83] and recycling centrifugal[84] extrusion have been described as extrusion microencapsulation techniques. In double capillary extrusion, the core substance and the wall material are fed through the inner and outer of a coaxial double capillary, respectively. The two fluids form a unified flow at the top of the coaxial nozzle to form microparticles.[5] The mechanism of multi-orifice centrifugal extrusion is similar to that of coaxial capillary extrusion. The wall material

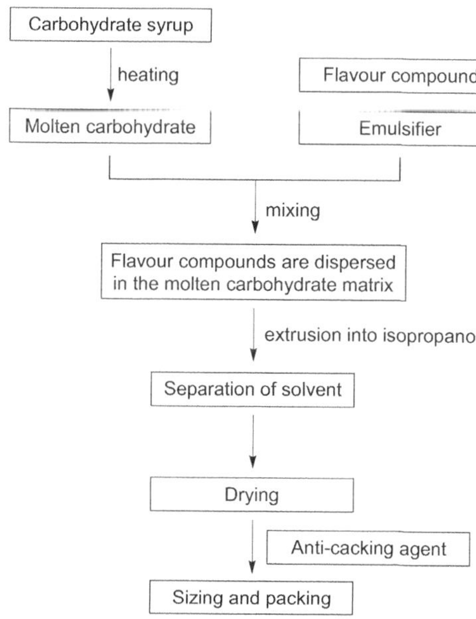

Scheme 12.1 *Flow diagram for the microencapsulation of flavours by extrusion*

solution is fed through the groove above and below the centrifugal rotating disc. The rotating disc pushes the core substance into the wall fluid at the centre of the orifice to form a microparticle. In recycling centrifugal extrusion, the suspension is extruded through the rotating disc in the usual way but the excess coating fluid is atomized and separated from the coated particles.

Orange oil flavour compounds were combined and blended with a sugar and starch hydrolysate mixture to form a homogenous melt.[19] The melt was then extruded into a cool solvent. Next, it was dried and combined with an anti-caking agent to produce the stable product. During the process, encapsulation efficiency was maintained at 60%. Cooking temperature for the process was maintained at or below a maximum of about 126 °C. Saleeb and Pickup utilized an extruder as a melting system to form the carrier melt in a continuous process.[22] A matrix composition consisted of mono- and disaccharides, corn syrup solids, organic acid and maltodextrin. The matrix base was then blended with flavour compounds and melted in a single screw extruder to yield a solid matrix.

Levine *et al.* utilized a composition consisting of a modified starch, maltodextrin, polyol and mono- and disaccharide components.[85] The balance of the composition comprised 10–40% maltodextrin, 5–20% corn syrup solids and 5–20% mono- or disaccharides. The resultant matrices are useful for encapsulating the flavour compounds. Porzio and Popplewell described the preparation of carbohydrate-based glassy matrices, which are stable in the glass state at ambient temperature, by using aqueous plasticizers with melt extrusion.[20] The glassy matrix was comprised of maltodextrin, corn syrup solids, organic acid, mono- or disaccharide, polyhydric alcohol, or sodium octenylsuccinate modified starch. The flavour compounds were then mixed with a glassy matrix in the presence of a plasticizer in an extruder to obtain a melted matrix.

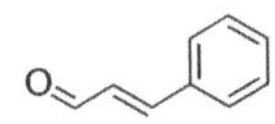

6
(Cinnamic aldehyde)

Figure 12.2 *Chemical structure of cinnamic aldehyde (**6**)*

Reichart *et al.* reported that a flavour encapsulated in a glassy carbohydrate matrix composed of polydextrose and lactitiol eliminates the need for adding water or other plasticizers in the extrusion.[21] When the matrix contains less than 3.5% water, it remains stable when stored at room temperature. McIver *et al.* utilized a matrix composition consisting of a maltodextrin, sucrose, or hydrogenated starch hydrolysate with from 1 to 7% prehydrated agar-agar for the encapsulation of cinnamic aldehyde (**6**, Figure 12.2) and orange oil.[86] The system is particularly stable in aqueous environments and is capable of providing the controlled release of flavour compounds.

The extrusion process has the following advantages:[12,87]

- It is a continuous and simple process.
- It uses minimal water, so drying conditions are greatly simplified.
- Flavour compounds are completely surrounded by the wall material.
- Any residual flavours are removed from the surface in an alcohol bath.
- The finished product can be shaped to any desired form,
- The release of the encapsulated agent is gradual, dependent on the hydration of the matrix.
- The product can be kept for one to two years without any substantial quality degradation.

However, there are several disadvantages to this technique. It is limited to the high boiling point flavours because most processes in extrusion are performed at relatively high temperature. To expand the application to low boiling point flavours, isopropanol is added to the process. Traces of isopropanol remaining in the final product can be hazardous. In addition, the microcapsule may contain microporosity when low boiling point components are present in the flavour.[88]

12.3.3 Cyclodextrin Inclusion Complexes

Cyclodextrins (CDs) are cyclic oligosaccharides with a hydrophilic exterior and a hydrophobic central cavity (Figure 12.3). The hydrophobic central cavity of CDs is torus (doughnut)-shaped, and their molecular dimensions allow total or partial inclusion of guest compounds.[89] The inclusion complexes are defined as a smaller guest molecule which fits into and is surrounded by the lattice of the larger host molecule. In these complexes, the flavour compounds are entrapped inside the cavity of the CDs. Complex formation is a dimensional fit between host cavity and guest molecule.[90] The formation and stability of the inclusion complexes depend on the binding forces between CDs and the flavour compounds. The binding forces include hydrogen bonds, van der Waals forces and dipole-dipole interactions.[14,91] No covalent bonds are formed during the preparation of inclusion complexes.[92] The main driving force of complex formation is the release of water molecules from the cavity. Water molecules are displaced by hydrophobic guest molecules to attain nonpolar-nonpolar association and to reach more stable lower energy states.[93]

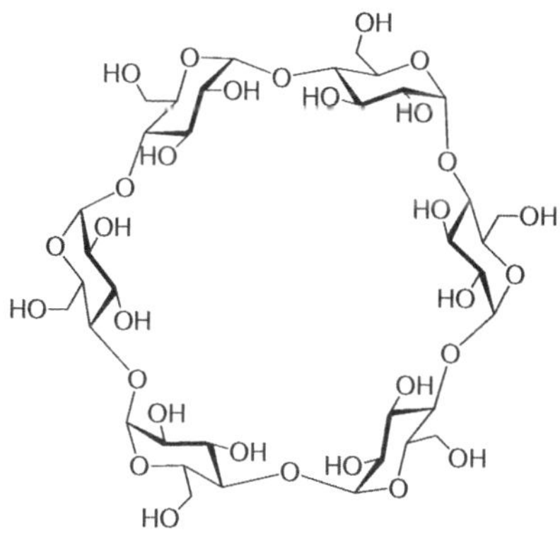

7
(α-Cyclodextrin)

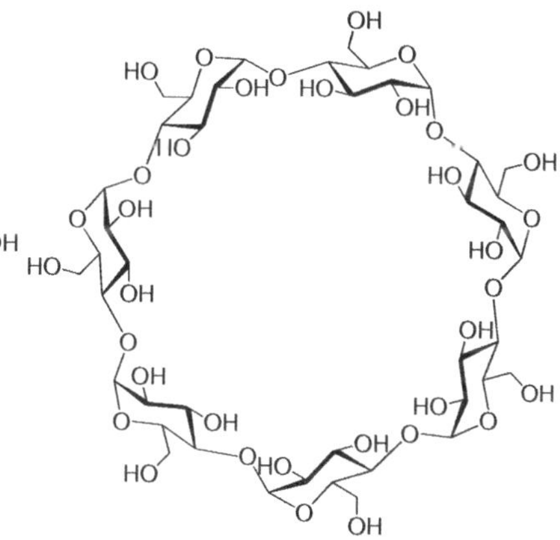

8
(β-Cyclodextrin)

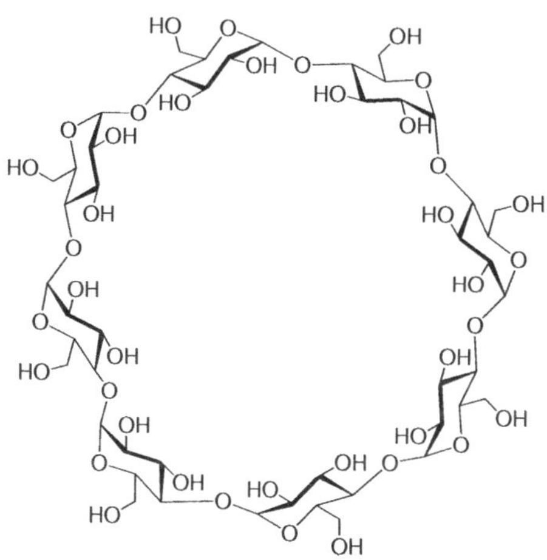

9
(γ-Cyclodextrin)

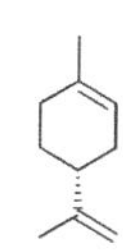

10
(D-Limonene)

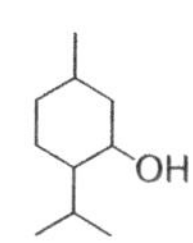

11
(Menthol)

Figure 12.3 *Chemical structures of α-cyclodextrin (**7**), β-cyclodextrin (**8**), γ-cyclodextrin (**9**) and flavour compounds D-limonene (**10**) and menthol (**11**)*

The commercially available CDs have 6, 7, or 8 glucopyranose units in the cycle and are referred to as α- (**7**), β- (**8**) and γ- (**9**) cyclodextrin, respectively (Figure 12.3). The inner cavities of α-, β- and γ-cyclodextrin, have reported diameters of 5.7, 7.8 and 9.5 Å and volumes of 174, 262 and 427 $Å^3$, respectively.[26,94] The formation of inclusion complexes depends on the size match between the guest flavour molecules and the host cyclodextrin. The maximum inclusion ratio of D-limonene (**10**) was about 0.65 for β-cyclodextrin and 1.2 for γ-cyclodextrin, while D-limonene was not included in α-cyclodextrin.[28] β-Cyclodextrin appeared to be a better encapsulant for menthol (**11**) than α- or γ-cyclodextrin.[25] Reineccius *et al.* demonstrated that γ-cyclodextrin generally achieved the highest initial flavour retention, while α- and β-cyclodextrin yielded lower but similar results to one another.[26] Once an inclusion complex is formed, some flavour compounds are much less readily released than others. Losses of volatiles during storage were greatest for γ-cyclodextrin and least in the case of α-cyclodextrin. From a practical standpoint, β-cyclodextrin is the cheapest and has the most applications among the CDs.[23,26,95]

β-Cyclodextrin forms inclusion complexes with flavour compounds in a process that only takes place in the presence of water.[93] Molecules that are less polar than water (i.e. most flavour substances) and that have suitable molecular dimensions to fit inside the cavity can be incorporated into the molecule. There are three methods to produce the flavour/β-cyclodextrin complexes using different amounts of water:[4,11,96] aqueous, concentrated suspension and slurry. In the first method, β-cyclodextrin is dissolved in enough water to form an aqueous solution. In the second method, β-cyclodextrin is dissolved in a lesser amount of water to form a concentrated suspension. In the third method, β-cyclodextrin is dissolved in a much lower water content to form a paste or slurry. Flavours are then added to form the inclusion complex. The D-limonene/β-cyclodextrin inclusion complex made using a twin screw kneader with low water content gave rise to a much higher inclusion fraction of the volatile in the complex than a sample made by the aqueous method.[27]

Among conventional microencapsulation methods, β-cyclodextrin molecular inclusion is the most effective for protecting the flavours. The advantages of cyclodextrin encapsulation are the unique release characteristics and the thermal/chemical stability. The cyclodextrin complexes are thermodynamically stable up to 200 °C[14] and encapsulated flavours can provide better protection from volatilization during extrusion; however, the flavour compounds in the cavity will be released whenever a better substrate becomes available. For example, release can be triggered by the presence of various biological components in the mouth.[97] Drawbacks include low flavour loading, the high cost of cyclodextrins (5.5–6.5 $ kg^{-1}) and regulatory requirements in a number of countries.[4,14] In addition, any flavour compounds larger than the cavity are left unencapsulated.

12.3.4 Helical Inclusion Complexes

Helical inclusion complexes, also known as starch (or amylose) inclusion complexes, are another type of inclusion complex. It has been thought that helical inclusion complexes are useful for food flavour retention.[29,98] Starch contains two polysaccharides: amylose and amylopectin. Amylose is a linear polymer with a very low degree of branching, while amylopectin is more highly branched. In the presence of suitable guest molecules, the amylase of starch adopts a single helix formation that is stabilized by hydrogen bonds.[33,99,100] The outside surface of the single helix is hydrophilic while the inner surface of the helical channel is hydrophobic. These helices are suitable for accommodating hydrophobic guest molecules of appropriate dimensions.[34,101] Amylose readily forms complexes with suitable guest molecules since helication is not hindered by side chains. However, amylose double helices hinder the formation of complexes due to the absence of a central channel.[100]

Starch can form inclusion complexes with various flavour compounds.[30,102] The microencapsulation process of flavour compounds into helical inclusion complexes is depicted in Scheme 12.2. A starch dispersion is prepared using distilled water. Specific amounts of flavour compounds are added to the starch dispersion. The water solubility of the flavours used is very low, therefore they are added in greater quantities than the starch solution. The mixture is then heated for 45–60 min at 95–121 °C.[31,103,104] When excess water is present,

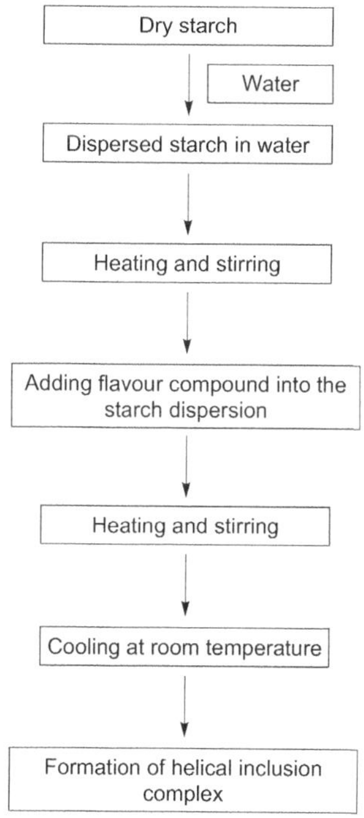

Scheme 12.2 *Flow diagram for the microencapsulation of flavours by helical inclusion complex formation*

heating leads to a swelling of the starch granules. The mobility of starch is related to its glass transition temperature and water is a good plasticizer for this polymer. However, water is a rather poor solvent for amylose. Therefore, the complexation of pure amylose in aqueous systems requires temperatures higher than 100 °C to avoid the spontaneous formation of double helices.[100] An essential feature of inclusion complexes is the presence of a left-handed amylose helix also known as V-amylose. The characterization of the crystalline packing of V-type amylose by wide-angle X-ray diffraction provides proof of complex formation.[100,103]

Arvisenet *et al.* measured the retention of three flavour compounds [isoamyl acetate (**12**), ethyl hexanoate (**13**), and linalool (**14**), Figure 12.4] in starch matrices.[105] Four different types of starches are used: a standard corn starch (26% amylose), a waxy corn starch (< 1% amylose), a stabilized and cross-linked waxy corn starch (< 1% amylose) and an amylase-rich corn starch (58% amylose). Because flavour compounds have a higher affinity for amylase than amylopectin, amylose-rich corn starch has the highest aroma retention and

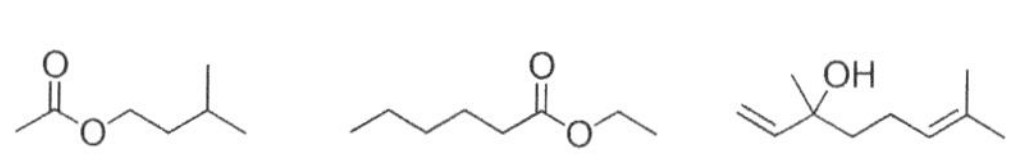

12 (Isoamyl acetate) **13** (Ethyl hexanoate) **14** (Linalool)

Figure 12.4 *Chemical structures of compounds encapsulated in starch matrices*

waxy corn starch has the lowest. Modified waxy corn starch and standard corn starch showed similar levels of aroma retention. Heinemann *et al.* demonstrated structural features of starch-lactone inclusion complexes using native potato starch and native amylopectin starch.[31] A variety of structures can be generated by selecting lactones with different chain lengths. Both amylase and amylopectin are able to interact with lactones. The amylase fraction dominates in the short term, while the amylopectin-lactone interactions develop over the course of several days. Itthisoponkul *et al.* were able to form inclusion complexes of tapioca starch (20.3% amylose) with flavour compounds, in particular with aliphatic alcohols.[32] A combination of differential scanning calorimetry (DSC) and wide-angle X-ray diffractometry was used to investigate the formation of complexes.

12.3.5 Fluidized Bed Coating

Numerous foods, food ingredients and food additives, have been coated using fluidized bed coating.[106] In fluidized bed coating, particles are introduced into a closed cell and fluidized by blowing air. The air used for atomization contributes to the evaporation of the coating solvent. The coating material is pumped to a nozzle and sprayed on the particles. During this process, there is a homogeneous layering of the coating material on the particles.[107] Coating droplet formation, contact, spreading, coalescence and evaporation occur almost simultaneously during the process.[108] There are four different types of fluidized bed coating caused by changing the position of the nozzle to be used for coating the solid particles: (*a*) top spray, (*b*) bottom spray, (*c*) wurster and (*d*) tangential spray fluidized bed coating methods (Figure 12.5).

12.3.6 Top Spray Fluidized Bed Coating

The top spray fluidized bed coating is widely used in pharmaceutical and food coating operations in pilot scale as well as in larger production scale.[109] The conventional top spray method has the greatest possibility of being introduced successfully in the food industry when compared to other methods.[110] This is due to its high versatility, relatively high batch size, and simplicity. In this technique, the spray nozzle is placed at the top of the fluid bed chamber and it sprays coating liquid counter currently onto the fluidized particles (Figure 12.5a). If processing conditions such as fluidization velocity, bed temperature and humidity of the air are incorrectly chosen, either excessive agglomeration or excessive spray drying of the spray feed will happen. In both cases, a poor product quality and loss will be expected.[111] Solis-Morales *et al.* prepared syrup coated puffed wheat by top spray fluidized bed and tumble coating.[112] The fluidized bed treated sample was considered crispier and more related to chocolate flavour than the commercial sample.

12.3.7 Bottom Spray System

The bottom spray system sprays the coating liquid from the bottom (Figure 12.5b). The path of the coating liquids concurrently towards the fluidized particle is extremely short so that premature droplet evaporation is almost absent.[113] The bottom spray also increases the collision between particles and droplets and results in a larger coating material efficiency and a reduction in spray drying. However, there were many possibilities of agglomeration, and the bottom spray system did not display a uniform thickness with respect to particle size.[114]

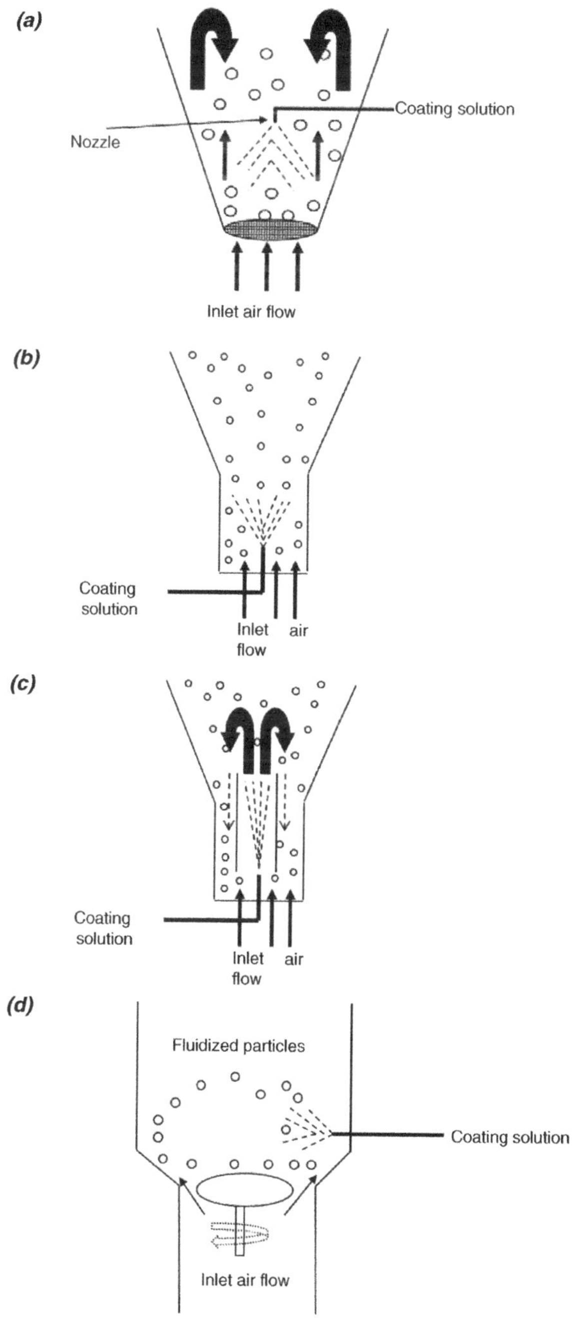

Figure 12.5 *Fluidized bed coating techniques: (a) top spray. [11] Reprinted by permission of Taylor & Francis Group, http://www.informaworld.com (b) bottom spray. [11] Reprinted by permission of Taylor & Francis Group, http://www.informaworld.com (c) wurster. Reprinted from [107], with permission from Elsevier (d) tangential spray. Reprinted from [113], with permission from Elsevier*

12.3.8 Wurster System

In the wurster system, the spray nozzle is placed at the bottom of the chamber like in the bottom spray system (Figure 12.5c). However, there is an air circulation in the wurster system, and the circulation of air increases the drying rate and reduces the potential for agglomeration.

12.3.9 Tangential Spray or Rotary Fluidized Bed Coating

The tangential spray or rotary fluidized bed coating has rotating disc and it sprays the coating solution tangentially to and concurrently with the flow of particles (Figure 12.5d). The combination of the rotation and the air flow provides a spherical shape and higher density to the resulting coated particles.[107] It can be expected that the coating film quality is similar to that obtained with a wurster system. Torres-Martinez *et al.* developed corn flavour coated puffed wheat using the tangential spray fluidized bed technique.[115] The fluidized bed sample presented a firmer consistency and a more uniform colour than tumble coating and commercial wheat.

12.3.10 Coacervation

Coacervation involves the separation of a coating material from a polymeric solution followed by coating that phase as a uniform layer around suspended core particles.[11] There are two types of coacervation: simple and complex, depending on the number of polymers. Simple coacervation involves only one type of polymer. In complex coacervation, however, two or more types of polymers are used.[1] Microcapsules produced by coacervation are divided into two types:[116] mononuclear and multinuclear capsules, depending on their internal structure. Mononuclear microcapsules are produced by encapsulation of only one droplet whereas multinuclear microcapsules are formed by aggregation of multiple mononuclear microcapsules.[117]

The ionic interaction between two oppositely charged polymers forms microcapsules in complex coacervation.[4] The coacervate is produced by neutralization of two opposite charges and can be cross-linked to form microcapsules. Various coating materials have been evaluated for use in the coacervation technique, such as gum acacia, heparin, pectin, carboxymethylcellulose and alginate as anionic polymers and gelatin or β-lactoglobulin as a cationic polymer.[4,118]

The most studied and best understood coating system is the gelatin/gum acacia system.[14,35,119] In this method, flavour oil was put into the gelatin solution and homogenized. Gum acacia was then added to the suspension and homogenized. Finally, the addition of acetic acid to reduce pH to 4.5 gave rise to the coacervation process.[38] Burgess and Carless noted that a maximum coacervate yield was obtained at the electrical equivalence pH of pure gelatin/acacia mixtures.[120] Daniels and Mittermaier also reported the influence of pH adjustment on microcapsules made by complex coacervation of gelatin/gum acacia.[35] Leclerq *et al.* characterized the formation of microcapsules by complex coacervation.[37] They produced coavervates using limonene (**10**) as a core material, gelatin/gum acacia as wall materials and glutaraldehyde as a cross-linker.

The coacervation method has some drawbacks. The complex coacervates are highly unstable and highly toxic compounds (e.g. formaldehyde or glutaraldehyde) are used as cross-linkers. In recent years, transglutaminase has been used as a cross-linker instead of glutaraldehyde.[24,36,121] Dong *et al.* reported the effect of processing conditions on the gelatin/gum acacia multinuclear microcapsules by coacervation.[116] They used peppermint oil as a core material and transglutaminase as a cross-linker. The microcapsules with desired particle mean size could be manufactured by modulating the pH value and stirring speed. The ideal conditions were pH 3.7 at a stirring speed of 400 rpm.

12.3.11 Double or Multiple Emulsion with Freeze Drying

Freeze drying or lyophilization is the process used for the dehydration of all heat-sensitive materials and flavour compounds.[11] In general, O/W/O or W/O/W multiple emulsions (O = oil, W = water) were prepared with or without ultrasonication and then stirred vigorously. The resulting microcapsules were filtered and freeze dried at a low temperature for 12–48 h. In the early 1970s, several studies were done concerning the retention of volatiles in freeze-dried carbohydrate (e.g. maltose, maltodextrin, glucose, starch, dextran, cellulose) matrices.[41,122]

There are some differences between microcapsules produced by hot air drying (e.g. spray drying) and those produced by freeze drying. Hot air drying produced hollows on the microcapsules,[123] while freeze drying maintained the form of microcapsules owing to the fixation by freezing.[124] Minemoto *et al.* compared the oxidation of methyl linoleate (**15**, Figure 12.6) encapsulated with gum arabic by hot air drying and freeze drying.[125] Methyl linoleate encapsulated by freeze drying was shown to be more resistant than that encapsulated by hot air drying and thus it was more slowly oxidized.

Kaushik and Roos described the preparation of limonene (**10**) microcapsules by freeze drying.[42] They used gum arabic, sucrose and gelatin matrices as wall materials. Gum arabic was shown to be an efficient encapsulation matrix for limonene encapsulation by freeze drying, while matrices consisting of gelatin collapsed during freeze drying processes. They also produced limonene microcapsules with a gum arabic:sucrose:gelatin (1 : 1 : 1) mixture. High retention of limonene is achieved by homogenization of the emulsion containing the gum arabic:sucrose:gelatin mixture.

Choi *et al.* compared the encapsulation of eugenol (**16**) by inclusion with β-cyclodextrin (**8**) and by the emulsion-diffusion method with polycaprolcatone using freeze drying.[40] The emulsion-diffusion methods showed higher encapsulation efficiencies and better oxidation stability than the molecular inclusion. They concluded that emulsion-diffusion with freeze drying was the most effective eugenol encapsulation technique to protect it from light oxidation during storage.

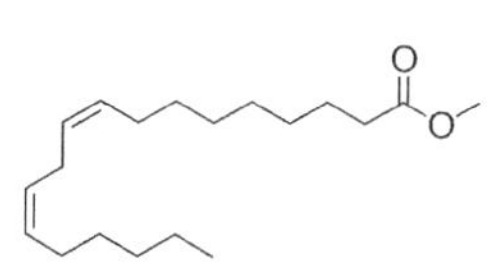

15
(Methyl linoleate)

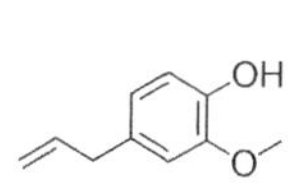

16
(Eugenol)

*Figure 12.6 Chemical structures of methyl linoleate (**15**) and eugenol (**16**)*

Freeze drying is a simple technique and sometimes it shows better retention of flavour compounds than spray drying;[39,76] however, it also has some drawbacks. The main disadvantages of freeze drying are high operation costs and long dehydration times. Additional disadvantages are that this technique generally results in losses of highly volatile compounds and a porous structure of microcapsules.

12.3.12 Co-Crystallization

Co-crystallization or co-precipitation utilize sucrose as a matrix for the incorporation of active compounds.[11] Compared with other flavour encapsulation techniques, it is a low-cost and simple procedure. Very few studies have been published on the use of co-crystallization in the encapsulation process,[43,44] although several patents have been reported on the co-crystallization technique.[45,47] Chen *et al.* showed that flavour compounds can be encapsulated by co-crystallization.[46]

In general, the sucrose syrup is concentrated and maintained at a high temperature (above 120 °C) and low moisture to prevent crystallization.[11] A core material is then added to the concentrated syrup with stirring. As the syrup reaches the crystallization temperature, a substantial amount of heat is emitted. Agitation is continued in order to promote and extend crystallization until the agglomerates are discharged from the vessel. The encapsulated products are then dried to a moisture content of less than 1% by weight and screened to ensure a uniform size.[126]

Beristain *et al.* described the microencapsulation of orange peel oil by co-crystallization.[43] They produced a granular shape of orange peel oil co-crystalliates having good flow characteristics that did not form aggregates. The crystals retained all the orange oil and the encapsulation capacity of sucrose syrups was found to be greater than 90%.

12.3.13 Spray Chilling and Spray Cooling

Spray chilling and spray cooling are similar to spray drying. However, the core and wall mixtures in spray chilling or spray cooling are atomized into chilled or cooled air, respectively. Therefore, unlike spray drying, these techniques do not involve the evaporation of water.[11]

The coating material in the spray cooling technique is typically vegetable oil or its derivatives; however, a wide range of other encapsulating materials may be employed. For instance, a molten fat is used as a wall material in which the core material is dispersed. These include fat and stearin with melting points of 45–122 °C as well as hard mono- and diacylglycerols with melting points of 45–65 °C.[3] In the spray chilling technique, the coating material is typically a fractionated or hydrogenated vegetable oil with a melting point within the range 32–42 °C.[48]

Microcapsules prepared by spray-chilling and spray-cooling are insoluble in water due to the lipid coating. Consequently, these techniques can be used for the encapsulation of water soluble flavours that may otherwise be volatilized during thermal processing.[49] These techniques have some advantages, such as lower processing costs, free flowing powder and delayed release of the encapsulated flavour compound in a wet environment. However, microcapsules prepared by spray chilling and spray cooling require special handling and storage conditions.[1,3] In addition, they have a relatively low payload of flavour compounds.[14]

12.3.14 Supercritical Fluids

The use of supercritical fluids in the microencapsulation of flavours is a relatively new technique. Supercritical fluids offer various possibilities in the production of microcapsules.[50,53] Supercritical carbon dioxide (sc-CO_2) is the most commonly used supercritical fluid for microencapsulation. There are various microencapsulation methods based on supercritical fluids: rapid expansion of supercritical solutions (RESS), gas (or supercritical fluids) anti-solvent (GAS or SAS), aerosol solvent extraction system (ASES), solution enhanced dispersion by supercritical fluids (SEDS), supercritical fluid extraction of emulsions (SFEE), supercritical solvent impregnation (SSI) and particles from gas-saturated solutions (PGSS).

In RESS, sc-CO_2 acts as the solvent (Scheme 12.3a). Flavour compounds are dissolved in a supercritical fluid under high pressure. Depressurizing this solution through a nozzle into a low pressure chamber resulted in the formation of microcapsules.[50,52,53] The morphology of microcapsules depends on the core material structure and the RESS processing parameters.[51,127] Microcapsules with a narrow size distribution can be prepared at relatively low temperatures by RESS. However, the main limitation of RESS is the low solubility of many substances in sc-CO_2. Additional disadvantages are: (*a*) RESS processing is extremely fast, so it is very difficult to control the morphology and loading of the flavour,[128] (*b*) elimination of the co-solvent is not simple and cheap.[50]

In GAS or SAS, a supercritical fluid acts as the anti-solvent (Scheme 12.3b). Flavour compounds are dissolved in a conventional organic solvent. The sc-CO_2 is then introduced into the vessel, preferably from the bottom to achieve a better mixing of the solvent and anti-solvent.[50,52,53] ASES and SEDS are based on the concept of GAS. ASES involves spraying the solution through an atomization nozzle. The sc-CO_2 is introduced to the top of the vessel. Next, the active compound solution is introduced into the vessel through a nozzle. Particles are collected at the bottom of the vessel.[50] SEDS offers a nozzle with two coaxial passages allowing for the introduction of the sc-CO_2 and an active compound solution into the vessel. The high velocity of the supercritical fluid allows for the solution to be broken into small droplets.[50]

In PGSS, the sc-CO_2 is dissolved in a solution of active compounds in solvent (Scheme 12.3c). It leads to gas-saturated solution/suspension; and depressurization of the mixture through a nozzle results in the formation of solid particles or droplets. The plasticizing and swelling effect caused by CO_2 dissolution can enhance the incorporation of the active compounds. The PGSS process has several advantages, such as low processing costs, simplicity of processing and the wide range of products that can be treated. However, it is difficult to monitor particle size. Based on the principles of PGSS, other processes have been developed, such as carbon dioxide assisted nebulization with a bubble dryer (CAN-BD), supercritical assisted atomization (SAA), depressurization of an expanded liquid organic solution (DELOS) and continuous powder coating spraying process (CPCSP).[52]

12.3.15 Other Techniques

Other microencapsulation techniques/products have been described, including: liposomes, centrifugal extrusion, centrifugal suspension separation and liposome entrapment.[11] However, these techniques/products are not widely used in the food industry. Recently,

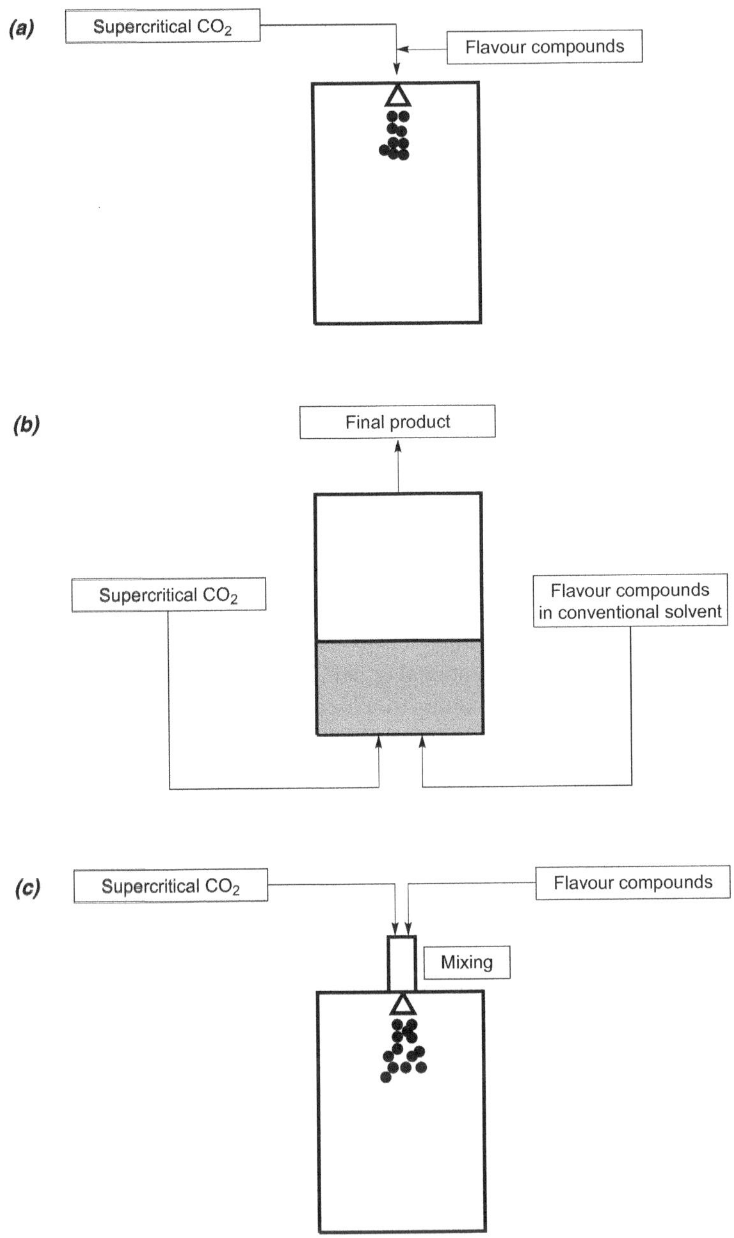

Scheme 12.3 *Schematic representation of microencapsulation equipment based on supercritical fluids: (a) RESS, (b) GAS, (c) PGSS*

new concepts for the microencapsulation of food flavours have been reported, including: sol-gel-made materials, yeasts, electrostatic extrusion and temperature-responsive gels.

Sol-gel-made materials are used as encapsulation matrices for flavours by embedding flavour compounds into the silica matrix with the silica nanoparticles enclosing the flavour molecules. The major mechanism of flavour retention on silica is physical adsorption by hydrogen bonding. Veith *et al.* demonstrated encapsulation and retention of decanoic acid in sol-gel-made silica.[129] Decanoic acid remains quite stable in the sol-gel-made silica at temperatures from 100 to 200 °C for a prolonged time.

Empty yeast cells are used in thermostable flavour retention.[130] They are stable up to temperatures of 250 °C and flavours encapsulated within the empty yeast cells leak out only at temperatures above 243 °C. Water is another important factor in yeast-based microencapsulation. Dry yeast is a strong barrier to active compounds whereas wet yeast promotes the release of active compounds. Dardelle *et al.* demonstrated that the flavour loading of yeasts is driven by the affinity between the flavour molecules and the hydrophobic bag formed by the phopholipid bilayer.[131]

The electrostatic extrusion technique has been developed for the encapsulation and production of uniform gel microbeads. It is based on the premise that electrostatic forces form a charged stream of small droplets. This technique is suitable for the production of very small particles. Bugarski *et al.* noted the size of microcapsules is a complex function of the operating parameters, the system properties and the properties of the polymer solution.[132] Manjlovic *et al.* demonstrated the application of electrostatic extrusion.[133] They used calcium alginate as the matrix and vanillin (**5**) as a flavour compound. The vanillin-loaded microcapsules were produced with regular spheres of a uniform size by electrostatic extrusion.

Recently, another flavour encapsulation technique has been developed. Heitfeld *et al.* used temperature-responsive gels (TRGs) for flavour encapsulation.[134] Modified hydroxypropyl cellulose TRGs change their volume in response to temperature. The gel shrinks and expels water above a critical temperature, and it swells and absorbs water below this critical temperature. Because of this transition, hydroxypropyl cellulose (HPC) encapsulation systems release flavour slowly and more flavour compounds stay in the matrix at increased temperatures.

12.4 Conclusion and Future Trends

Several microencapsulation techniques have been developed within the food flavour industry. The choice of appropriate encapsulation techniques depends on the properties of food flavour compounds and wall materials, desired release properties, maximum payloads and processing costs. Among those factors, processing cost is the main concern in the food flavour industry. Cost pressures will remain an issue and cost control will continue to be a critical factor for the next few years in that industry. From an industrial point of view, spray drying is recommended as the preferred technique for the microencapsulation of food flavour. It is the most economical and flexible of the encapsulation techniques. Further research on the use of modified starches as wall materials can eliminate the limitation of its application, particularly the limited number of wall materials. In addition, the use of modified starches when preparing inclusion complexes results in the formation of helical inclusion complexes, a possible microencapsulation product that can be used along with cyclodextrin inclusion complexes. More in-depth research on helical inclusion complexes will be performed within the next five to ten years. Microencapsulation techniques based on supercritical fluids are also promising for the food industry.

References

1. A. Madene, M. Jacquot, J. Scher, S. Desobry, Flavor encapsulation and controlled release – a review, *Int. J. Food Sci. Technol.*, **41**, 1–21 (2006).
2. G. A. Reineccius, Flavor encapsulation, *Food Rev. Int.*, **5**, 147–150 (1989).
3. A. H. Taylor, Encapsulation systems and their applications in the flavor industry, *Food Flavor Ingredients Packag. Process.*, **5** (9), 48–51 (1983).
4. S. Yuliani, B. Bhandari, R. Rutgers, B. D'Arcy, Application of microencapsulated flavor to extrusion product, *Food Rev. Int.*, **20**, 163–185 (2004).
5. R. Arshady, Microcapsules for food, *J. Microencapsulation*, **10**, 413–435 (1993).
6. (*a*) X. G. Chen, C. S. Liu, C. G. Liu, X. H. Meng, C. M. Lee, H. J. Park, Preparation and biocompatibility of chitosan microcarriers as biomaterial, *Biochem. Eng. J.*, **27**, 269–274 (2007); (*b*) K. G. H. Desai, H. J. Park, Effect of manufacturing parameters on the characteristics of vitamin C encapsulated tripolyphosphate-chitosan microspheres prepared by spray drying, *J. Microencapsulation*, **23**, 91–103 (2006).
7. K. G. H. Desai, H. J. Park, Study of gamma-irradiation effects on chitosan microparticles, *Drug Delivery*, **13**, 39–50 (2006).
8. (*a*) B. K. Kim, S. J. Hwang, J. B. Park, H. J. Park, Preparation and characterization of drug-loaded microspheres by an emulsion solvent evaporation method, *J. Microencapsulation*, **19**, 811–822 (2002); (*b*) B. K. Kim, S. J. Hwang, J. B. Park, H. J. Park, Characteristics of felodipine-loaded poly(ε-caprolactone) microspheres, *J. Microencapsulation*, **22**, 193–203 (2005); (*c*) S.-J. Kim, S. Y. Cho, S. H. Kim, O.-J. Song, I.-S. Shin, D. S. Cha, H. J. Park, Effect of microencapsulation on viability and other characteristics in *Lactobacillus acidophilus* ATCC 43121, *LWT-Food Sci. Technol.*, **41**, 493–500 (2008); (*d*) C. J. Kirby, Microencapsulation and controlled delivery of food ingredients, *Food Sci. Technol. Today*, **5** (2), 74–80 (1991); (*e*) J. A. Ko, H. J. Park, S. J. Hwang, J. B. Park, J. S. Lee, Preparation and characterization of chitosan microparticles intended for controlled drug delivery, *Int. J. Pharm.*, **249**, 165–174 (2002); (*f*) J. A. Ko, S. Y. Koo, H. J. Park, Effects of alginate microencapsulation on the fibrinolytic activity of fermented soybean paste (Cheonggukjang) extract, *Food Chem.*, **111**, 921–924 (2008); (*g*) D. W. Lee, S. J. Hwang, J. B. Park, H. J. Park, Preparation and release characteristics of polymer-coated and blended alginate microspheres, *J. Microencapsulation*, **20**, 179–192 (2003).
9. M. Augustin, L. Sanguansri, C. Margetts, B. Young, Microencapsulation of food ingredients, *Food Aust.*, **53**, 220–223 (2001).
10. (*a*) J. A. Bakan, Microencapsulation of food and related products, *Food Technol.*, **27** (11), 34–38 (1973); (*b*) L. L. Balsa, G. O. Fanger, Microencapsulation in food industry, *Crit. Rev. Food Technol.*, **2**, 245–249 (1971); (*c*) J. D. Dziezak, Microencapsulation and encapsulated food ingredients, *Food Technol.*, **42** (4), 136–151 (1998); (*d*) R. D. Todd, Microencapsulation and food industry, *Flavour Ind.*, **1**, 78–81 (1970).
11. K. G. H. Desai, H. J. Park, Recent developments in microencapsulation of food ingredients, *Drying Technol.*, **23**, 1361–1394 (2005).
12. B. F. Gibbs, S. Kermasha, I. Alli, C. N. Mulligan, Encapsulation in food industry: a review, *Int. J. Food Sci. Nutr.*, **50**, 213–234 (1999).
13. F. Shahidi, X. Q. Han, Encapsulation of food ingredients, *Crit. Rev. Food Sci. Nutr.*, **33**, 501–504 (1993).
14. S. Gouin, Microencapsulation: industrial appraisal of existing technologies and trends, *Trends Food Sci. Technol.*, **15**, 330–347 (2004).
15. S. Chattopadhyaya, R. S. Singhal, P. R. Kulkarni, Oxidized starch as gum arabic substitute for encapsulation of flavors, *Carbohydr. Polym.*, **37**, 143–144 (1998).
16. S. Krishnan, R. Bhosale, R. S. Singhal, Microencapsulation of cardamom oleoresin: evaluation of blends of gum arabic, maltodextrin and a modified starch as wall materials, *Carbohydr. Polym.*, **61**, 95–102 (2005).
17. S. Varavinit, N. Chaokasem, S. Shobsngob, Studies of flavor encapsulation by agents produced from modified sago and tapioca starches, *Starch/Stärke*, **53**, 281–287 (2001).

18. B. Murúa-Pagola, C. I. Beristain-Guevara, F. Martínez-Bustos, Preparation of starch derivatives using reactive extrusion and evaluation of modified starches as shell materials for encapsulating of flavoring agents by spray drying, *J. Food Eng.*, **91**, 380–386 (2009).
19. D. H. Miller, J. R. Mutka, Solid essential oil flavor composition (to Sunkist Growers Inc.), US 4,610,890 (1986), US 4,707,367 (1987); *Chem. Abstr.*, **105**, 5420 (1986).
20. M. A. Porzio, L. M. Popplewell, Encapsulation compositions (to McCormick and Co.), US 5,603,971 (1997), US 6,416,799 (2002); *Chem. Abstr.*, **122**, 54700 (1995).
21. G. Reichart, K. W. Hyung, N. Prasad, Flavor composition (to Nestac SA), US 6,413,573 (2002); *Chem. Abstr.*, **136**, 69090 (2002).
22. F. Z. Saleeb, J. G. Pickup, Fixation of volatiles in extruded glass substrates (to General Foods Corp.), US 4,820,534 (1989); *Chem. Abstr.*, **104**, 128626 (1986).
23. P. Gornas, G. Neunert, K. Baczynski, K. Polewski, Beta-cyclodextrin complexes with chlorogenic and caffeic acids from coffee brew: spectroscopic, thermodynamic and molecular modeling study, *Food Chem.*, **114**, 190–196 (2009).
24. W. Li, G. Wu, H. Chen, M. Wang, Preparation and characterization of gelatin/SDS/NaCMC microcapsules with compact wall structure by complex coacervation, *Colloids Surf., A*, **333**, 133–137 (2009).
25. X. Liu, T. Furuta, H. Yoshi, P. Linko, W. J. Coumans, Cyclodextrin encapsulation to prevent the loss of L-menthol and its retention during drying, *Biosci. Biotechnol. Biochem.*, **64**, 1608–1613 (2000).
26. T. A. Reineccius, G. A. Reineccius, T. L. Peppard, Encapsulation, Encapsulation of flavors using cyclodextrins: comparison of flavor retention in alpha, beta, and gamma types, *Food Chem. Toxicol.*, **67**, 3271–3279 (2002).
27. H. Yoshi, T. Furuta, E. Okita, A. Toyomi, Y. Linko, P. Linko, The increased effect of kneading on the formation of inclusion complexes between D-limonene and β-cyclodextrin at low water content, *Biosci. Biotechnol. Biochem.*, **62**, 464–468 (1998).
28. H. Yoshi, T. Kometani, T. Furuta, Y. Watanabe, Y. Linko, P. Linko, Formation of inclusion complexes of cyclodextrin with ethanol under anhydrous conditions, *Biosci. Biotechnol. Biochem.*, **62**, 2166–2170 (1998).
29. M. C. Godet, V. Tran, P. Colonna, A. Buleon, M. Pezolet, Inclusion/exclusion of fatty acids in amylose complexes as a function of the fatty acid chain length, *Int. J. Biol. Macromol.*, **17**, 405–408 (1995).
30. (*a*) C. Heinemann, B. Conde-Petit, J. Nuessli, F. Escher, Evidence of starch inclusion complexation with lactones, *J. Agric. Food Chem.*, **49**, 1370–1376 (2001); (*b*) C. Heinemann, M. Zinsli, A. Renggli, F. Escher, B. Conde-Petit, Influence of amylose-flavor complexation on build-up and breakdown of starch structures in aqueous food model systems, *LWT-Food Sci. Technol.*, **38**, 885–894 (2005).
31. C. Heinemann, F. Escher, B. Conde-Petit, Structural features of starch-lactone inclusion complexes in aqueous potato starch dispersions: the role of amylose and amylopection, *Carbohydr. Polym.*, **51**, 159–168 (2003).
32. T. Itthisoponkul, J. R. Mitchell, A. J. Taylor, I. A. Farhat, Inclusion complexes of tapioca starch with flavor compounds, *Carbohydr. Polym.*, **69**, 106–115 (2007).
33. J. Nuessli, S. Handschin, B. Conde-Petit, F. Escher, Rheology and structure of amylopectin potato starch dispersions without and with emulsifier addition, *Starch/Stärke*, **52**, 22–27 (2000).
34. J. Nuessli, J. L. Putaux, P. Le-Bail, A. Buleon, Crystal, structure of amylose complexes with small ligands, *Int. J. Biol. Macromol.*, **33**, 227–234 (2003).
35. R. Daniels, E. M. Mittermaier, Influence of pH adjustment on microcapsules obtained from complex coacervation of gelatin and acacia, *J. Microencapsulation*, **12**, 591–599 (1995).
36. Z. Dong, S. Xia, S. Hua, K. Hayat, X. Zhang, S. Xu, Optimization of cross-linking parameters during production of transglutaminase-hardened spherical multinuclear microcapsules by complex coacervation, *Colloids Surf., B*, **63**, 41–47 (2008).
37. S. Leclercq, K. R. Harlander, G. A. Reineccus, Formation and characterization of microcapsules by complex coacervation with liquid or solid aroma cores, *Flavour Fragr. J.*, **24**, 17–24 (2008).

38. G. F. Palmieri, S. Martell, D. Lauri, P. Wehrle, Gelatin-acacia complex coacervation as a method for ketoprofen microencapsulation, *Drug Dev. Ind. Pharm.*, **22**, 951–957 (1996).
39. Y. H. Cho, J. Park, Evaluation of process parameters in the O/W/O multiple emulsion method for flavor encapsulation, *J. Food Sci.*, **68**, 534–538 (2003).
40. M. Choi, A. Soottitantawat, O. Nuchuchua, S. Min, U. Ruktanonchai, Physical and light oxidative properties of eugenol encapsulated by molecular inclusion and emulsion-diffusion method, *Food Res. Int.*, **42**, 148–156 (2009).
41. J. Flink, F. Gejl-Hansen, Retention of organic volatiles in freeze-dried solutions of carbohydrates. Solution: microscopic observations, *J. Agric. Food Chem.*, **20**, 691–694 (1972).
42. V. Kaushik, Y. H. Roos, Limonene encapsulation in freeze drying of gum arabic-sucrose-gelatin systems, *LWT-Food Sci. Technol.*, **40**, 1381–1391 (2007).
43. C. I. Beristain, A. V. Vazquez, H. S. Garcia, E. J. Vernon-Carter, Encapsulation of orange peel oil by co-crystallization, *LWT-Food Sci. Technol.*, **29**, 645–647 (1996).
44. D. R. Bhandari, N. Datta, B. R. D'Arcy, G. B. Rintoul, Co-crystallization of honey with sucrose, *LWT-Food Sci. Technol.*, **31**, 138–142 (1998).
45. A. C. C. Chen, S. J. Drescher, C. P. Graham, Maple sugar product and method of preparing and using same (to Amstar Corp.), US 4,159,210 (1979); *Chem. Abstr.*, **91**, 93330 (1979).
46. A. C. C. Chen, C. E. Lang, C. P. Graham, A. B. Rizzutto,Crystallized, readily water dispersible sugar product containing heat sensitive, acidic or high invest sugar substances (to Amstar Corp.), US 4,362,757 (1982); *Chem. Abstr.*, **97**, 74280 (1982).
47. M. D. Miller, C. P. Graham, Sugar process and product (to American Sugar Co.), US 3,365,331 (1968); *Chem. Abstr.*, **69**, 42913 (1968).
48. D. Blenford, Fully protected, *Food Flavor Ingredients Packag. Process.*, **8** (7), 43–45 (1986).
49. R. Lamb, Spray chilling, *Food Flavor Ingredients Packag. Process.*, **9** (12), 39–42 (1987).
50. J. Jung, M. Perrut, Particle design using supercriticial fluids: literature and patent survey, *J. Supercrit. Fluids*, **20**, 179–219 (2001).
51. E. M. Pillips, V. J. Stella, Rapid expansion from supercritical solutions: application to pharmaceutical processes, *Int. J. Pharm.*, **94**, 1–10 (1993).
52. A. Shariati, C. J. Peters, Recent developments in particle design using supercritical fluids, *Curr. Opin. Solid Mater. Sci.*, **7**, 371–383 (2003).
53. M. J. Cocero, A. Martin, F. Mattea, S. Varona, Encapsulation and co-precipitation processes with supercritical fluids: fundamentals and applications. *J. Supercrit. Fluids*, **47**, 546–555 (2009).
54. M. A. Godshall, The role of carbohydrates in flavor development, *Food Technol.*, **42** (11), 71–74 (1988).
55. B. R. Bhandari, B. R. D'Arcy, I. Padukka, Encapsulation of lemon oil by paste method using β-cyclodextrin: encapsulation efficiency and profile of oil volatiles, *J. Agric. Food Chem.*, **47**, 5194–5197 (1999).
56. (*a*) I. K. Greener, O. Fennema, Barrier properties and surface characteristics of edible, bilayer films, *J. Food Sci.*, **54**, 1393–1395 (1989); (*b*) I. K. Greener, O. Fennema, Evaluation of edible, bilayer films for use as moisture barriers for food, *J. Food Sci.*, **54**, 1400–1403 (1989).
57. M. Rosenberg, I. J. Kopelman, Y. Talmon, Factors affecting retention in spray drying microencapsulation of volatile materials, *J. Agric. Food Chem.*, **28**, 1288–1294 (1990).
58. H. Shiga, H. Yoshii, T. Nishiyama, T. Furuta, P. Forssele, K. Poutanen, P. Linko, Flavor encapsulation and release characteristics of spray-dried powder by the blended encapsulant of cyclodextrin and gum arabic, *Drying Technol.*, **19**, 1385–1395 (2001).
59. H. Yoshii, M. Yasuda, T. Furuta, H. Kuwahara, M. Ohkawara, P. Linko, Retention of cyclodextrin complexed shiitake (*Lentinus edodes*) flavors with spray drying, *Drying Technol.*, **23**, 1205–1215 (2005).
60. R. Baranauskiene, P. R. Venskutonis, K. Dewettinck, R. Vershe, Properties of oregano (*Orignum vulgare* L.), citronella (*Cymbopogon nardus* G.), and marjorana (*Majornan hortensis* L.) flavors encapsulated into milk protein-based matrices, *Food Res. Int.*, **39**, 413–425 (2006).
61. F. Ono, New encapsulation technique with protein-carbohydrate matrix, *J. Jpn. Food Sci. Technol.*, **27**, 529–535 (1980).

62. M. Rosenberg, T. Y. Sheu, Microencapsulation of volatiles by spray drying in whey protein based wall systems, *Int. Dairy J.*, **6**, 273–284 (1996).
63. S. L. Kamper, O. Fennema, Water vapor permeability of an edible, fatty acid, bilayer film, *J. Food Sci.*, **49**, 1482–1485 (1984).
64. H.-H. Y. Kim, I. C. Baianu, Novel liposome microencapsulation techniques for food applications, *Trends Food Sci. Technol.*, **2**, 55–60 (1991).
65. L. M. Popplewell, M. A. Porzio, Fat-coated encapsulation compositions and method for preparing the same (to McCormick & Co.), US 6,245,366 (2001); *Chem. Abstr.*, **129**, 4018 (1998).
66. L. G. H. Desai, C. Liu, H. J. Park, Characteristics of vitamin C encapsulated tripolyphosphate-chitosan microspheres as affected by chitosan molecular weight, *J. Microencapsulation*, **23**, 79–90 (2006).
67. K. G. H. Desai, H. J. Park, Encapsulation of vitamin C in tripolyphosphate crosslinked chitosan microspheres by spray drying, *J. Microencapsulation*, **22**, 179–192 (2005).
68. S. A. Desobry, F. M. Netto, T. B. Labuza, Comparison of spray drying, drum drying and freeze drying for ($1 \rightarrow 3$, $1 \rightarrow 4$)-β-carotene encapsulation and preservation, *J. Food Sci.*, **62**, 1158–1162 (1997).
69. A. Gharsallaoui, G. Roudaut, O. Chambin, A. Voilley, R. Saurel, Applications of spray drying in microencapsulation of food ingredients: an overview, *Food Res. Int.*, **40**, 1107–1121 (2007).
70. Z. Liu, J. Zhou, Y. Zeng, X. Ouyang, The enhancement and encapsulation of *Agaricus biporus* flavour, *J. Food Eng.*, **65**, 391–396 (2004).
71. (*a*) Y. Beatus, A. Raziel, M. Rosenberg, I. J. Kopelman, Spary drying microencapsulation of paprika oleoresin, *LWT-Food Sci. Technol.*, **18**, 28–34 (1985); (*b*) A. J. Zakarian, C. J. King, Volatiles loss in the zone during spray drying of emulsions, *Ind. Eng. Chem. Process Des. Dev.*, **21**, 107–113 (1982).
72. F. M. Goycoolea, A. M. Calderon De La Barca, J. R. Balderrama, Immunological and functional properties of the exudate gum from northwestern Mexican mesquite (*Prosopis spp.*) in comparison with gum arabic, *Int. J. Biol. Macromol.*, **21**, 29–36 (1997).
73. C. I. Beristain, E. J. Vernon-Carter, Utilization of mesquite (*Prosopis juliflora*) gum as emulsion stabilizing agent for spray dried encapsulated orange peel oil, *Drying Technol.*, **12**, 1727–1733 (1994).
74. C. I. Beristain, H. S. Garcia, E. J. Vernon-Carter, Mesquite gum (*Prosopis juliflora*) and maltodextrin blends as wall materials for spray-dried encapsulated orange peel oil, *Food Sci. Technol. Int.*, **5**, 353–356 (1999).
75. C. I. Beristain, H. S. Garcia, E. J. Vernon-Carter, Spray-dried encapsulation of cardamom (*Elettaria cardamomum*) essential oil with mesquite (*Prosopis juliflora*) gum, *LWT-Food Sci. Technol.*, **34**, 398–401 (2001).
76. Y. H. Cho, J. Park, Characteristics of double-encapsulated flavor powder prepared by secondary fat coating process, *J. Food Sci.*, **67**, 968–972 (2002).
77. J. Finney, R. Buffo, G. A. Reineccius, Effects of type of atomization and processing temperatures on the physical properties and stability of spray dried flavors, *J. Food Sci.*, **67**, 1108–1114 (2002).
78. M. I. Re, Microencapsulation by spray drying, *Drying Technol.*, **16**, 1195–1196 (1998).
79. C. I. Beristain, E. J. Vernon-Carter, Studies on the interaction of arabic (*Acacia senegal*) and mesquite (*Prosopis juliflora*) gum as emulsion stabilizing agents for spray dried encapsulated orange peel oil, *Drying Technol.*, **29**, 645–667 (1995).
80. H. E. Swisher, Solid flavoring composition and method of preparing the same (to Sunkist Growers, Inc.), US 2,809,895 (1957); *Chem. Abstr.*, **52**, 8231 (1958).
81. (*a*) J. M. Barnes, J. A. Steinke, Encapsulation matrix composition and encapsulate containing same (to SCM Corp.), US 4,689,235 (1987); *Chem. Abstr.*, **103**, 213769 (1985); (*b*) E. E. Beck, Essential oil composition and method of preparing the same, US 3,704,137 (1972); *Chem. Abstr.*, **78**, 47659 (1973); (*c*) L. Sair, R. A. Sair, Food supplement concentrate in a dense glasseous extrudate (to Griffith Laboratories Inc.), US 4,232,047 (1980); (*d*) L. Sair, R. A. Sair, Encapsulation of active agents as microdispersions in homogeneous natural polymeric matrices (to Griffith Laboratories Inc.), US 4,230,687 (1980); (*e*) T. H. Schultz, Preparation of solid flavoring

compositions (to US Department of Agriculture), US 2,856,291 (1958); *Chem. Abstr.*, **53**, 8426 (1959); (*f*) T. H. Schultz, A. H. Brown, L. H. Walker, Production of flavoring oil compositions (to US Department of Agriculture), US 2,857,281 (1958); *Chem. Abstr.*, **53**, 36185 (1959).

82. S. Matsumoto, H. Kobayashi, Y. Takashima, Production of monodisperse capsules, *J. Microencapsulation*, **3**, 25–31 (1986).
83. G. R. Somerville Jr., Encapsulating apparatus (to Southwest Research Institute), US 3,015,128 (1962).
84. R. E. Sparks, N. S. Mason, Method for coating particles or liquid droplets (to Washington University Technology Associates, Inc.), US 4,675,140 (1987); *Chem. Abstr.*, **104**, 132135 (1986).
85. (*a*) H. Levine, B. van Lengerich, J. G. Pickup, Glassy matrices containing volatile and/or labile components, and processes for preparation and use thereof (to Nabisco Brands Inc.), US 5,009,900 (1991); (*b*) H. Levine, B. van Lengerich, J. G. Pickup, Double encapsulated compositions containing volatile and/or labile components, and processes for preparation and use thereof (to Nabisco Brands Inc.), US 5,087,461 (1992).
86. R. C. McIver, F. Vlad, R. A. Golding Jr., T. D. Leichssenring, D. Benczédi, Encapsulated flavor and/or fragrance composition (to Firmenich SA), US 6,932,982 (2005); *Chem. Abstr.*, **137**, 184789 (2002).
87. (*a*) F. Thevenet, Acacia gums: natural encapsulation agent for food ingredients, in *Encapsulation and Controlled Release of Food Ingredients*, S. J. Risch, G. A. Reineccius (Eds), ACS Symp. Ser., Vol. 590, American Chemical Society, Washington D. C., 1995, pp. 51–59; (*b*) D. M. Jones, Controlling particle size and release properties. Secondary processing techniques, in *Flavor Encapsulation*, S. J. Risch, G. A. Reineccius (Eds), ACS Symp. Ser., Vol. 370, American Chemical Society, Washington, D. C., 1988, pp. 158–175; (*c*) T. J. Dezarn, Food ingredient encapsulation, in *Encapsulation and Controlled Release of Food Ingredients*, S. J. Risch, G. A. Reineccius (Eds), ACS Symp. Ser., Vol. 590, American Chemical Society, Washington D. C., 1995, pp. 74–86.
88. C. V. Fulger, L. M. Popplewell, Flavor encapsulation (to McCormick & Company Inc.), US 5,958,502 (1999); *Chem. Abstr.*, **126**, 198783 (1997).
89. G. Astray, C. Gonzalez-Barreiro, J. C. Mejuto, R. Rial-Otero, J. Simal-Gándara, A review on the use of cyclodextrins in foods, *Food Hydrocolloids*, **23**, 1631–1640 (2009).
90. S. Muñoz-Botella, B. del Castillo, M. A. Martin, 1995, Cyclodextrin properties and applications of inclusion complex formation, *Ars. Pharm.*, **36**, 187–198 (1995).
91. M. Singh, R. Sharma, U. C. Banerjee, Biotechnological applications of cyclodextrins, *Biotechnol. Adv.*, **20**, 341–359 (2002).
92. E. Schneiderman, A. M. Stalcup, Cyclodextrin: a versatile tool in separation science, *J. Chromatogr. B*, **745**, 83–102 (2000).
93. J. Szejtli, Introduction and general overview of cyclodextrin chemistry, *Chem. Rev.*, **98**, 1743–1753 (1998).
94. A. R. Hedges, W. J. Shieh, C. T. Sikorski, Use of cyclodextrin for encapsulation in the use and treatment of food products, in *Encapsulation and Controlled Release of Food Ingredients*, S. J. Risch, G. A. Reineccius (Eds), ACS Symp. Ser., Vol. 590, American Chemical Society, Washington D. C., 1995, pp. 60–71.
95. (*a*) T. A. Reineccius, G. A. Reineccius, T. L. Peppard, Flavor release from cyclodextrin complexes: comparison of alpha, beta, and gamma types, *Food Chem. Toxicol.*, **68**, 1234–1239 (2003); (*b*) T. A. Reineccius, G. A. Reineccius, T. L. Peppard, Utilization of β-cyclodextrin for improved flavor retention in thermally processed foods, *Food Chem. Toxicol.*, **69**, FCT58–FCT62 (2004); (*c*) T. A. Reineccius, G. A. Reineccius, T. L. Peppard, The effect of solvent interactions on α-, β-, γ-cyclodextrin/flavor molecular inclusion complexes, *J. Agric. Food Chem.*, **53**, 388–392 (2005).
96. (*a*) T. Furuta, H. Yoshi, T. Kobayashi, T. Nishitarumi, A. Yasunishi, Powdery encapsulation of D-limonene by kneading with mixed powders of β-cyclodextrin and maltodextrin at low water content, *Biosci. Biotechnol. Biochem.*, **58**, 847–850 (1994); (*b*) J. S. Pagington, β-Cyclodextrin

and its uses in the flavour industry, in *Developments in Food Flavours*, G. G. Birch, M. G. Lindley (Eds), Elsevier Applied Science, London, 1986, pp. 31–150.

97. J. S. Pagington, Molecular encapsulation with β-cyclodextrin, *Food Flavor Ingredients Packag. Process.*, **7** (9), 51–55 (1985).
98. (*a*) C. G. Biliaderis, G. Galloway, Crystallization behavior of amylose-V complexes: structure–property relationships, *Carbohydrate Res.*, **189**, 31–48 (1989); (*b*) C. G. Biliaderis, Structures and phase transitions of starch in food systems, *Food Technol.*, **46** (6), 98–100 (1992); (*c*) M. C. Godet, A. Buleon, V. Tran, P. Colonna, Structural features of fatty acid-amylose complexes, *Carbohydrate Polym.*, **21**, 91–95 (1993).
99. G. Arvisenet, A. Voilley, N. Cayot, Retention of aroma compounds in starch matrices: competitions between aroma compounds toward amylose and amylopectin, *J. Agric. Food Chem.*, **50**, 7345–7349 (2002).
100. B. Conde-Petit, F. Escher, J. Nuessli, Structural features of starch–flavor complexation in food model systems, *Trends Food Sci. Technol.*, **17** (5), 227–235 (2006).
101. S. Immel, F. W. Lichtenthaler, The hydrophobic topographies of amylose and its blue iodine complex, *Starch/Stärke*, **52**, 1–8 (2000).
102. (*a*) O. Tapanapunnitikul, S. Chaiseri, D. G. Peterson, D. B. Thompson, Water solubility of flavor compounds influences formation of flavor inclusion complexes from dispersed high-amylose maize starch, *J. Agric. Food Chem.*, **56**, 220–226 (2008); (*b*) G. Wulff, G. Avgenaki, M. S. P. Guzmann, Molecular encapsulation of flavors as helical inclusion complexes of amylose, *J. Cereal Sci.*, **41**, 239–249 (2005).
103. P. Le-Bail, C. Rondeau, A. Blueon, Structural investigation of amylose complexes with small ligands: helical conformation, crystalline structure and thermostability, *Int. J. Biol. Macromol.*, **35**, 1–7 (2005).
104. C. Jouquand, V. Ducruet, P. Le-Bail, Formation of amylose complexes with C6-aroma compounds in starch dispersions and its impact on retention, *Food Chem.*, **96**, 461–470 (2006).
105. G. Arvisenet, P. Le-Bail, A. Voilley, N. Cayot, Influence of physicochemical interactions between amylose and aroma compounds on the retention of aroma in food-like matrices, *J. Agric. Food Chem.*, **50**, 7088–7093 (2002).
106. P. G. Smith, *Application of Fluidization to Food Processing*, Blackwell Publishing, Oxford, 2007, pp. 8–9.
107. E. Teunou, D. Poncelet, Batch and continuous fluid bed coating – review and state of the art, *J. Food Eng.*, **53**, 325–340 (2002).
108. D. M. Jones, Factors to consider in fluid-bed processing, *Pharm. Technol.*, **9** (4), 50–62 (1985).
109. O. P. Rubino, Fluid-bed technology. Overview and criteria for process selection, *Pharm. Technol.*, **23**, 104–117 (1999).
110. K. Dewettinck, A. Huyghebaert, Top-spray fluidized bed coating: effect of process variables on coating efficiency, *LWT-Food Sci. Technol.*, **31**, 568–575 (1998).
111. P. D. Hede, P. Bach, A. D. Jensen, Small scale top-spray fluidized bed coating: granule impact strength, agglomeration tendency and coating layer morphology, *Powder Technol.*, **176**, 156–167 (2007).
112. D. Solis-Morales, C. M. Saenz-Hernadez, E. Ortega-Rivas, Attrition reduction and quality improvement of coated puffed wheat by fluidized bed technology, *J. Food Eng.*, **93**, 236–241 (2009).
113. K. Dewettinck, A. Huyghebaert, Fluidized bed coating in food technology, *Food Sci. Technol.*, **10**, 163–168 (1999).
114. R. Wesdyk, Y. M. Joshi, J. de Vincentis, A. W. Newman, N. N. Jain, Factors affecting differences in film thickness of beads coated in fluidized bed unit, *Int. J. Pharm.*, **93**, 101–109 (1993).
115. J. L. Torres-Martinez, H. Mujica-Paz, A. Valdez-Fragoso, E. Ortega-Rivas, Coating of puffed wheat by a tumbling method and a fluidized bed technique, *Part. Sci. Technol.*, **25**, 549–554 (2007).
116. Z. J. Dong, A. Toure, C. S. Jia, X. M. Zhang, S. Y. Xu, Effect of processing parameters on the formation of spherical multinuclear microcapsules encapsulationg peppermint oil by coacervation, *J. Microencapsulation*, **24**, 634–646 (2007).

117. (*a*) H. Matsukawa, K. Saeki, T. Shimada, Process for preparing microcapsules containing hydrophobic oil drops (to Fuji Photo Film Co.), US 3,944,502 (1976); *Chem. Abstr.*, **76**, 145304 (1972); (*b*) T. Matsushita, M. Miyake, S. Shioi, T. Shiozaki, T. Shiraishi, Method of making oil containing microcapsules (to Kanzaki Paper Mfg. Co.), US 4,219,439 (1980); *Chem. Abstr.*, **94**, 74772 (1981).
118. M. Rabiskova, J. Valaskova, The influence of HLB on the encapsulation of oils by complex coavervation, *J. Microencapsulation*, **15**, 747–751 (1998).
119. V. B. Junyaprasert, A. Mitrevej, N. Sinchaipanid, P. Boonme, D. E. Wurster, Effect of process variables on the microencapsulation of vitamin A palmitate by gelatin-acacia coacervation, *Drug Dev. Ind. Pharm.*, **27**, 561–566 (2001).
120. D. J. Burgess, J. E. Carless, Microelectrophoretic studies of gelatin and acacia for the prediction of complex coacervation, *J. Colloid Interface. Sci.*, **98**, 1–8 (1984).
121. J. C. Soper, M. T. Thomas, Enzymatically protein encapsulating oil particles by complex coacervation (to Givaudan SA and Givaudan-Roure SA), US 6,039,901 (1997); *Chem. Abstr.*, **129**, 172773 (1998).
122. (*a*) J. Chirife, M. Karel, Volatile retention during freeze drying of aqueous suspensions of cellulose and starch, *J. Agric. Food Chem.*, **21**, 936–938 (1973); (*b*) J. Flink, M. Karel, Retention of organic volatiles in freeze dried solutions of carbohydrates, *J. Agric. Food Chem.*, **18**, 295–297 (1970).
123. (*a*) M. Rosenberg, I. J. Kopelman, Y. Talmon, A scanning electron microscopy study of microencapsulation, *J. Food Sci.*, **50**, 139–144 (1985); (*b*) J. Imagi, H. Kako, K. Nakanishi, R. Matsuno, Entrapment of liquid lipids in matrices of saccharides, *J. Food Eng.*, **12**, 207–222 (1990).
124. T. Nagata, Techniques and application of electron microscopic radioautography. *J. Electron Microsc.*, **45**, 258–274 (1996).
125. Y. Minemoto, S. Adachi, R. Matsuno, Comparison of oxidation of methyl linoleate encapsulated with gum arabic by hot air drying and freeze drying, *J. Agric. Food Chem.*, **45**, 4530–4534 (1997).
126. A. B. Rizzuto, A. C. Chen, M. F. Veiga, Modification of the sucrose crystal structure to enhance pharmaceutical properties of excipient and drug substances, *Pharm. Technol.*, **8** (9), 32–39 (1984).
127. (*a*) R. D. Smith, Supercritical fluid molecular spray thin films and fine powders (to Battelle Memorial Institute), US 4,734,451 (1988); *Chem. Abstr.*, **109**, 152362 (1988); (*b*) R. E. Sievers, B. M. Hybertson, B. N. Hansen, Methods and apparatus for drug delivery using supercritical solutions, US 5,301,664 (1994).
128. S. D. Yeo, E. Kiran, Formation of polymer particles with supercritical fluids: a review, *J. Supercrit. Fluids*, **34**, 287–308 (2005).
129. S. R. Veith, M. Perron, S. E. Pratsinis, Encapsulation and retention of decanoic acid in sol-gel-made silicas, *J. Colloid Interface Sci.*, **283**, 495–502 (2005).
130. B. F. Sagar, A. J. G. Sagar, S. G. Graham, R. T. Wragg, Method for encapsulating substances in biocapsules (to British Textile Technology Group), US 5,660,769 (1997); *Chem. Abstr.*, **121**, 296642 (1994).
131. G. Dardelle, V. Normand, M. Steenhoudt, P.-E. Bouquerand, M. Chevalier, P. Baumgartner, Flavour encapsulation and flavour release performances of a commercial yeast based delivery system, *Food Hydrocolloids*, **21**, 953–960 (2007).
132. B. M. Bugarski, B. Obradovic, V. A. Nedovic, D. Poncelet, Immobilization of cells and enzymes using an electrostatic droplet generator, in *Focus on Biotechnology*, Vol. 8a (Fundamentals of Cell Immobilization Biotechnology), V. A. Nedovic, R. G. Willaert (Eds), Kluwer Academic Publishers, Dordrecht, 2004, pp. 274–294.
133. V. Manojlovic, N. Rajic, J. Djonlagic, B. Obradovic, V. Nedovic, B. Bugarski, Application of electrostatic extrusion – flavor encapsulation and controlled release, *Sensors*, **8**, 1488–1496 (2008).
134. K. A. Heitfeld, T. Guo, G. Yang, D. W. Schaefer, Temperature responsive hydroxypropyl cellulose for encapsulation, *Mater. Sci. Eng., C*, **28**, 374–379 (2008).

13

Profragrances and Properfumes

Andreas Herrmann

Firmenich SA, Division Recherche et Développement,
1 Route des Jeunes, B. P. 239, 1211 Genève 8, Switzerland

13.1 Introduction

To fulfil their biological activity, volatile organic molecules have to be efficiently evaporated and transported by diffusion through the air to eventually reach a specific target where they are perceived as a selective signal. Bioactive volatiles are therefore generally characterized by a unique molecular structure as a prerequisite for selective recognition, combined with relatively low molecular weights, typically correlating with high vapour pressures (volatilities), to allow for an efficient evaporation into the environment. Once released into the air, the high vapour pressures of volatile compounds result in a limited duration of their biological activity as a consequence of rapid diffusion. Nevertheless, by generating volatiles from nonvolatile precursors, nature has set up mechanisms for the selective and timely release of volatile compounds to ensure their structure-related perception after delivery to the biological target via transport through the air.

As a consequence of their pleasant taste or smell to humans, many bioactive volatiles have been used since antiquity as flavours and fragrances in our everyday life.[1,2] Due to their low olfactory thresholds,[3] flavours and fragrances can easily be detected by humans, even at very low concentrations.

Since the performance of perfumed consumer articles is often judged by the duration of fragrance perception, the development of delivery systems to control the release of natural and synthetic fragrance raw materials has become an important research area in the flavour and fragrance industry. Apart from the development of specific matrices or capsules for the diffusion-controlled release of volatiles,[4] the preparation of fragrance precursors, so-called 'profragrances' (if one single perfumery raw material is released) or

The Chemistry and Biology of Volatiles Edited by Andreas Herrmann

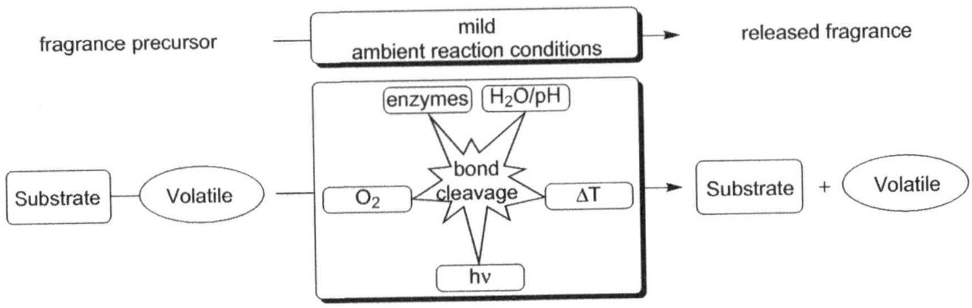

Scheme 13.1 *Triggers used for the controlled release of bioactive volatiles by covalent bond cleavage under ambient reaction conditions*

'properfumes' (if several different volatiles can be generated from the same molecule) has systematically been investigated.[5] The active volatile is covalently bound to a well defined substrate to form an ideally nonvolatile and odourless precursor, the size of which can range from small molecular systems to polymer structures[5,6] and which, especially in the context of macromolecular structures, is often denominated as 'conjugate'. For fragrance release, the covalent bond has to be cleaved under mild reaction conditions which are prevalent in the environment during the application of the product. Typical triggers that have successfully been used to release volatiles from their corresponding precursors comprise variations of temperature, exposure to light, aqueous hydrolysis (including change in pH), oxidation and the action of enzymes from different microorganisms or bacteria (Scheme 13.1).[5]

Whereas the encapsulation and diffusion-controlled release of volatiles from capsules or matrices mainly depends on the physicochemical parameters of the compounds, covalent bond cleavage reactions depend on the structure of the precursors which selectively generate specific chemical functionalities such as alcohols, carbonyl compounds, amines, alkenes, esters and lactones. This means that a given precursor generally releases only one class of compounds. In the case of insect attractants or repellents, this is not a problem, as usually one single compound with a well defined structure has to be released to interact with one specific species. However, in perfumery, where mixtures of many structurally different volatiles are used, the release of a single compound might seem to be an important limitation of the technique. Nevertheless, it has been shown in many cases that it is sufficient to select one, usually highly volatile, key note of a perfume composition and to prolong its effect in a perfume mixture. It is interesting to note that, although it would theoretically be possible to encapsulate an entire perfume, most fragrance capsules are used to release only a few selective compounds of the perfume mixture. Despite the fact that the concept of releasing volatile compounds by covalent bond cleavage of a nonvolatile precursor is generally applicable for all types of volatiles, almost all developments of this technique have been devoted to the flavour and fragrance industry.[5–7]

In view of the considerable commercial interest in these systems, most of the literature on properfumes and profragrances consists of patent applications.[5] However, as the goal of this contribution is not to give a complete overview of the topic, but rather illustrate the main concepts of profragrance and properfume technologies using selected examples, this review is mainly based on some recent developments reported in the scientific literature. Furthermore, as alcohols and carbonyl compounds are the most important functionalities released by covalent bond cleavage reactions,[7] we mainly focus on the release of these two classes of compounds. In the following discussion, the structures of the volatiles released from a given precursor are highlighted with a frame in the corresponding figures and schemes.

Although this chapter mainly concentrates on fragrance delivery, it is quite evident that the described release strategies can generally be applied to control the evaporation of bioactive volatiles for other purposes.

13.2 Release of Alcohols

Despite their polar hydroxyl group, most flavour and fragrance alcohols are rather hydrophobic and are only sparingly soluble in water. To store, transport and release hydrophobic volatile compounds, nature has thus generated a broad variety of nonvolatile bioconjugates. These natural, nonvolatile precursors, or their synthetic analogues, are potential replacements for the free fragrance compound in practical applications, if suitable reaction conditions are available which cleave the precursors and thus release the corresponding flavour or fragrance compound at a desired rate. In the case of natural precursors, the most evident approach to control the release of volatiles is to use the enzymes and microorganisms which are present in our everyday environment.

13.2.1 Enzymatic Hydrolysis

Together with phosphate esters, glycosides represent one of the most important classes of precursors for natural flavour and fragrance alcohols in fruits or other plant tissues.[8] Glycosylated volatiles display an increased water solubility for storage and transport within the plant, and are less reactive than the corresponding free aglycones which might damage the relatively labile cell membrane structure as a consequence of their high lipophilicity. Typical volatiles which are bound to glycosides and liberated by the action of glycoside hydrolases or glycoside transferases comprise mono- or sesquiterpenols, aliphatic alcohols, phenylpropanol derivatives and a variety of norisoprenoids.[8] The precursors can be isolated from suitable plant material, or prepared in enzymatic reactions by a variety of glycosyltransferases.[9] In general, the aglycone is directly attached to a β-D-glucose moiety, the latter of which may be further linked to a second carbohydrate. While monosaccharides are directly cleaved by β-glucosidases, the release of the volatiles from disaccharides occurs in a two-step sequence with the first step being the cleavage of the two carbohydrate units.[8] Furthermore, the structure of the volatile has an influence on the activity of the β-glucosidases. Primary and secondary alcohols are successfully hydrolysed by plant and microbial enzymes, whereas tertiary alcohols are efficiently released by glucosidases from fungi.

The presence of glycosidases on the skin or in skin bacteria allows the cleavage of a broad variety of mono- and disaccharides and thus the release of fragrances from natural or synthetic glycoconjugates. Ikemoto *et al.* showed that incubation of several types of skin bacteria with natural β-D-glucosides such as **1** and **2** (Figure 13.1) resulted in the release of 4-(4-hydroxyphenyl)butan-2-one (raspberry ketone) and 2-phenylethanol (phenethylol), respectively, as fragrance materials.[10] Besides providing an increased duration of fragrance

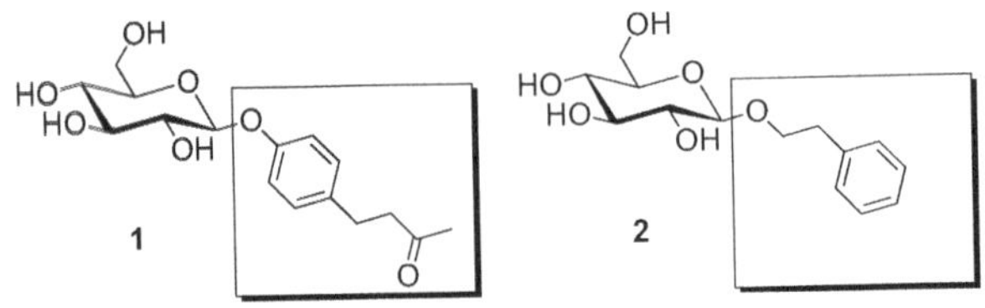

Figure 13.1 *β-D-Glucosides **1** and **2** as precursors for β-glucosidase-mediated alcohol release*

perception, the profragrances also helped in covering body odours, as for example those formed in the human axillary region.

Human axillar malodours originate from initially odourless sweat which is secreted by the apocrine glands and transformed by bacterial enzymes of the skin microflora.[11,12] A series of sulfanylalkanols[13] and structurally related branched hexanoic and hexenoic acids[14] have been identified as the most important compounds contributing to the typical human axillary malodour. These compounds are generated from amino acid or peptide conjugates such as **3**[15] or **4**[16] by pyridoxal phosphate-dependent β-lyases or Zn^{2+}-dependent aminoacylases from *Staphylococci* or *Corynebacteria* (Scheme 13.2). The structure elucidation of malodour precursors and the determination of their enzymatic degradation by skin bacteria are the key steps for the development of new deodorant strategies, either by developing efficient enzyme inhibitors or by preparing competitive substrates which release volatiles with a neutral or pleasant smell.[17]

Based on the fact that the same Zn^{2+}-dependent aminoacylase from *Corynebacteria* releases a variety of hexenoic acid derivatives,[16] Natsch and coworkers showed that the enzyme is indeed strongly selective to the L-glutamine residue but tolerates structural variations at the acyl part of the substrate. When fed to the *Corynebacteria* at the moment of the malodour formation, carbamate **5** releases 3-methyl-5-phenylpentanol (Phenoxanol) as

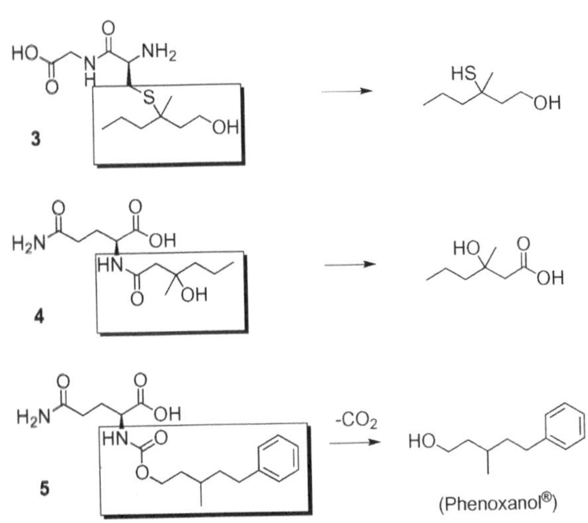

Scheme 13.2 *Generation of volatiles by bacterial enzymes on the skin*

a masking fragrance alcohol together with CO_2 and glutamine (Scheme 13.2).[18] Furthermore, *in vitro* measurements showed that **5** is a valuable competitive substrate to malodour-generating conjugate **4**, as in its presence less malodour acids were formed. Carbamate **5** was also successfully cleaved *in vitro* upon incubation with *Corynebacteria*, or after application to human underarm skin in the form of a deodorant.[19]

Lipases are another important class of enzymes commonly encountered in association with living organisms. They are for example produced and secreted by bacteria,[20] and can thus be found as extracellular enzymes on the stratum corneum of the skin.[12] They primarily hydrolyse glycerol esters of long-chain fatty acids and have therefore been developed as ingredients for stain removal in detergents,[21] but they are in general relatively tolerant towards a broad variety of carboxylic esters. Due to their ability to support elevated temperatures, high pH values and the presence of surfactants, lipases are nowadays present in many consumer products. When added to laundry detergents, modern enzymes survive a typical machine washing cycle and are, in the presence of ambient humidity, still active on dry fabric. Detergent enzymes can thus be used to trigger the cleavage of carboxylates in laundry applications.

Geraniol-releasing carboxylates, such as digeranyl succinate (**6**) – or its analogue obtained from a mixture of geraniol and nerol – and geranyl palmitate (Hexarose, **7**, Figure 13.2) have thus been successfully commercialized as ingredients in fabric softener formulations.[22,23] Addition of the profragrances to the fabric softener rather than to the enzyme-containing detergent has the advantage that the precursor is separated from the enzyme and thus not degraded during product storage. In the washing cycle, the two compounds are consecutively deposited onto the fabric where the enzyme starts to cleave the ester bond of the profragrance. Panel evaluations carried out on dried textiles that had been washed with a lipase (Lipolase 100T) containing detergent powder and then rinsed with a fabric softener which contained profragrance **6** showed an increased duration of fragrance release in comparison to textiles treated with the free fragrance alcohol.[22]

Similarly, properfume **8** generated an equimolar mixture of geraniol and Hedione (methyl 3-oxo-2-pentyl-1-cyclopentaneacetate) as a binary fragrance accord in a stepwise reaction sequence. In a first step, enzymatic cleavage afforded geraniol and a sterically hindered enol ester of succinic acid which then, in a second step, releases Hedione by hydrolysis or intramolecular substitution, while forming succinic anhydride.[19] Thin-layer chromatography (TLC) of buffered solutions at different pH values showed that the release of the

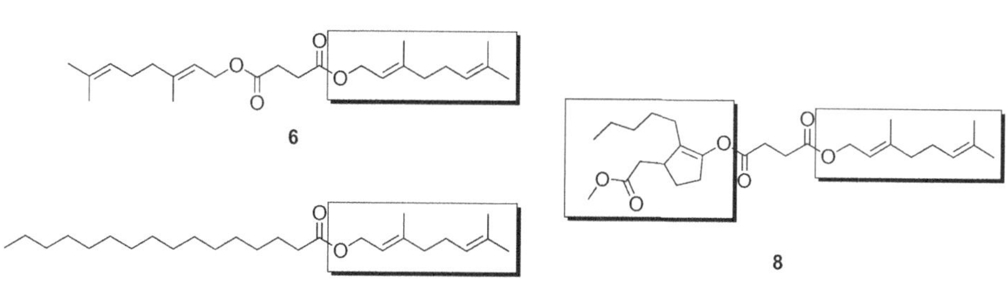

7 (Hexarose®)

Figure 13.2 *Carboxylates (**6–8**) as precursors for lipase-activated release of alcohols*

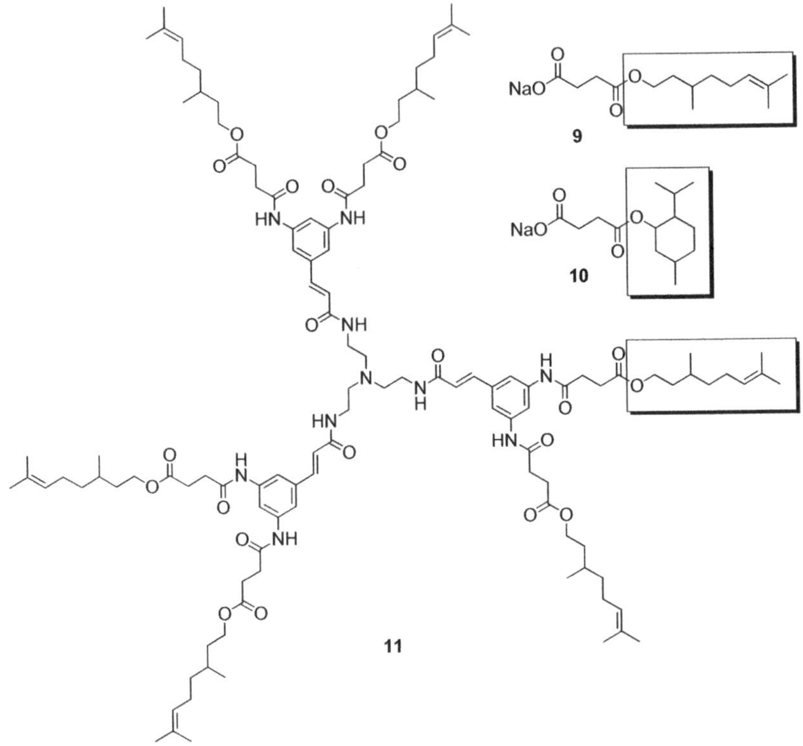

Figure 13.3 *Linear and dendritic succinates (**9–11**) as lipase-sensitive profragrances*

fragrance accord occurred selectively in the presence of the enzyme (Lipolase), while in its absence precursor **8** remained practically unchanged.

Monosuccinates of primary and secondary fragrance alcohols such as **9** or **10** (Figure 13.3) were found to form micelles in aqueous solutions. In the absence of enzymes, the compounds are relatively stable, only undergoing hydrolysis in strongly alkaline conditions.[24] To investigate the enzyme-catalysed cleavage of carboxylates from larger structures such as polymers or dendrimers, Hayes and coworkers prepared a series of branched polyamides such as **11** by grafting the acid form of monosuccinates **9** or **10** to the primary amines of a dendrimer substrate.[25] Hydrolysis experiments were performed in an aqueous buffer at neutral pH using a lipase (from *Candida cylindracea*) or a cutinase (from *Fusarium solani pisii*). The enzymes were found to be highly selective for the hydrolysis of primary alcohols. Furthermore, it was shown that the efficiency of ester cleavage decreases with an increase of bulkiness and rigidity of the polyamide substrate and, at the same time, depends on the type of enzyme used. Whereas the cutinase cleaved the ester bonds in **11**, no citronellol was released in the presence of the lipase.[25]

Lipases were also found to hydrolyse carbonates such as (*Z*)-3-hexenol (leaf alcohol) precursor **12** (Figure 13.4).[19] When incubated with buffer solutions at different pH, profragrance **12** was found to be stable over a pH range from 2.5 to 9.0, as shown by

TLC analysis. However, in the presence of lipase the alcohol was successfully released. Nevertheless, after prewashing with a lipase-containing detergent and adding **12** together with the fabric softener in the rinsing cycle, the typical fresh, green note of the leaf alcohol was only perceived for a limited period of time on dry fabric.

To increase the duration of fragrance perception, Flachsmann *et al.* investigated the possibility of using anchimeric assistance from free OH, COOH or NH_2 groups in proximity to the carbonate, by preparing a series of commercially available polyol derivatives.[19] Panel evaluations and headspace analysis on dry fabric showed that polycarbonate **13** efficiently released (*Z*)-3-hexenol during more than three weeks, independently of the fact whether or not the fabric was washed with a lipase-containing detergent. Precursor **13** was obtained as the main product of a compound mixture, in which symmetrical carbonate **14**, tricarbonates **15**–**17** and tetracarbonate **18** could be identified as side-products (Figure 13.4). Furthermore, upon ageing, a freshly prepared sample of **13** spontaneously changed its composition, and the stepwise formation of a series of cyclic carbonates was observed.[19] With the exception of carbonate **14** (which was found to be stable), the reactions and rearrangements of the other compounds liberate (*Z*)-3-hexenol, and may thus explain the observed duration of the olfactive effect in application.

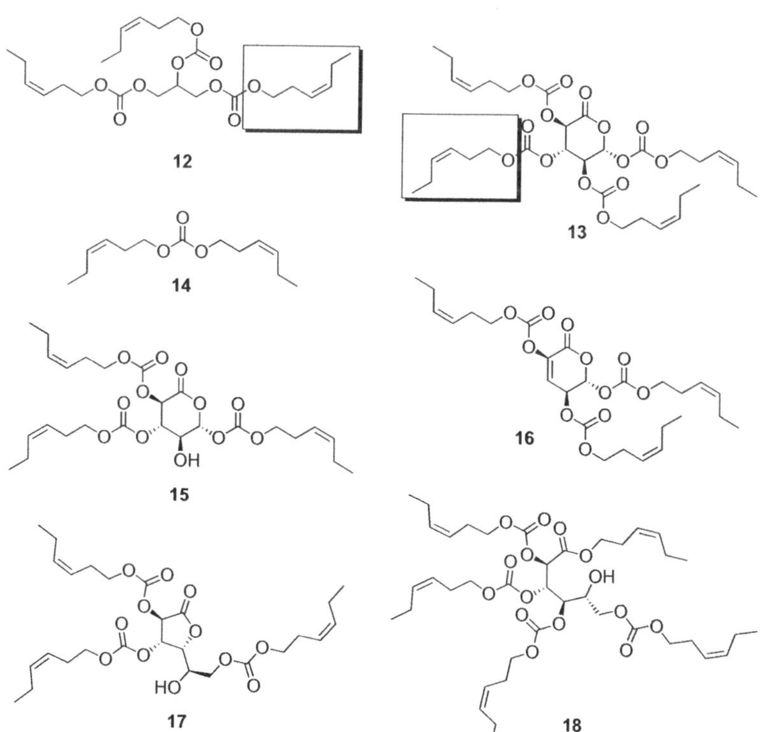

Figure 13.4 *Polyol carbonates **12–18** as hydrolytically cleavable precursors in the presence or absence of enzymes*

13.2.2 Neighbouring-Group-Assisted, Non-Enzymatic Hydrolysis

Despite their high efficiency to cleave covalent bonds under mild reaction conditions at neutral pH, several drawbacks limit the use of enzymes in practical applications. The high structural selectivity of enzymes and their sensitivity towards bulky and rigid substrates might result in insufficient enzyme activity with respect to certain substrates under various application conditions. Furthermore, as a consequence of possible side effects such as skin irritations and allergic reactions which may be related to enzyme activity, consumers are more and more concerned about their presence in consumer products.[5]

Taking inspiration from enzymatic processes, where the spatial arrangement of specific functionalities in close proximity to the substrate to be cleaved favours a given transformation, the concept of intramolecular catalysis[26] or neighbouring-group participation[27] has been developed. This principle generally covers intramolecular acid and base catalysis or nucleophilic participation (with or without anchimeric assistance) and allows the cleavage of covalent bonds under relatively mild reaction conditions.

The neighbouring-group-assisted hydrolysis of benzoates (and some related structures) is a typical example of a mild reaction using the participation of nucleophilic intermediates that has successfully been explored for the controlled release of volatiles at neutral pH.[5] The general principle of the concept is illustrated in Scheme 13.3.

Although stable under acidic conditions, the precursors are deprotonated under neutral or slightly alkaline conditions to generate the active nucleophilic species in close proximity to the ester bond to be cleaved. The release of the volatile is triggered by a change of pH from acidic to neutral or slightly alkaline conditions, as for example by dilution in a typical fabric softener application. Intramolecular cyclization by attack of the nucleophile (Nu^-) on the carboxyl group then releases the volatile alcohol. In some cases, hydrolysis of the cyclic species regenerates the open form of the protonated nucleophile (Scheme 13.3). This concept is applicable to a broad variety of structures. The rate of fragrance release depends on the structure of the alcohol to be generated in the intramolecular reaction (primary alcohols are generally more efficiently released than secondary or tertiary alcohols), the nature of the nucleophile and the orientation of the nucleophile towards the carboxyl group.

Figure 13.5 lists the second-order rate constants measured in buffered solutions of water/acetonitrile 2:1 for geraniol releasing precursors **19–27**.[28] According to the structure of the substrates, the rate constants were found to vary between 1735 (**19**) and 0.2 l and 0.2

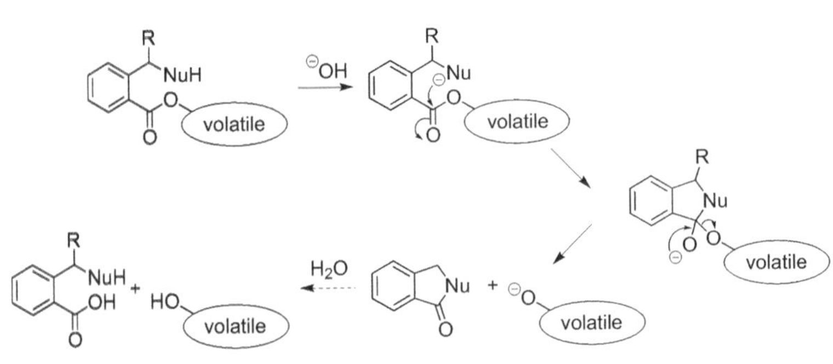

__Scheme 13.3__ Mechanism for the neighbouring-group-assisted hydrolysis of benzoates for the controlled release of volatile alcohols (HO-volatile; Nu = nucleophile)

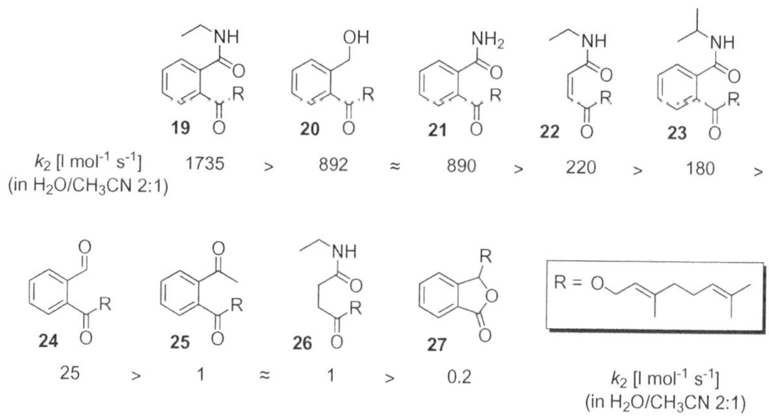

Figure 13.5 *Structures and second-order rate constants for the neighbouring-group-assisted release of geraniol*

l mol^{-1} s^{-1} (**27**), thus spanning a range of almost four orders of magnitude. For practical applications this is particularly interesting, as the substrate structure can be adapted depending on the desired release rate.

2-(Ethylaminocarbonyl)benzoate **19** reacts faster than the unsubstituted 2-(aminocarbonyl) derivative **21** or the sterically more hindered *iso*-propyl analogue **23**. Forcing the nucleophile into a favourable position for intramolecular cyclization increases the rate of hydrolysis as shown for the series of profragrances **19**, **22** and **26**. The presence of a double bond in the (*Z*)-configuration increases the rate constant for the cyclization of maleate **22** by a factor of more than 200 with respect to the corresponding succinate **26**, where the carbamoyl moiety can freely rotate around the central carbon-carbon bond of the molecule.

The kinetic measurements showed that 2-formyl and 2-acylbenzoate **24** and **25** cyclized much slower than precursors **19**–**23** (Figure 13.5).[29] This is presumably due to the fact that **24** and **25** need to be hydrated prior to the formation of the nucleophile in alkaline media.[30] The smallest rate constants were measured for phthalide **27**,[29] where lactone opening with an external hydroxide ion is the rate-determining step and therefore relatively slow.

Due to their ease of preparation in a two-step reaction sequence and the fact that they successfully release tertiary alcohols at interesting rates for practical applications,[28,31] 2-carbamoylbenzoates releasing 2-methyl-1-phenyl-2-propanol were chosen as substrates to investigate the physicochemical parameters influencing release of tertiary alcohols as a function of molecular size. At the example of 'dimer' **28** as substrate, it was shown that the release of the tertiary alcohol occurs via a multistep reaction sequence consisting of two cyclization and two reopening steps of the precursor (Scheme 13.4), the latter two being without consequence for the release of the volatile alcohol.[31] The individual reaction intermediates (mono- and diphthalimides as well as the reopened species) were easily separated by analytical high performance liquid chromatography (HPLC) in buffered solutions of water/acetonitrile 2 : 1 at neutral pH, and the rate constants of the individual steps could be determined. The cyclization reactions were found to be faster than the reopening of the phthalimides. Baseline separation of quite large linear or spherical

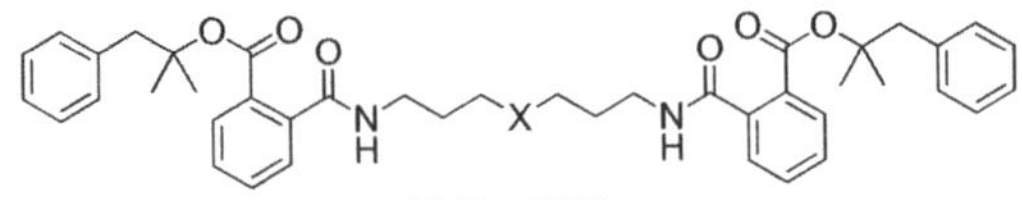

28: X = NCH_3
31: X = CH_2

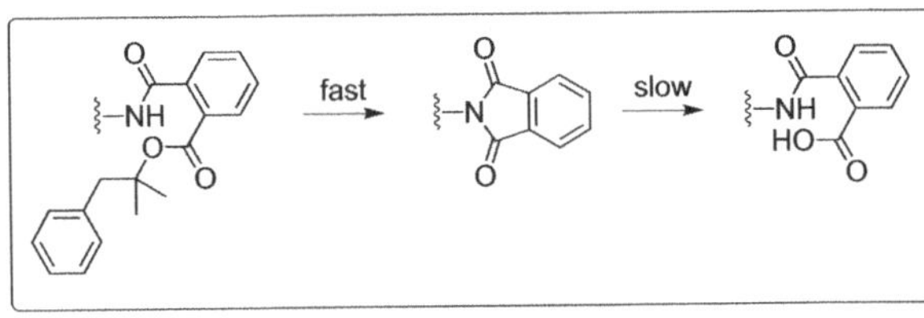

Scheme 13.4 *Stepwise cyclization and re-opening of 2-carbamoylbenzoates*

structures such as stylomer **29** or dendrimer **30** (Figure 13.6) from their corresponding mono-phthalimide derivatives by analytical HPLC allowed the comparison of the rate constants for a series of compounds with different shape, size and number of 2-carbamoylbenzoate units per molecule.[32,33] Due to the limited solubility of the larger structures in the reaction medium (and to take into account the presence of surfactants in many practical applications) the rate constants for the neighbouring-group-assisted cyclizations were measured in the presence of a non-ionic surfactant.

Comparison of the rate constants measured for a series of molecules with an increasing number of 2-carbamoylbenzoate end-groups showed that, from a certain size on, there seemed to be no significant size dependence for the release rates, if the total amount of alcohol to be released was kept constant in all samples.[33] The kinetic data show that the polarity of the precursors, which is related to their solubility in the aqueous solution, has a stronger influence on the rates of alcohol release than the shape (linear or spherical) or the size (number of release units) of the conjugates.

It was also noticed that the presence of catalysing groups in close proximity to the release unit has a strong influence on the release rates. In the case of 2-carbamoylbenzoates, the presence of the polar, tertiary amine function in the spacer between the alcohol-releasing moieties in **28**, as well as in macromolecules **29** and **30**, considerably accelerates the rate of ester hydrolysis. Comparison with precursor **31**, which is lacking the tertiary amino function, shows that the release of the tertiary alcohol is slowed down by several orders of magnitude.[33] The understanding of the physicochemical parameters which have an impact on the environment close to the covalent bond cleavage site plays an important role in the conception of polymeric delivery systems.

Kamogawa *et al.* reported that the release of fragrance alcohols from 4-styrenesulfonate/*N*-vinyl-2-pyrrolidone co-polymers strongly depends on the structure of the leaving alcohol.[34] Figure 13.7 indicates the extent of hydrolysis of co-polymers **32**–**35** determined in an aqueous solution of dioxane at 20 °C after 24 h. The different amounts of alcohols released from the polymer conjugates were attributed to the possible stabilization of an intermediately formed carbocation by electron-donating groups such as the double bond in **32**, or by inductive effects of secondary alkyl groups as in **33**, which increase the rates of hydrolysis with respect to the nonconjugated primary alcohol-releasing polymer **34**. A particular case is co-polymer **35** which readily released citronellol under the given reaction

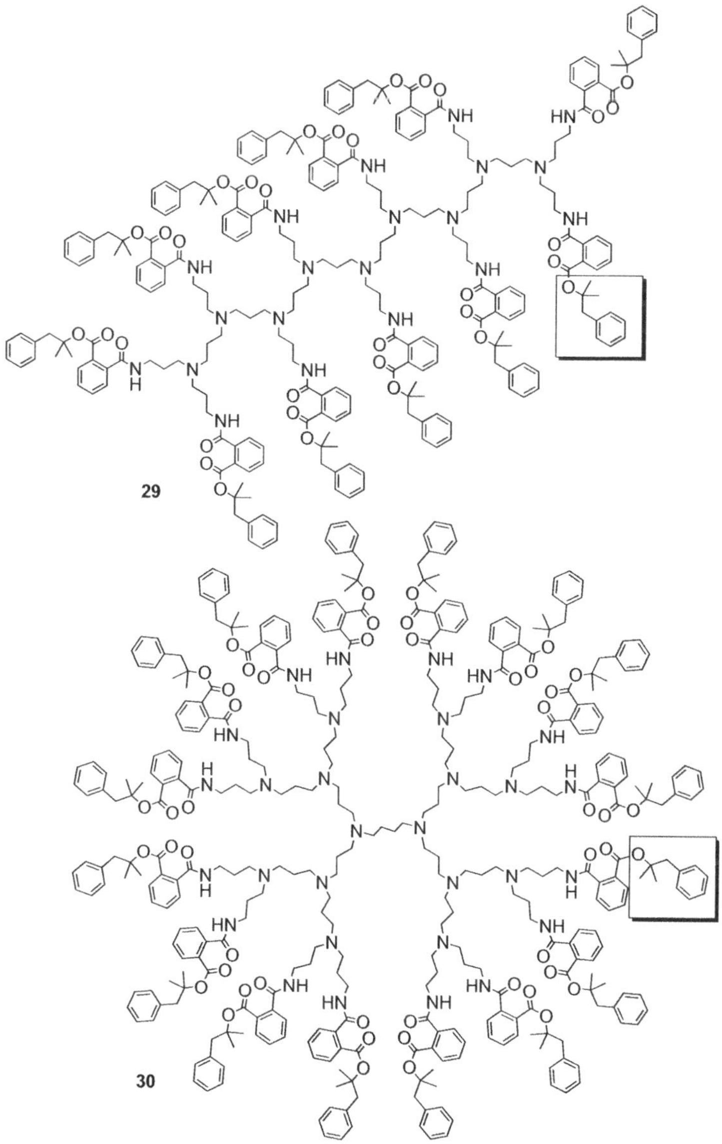

29

30

Figure 13.6 *Linear stylomer **29** and spherical dendrimer **30** as precursors for the neighbouring-group-assisted release of tertiary alcohols*

conditions, presumably as a result of the neighbouring-group effect of the π-electrons of the alkyl group in the leaving alcohol (Figure 13.7).[34]

Some additional aspects arising from the polymer structure itself are further discussed below.

An alternative method to efficiently release volatile alcohols by a neighbouring-group-assisted intramolecular cyclization reaction was reported by Anderson and Fráter,[35] and

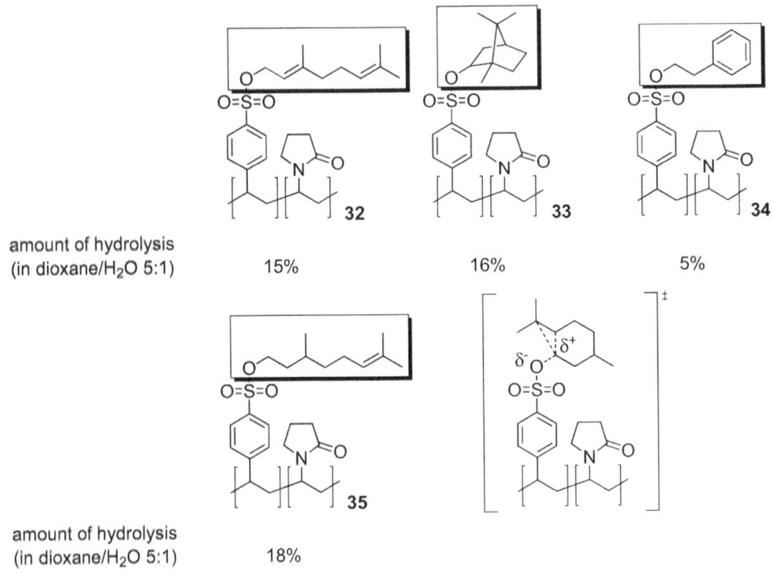

Figure 13.7 *Poly(styrenesulfonate) co-polymers (**32–35**) as hydrolytically cleavable alcohol precursors*

further developed by Flachsmann and coworkers.[36] Upon exposure to daylight, the double bond of *o*-hydroxy cinnamates isomerizes to the (*Z*)-isomer, which then spontaneously cyclizes into coumarin (2-chromenone) while releasing an alcohol (Scheme 13.5). In general, a low-energy photon is sufficient to induce the required (*E*/*Z*)-isomerization of the cinnamate. As coumarin and its derivatives are important perfumery ingredients, this process is highly atom-economic as all precursor atoms are used for the generation of an equimolar mixture of two different fragrance raw materials in a fixed 1:1 ratio. Properfume **36**, releasing coumarin and 9-decenol (Rosalva), was recently commercialized under the trade name Tonkarose.

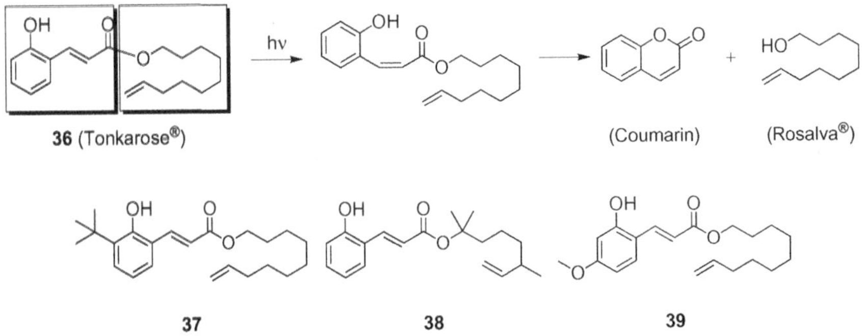

Scheme 13.5 *Reaction pathway for the light-induced formation of coumarin from o-hydroxy cinnamate **36***

As seen above for the 2-carbamoylbenzoates, structural variations of the precursor structure influence the rate of fragrance release. Sterically demanding substituents either close to the reactive phenyl group (as in **37**) or next to the ester to be hydrolysed (as in **38**) considerably decreased the rate of generation of the volatiles (Scheme 13.5). Irradiation of **38** (0.5 mM in acetonitrile/water 9 : 1) for 30 min with a mercury medium pressure lamp yielded a mixture of (*Z*)- and (*E*)-coumarates, the former of which, when isolated in pure form and kept in the dark, released dihydromyrcenol (2,6-dimethyl-7-octen-2-ol) during an entire week.[36] In the case of the 7′-methoxy analogue **39**, the observed decrease in the rate of lactonization was attributed to the reduced electrophilicity of the ester function due to the presence of the methoxy substituent and a possible quenching of the excited states involved in the photoisomerization by 7-methoxycoumarin (methylumbelliferon) formed during the reaction.[36]

Controlling the release of volatiles by using light as the trigger is very attractive for practical applications, as the photosensitive precursors can easily be protected from light by storage in opaque packing materials. To be perceived by their biological target, volatiles have to be deposited onto a surface from which they evaporate into the air. As the surfaces are generally exposed to natural daylight, the photochemically induced release starts with the deposition of the precursor onto the target surface.

Besides effecting the isomerization of double bonds, natural daylight is one of the most important energy sources for many biological processes involving the formation and breaking of covalent bonds. Photocleavable molecules, so-called 'caged compounds', have found numerous applications in a variety of bioorganic systems.[5,37,38] To be effective under realistic application conditions, photochemical reactions have to proceed in daylight in a polar solution, preferably in water, and tolerate the presence of molecular oxygen,[5] which has been reported to efficiently quench the formation of intermediate biradicals in photochemical reactions.

Several photochemical reactions fulfilling these criteria have been described. Pika *et al.*, for example, investigated the light-induced cleavage of 2-benzoylbenzoates for the controlled release of primary and secondary fragrance alcohols.[39] This reaction, previously reported by Porter and coworkers,[40] proceeds via an excited triplet state by hydrogen abstraction and intramolecular lactonization of the intermediate 1,4-biradical, to generate the alcohol together with a cyclized product arising from the photoreactive 2-benzoylbenzoate moiety (Scheme 13.6). Unsubstituted 2-benzoylbenzoates, such as **40**, react only

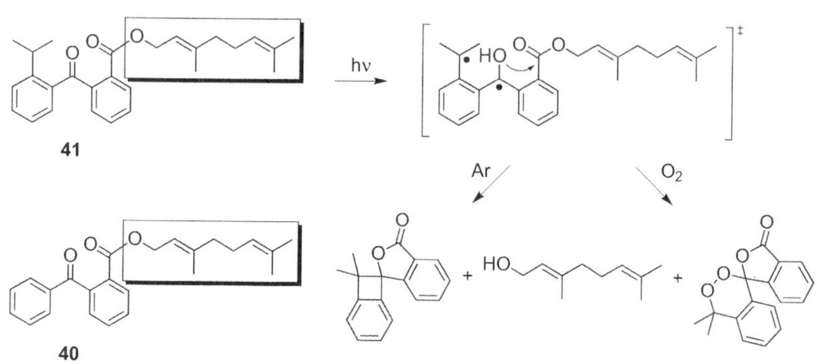

Scheme 13.6 *Mechanism for the photochemical reaction of 2-benzoylbenzoates in the presence and absence of oxygen*

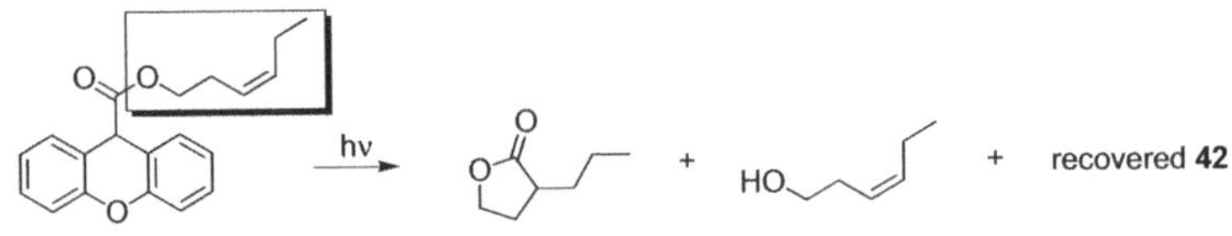

Scheme 13.7 *Photofragmentation of xanthenoic ester* ***42***

in the presence of an electron-donating molecule or in solvents bearing abstractable hydrogen atoms.[39] This problem could be circumvented by placing a secondary alkyl group in close proximity to the photoactive carbonyl group as in **41**, which readily released geraniol, independently of the choice of the solvent. The reaction was found to proceed in the presence and absence of oxygen; however, depending on the reaction conditions (degassed or undegassed solution), two different polycyclic reaction products were formed in addition to the expected alcohol (Scheme 13.6).

The investigation of photofragmentation mechanisms is an important prerequisite for understanding the possible formation of side-products and thus to enhance the efficiency of light-induced release reactions. Alkenyl xanthenoic esters, such as **42**, form small amounts of volatile alcohols as a result of the homolytic cleavage of the ester bond (Scheme 13.7). After irradiation at wavelengths above 300 nm for two to four days, lactones with different ring sizes were obtained as the major reaction products, by intramolecular cyclization of photochemically generated formyl radicals.[41] The ring size was found to depend on the position of the double bond in the ester side chain. However, due to the limited performance of the reaction to release olfactively interesting compounds, xanthenoic esters were not further investigated as fragrance precursors in practical applications.[36]

13.3 Release of Carbonyl Derivatives

Precursors of volatile aldehydes are therefore not only interesting for a control of their evaporation, but also as a means to increase their lifetime in different media. The carbonyl group of aldehydes and ketones can be generated from different structures under various reaction conditions, as for example by mild oxidation.[5] Whereas ketones are generally stable, aldehydes are readily degraded by oxidation or polymerization during storage in various consumer product formulations.

13.3.1 Oxidations

Analogous to the light-induced release of alcohols described in the previous section, photochemical pathways also allow the generation of carbonyl derivatives, especially if the photoreaction contains an oxidative step. Upon irradiation at wavelengths between 350 and 370 nm, 2-oxoacetates (α-keto esters) undergo photo-oxidation to generate aldehydes and ketones in good yields.[42,43] This reaction, generally referred to as a Norrish type II photofragmentation, proceeds in polar solvents and in the presence of oxygen[44] and is particularly suitable for the controlled release of volatile carbonyl compounds under mild everyday conditions.[45] According to the general mechanism described in Scheme 13.8,

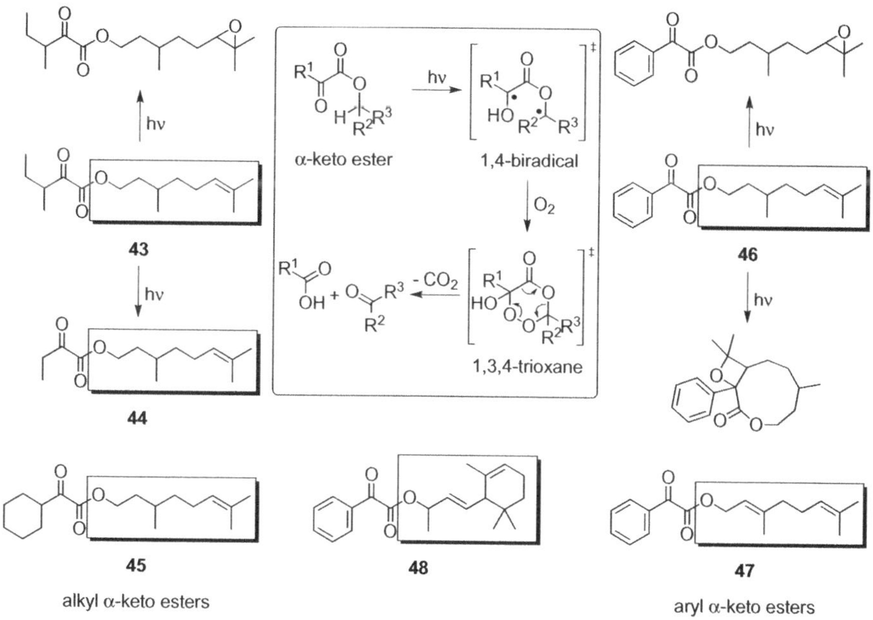

Scheme 13.8 *Mechanism for the Norrish type II photofragmentation of alkyl (**43–45**) and aryl (**46–48**) α-keto esters in the presence of oxygen*

photoirradiation of α-keto esters generates (by intramolecular hydrogen abstraction from the carbon atom next to the ester alkyl oxygen atom) a 1,4-biradical which can react with oxygen to give an intermediate 1,3,4-trioxane species, rearrangement of which then affords the desired carbonyl compound together with a carboxylic acid. Both alkyl and aryl 2-oxoacetates were successfully used to release a series of fragrance aldehydes and ketones under realistic application conditions.[43,46] Understanding the mechanistic pathway of the photoreaction allows an optimization of the structure of the profragrance substrates on both sides of the photoreactive keto group, and thus to minimize side reactions. In addition to the desired ester chain fragmentation to generate the carbonyl compounds, alkyl α-keto esters such as **43** can also undergo alkyl chain fragmentation, resulting in another keto ester with a shorter alkyl chain (**44**). Although this reaction does not suppress the desired photo-oxidation, it can be avoided by choosing a cyclic ring system (such as in **45**), where hydrogen abstraction is sterically impossible. Intramolecular cyclization or epoxidation of aryl keto ester **46** also results in the formation of unwanted side-products which, however, can be prevented by choosing an aldehyde to be released other than citronellal.[42,43] Due to their easy preparation from commercially available 2-oxo-(2-phenyl)acetic acid (phenylglyoxylic acid) and its methyl or ethyl esters, aryl α-keto esters such as **47** and **48** are particularly interesting for practical applications.[47] Besides the desired aldehydes, the formation of the corresponding alcohol was also observed in some cases. Although partial hydrolysis of the keto ester cannot be ruled out, it appears that alcohol formation is the result of the light-induced reaction pathway.[47] Nevertheless, the mixture of citral and geraniol released from **47** is appreciated in practical applications.

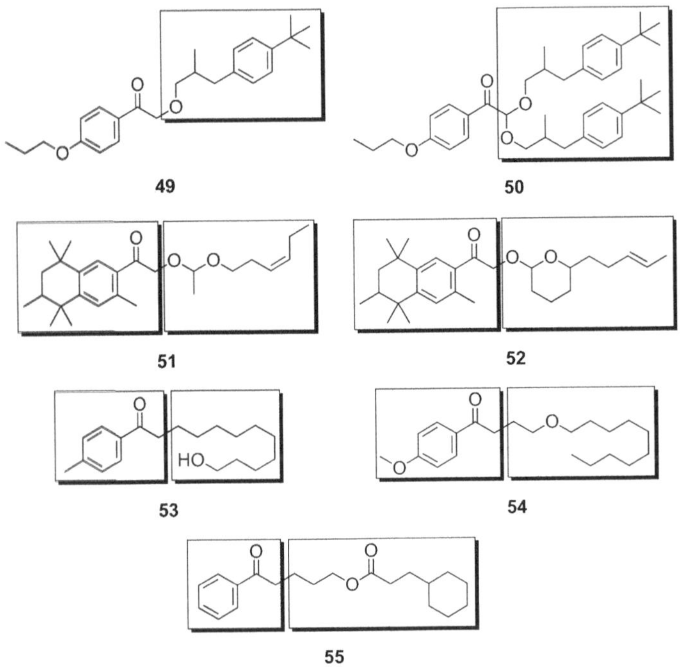

Figure 13.8 *Photolabile profragrances and properfumes for the release of aldehydes, esters, lactones and terminal alkenes*

The Norrish type II photoreaction is very general and allows for numerous structural variations of the substrate. Besides α-keto esters, a series of alkyl phenyl ketones were found to be efficient for the light-induced generation of carbonyl compounds.[45] Derrer *et al.* reported α-alkoxyacetophenone derivatives **49** and **50** (Figure 13.8) as examples for photolabile properfumes releasing 3-(4-*tert*-butylphenyl)-2-methylpropanal (Lilial) and an acetophenone derivative, both of which are olfactively interesting volatiles.[36] It was found that the choice of the substituents on the phenyl ring has a direct influence on the efficiency of the photoreaction. For the optimization of the photochemical reaction, it might therefore be desirable to select an odourless phenylketone chromophore and focus on the release of the volatile carbonyl compound. This is particularly important, as the photolabile precursor can be cleaved under natural daylight conditions. Indeed, using electron donating substituents on the phenyl ring, such as alkoxy or dialkylamino groups, the absorption maximum responsible for the photoreaction is shifted by about 30–100 nm towards the visible region of the spectrum.[36] Further structural modifications of the alkyl phenyl ketone structure allowed the light-induced release of esters and lactones (**51**, **52**)[36] as well as of terminal alkenes (**53**), allyl ethers (**54**) or vinyl esters (**55**; Figure 13.8).[48] Again, oxygen can react with the photochemically formed biradical intermediate and form a series of different side-products,[48] which are not further discussed here. Dependent on the choice of the substrate, all of the precursor atoms can be transformed into fragrance compounds. For example, properfumes **51–55** each generate a 1 : 1 mixture of two bioactive volatiles as the main fragmentation products of the photoreaction.

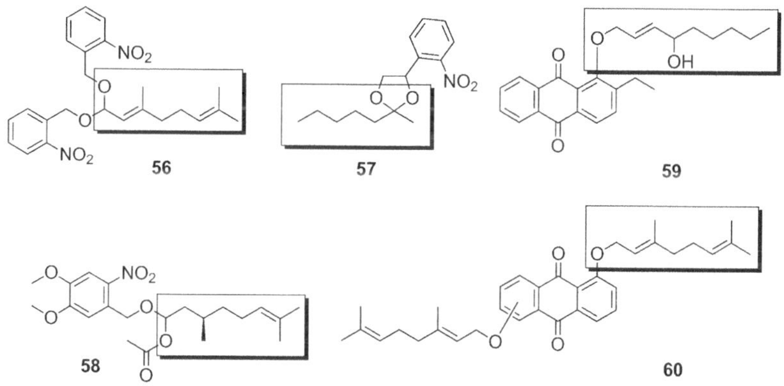

56 57 59

58 60

Figure 13.9 Photolabile aldehyde precursors **56–60**

Other recently reported light-sensitive precursors which release carbonyl derivatives, but are not based on a Norrish type II process, comprise *o*-nitrobenzyl derivatives **56** and **57**,[49] α-acetoxy nitroveratryl ether **58**[50] and 1-alkoxy-9,10-anthraquinone derivatives **59**[51] and **60**[52] (Figure 13.9). Nitrobenzyl groups were efficiently cleaved upon irradiation around 350 nm and efficiently released the corresponding pheromones and fragrances in polar and apolar solution.[49,50] 1-Alkoxy-9,10-anthraquinones were originally reported for the photochemical preparation of aldehydes and ketones.[53] They can be prepared from commercially available intermediates used as colorants in the dyestuff industry. However, in practical applications of functional perfumery, they turned out to be less efficient than the Norrish type II photorelease systems described above.[52]

The kinetics of the light-induced delivery systems for the controlled release of volatiles discussed so far (alcohols and carbonyl compounds) directly depend on the light intensity of the irradiating ambient daylight.[5,45,46] This means that irradiation of the precursor with (unclouded) outdoor sunlight can release the volatiles within several hours, whereas exposure to indoor daylight slows down the photoreaction to last for several days or even weeks until the precursor has entirely been consumed.

Other oxidation reactions, which use molecular oxygen but are not based on photochemical processes, are more complicated to be used in practical applications because the presence of oxygen in the final consumer article cannot be avoided and therefore slowly degrades the precursor during storage.[5] Reymond and coworkers described β-amino alcohol derivatives, as for example **61** and **62** (Figure 13.10), which released the corresponding aldehydes or ketones by oxidation with $NaIO_4$ or nontoxic $NaBiO_3$.[54] Grinding the profragrances (1% by weight) with anhydrous Na_2SO_4 or MgO and the solid oxidant

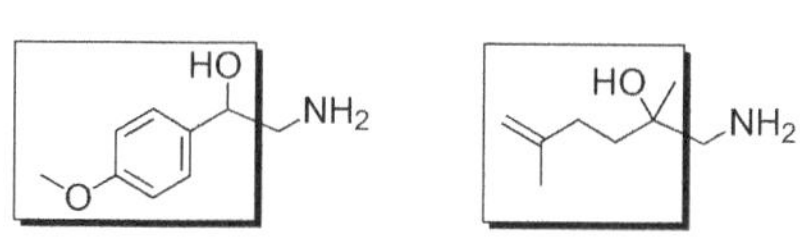

61 62

Figure 13.10 Oxidation sensitive β-amino alcohols for the release of aldehydes and ketones resulted in an odourless, colourless powder, which, upon exposure to air and uptake of ambient humidity, slowly released the volatile carbonyl compounds.

Up to now, all the precursors described above for the release of carbonyl compounds were prepared from their corresponding alcohols. Despite the fact that this might be quite practical from a synthetic point of view, the following discussion illustrates a series of carbonyl group-releasing precursors which are directly prepared from the corresponding aldehydes or ketones.

13.3.2 Reversible Systems

One possibility to form volatile carbonyl compounds (and/or volatile alcohols) is by acidic hydrolysis of acetals and ketals. This reaction has been studied for both the release of fragrance alcohols and aldehydes, exemplified by a variety of monomeric[55] or polymeric substrates.[56] However, the efficient hydrolysis of acetals and ketals requires relatively strong acidic conditions, which are often not attainable in many practical applications. This limits the applicability of these precursors as fragrance delivery systems.

In contrast, imines (Schiff bases) are known to be rather unstable in aqueous media and should therefore be readily cleaved upon hydrolysis under mild reaction conditions. In their pioneering work on polymeric fragrance release systems, Kamogawa *et al.* prepared a series of monomeric Schiff bases (**63–66**) by reaction of fragrance aldehydes with *m*- or *p*-aminostyrene in ethanol, which were then co-polymerized by radical polymerization with *N*-vinyl-2-pyrrolidone to ensure water solubility of the resulting random co-polymers (**67–70**) for the hydrolysis measurements (Figure 13.11).[57] Hydrolysis in pure water (co-

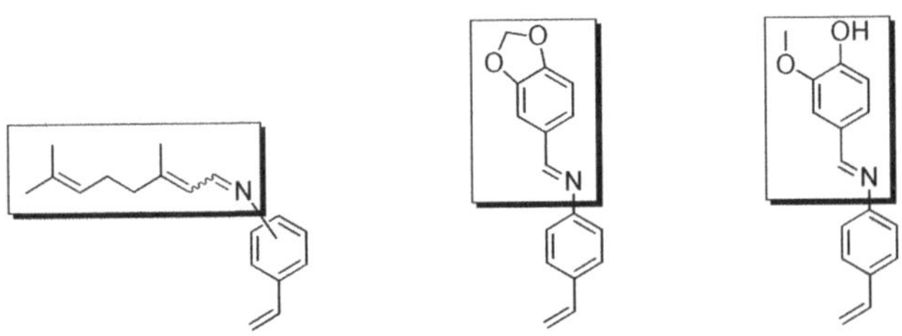

amount of hydrolysis (in (THF)/H_2O/acetic acid) **63** 36% (para) **64** 40% (meta) **65** 50% **66** 24%

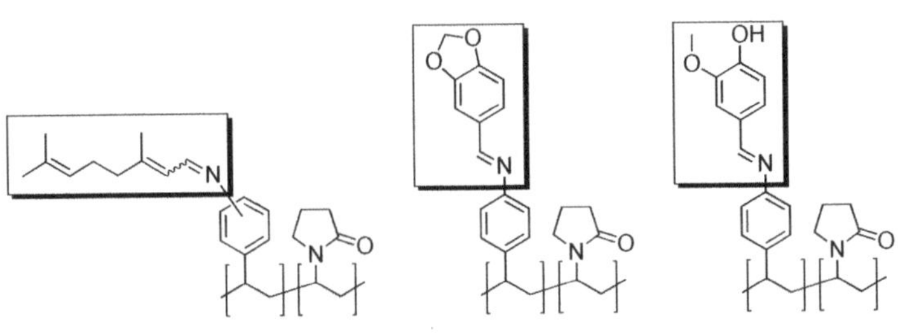

amount of hydrolysis (in (THF)/H_2O/acetic acid) **67** 19% (para) **68** 8% (meta) **69** 50% **70** 7%

Figure 13.11 *Monomeric (**63–66**) and polymeric (**67–70**) Schiff bases as hydrolytically cleavable aldehyde precursors*

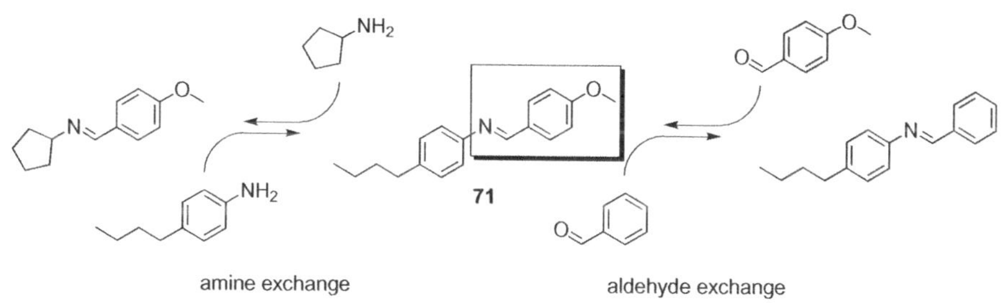

Scheme 13.9 *Amine and aldehyde exchange reactions of liquid crystalline film forming imine **71** triggered by the application of an electric field*

polymers) or THF/water mixtures (monomers) in the presence of acetic acid for 24 h showed that, with the exception of **69**, polymeric profragrances hydrolysed more slowly than the corresponding monomers, whereas no systematic difference was observed for the hydrolysis of the different *m*- or *p*-isomers (Figure 13.11).

Imines of aromatic amines and benzaldehyde derivatives with negative dielectric anisotropy, such as *N*-(4-methoxybenzylidene)-4-butylamine (**71**), form liquid crystalline phases upon application of an electric field (Scheme 13.9).[58] Volatile compounds which do not participate in the liquid crystalline phase formation, such as alcohols, esters, lactones or nitriles, were found to be efficiently expelled from the system. The rate of evaporation was dependent on the vapour pressure of the compounds to be expelled as well as on the strength of the applied electric field. In contrast, amines or aromatic aldehydes were found to react with the liquid crystalline phase-forming imine upon component exchange by the generation of constitutional dynamic libraries (Scheme 13.9). The evaporation of volatiles from the reversible dynamic system was modulated by the strength of the applied electric field.[58] The efficient expulsion of compounds from the liquid crystalline phase allows the construction of devices for the release of bioactive volatiles as a function of an external electric field.

Birkbeck *et al.* described the use of aromatic amines as reversible trapping agents for the controlled release of volatile carbonyl compounds by the formation of a reversible equilibrium between the amine, the carbonyl compound and the corresponding imine in aqueous media.[59] However, in the case of imines formed from aromatic amines, the equilibrium of the reaction at neutral pH is mainly on the side of the unreacted compounds; more stable imines are formed by reacting carbonyl compounds with hydrazides.

The formation of dynamic libraries by reversible reactions of hydrazides with carbonyl compounds was previously developed for drug discovery in a combinatorial manner[60] and prompted us to investigate the potential of these systems to modulate the evaporation of fragrances.[5,61] The concept of using 'dynamic mixtures' for the controlled release of volatile aldehydes and ketones is quite straightforward, as shown in Scheme 13.10. Addition of a hydrazide derivative to a carbonyl compound results in the formation of an equilibrium consisting of the hydrazone (formed upon condensation of the carbonyl compound with the hydrazide) together with a certain amount of the two unreacted species. In aqueous systems, the hydrazone formation is entirely reversible. With respect to 'classical' profragrance techniques described so far, the use of dynamic mixtures has several advantages. First of all, the fragrance precursor, in this case the hydrazone, is formed *in situ* and has therefore not to be prepared separately prior to its use. Furthermore, if a hydrazide derivative is added to a

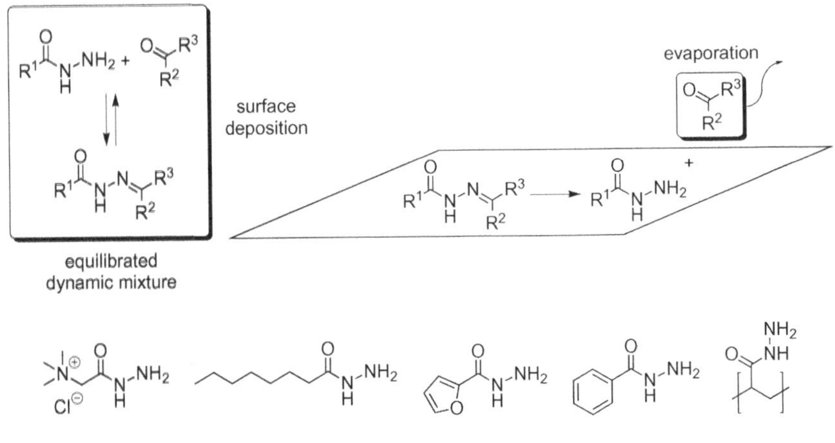

Scheme 13.10 Principle of the controlled release of volatile aldehydes and ketones from dynamic mixtures formed in the presence of hydrazides

mixture of carbonyl compounds, several hydrazones are formed simultaneously, thus influencing the release of many different carbonyl compounds at the same time without requiring the synthesis of individual precursors for each molecule to be released.[61] Because the formation of the hydrazones is an equilibrium process, the stability of the profragrances is not a concern; as long as the fragrance is not removed from the equilibrium through evaporation, the composition of the equilibrium only depends on external parameters such as concentration, temperature, or pH. After deposition of the dynamic mixture onto the target surface, the volatiles in the mixture slowly evaporate, which shifts the equilibrium towards the hydrolysis of the hydrazone while giving rise to the desired long-lasting effect of fragrance release (Scheme 13.10).[61]

Kinetic measurements, carried out by UV/Vis spectroscopy for the formation of a single hydrazone by reaction of one carbonyl derivative with one hydrazide, show that the rate constants for reversible hydrazone formation strongly depend on the pH and decrease with increasing pH. The influence of the pH on the release rates was found to be more important than structural variations of the hydrazide or carbonyl compound, and similar results were observed for a broad variety of different alkyl or aryl hydrazides (Scheme 13.10). Furthermore, the presence of surfactants, encountered in most practical applications, reduced the rate of equilibration.[61]

The presence of a hydrazide in a mixture of carbonyl compounds has a strong effect on their evaporation profile. This was shown by equilibrating a mixture of six fragrance aldehydes or ketones in the presence or absence of adipic dihydrazide (**72**) in a fabric softener formulation. After deposition of the mixture onto the cotton surface, the sample was dried and analysed by dynamic headspace analysis. Figure 13.12 shows the evaporation profiles recorded for two of the six compounds (one aldehyde, one ketone) in the presence or absence of **72**. In this particular example, the presence of the hydrazide increased the headspace concentrations of the two compounds by factors of 30 and 350, as compared to the reference sample without hydrazide. The highest increase in headspace concentrations was observed for volatiles with high vapour pressures and low water solubilities, and the effect was generally more pronounced for ketones than for aldehydes.[61] The set-up of

72

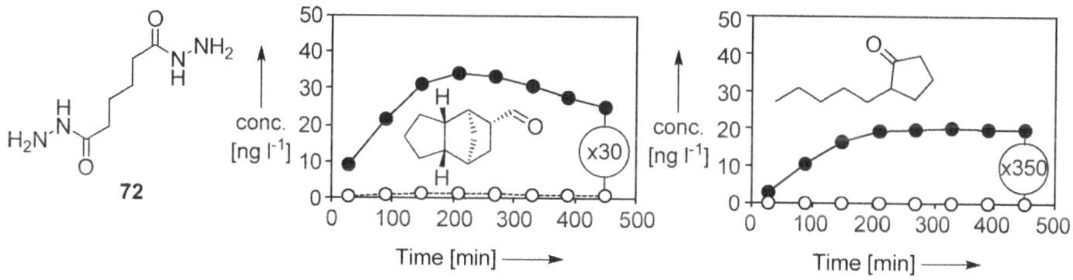

Figure 13.12 *Amounts of aldehydes and ketones released from a dynamic mixture in the presence (–●–) or absence (–○–) of dihydrazide* ***72***

dynamic mixtures using alternative systems, such as the reversible formation of aminals,[62] are currently under investigation.

The use of dynamic mixtures was further expanded to supramolecular systems, by combining the possibility of reversible covalent bond formation with the generation of larger structural assemblies. A typical example of such a combined system is represented by guanosine-5′-hydrazide (**73**), which was recently reported by Sreenivasachary and Lehn.[63] Besides the hydrazide function for reversible hydrazone formation, **73** possesses a guanosine moiety which allows formation of stable hydrogels by self-assembly to guanosine quartets (G-quartets)[64] in the presence of alkali metal cations (Figure 13.13). The release of bioactive volatile carbonyl compounds can thus be controlled not only by physical entrapment inside the supramolecular hydrogel structure, but also by reversible hydrazone formation. Dynamic headspace analysis showed an increased duration of fragrance release from hydrogels composed of **73** as compared to those formed from guanosine, which cannot interact with the carbonyl compounds by reversible covalent bond formation.[65]

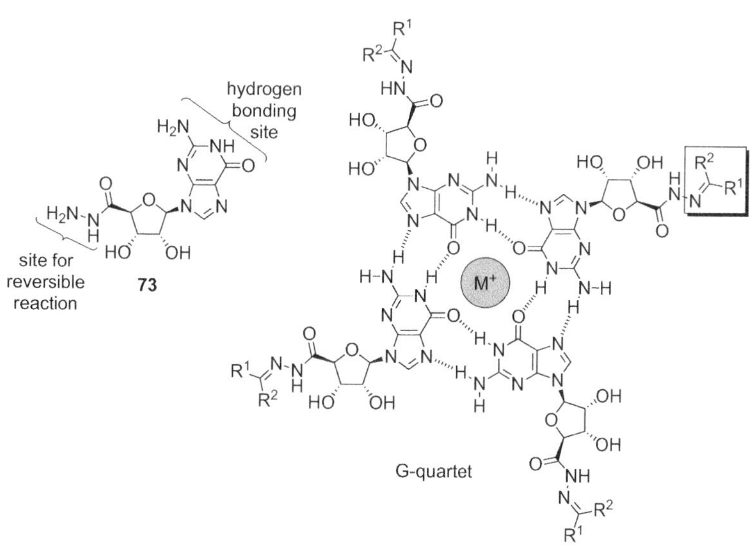

Figure 13.13 *Hydrogel formation of dynamic mixtures containing hydrazide* ***73*** *by self-assembly to guanosine quartets*

The simplicity of forming dynamic mixtures by adding a reactive substrate to a mixture of carbonyl compounds, combined with a highly efficient, long-lasting fragrance perception, makes these systems particularly suitable for practical applications.[61,62]

13.3.3 Retro 1,4-Additions

If primary amines are reacted with enones, such as ionones or damascones and damascenones (the so-called 'rose ketones'), they form stable β-amino ketones by Michael-type 1,4-addition to the enone double bond, rather than less stable imines by reaction with the carbonyl group. By heating a mixture of δ-damascone [(±)-(*E*)-1-(2,6,6-trimethyl-3-cyclohexen-1-yl)-2-buten-1-one] and poly(ethylene imine), Busch *et al.* prepared polymeric reaction product **74** (Figure 13.14),[66] which was commercialized as an ingredient of detergent powders, to increase the duration of δ-damascone in laundry applications.

Fehr and Galindo showed that 1,4-additions to hindered enones (such as damascones or ionones) also occur with *O*- and *S*-nucleophiles.[67] Under basic conditions, alcohols, carboxylic acids and alkyl thiols added to the enone double bond of rose ketones and allowed the preparation of a series of different addition products, such as **75–77** (Figure 13.14). Thioether **77**, which has recently been commercialized, was found to be particularly efficient for the release of δ-damascone in fabric softener applications.

As an extension of this work, amphiphilic polymethacrylate- and polystyrene-based β-acyloxy ketones **78** and **79**, with different stoichiometric ratios between the hydrophobic damascone release unit and the corresponding hydrophilic carboxylic acid in the polymer backbone, were prepared (Figure 13.15).[68] The presence of the hydrophilic carboxylic acid co-monomers allows for a better dispersion of the polymer in aqueous media, and was expected to enhance the release rates of the rose ketone. Furthermore, as a result of increasing deprotonation of the carboxylic acid moieties at higher pH, the random co-polymers were expected to change their three-dimensional structure from a strongly coiled arrangement at low pH to an unfolded structure at higher pH. In their protonated state at an acidic pH, the co-polymers were found to be stable, whereas under neutral or alkaline conditions the amount of δ-damascone released from the co-polymers increased over time, thus indicating that polar groups in close proximity to the release unit have a strong influence

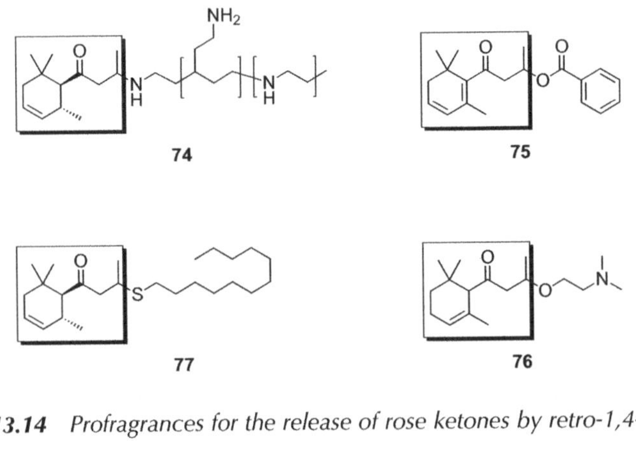

Figure 13.14 *Profragrances for the release of rose ketones by retro-1,4-addition*

on the rate of fragrance release. Furthermore, as was observed for other macromolecular or polymeric systems discussed above, fragrance release from the polymers was found to be considerably slower than from the corresponding monomers, indicating a stabilizing effect of the polymer backbone. It was also found that the structure of the polymer backbone has a higher influence on the release of the volatiles than the ratio of hydrophilic and hydrophobic monomers within the polymer chain.[68]

This general trend was confirmed by the release of δ-damascone from polymeric β-mercapto ketones **80–82** (Figure 13.15), which were obtained by ring-opening of commercially available poly(maleic anhydride) based co-polymers, giving rise to varying polymer backbone polarities.[69] For each β-mercapto ketone which was grafted onto the polymer backbone, one carboxylic acid function was generated next to it. Data from dynamic headspace analysis on dry cotton showed that the highest amount of δ-damascone was released from co-polymer **81**, whose global structure appears to represent an ideal

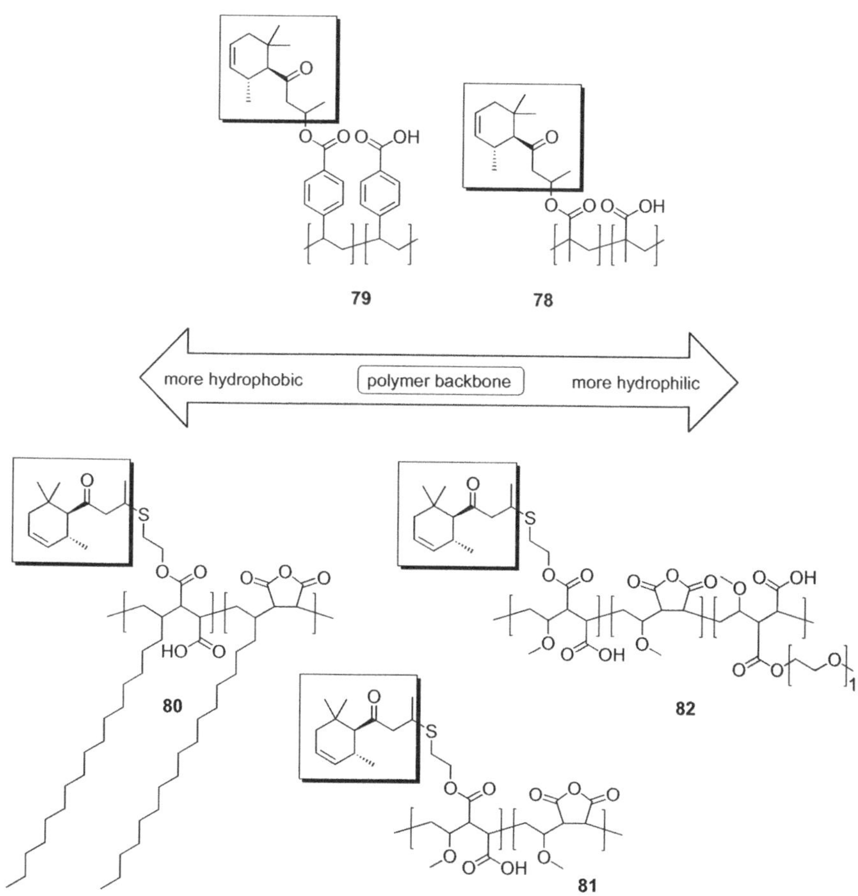

Figure 13.15 *Polymer-based precursors of δ-damascone with variable hydrophilicity of the polymer backbone*

compromise between hydrophilicity (to accelerate the release of the active volatile) and hydrophobicity of the polymer backbone (which enables good surface deposition from aqueous solution). The more hydrophobic polymer conjugate **80** was relatively well deposited onto the cotton surface, but displayed only a modest release of the active compound. However, the more polar, and thus more water soluble, co-polymer **82** with lateral poly(ethylene glycol) side chains, was deposited less, but released the volatile at higher rates. In an aqueous environment, there should be a judicious balance between hydrophobic and hydrophilic units in the polymer backbone in order to optimize the deposition of polymers onto the target surface, whilst maintaining an efficient release of the active volatile.[69]

13.4 Profragrance and Properfume Strategies

13.4.1 Performance and Cost Efficiency

The most important criterion for choosing a particular profragrance technology is the performance of the precursor in application. To efficiently release bioactive volatiles, several parameters have to be considered and optimized. Apart from the choice of suitable reaction conditions, allowing the cleavage of one (or several) covalent bonds under the conditions of use of the delivery system, several physicochemical parameters of the precursors and the volatile to be released have to be taken into account. Ideally, the volatile has a low olfactory threshold for ready detection, especially if the release reaction is slow or if only small amounts of volatile are released.

One of the most important physicochemical parameters influencing the precursor's efficiency is its hydrophobicity, often expressed as the logarithm of its octanol/water partition coefficient ($\log P_{o/w}$).[70] As hydrophobic materials (with a high $\log P$ value) are often preferentially deposited on surfaces, especially in aqueous systems such as laundry applications,[71] the choice of the precursor structure influences the amount of profragrance deposited on the target surface and thus determines, at least partially, the quantity of volatile to be released at the end. The possibility of modifying the deposition and release of different precursors according to their hydrophilicity or hydrophobicity is particularly interesting for the controlled release of hydrophilic volatiles in aqueous media, which otherwise would be rinsed off the target surface by the washing water.

The successful commercialization of profragrance technologies is also dependent on the cost efficiency of the delivery system.[5] Not only are certain key ingredients used in perfumery relatively inexpensive, which means that the performance of the corresponding precursor has to justify an increased price of the delivery technology, but also many consumer articles using the release technology are mass market products with low profit margins. This limits the development of profragrance or properfume technologies to compounds that can be efficiently prepared at low cost using only a few reaction steps. Cost efficiency is usually also achieved for precursors containing a high percentage of volatiles to be released with respect to the total mass of the delivery system. Although this aspect seems to favour properfumes which give rise to one or more volatile compounds, it should be noted that these compounds can only be released in a fixed stoichiometric ratio, which results in a considerable limitation with respect to perfume creation.

13.4.2 Stability

Another important aspect of profragrance design is to ensure sufficient chemical stability of the precursors with regard to decomposition during product storage.[5] As the precursors are usually designed to be efficiently cleaved under mild reaction conditions, a reasonable compromise between release efficiency and precursor stability has to be found. Limited precursor stability is often observed with hydrolytically cleavable profragrances, especially if these are kept in an aqueous environment. The most efficient strategy to ensure high product stability during storage is to separate the profragrance from the trigger which is used for the release of the volatiles. This can be achieved by incorporating hydrolytically labile profragrances or properfumes into solid product formulations (such as soaps, detergent powders or tablets), by compounding light-sensitive precursors into opaque packing materials, or by using combinations of different products to release the active volatiles, such as an enzyme-containing detergent powder with a fabric softener that contains the corresponding fragrance precursor.

An easy and quite efficient way to circumvent the instability of hydrolytically labile precursors in an aqueous environment is to use reversible systems such as dynamic mixtures. As discussed above, if the volatile product is not removed from the mixture by evaporation, the position of such an equilibrium only depends on external parameters such as pH, concentration and temperature, and a shift of the equilibrium during product storage can easily be corrected by re-establishing the original conditions.

13.5 Conclusion

The high volatility of bioactive volatile organic compounds requires ways to control their release over time in order to prolong their duration of perception. Covalent bond cleavage of suitably designed profragrances or properfumes under mild ambient conditions is an efficient strategy to increase the durability of the evaporation of volatile compounds from various surfaces. Volatile alcohols, aldehydes and ketones are typically released by enzymatic reactions, aqueous hydrolysis, photolysis or oxidation. Despite the limited choice of reactions allowing facile covalent bond cleavage, combined with additional requirements such as high product stability during storage, cost efficiency and biocompatibility of the precursors, a number of efficient profragrances has been developed over the past decade. The successful commercialization of several precursors has established profragrance technologies as a valuable alternative to encapsulation techniques.

Besides the investigation of a large variety of chemical reactions occurring under conditions that are entirely defined by the targeted application, progress has also been made in the understanding and control of the physicochemical parameters which have a direct influence on the release efficiency of the target compound. In particular, the use of polymer-based systems, which allows a selective modification of the precursor structures with respect to efficient surface deposition, and the use of dynamic mixtures to influence the simultaneous evaporation of several volatiles offer promising developments for the controlled release of bioactive volatile compounds. Furthermore, the exploration of covalent bond cleavage reactions within supramolecular entities is expected to give more insight into the stability and reactivity of profragrances in complex media.

Acknowledgements

Many thanks go to my coworkers and colleagues at Firmenich for their valuable contributions to the development of profragrance technologies, in particular to Barbara Buchs (Levrand) and Alain Trachsel for their long-standing collaboration, as well as to Roger Snowden for constructive comments on the manuscript.

References

1. See for example: (*a*) *Chemistry and Technology of Flavors and Fragrances*, D. J. Rowe (Ed.), Blackwell Publishing Ltd., Oxford, 2005; (*b*) *The Chemistry of Fragrances*, D. Pybus, C. Sell (Eds), The Royal Society of Chemistry, Cambridge, 1999; (*c*) G. Ohloff, *Scent and Fragrances. The Fascination of Odors and their Chemical Perspectives*, Springer Verlag, Berlin, 1994; (*d*) T. Curtis, D. G. Williams, *Introduction to Perfumery*, Ellis Horwood Ltd., Hemel Hempstead, 1994; (*e*) P. J. Teisseire, *Chemistry of Fragrant Substances*, VCH Publishers, New York, 1994.
2. (*a*) S. Arctander, *Perfume and Flavor Chemicals*, published by the author, Montclair, 1969; (*b*) H. Surburg, J. Panten, *Common Fragrance and Flavor Materials*, 5th Ed., Wiley-VCH, Weinheim, 2006.
3. *Standardized Human Olfactory Thresholds*, M. Devos, F. Patte, J. Rouault, P. Laffort, L. J. Van Gemert (Eds), Oxford University Press, Oxford, 1990.
4. (*a*) M. A. Augustin, Y. Hemar, Nano- and micro-structured assemblies for encapsulation of food ingredients, *Chem. Soc. Rev.*, **38**, 902–912 (2009); (*b*) *Delivery System Handbook for Personal Care and Cosmetic Products: Technology, Applications, and Formulations*, M. R. Rosen (Ed.), William Andrew Publishing, Norwich, 2005; (*c*) K. G. H. Desai, H. J. Park, Recent developments in microencapsulation of food ingredients, *Drying Technol.*, **23**, 1361–1394 (2005); (*d*) S.-J. Park, R. Arshady, Microcapsules for fragrances and cosmetics, *Microspheres, Microcapsules, Liposomes*, **6**, 157–198 (2003); (*e*) J. Ness, O. Simonsen, K. Symes, Microcapsules for household products, *Microspheres, Microcapsules, Liposomes*, **6**, 199–234 (2003); (*f*) C. Quellet, M. Schudel, R. Ringgenberg, Flavors & fragrance delivery systems, *Chimia*, **55**, 421–428 (2001).
5. A. Herrmann, Controlled release of volatiles under mild reaction conditions: from nature to everyday products, *Angew. Chem.*, **119**, 5938–5967 (2007); *Angew. Chem. Int. Ed.*, **46**, 5836–5863 (2007) and references cited therein.
6. R. J. Perry, 'Pro-fragrant' silicone delivery polymers, in *Delivery System Handbook for Personal Care and Cosmetic Products: Technology, Applications, and Formulations*, M. R. Rosen (Ed.), William Andrew Publishing, Norwich, 2005, pp. 667–682.
7. V. Rataj, F. Ruyffelaere, J.-M. Aubry, Libération contrôlée de molécules de parfum à partir de précurseurs, *Cah. Formulation*, **12**, 82–96 (2005).
8. (*a*) E. Stahl-Biskup, F. Intert, J. Holthuijzen, M. Stengele, G. Schulz, Glycosidically bound volatiles – a review 1986–1991, *Flavour Fragr. J.*, **8**, 61–80 (1993); (*b*) P. Winterhalter, G. K. Skouroumounis, Glycoconjugated aroma compounds: occurrence, role and biotechnological transformation, *Adv. Biochem. Eng. Biotechnol.*, **55**, 73–105 (1997); (*c*) J.-E. Sarry, Z. Günata, Plant and microbial glycoside hydrolases: volatile release from glycosidic aroma precursors, *Food Chem.*, **87**, 509–521 (2004).
9. L. Caputi, E.-K. Lim, D. J. Bowles, Discovery of new biocatalysts for the glycosylation of terpenoid scaffolds, *Chem. Eur. J.*, **14**, 6656–6662 (2008).
10. (*a*) T. Ikemoto, B.-I. Okabe, K. Mimura, T. Kitahara, Formation of fragrance materials from odourles glycosidically-bound volatiles by skin microflora (Part 1), *Flavour Fragr. J.*, **17**, 452–455 (2002); (*b*) T. Ikemoto, K. Mimura, T. Kitahara, Formation of fragrant materials from odourles glycosidically-bound volatiles on skin microflora (Part 2), *Flavour Fragr. J.*, **18**, 45–47 (2003).
11. (*a*) W. B. Shelley, H. J. Hurley Jr., A. C. Nichols, Axillary odor: experimental study of the role of bacteria, apocrine sweat, and deodorants, *Arch. Dermatol. Syphiol.*, **68**, 430–446 (1953); (*b*) J. J.

Leyden, Bacteriology of the human axilla: relationship to axillary odor, *Cosmet. Sci. Technol. Ser.*, **7**, 311–320 (1988).
12. J. P. Forestier, Les enzymes de l'espace extra-cellulaire du stratum corneum, *Int. J. Cosmet. Sci.*, **14**, 47–63 (1992).
13. (*a*) M. Troccaz, C. Starkenmann, Y. Niclass, M. van de Waal, A. J. Clark, 3-Methyl-3-sulfanylhexan-1-ol as a major descriptor for the human axilla-sweat odour profile, *Chem. Biodivers.*, **1**, 1022–1035 (2004); (*b*) A. Natsch, J. Schmid, F. Flachsmann, Identification of odoriferous sulfanylalkanols in human axilla secretions and their formation through cleavage of cysteine precursors by a C–S lyase isolated from axilla bacteria, *Chem. Biodivers.*, **1**, 1058–1072 (2004); (*c*) Y. Hasegawa, M. Yabuki, M. Matsukane, Identification of new odoriferous compounds in human axillary sweat, *Chem. Biodivers.*, **1**, 2042–2050 (2004).
14. (*a*) X.-N. Zeng, J. J. Leyden, H. J. Lawley, K. Sawano, I. Nohara, G. Preti, Analysis of characteristic odors from human male axillae, *J. Chem. Ecol.*, **17**, 1469–1492 (1991); (*b*) A. Natsch, S. Derrer, F. Flachsmann, J. Schmid, A broad diversity of volatile carboxylic acids, released by a bacterial aminoacylase from axilla secretions, as candidate molecules for the determination of human-body odor type, *Chem. Biodivers.*, **3**, 1–20 (2006).
15. C. Starkenmann, Y. Niclass, M. Troccaz, A. J. Clark, Identification of the precursor of (*S*)-3-methyl-3-sulfanylhexan-1-ol, the sulfury malodour of human axilla sweat, *Chem. Biodivers.*, **2**, 705–716 (2005).
16. A. Natsch, H. Gfeller, P. Gygax, J. Schmid, G. Acuna, A specific bacterial aminoacylase cleaves odorant precursors secreted in the human axilla, *J. Biol. Chem.*, **278**, 5718–5727 (2003).
17. M. Gautschi, A. Natsch, F. Schröder, Biochemistry of human axilla malodor and chemistry of deodorant ingredients, *Chimia*, **61**, 27–32 (2007).
18. A. Natsch, H. Gfeller, P. Gygax, J. Schmid, Isolation of a bacterial enzyme releasing axillary malodor and its use as screening target for novel deodorant formulations, *Int. J. Cosmet. Sci.*, **27**, 115–122 (2005).
19. F. Flachsmann, M. Gautschi, J.-P. Bachmann, G. Brunner, Enzyme-triggered and self-cleaving fragrant alcohol precursors, *Chem. Biodivers.*, **5**, 1115–1136 (2008).
20. K.-E. Jaeger, S. Ransac, B. W. Dijkstra, C. Colson, M. van Heuvel, O. Misset, Bacterial lipases, *FEMS Microbiol. Rev.*, **15**, 29–63 (1994).
21. (*a*) *Enzymes in Detergency*, J. H. van Ee, O. Misset, E. J. Baas (Eds), Surfactant Sci. Ser., Vol. 69, Marcel Dekker, New York, 1997; (*b*) A. Crutzen, M. L. Douglass, Detergent enzymes: a challenge, in *Handbook of Detergents*, G. Brose (Ed.), Surfactant Sci. Ser., Vol. 82, Marcel Dekker, New York, 1999, pp. 639–690.
22. W. Paget, D. Reichlin, R. L. Snowden, E. C. Walborsky, C. Vial, Fabric scenting method (to Firmenich SA), WO 95/04809 (1995); *Chem. Abstr.*, **123**, 116298 (1995).
23. F. A. Hartman, M. R. Sivik, J. C. Severns, S. W. Waite, C. L. Eddy, Perfumes for laundry and cleaning compositions (to The Procter & Gamble Company), WO 96/02625 (1996); *Chem. Abstr.*, **124**, 320199 (1996).
24. M. Avendaño, V. Nardello, J.-M. Aubry, F. Marchal, G. Tiddy, New monoester surfactants as potential fragrance releasers, *Comun. Jorn. Com. Esp. Deterg.*, **37**, 291–300 (2007).
25. F. Aulenta, M. G. B. Drew, A. Foster, W. Hayes, S. Rannard, D. W. Thornthwaite, T. G. A. Youngs, Fragrance release from the surface of branched poly(amide)s, *Molecules*, **10**, 81–97 (2005).
26. (*a*) A. J. Kirby, A. R. Fersht, Intramolecular catalysis, *Prog. Bioorg. Chem.*, **1**, 1–82 (1971); (*b*) L. Mandolini, Intramolecular reactions of chain molecules, *Adv. Phys. Org. Chem.*, **22**, 1–111 (1986).
27. B. Capon, S. P. McManus, *Neighboring Group Participation*, Plenum Press, New York, 1976.
28. J.-Y. de Saint Laumer, E. Frérot, A. Herrmann, Controlled release of perfumery alcohols by neighboring-group participation. Comparison of the rate constants for the alkaline hydrolysis of 2-acyl-, 2-(hydroxymethyl)-, and 2-carbamoylbenzoates, *Helv. Chim. Acta*, **86**, 2871–2899 (2003).
29. P. Enggist, S. Rochat, A. Herrmann, Controlled release of perfumery alcohols by alkaline hydrolysis of 2-formyl- and 2-acetylbenzoates and their corresponding phthalides, *J. Chem. Soc., Perkin Trans. 2*, 438–440 (2001).

30. (*a*) K. Bowden, Neighbouring group participation by carbonyl groups in ester hydrolysis, *Adv. Phys. Org. Chem.*, **28**, 171–206 (1993); (*b*) K. Bowden, Intramolecular catalysis: carbonyl groups in ester hydrolysis, *Chem. Soc. Rev.*, **24**, 431–435 (1995).
31. A. Trachsel, A. Govoni, J.-Y. de Saint Laumer, E. Frérot, A. Herrmann, Controlled release of volatile secondary and tertiary alcohols by neighboring group participation: stepwise cyclization and re-opening of 2,2′-bis(carbamoyl)dibenzoates at neutral pH, *Chem. Biodivers.*, **5**, 2621–2639 (2008).
32. E. Frérot, K. Herbal, A. Herrmann, Controlled stepwise release of fragrance alcohols from dendrimer-based 2-carbamoylbenzoates by neighbouring group participation, *Eur. J. Org. Chem.*, 967–971 (2003).
33. A. Trachsel, J.-Y. de Saint Laumer, O. P. Haefliger, A. Herrmann, Parameters influencing the release of tertiary alcohols from the surface of 'spherical' dendrimers and 'linear' stylomers by neighbouring-group-assisted hydrolysis of 2-carbamoylbenzoates, *Chem. Eur. J.*, **15**, 2846–2860 (2009).
34. H. Kamogawa, Y. Haramoto, Y. Misaka, Y. Asada, Y. Ohno, M. Nanasawa, Chemical release control: sulfonate esters from perfume and herbicide alcohols and *p*-styrenesulfonyl chloride, *J. Polym. Sci., Polym. Chem. Ed.*, **23**, 1517–1526 (1985).
35. D. Anderson, G. Frater, Aryl-acrylic acid esters useful as precursors for organoleptic compounds (to Givaudan Roure SA), EP 0 936 211 (1999); *Chem. Abstr.*, **131**, 170173 (1999).
36. S. Derrer, F. Flachsmann, C. Plessis, M. Stang, Applied photochemistry – light controlled perfume release, *Chimia*, **61**, 665–669 (2007).
37. M. Goeldner, R. Givens, *Dynamic Studies in Biology: Phototriggers, Photoswitches and Caged Biomolecules*, Wiley-VCH, Weinheim, 2005.
38. (*a*) D. D. Young, A. Deiters, Photochemical control of biological processes, *Org. Biomol. Chem.*, **5**, 999–1005 (2007); (*b*) G. Mayer, A. Heckel, Biologically active molecules with a 'light switch' *Angew. Chem.*, **118**, 5020–5042 (2006); *Angew. Chem. Int. Ed.*, **45**, 4900–4921 (2006); (*c*) G. Dormán, G. D. Prestwich, Using photolabile ligands in drug discovery and development, *Trends Biotechnol.*, **18**, 64–77 (2000).
39. (*a*) J. Pika, A. Konosonoks, R. M. Robinson, P. N. D. Singh, A. D. Gudmundsdottir, Photoenolization as a means to release alcohols, *J. Org. Chem.*, **68**, 1964–1972 (2003); (*b*) J. Pika, A. Konosonoks, P. Singh, A. D. Gudmundsdottir, Designing esters which release alcohols upon exposure to light, *The Spectum (Bowling Green)*, **16** (4), 12–17 (2003); (*c*) A. Konosonoks, P. J. Wright, M.-L. Tsao, J. Pika, K. Novak, S. M. Mandel, J. A. Krause Bauer, C. Bohne, A. D. Gudmundsdóttir, Photoenolization of 2-(2-methyl benzoyl) benzoic acid, methyl ester: effect of E photoenol lifetime on the photochemistry, *J. Org. Chem.*, **70**, 2763–2770 (2005).
40. P. B. Jones, M. P. Pollastri, N. A. Porter, 2-Benzoylbenzoic acid: a photolabile mask for alcohols and thiols, *J. Org. Chem.*, **61**, 9455–9461 (1996).
41. C. Plessis, S. Derrer, Novel photolactonisation from xanthenoic esters, *Tetrahedron Lett.*, **42**, 6519–6522 (2001).
42. (*a*) S. Hu, D. C. Neckers, Photochemical reactions of alkyl phenylglyoxylates, *J. Org. Chem.*, **61**, 6407–6415 (1996); (*b*) S. Hu, D. C. Neckers, Photochemical reactions of alkenyl phenylglyoxylates, *J. Org. Chem.*, **62**, 6820–6826 (1997).
43. S. Rochat, C. Minardi, J.-Y. de Saint Laumer, A. Herrmann, Controlled release of perfumery aldehydes and ketones by Norrish type-II photofragmentation of α-keto esters in undegassed solution, *Helv. Chim. Acta*, **83**, 1645–1671 (2000).
44. (*a*) S. Hu, D. C. Neckers, Mechanism of alkyl phenylglyoxylates photoreaction in the presence of oxygen, *J. Photochem. Photobiol. A: Chem.*, **118**, 75–80 (1998); (*b*) M. C. Pirrung, R. J. Tepper, Photochemistry of substituted benzoylformate esters. A convenient method for the photochemical oxidation of alcohols, *J. Org. Chem.*, **60**, 2461–2465 (1995).
45. A. Herrmann, Photochemical fragrance delivery systems based on the Norrish type II reaction – a review, *The Spectrum (Bowling Green)*, **17** (2), 10–13, 19 (2004).
46. A. Herrmann, C. Debonneville, V. Laubscher, L. Aymard, Dynamic headspace analysis of the light-induced controlled release of perfumery aldehydes and ketones from α-keto esters in bodycare and household applications, *Flavour Fragr. J.*, **15**, 415–420 (2000).

47. B. Levrand, A. Herrmann, Controlled light-induced release of volatile aldehydes and ketones by photofragmentation of 2-oxo-(2-phenyl)acetates, *Chimia*, **61**, 661–664 (2007).
48. B. Levrand, A. Herrmann, Light induced controlled release of fragrances by Norrish type II photofragmentation of alkyl phenyl ketones, *Photochem. Photobiol. Sci.*, **1**, 907–919 (2002).
49. (*a*) X. Liu, E. D. M. Macaulay, J. A. Pickett, Propheromones that release pheromonal carbonyl compounds in light, *J. Chem. Ecol.*, **10**, 809–822 (1984); (*b*) J. A. Pickett, G. W. Dawson, D. C. Griffiths, L. Xun, E. D. M. Macaulay, C. M. Woodcock, Propheromones: an approach to the slow release of pheromones, *Pestic. Sci.*, **15**, 261–264 (1984).
50. J. Lage Robles, C. G. Bochet, Photochemical release of aldehydes from α-acetoxy nitroveratryl ethers, *Org. Lett.*, **7**, 3545–3547 (2005).
51. (*a*) R. G. Brinson, P. B. Jones, Caged *trans*-4-hydroxy-2-nonenal, *Org. Lett.*, **6**, 3767–3770 (2004); (*b*) P. B. Jones, R. G. Brinson, S. J. Sarma, S. Elkazaz, Observation of heavy atom effects in the development of water soluble caged 4-hydroxy-*trans*-2-nonenal, *Org. Biomol. Chem.*, **6**, 4204–4211 (2008).
52. B. Levrand, A. Herrmann, Light-induced controlled release of fragrance aldehydes from 1-alkoxy-9,10-anthraquinones for applications in functional perfumery, *Flavour Fragr. J.*, **21**, 400–409 (2006).
53. (*a*) R. L. Blankespoor, R. P. Smart, E. D. Batts, A. A. Kiste, R. E. Lew, M. E. Vander Vliet, Photochemistry of 1-alkoxy- and 1-(benzyloxy)-9,10-anthraquinones in methanol: a facile process for the preparation of aldehydes and ketones, *J. Org. Chem.*, **60**, 6852–6859 (1995); (*b*) R. L. Blankespoor, T. DeVries, E. Hansen, J. M. Kallemeyn, A. M. Klooster, J. A. Mulder, R. P. Smart, D. A. Vander Griend, Photochemical synthesis of aldehydes in the solid phase, *J. Org. Chem.*, **67**, 2677–2681 (2002).
54. Y. Yang, D. Wahler, J.-L. Reymond, β-Amino alcohol properfumes, *Helv. Chim. Acta*, **86**, 2928–2936 (2003).
55. See for example: A. Sharma, N. Nagarajan, K. N. Gurudutt, Stabilization of aldehydes as propylene glycol acetals, *J. Agric. Food Chem.*, **46**, 654–656 (1998).
56. (*a*) H. Kamogawa, S. Okabe, M. Nanasawa, Synthesis of polymerizable acetals. I. Vinylbenzaldehyde acetals with perfume alcohols, *Bull. Chem. Soc. Jpn.*, **49**, 1917–1919 (1976); (*b*) H. Kamogawa, Y. Haramoto, M. Nanasawa, Synthesis of polymerizable acetals. II. Readily hydrolyzable acetals from citronellol and vitamins, *Bull. Chem. Soc. Jpn.*, **52**, 846–848 (1979); (*c*) H. Kamogawa, Y. Haramoto, T. Nakazawa, H. Sugiura, M. Nanasawa, 1,3-Dioxolane bearing perfume and herbicide aldehyde residues, *Bull. Chem. Soc. Jpn.*, **54**, 1577–1578 (1981); (*d*) H. Morinaga, H. Morikawa, Y. Wang, A. Sudo, T. Endo, Amphiphilic copolymer having acid-labile acetal in the side chain as a hydrophobe: controlled release of aldehyde by thermoresponsive aggregation-dissociation of polymer micelles, *Macromolecules*, **42**, 2229–2235 (2009).
57. H. Kamogawa, H. Mukai, Y. Nakajima, M. Nanasawa, Chemical release control – Schiff bases of perfume aldehydes and aminostyrenes, *J. Polym. Sci., Polym. Chem. Ed.*, **20**, 3121–3129 (1982).
58. (*a*) N. Giuseppone, J.-M. Lehn, Electric-field modulation of component exchange in constitutional dynamic liquid crystals, *Angew. Chem.*, **118**, 4735–4740 (2006); *Angew. Chem. Int. Ed.*, **45**, 4619–4624 (2006); (*b*) A. Herrmann, N. Giuseppone, J.-M. Lehn, Electric-field triggered controlled release of bioactive volatiles from imine-based liquid crystalline phases, *Chem. Eur. J.*, **15**, 117–124 (2009).
59. A. A. Birkbeck, O. Moulin, C. Nagel, K. D. Perring, C. S. Sell, K. M. Tuck, Improvements in or relating to perfumes (to Quest International B.V.), WO 01/93823 (2001); *Chem. Abstr.*, **136**, 24978 (2002).
60. (*a*) J.-M. Lehn, Dynamic combinatorial chemistry and virtual combinatorial libraries, *Chem. Eur. J.*, **5**, 2455–2463 (1999); (*b*) K. R. West, S. Otto, Reversible covalent chemistry in drug delivery, *Curr. Drug Discovery Technol.*, **2**, 123–160 (2005); (*c*) P. T. Corbett, J. Leclaire, L. Vial, K. R. West, J.-L. Wietor, J. K. M. Sanders, S. Otto, Dynamic combinatorial chemistry, *Chem. Rev.*, **106**, 3652–3711 (2006); (*d*) J.-M. Lehn, From supramolecular chemistry towards constitutional dynamic chemistry and adaptive chemistry, *Chem. Soc. Rev.*, **36**, 151–160 (2007); (*e*) S. Ladame, Dynamic combinatorial chemistry: on the road to fulfilling the promise, *Org. Biomol. Chem.*, **6**, 219–226 (2008); (*f*) A. Herrmann, Dynamic mixtures and combinatorial libraries: imines as

probes for molecular evolution at the interface between chemistry and biology, *Org. Biomol. Chem.*, **7**, 3195–3204 (2009).
61. (*a*) B. Levrand, Y. Ruff, J.-M. Lehn, A. Herrmann, Controlled release of volatile aldehydes and ketones by reversible hydrazone formation – 'classical' profragrances are getting dynamic, *Chem. Commun.*, 2965–2967 (2006); (*b*) B. Levrand, W. Fieber, J.-M. Lehn, A. Herrmann, Controlled release of volatile aldehydes and ketones from dynamic mixtures generated by reversible hydrazone formation, *Helv. Chim. Acta*, **90**, 2281–2314 (2007).
62. G. Godin, B. Levrand, A. Trachsel, J.-M. Lehn, A. Herrmann, Reversible formation of aminals: a new strategy to control the release of bioactive volatiles from dynamic mixtures, *Chem. Commun.*, **46**, 3125–3127 (2010).
63. (*a*) N. Sreenivasachary, J.-M. Lehn, Gelation-driven component selection in the generation of constitutional dynamic hydrogels based on guanine-quartet formation, *Proc. Natl. Acad. Sci. USA*, **102**, 5938–5943 (2005); (*b*) N. Sreenivasachary, J.-M. Lehn, Structural selection in G-quartet-based hydrogels and controlled release of bioactive molecules, *Chem. Asian J.*, **3**, 134–139 (2008).
64. See for example: J. T. Davis, G. P. Spada, Supramolecular architectures generated by self-assembly of guanosine derivatives, *Chem. Soc. Rev.*, **36**, 296–313 (2007).
65. J.-M. Lehn, N. Sreenivasachary, A. Herrmann, Hydrogels for the controlled release of bioactive materials (to Firmenich SA, Université Louis Pasteur and CNRS), WO 2006/100647 (2006); *Chem. Abstr.*, **145**, 383495 (2006).
66. A. Busch, M. Homble, C. Laudamiel, J. Smets, R. Trujillo, J. Wevers, Process for producing particles of amine reaction product (to The Procter&Gamble Company), EP 0 971 021 (2000); *Chem. Abstr.*, **132**, 80090 (2000).
67. C. Fehr, J. Galindo, Aldols by Michael addition: application of the retro-Michael addition to the slow release of enones, *Helv. Chim. Acta*, **88**, 3128–3136 (2005).
68. D. Berthier, A. Trachsel, C. Fehr, L. Ouali, A. Herrmann, Amphiphilic polymethacrylate- and polystyrene-based chemical delivery systems for damascones, *Helv. Chim. Acta*, **88**, 3089–3108 (2005).
69. (*a*) D. Berthier, A. Herrmann, Polymer conjugates for a controlled release of active molecules (to Firmenich SA), WO 2008/044178 (2008); *Chem. Abstr.*, **148**, 472574 (2008); (*b*) D. L. Berthier, N. Paret, A. Trachsel, A. Herrmann, Influence of the backbone structure on the release of bioactive volatiles from maleic acid-based polymer conjugates, *Bioconjugate Chem.*, submitted.
70. (*a*) A. Leo, C. Hansch, D. Elkins, Partition coefficients and their uses, *Chem. Rev.*, **71**, 525–616 (1971); (*b*) A. J. Leo, Calculating log P_{oct} from structures, *Chem. Rev.*, **93**, 1281–1306 (1993).
71. (*a*) S. D. Escher, E. Oliveros, A quantitative study of factors that influence the substantivity of fragrance chemicals on laundered and dried fabrics, *J. Am. Oil. Chem. Soc.*, **71**, 31–40 (1994); (*b*) T. Stora, S. Escher, A. Morris, The physicochemical basis of perfume performance in consumer products, *Chimia*, **55**, 406–412 (2001); (*c*) S. K. Obendorf, H. Liu, K. Tan, M. J. Leonard, T. J. Young, M. J. Incorvia, Adsorption of aroma chemicals on cotton fabric in different aqueous environments, *J. Sufactants Deterg.*, **12**, 43–58 (2009).

14

Reactions of Biogenic Volatile Organic Compounds in the Atmosphere

Russell K. Monson

Department of Ecology and Evolutionary Biology, and Cooperative Institute for Research in Environmental Sciences, University of Colorado, Boulder, CO, 80309, USA

14.1 Introduction

Plants and microorganisms produce a variety of organic and inorganic compounds that react with photochemical components of the atmosphere and influence the production and turnover of atmospheric oxidants.[1] It is this continuous processing of reactive compounds from various terrestrial and marine ecosystems that maintains the atmosphere in a state of chemical disequilibrium and sustains some key atmospheric components that are required for life to persist. For example, most of the molecular oxygen (O_2) that exists in the atmosphere, and is used as an electron acceptor in aerobic respiration, is derived from autotrophic photosynthesis. Tropospheric O_2, in turn, sustains the production of stratospheric ozone (O_3), a compound that protects nucleic acids, the compounds that carry the genetic code for life, from potentially harmful fluxes of ultraviolet radiation. There is clearly need for a coupled, synthetic view of biology and chemistry in order to completely understand atmospheric biogeochemical dynamics and their role in the sustenance of life.

In this chapter, I focus on the emission of biogenic volatile organic compounds (BVOCs), principally by plants, and their photochemical fate in the atmosphere. The diversity of BVOCs capable of fuelling atmospheric chemistry is immense, including alkenes, alkanes, alcohols, aldehydes, ketones and organic acids. These compounds exhibit a broad range of atmospheric lifetimes, ranging from a few seconds to several years. Ultimately, they all represent sources of electrons that exist at higher potential energy levels than those in the fundamental inorganic products to which they are oxidized (e.g. CO_2) and, thus, thermo-

The Chemistry and Biology of Volatiles Edited by Andreas Herrmann

dynamic forcings favour the spontaneous, photochemical reactions caused by their emission. Prior to the past decade, most research on BVOC oxidation focused on gas phase reactions initiated by the hydroxyl radical (•OH); commonly called the 'detergent of the atmosphere'. It was thought that gas phase reactions accounted for most BVOC oxidation. In the past few years, however, several discoveries have led us to re-evaluate the conventional wisdom on this matter. The discoveries can be used to build a compelling case for high rates of BVOC oxidation through heterogeneous chemistry involving atmospheric aerosols. In fact, the heterogeneous processing of BVOCs is likely to be the source of most of the secondary aerosol formation above terrestrial ecosystems, influencing radiative transfer in the atmosphere and surface energy budgets.[2] I use these two distinct phases of research as the framework around which to organize the discussions in this chapter. In the initial sections of the chapter I focus on the types of BVOCs emitted to the atmosphere, making a key distinction between the atmospheric fates of anthropogenic versus biogenic VOCs. In the subsequent sections I take up the topic of gas phase oxidation of BVOCs, with a focus on newer discoveries, made over the past decade and the topic of BVOC oxidation and aerosol formation, which by nature is also focused on research reported during the past decade. Finally, I conclude with some general thoughts about current and future research trends concerning these issues.

14.2 The Relative Importance of Anthropogenic Versus Biogenic VOC Emissions to Atmospheric Chemistry

At the global scale, most volatile organic compound (VOC) emissions occur from natural ecosystems. Annual global VOC emissions from anthropogenic sources are in the range of 110–186 Tg (expressed as total compound mass) and include emissions from managed biomass burning (e.g. clearing fields and forests for agriculture) and natural forest fires.[3] This range represents a projected increase in emissions from approximately 100 Tg in 1990 to approximately 186 Tg in 2000, demonstrating the effect of continued economic development and changes in the way that emissions from biomass burning were estimated (favouring higher emissions from biomass burning in inventories constructed after 1995). Annual global emissions from biogenic sources are estimated as 1000–1200 Tg, expressed per unit of carbon mass and not including biomass burning.[4,5] It is difficult to compare the magnitude of anthropogenic versus biogenic emissions because of the use of non-equivalent units, lack of specificity of compounds in many of the BVOC inventories, cross emissions from both natural and human-caused sources (e.g. emissions of acetone) and varying designations as to which emissions qualify as anthropogenic and which as natural (e.g. forest fires).[6] However, we know that most of the BVOC emissions occur as isoprene and compounds constructed from multiple isoprene units (e.g. monoterpenes and sequiterpenes).[7] If we use 68 : 60 as the ratio of compound mass to carbon mass in isoprene, then global BVOC emissions on a compound basis are approximately 13% higher than those estimated on a carbon atom basis, bringing estimates of global BVOC emissions into the 1300–1500 Tg range. Thus, as a general estimate, it is likely that global BVOC emissions are at least tenfold higher in magnitude than global anthropogenic VOC emissions.

Within the class of anthropogenic VOCs, most are alkanes (paraffins), whereas within the biogenic VOCs, most are alkenes (olefins). The Emissions Database for Global

Atmospheric Research (EDGAR),[8] which has been used as the basis for much of the global atmospheric chemistry modelling over the past couple of decades, estimates anthropogenic VOCs as approximately 78% paraffins, 13% non-ethene olefins, 9% ethene and the remainder as aldehydes, methylglyoxal, formaldehydes, halogenated hydrocarbons and some additional trace species. Approximately 10% of global anthropogenic VOC emissions are aromatic compounds, including benzene, toluene and xylene.[9] The strongest source for many of the anthropogenic VOCs is evaporation from petroleum-based fuel sources and industrial processes.[10] Emitted BVOCs are dominated by the terpenes, which comprise approximately 55% of the total global emissons.[4] Isoprene (2-methyl-1,3-butadiene, **1**) makes up the greatest fraction of the BVOC terpenes, as it is emitted at relatively high rates from most forest ecosystems, especially those dominated by broad-leaved tree species.[11]

The ratio of anthropogenic to biogenic VOCs varies considerably depending on proximity to urban and suburban sources. In the northern hemisphere, where developed cities and towns cover more area, anthropogenic VOCs are proportionately more abundant.[12] The prevalence of anthropogenic VOCs in the northern hemisphere, where much of the atmospheric chemistry research community resides, has historically skewed the discussion on surface-atmosphere emissions and atmospheric quality. Only in the past three decades has the general recognition that BVOCs are an important component of global atmospheric chemistry caused a re-evaluation of the causes and aims of deterioration of regional atmospheric quality.[13]

The presence of double bonds in the olefin compounds that dominate BVOC emissions make them more reactive in atmospheric gas phase chemistry and accordingly they have shorter atmospheric lifetimes. Rate constants for the reactions of biogenic and anthropogenic VOCs with hydroxyl radical (•OH) vary by several orders of magnitude.[14] As a result of differential atmospheric lifetimes, regional and local concentrations of compounds can also vary widely; in urban air, alkanes and aromatic paraffins build-up to relatively high concentrations. In rural air, biogenic olefins are present at relatively low concentrations (a few ppbv), having been oxidized to various peroxy compounds and organic acids; the higher order products of olefin oxidation tend to have longer atmospheric lifetimes, though they are also more susceptible to atmospheric loss due to dry and wet deposition and partitioning into aerosol surfaces.[15]

14.3 Overview of BVOC Oxidation

The principal oxidation pathways that lead to the processing of BVOCs are shown in Scheme 14.1. The oxidative power that we recognize in the atmosphere is due to the reactive tendencies driven by O_x-HO_x-NO_x chemistry. The principal oxidative species within these pathways are O_2, O_3, •OH, HO_2•, NO, NO_2 and NO_3•. The left side of Scheme 14.1 shows the gas phase chemistry initiated by •OH oxidation. Hydroxyl radicals (•OH) are formed from the photolysis of O_3 in the presence of UV radiation with wavelengths less than ca. 320 nm:

$$O_3 + h\nu(<320\,\text{nm}) \rightarrow O\bullet + O_2 \quad (14.1)$$

$$O_3 + h\nu(<320\,\text{nm}) \rightarrow O\bullet(^1D) + O_2 \quad (14.2)$$

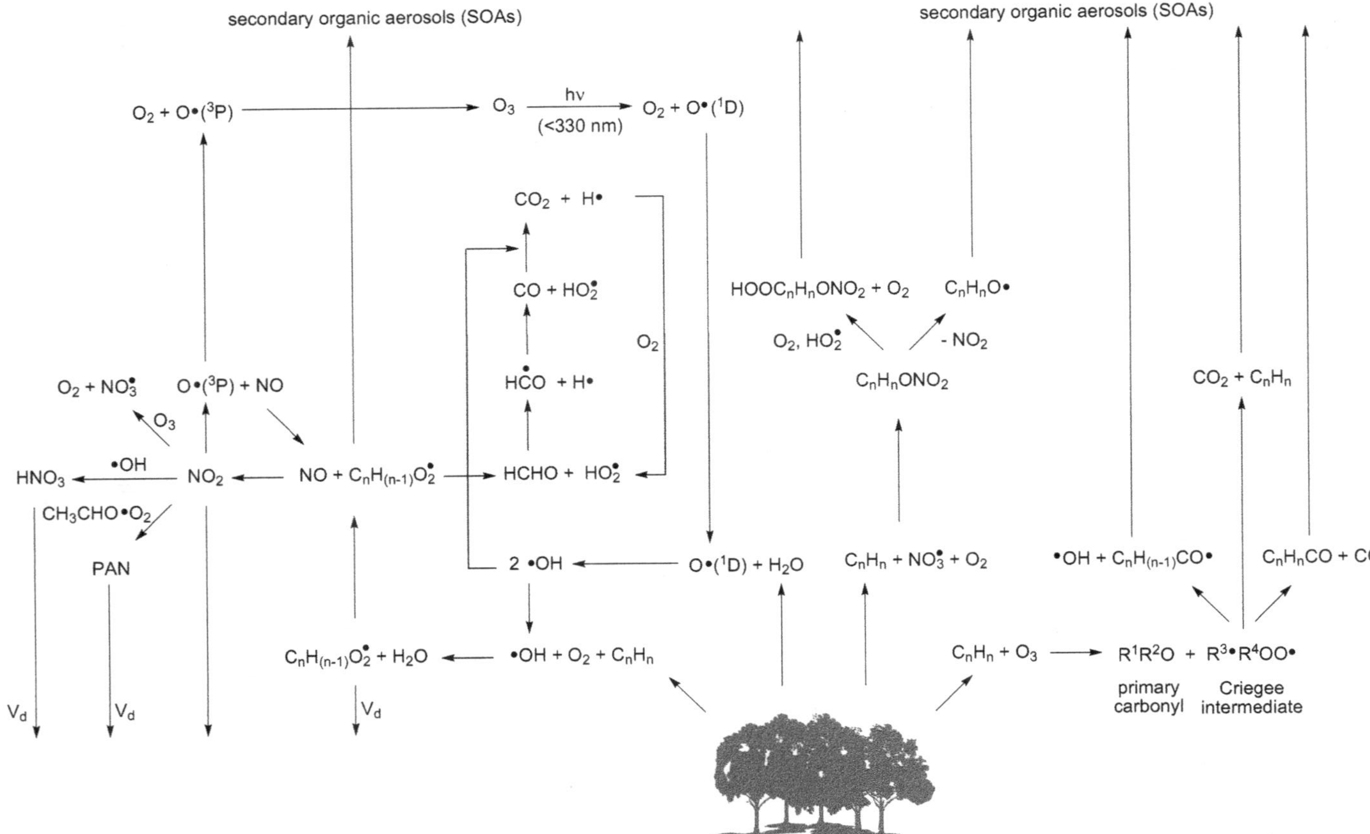

Scheme 14.1 *Major gas phase oxidative reactions in the troposphere associated with the emissions of BVOCs from trees. Parts of this scheme were taken from ref. 1*

The products of O_3 photolysis are both ground state singlet oxygen (Reaction 14.1) and excited singlet oxygen (Reaction 14.2), although •OH is only formed from the excited singlet, O•(^{1}D), according to:

$$O\bullet(^1D) + H_2O \rightarrow 2\ \bullet OH \qquad (14.3)$$

It is with Reaction 14.3 that we begin the photochemical sequences shown on the left of Scheme 14.1.

The principal gas phase oxidation route for both alkanes and alkenes in the atmosphere is through •OH. In the case of alkanes (including methane, CH_4), •OH attacks a carbon-bound H atom (generally the most weakly bound H in the molecule), and abstracts it to form H_2O, leaving behind an organic radical on the H-deficient carbon (represented as C•), which quickly reacts with O_2 to form an organic peroxy radical ($RO_2\bullet$):

$$C_nH_n + \bullet OH \rightarrow C\bullet C_{(n-1)}H_{(n-1)} + H_2O \qquad (14.4)$$

$$C\bullet C_{(n-1)}H_{(n-1)} + O_2 + M \rightarrow COO\bullet C_{(n-1)}H_{(n-1)} \qquad (14.5)$$

where we represent $COO\bullet C_{(n-1)}H_{(n-1)}$ generally as $RO_2\bullet$. In Reaction 14.5, *M* stands for a third molecule (e.g. O_2 or N_2) or surface which is required to take on some of the energy of the reaction and thus stabilize the reactive species to the point where a successful reaction can occur. The gas phase oxidation of alkenes also occurs predominantly through •OH, although the •OH radical attacks a double bond in the molecule, and adds itself to one of the carbons participating in the double bond, rather than abstracting an H atom:

$$C_nH_n + \bullet OH \rightarrow C\bullet C_{(n-1)}H_nOH \qquad (14.6)$$

$$C\bullet C_{(n-1)}H_nOH + O_2 \rightarrow COO\bullet C_{(n-1)}H_nOH \qquad (14.7)$$

As in the case for CH_4, these reactions result in formation of an organic peroxy radical ($RO_2\bullet$). In the presence of nitric oxide (NO), organic peroxy radicals are chemically reduced to form an organic alkoxy radical (RO•), and NO is oxidized to form nitrogen dioxide (NO_2):

$$RO_2\bullet + NO \rightarrow RO\bullet + NO_2 \qquad (14.8)$$

The NO_2 formed from this reaction undergoes photolysis to form NO and ground-state singlet oxygen, O•(^{3}P). The O•(^{3}P) species has the potential to react with molecular oxygen (O_2) to form O_3:

$$O\bullet(^3P) + O_2 \rightarrow O_3 \qquad (14.9)$$

The NO-NO_2 loop is cyclic, as NO is reformed after the photolysis of NO_2, and is available to react with additional $RO_2\bullet$ radicals.

The alkoxy radicals produced through Reaction 14.8 tend to react with O_2 to form aldehyde compounds (the simplest being formaldehyde, HCHO) and the hydroperoxy radical ($HO_2\bullet$):

$$RO\bullet + O_2 \rightarrow HRHO + HO_2\bullet \qquad (14.10)$$

$$HO_2\bullet + NO \rightarrow NO_2 + \bullet OH \qquad (14.11)$$

Thus, from Reactions 14.4 to 14.9 an •OH radical is consumed (in Reaction 14.4) and then regenerated (in Reaction 14.11). In the presence of NO, the reactions between Reactions

14.4 and 14.11 generate 'ozone-producing power' by converting NO to NO_2. In the case of formaldehyde (the product of Reaction 14.10), photolysis occurs with subsequent oxidation, represented as:

$$HCHO + h\nu \ (+ O_2) \rightarrow 2\,HO_2\bullet + CO \quad (14.12)$$

$$CO + \bullet OH \ (+ O_2) \rightarrow CO_2 + HO_2\bullet \quad (14.13)$$

In the case of more complex aldehydes which might be produced from Reaction 14.10, photolysis occurs with subsequent oxidation, represented as:

$$RCHO + \bullet OH \rightarrow RC\bullet O + H_2O \quad (14.14)$$

$$RC\bullet O + O_2 \rightarrow RC(O)O_2\bullet \quad (14.15)$$

where $RC\bullet O$ (in Reaction 14.14) is the acetyl radical and $RC(O)O_2\bullet$ (in Reaction 14.15) is the peroxyacetyl radical. The peroxyacetyl radical can potentially react with NO and NO_2 to form peroxyacetyl nitrate (PAN), a common pollutant in urban and suburban airsheds. The peroxyacetyl radical can also be oxidized to form CO_2 according to:

$$RC(O)O_2\bullet + NO \rightarrow NO_2 + RC(O)O\bullet \quad (14.16)$$

$$RC(O)O\bullet + O_2 \rightarrow ROO\bullet + CO_2 \quad (14.17)$$

All of the carbon atoms of the original hydrocarbon are potentially oxidized through Reactions 14.14 to 14.17 to form CO_2.

Reactions 14.1 to 14.9 take us through a cycle of reactions that begin and end with O_3. Given the multiply reduced carbon atoms that often occur in BVOCs, however, there is much more potential to produce O_3, than is consumed in the original production of •OH. Thus, it is accurate to say 'it takes a little O_3 to produce a lot of O_3 in the presence of organic compounds and NO_x'. We can appreciate the potential for O_3 production by examination of the accounting sequence for the case of the simplest oxidation path, that for CH_4:

$$CH_4 + \bullet OH \ (+ O_2) \rightarrow CH_3O_2\bullet + H_2O \quad (14.18)$$

$$CH_3O_2\bullet + NO \rightarrow CH_3O\bullet + NO_2 \quad (14.19)$$

$$CH_3O\bullet + O_2 \rightarrow HCHO + HO_2\bullet \quad (14.20)$$

$$HO_2\bullet + NO \rightarrow \bullet OH + NO_2 \quad (14.21)$$

$$2\,NO_2 + h\nu \rightarrow 2\,NO + 2\,O\bullet(^3P) \quad (14.22)$$

$$2\,O\bullet(^3P) + 2\,O_2 + M \rightarrow 2\,O_3 + M \quad (14.23)$$

$$\text{Net}: \ CH_4 + 4\,O_2 + h\nu \rightarrow HCHO + 2\,O_3 + H_2O \quad (14.24)$$

The summation given in Equation (14.24) makes clear that hydrocarbons in the presence of NO drive a photochemically energized oxidation process capable of producing two O_3 molecules. The HCHO can be oxidized further to produce $HO_2\bullet$ plus CO; the $HO_2\bullet$ reacts with NO to produce NO_2, and thus one more O_3 equivalent. Finally, the CO can be oxidized to form CO_2 and in the process generate an additional HO_2•, capable of producing an additional O_3 molecule. Thus, the complete oxidation of CH_4 to CO_2 in the presence of sufficient NO has the potential to produce four O_3 molecules. Note that these processes do not result in the net production or loss of •OH or HO_2•; thus, this process is truly a 'catalytic oxidation'.

It should be noted that the oxidized products of the original hydrocarbon are not necessarily oxidized to completion in the gas phase. Depending on their polarity, and thus volatility, they can also partition into organic aerosol particles, or be deposited back to the surface through wet or dry deposition (designated as V_d in Scheme 14.1). Reactions in the liquid phase, collectively referred to as heterogeneous chemistry, can further oxidize higher-generation products of hydrocarbon chemistry. On the right side of Scheme 14.1, is shown the oxidative pathways (in summarized form) for reaction with NO_3• and O_3, two additional oxidants that potentially react with emitted hydrocarbons. These reactions also often produce products of lower volatility that are likely to partition into the liquid phase of secondary organic aerosol (SOA) particles. Estimates of atmospheric lifetime for many of the species shown in Scheme 14.1 are provided in Table 14.1.[16–20]

Table 14.1 *Atmospheric lifetimes (τ) for some of the important species in the oxidative chemical schemes shown in Scheme 14.1 (modified from ref. 1)*

Chemical species	Chemical formula	Atmospheric lifetime
Hydroxyl radical	•OH	0.2–1.0 s[a]
Peroxy radical	RO_2•	5–900 s[b]
Nitrate radical	NO_3•	5–6 s (daytime)[c] $>$ 1000 s (nighttime)
Nitrogen dioxide	NO_2	143 s (daytime)[d] 7 h (nighttime)
Isoprene (**1**)	C_5H_8	0.06–1.5 days[e]
Monoterpenes	$C_{10}H_n$	0.06–20 days[f]
Sesquiterpenes	$C_{15}H_n$	2 min to 3 h[g]
Acetone	C_3H_6O	15–61 days (•OH)[g]
Methanol	CH_3OH	12 days (•OH)[g]
Ozone	O_3	5–300 days[h]

Note: [a] Calculated with respect to global average reaction with CH_4 (at total global content of 5000 Tg) and CO (at total global content of 360 Tg).

[b] Calculated for the methyl peroxy radical when reaction is with NO or HO_2•; reaction coefficients were taken from ref. 19.

for NO + CH_3O_2	10 pptv NO	$\tau = 540$ s
	1 ppbv NO	$\tau = 5.4$ s
for HO_2• + CH_3O_2	8 pptv HO_2	$\tau = 891$ s.

[c] Daytime zenith angle = 0°; nighttime τ calculated for reactions with NO_2 and VOCs in an unpolluted atmosphere, from ref. 20.

[d] Reaction coefficients were taken from ref. 19.

Daytime, zenith angle = 50° $\tau = 143$ s.
Nighttime, 50 ppb O_3 $\tau = 7$ h.

[e] From ref. 16 with isoprene lifetimes for individual oxidants as:

$\tau(\bullet OH) = 1.4$ h; $\tau(O_3) = 1.3$ days; $\tau(NO_3\bullet) = 1.6$ h.

[f] From ref. 16 with representative monoterpene lifetimes for individual oxidants as:

α-Pinene : $\tau(\bullet OH) = 2.6$ h; $\tau(O_3) = 4.6$ h; $\tau(NO_3\bullet) = 11$ min
β-Pinene : $\tau(\bullet OH) = 1.8$ h; $\tau(O_3) = 1.1$ days; $\tau(NO_3\bullet) = 27$ min.

[g] From refs. 16, 17 (in the case of acetone) and 18 (in the case of methanol).

[h] Photolysis and chemical reaction (surface deposition causes a reduced lifetime); from ref. 20:

Lower troposphere; 0-3 km $\tau = 5-8$ days (summer); $\tau = 17-100$ days (winter)
Upper troposphere; 6-10 km $\tau = 30-40$ days (summer); $\tau = 90-300$ days (winter).

14.4 The Types of Emitted BVOCs and General Roles in Atmospheric Chemistry

In terms of oxidative photochemistry, the most important BVOC emitted from terrestrial ecosystems is isoprene (**1**). Global isoprene emissions are likely to be in the range of 500–550 Tg C year^{-1}.[4,5,21,22] Isoprene has an atmospheric lifetime of one to two hours,[16,23] reacting relatively quickly with •OH radical. Typical isoprene concentrations in the few meters above isoprene-emitting ecosystems are 1–30 ppbv.[24–26] The oxidation of isoprene by •OH yields two primary products, methacrolein (MACR, **2**) and methyl vinyl ketone (MVK, **3**; Scheme 14.2), with fractional yields of 0.20–0.28 and 0.31–0.44, respectively.[27–30] Thus, a total of about 60% of the oxidation products can be accounted for as MVK and MACR; the remaining 40% of the products include various organic nitrates (for reactions in the presence of NO concentration above about 30 pptv) and various carbonyl compounds.[31] The primary gas phase oxidation of MVK and MACR also occurs through the •OH reaction channel, though estimated lifetimes of these products are a bit higher than those of isoprene, being approximately 9 h and 6 h, respectively.[30]

The emissions of C_{10} monoterpenes from plants have been studied for nearly as long as those for isoprene. In 1960, Went published an intriguing analysis in which he used the concentration of terpene-containing oils from sagebrush to estimate global terpene emissions as 175 Tg year^{-1}.[32] The most abundant terpene in sagebrush oils is camphor (**4**, Figure 14.1), a monoterpene estimated to compose 40–45% of the total oil.[33] Went's coarse estimate of global monoterpene emissions was not too different from more sophisticated recent estimates made with 3D emissions models, with most models predicting global monoterpene fluxes between 100–120 Tg year^{-1}.[22] Like isoprene, monoterpenes are oxidized mostly through reaction with •OH radicals. However, their internal double bonds

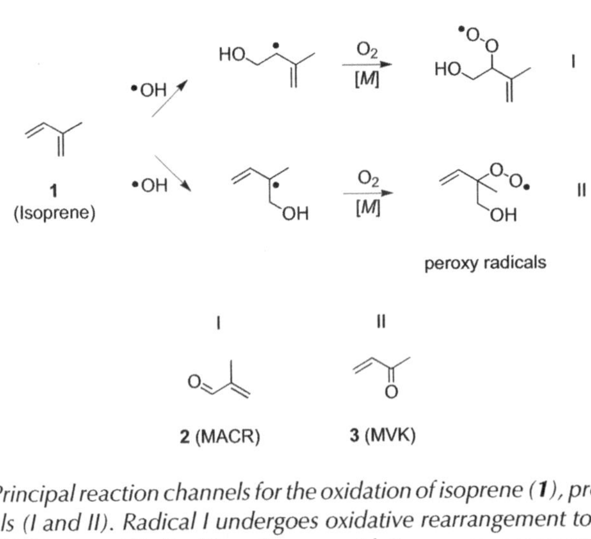

Scheme 14.2 *Principal reaction channels for the oxidation of isoprene (**1**), producing two types of peroxy radicals (I and II). Radical I undergoes oxidative rearrangement to form the product methacrolein (MACR, **2**) and Radical II undergoes oxidative rearrangement to form the product methyl vinyl ketone (MVK, **3**). Typical ratios of product formation are 6 : 4 favouring MVK*

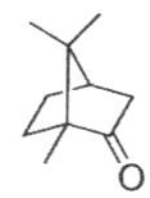

(1*R*)-**4**
(Camphor)

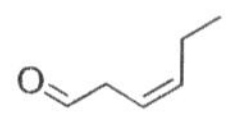

5
((*Z*)-3-Hexenal)

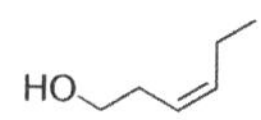

6
((*Z*)-3-Hexen-1-ol)

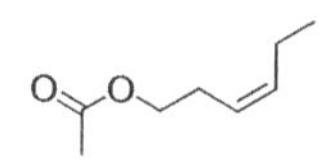

7
((*Z*)-3-Hexenyl acetate)

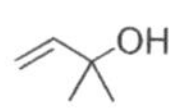

8
(Methylbutenol)

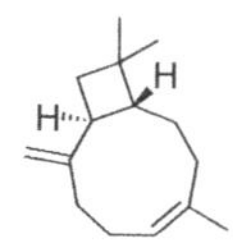

9
(*β*-Caryophyllene)

Figure 14.1 *Structures of camphor (**4**), (Z)-3-hexenyl derivatives **5–7**, methylbutenol (**8**) and β-caryophyllene (**9**)*

render them highly reactive with O_3, as well, and the O_3 reaction is more important in the overall oxidation of monoterpenes than it is with isoprene.[34] The primary oxidation products from monoterpenes tend to be of lower volatility than those for isoprene, and are more important as precursors to the formation of SOAs.[35] Because monoterpene emission rates tend to be lower than those for isoprene, monoterpenes are also present at lower atmospheric concentrations, typically in the range 50–100 pptv above actively emitting ecosystems.[36,37] Monoterpene lifetimes are in the range of 3–6 h, a bit longer than isoprene, but nonetheless short.[36] Most recently, tropical ecosystems have been identified as strong monoterpene sources, with potentially profound effects on regional atmospheric chemistry.[25,38]

Isoprene and the monoterpenes are the traditional foci of BVOC emissions research.[1,39] However, new classes of emitted compounds are being discovered continuously. It is likely that our views of how plant emissions influence the atmosphere will continue to change as more of these discoveries are made. Recently, the oxygenated BVOCs have been recognized as an important class of emitted compounds.[40] Oxygenated BVOC emissions are especially high from coniferous forests (acetone and methylbutenol), plants undergoing active leaf expansion (methanol), freshly cut turf and agricultural ecosystems [(*Z*)-3-hexenal (**5**), (*Z*)-3-hexen-1-ol (**6**) and (*Z*)-3-hexenyl acetate (**7**), Figure 14.1].[41–43] Recently, high rates of salicylic ester emissions have been detected from a desert ecosystem, potentially contributing to organic aerosol formation.[44] Globally, the importance of oxygenated BVOCs is seen in the observation of high concentrations in the remote free troposphere of the southern hemisphere, including identification of carbonyls, alcohols, organic nitrates, organic pernitrates and peroxides; these compounds as a group can be present at five times the concentration of non-oxygenated VOCs.[45] High concentrations of oxygenated VOCs in these remote areas are likely due to a combination of higher tropospheric lifetimes, compared to non-oxygenated species, and previously unrecognized sources, such as the photochemical oxidation of hydrocarbons in terrestrial air masses that drift to remote marine locations. Because there is a significant recent history of published papers on the topic of oxygenated BVOCs, I next spend some words briefly describing the main classes of compounds.

Acetone is emitted at global rates of approximately 95 Tg $year^{-1}$ with an estimated tropospheric lifetime of 15–60 days.[16,17] Tropospheric acetone is a source of peroxyacetyl radicals, which react with NO_x to form PAN, and is a primary source for the formation of •OH and HO_2• in the upper troposphere.[46,47] Methanol is emitted from vegetation[18,42,48] at the rate of 80–130 Tg $year^{-1}$ and from oceans[18] at the rate of 85 Tg $year^{-1}$. Methanol is oxidized in the troposphere through reaction with •OH, and exhibits an average lifetime of approximately five to ten days;[18] it is also lost to an ocean sink at the same approximate rate as atmospheric oxidation.[18,49] The products of methanol oxidation include formaldehyde (CH_2O) and hydroperoxy (HO_2•) radicals.[50] Methylbutenol [2-methyl-3-buten-2-ol (**8**), Figure 14.1] is an oxidized hemiterpene, similar in structure and biochemical origin to isoprene, and emitted at relatively high rates from North American pine forests.[41,51] Estimates of global methylbutenol emissions are not available; however regional estimates for North America suggest emissions of 3.2 Tg C $year^{-1}$.[52] Methylbutenol has a tropospheric lifetime of approximately 2 h with the primary loss due to oxidation by •OH.[53] Oxidation of methylbutenol contributes to tropospheric acetone and H_2O_2 production.[47,53,54] Acetaldehyde emissions occur from numerous different types of plants, and may be associated with vegetative stress.[40] In the troposphere, acetaldehyde is oxidized by •OH, exhibiting an average lifetime of one day, and feeding the production of PAN.[55] Short-chain organic acids are emitted to the atmosphere from plants and soil microbes,[56] and potentially produced photochemically from peroxy radicals.[57] Atmospheric concentrations of formic and acetic acid above several different ecosystems are in the range of 0.5–2.5 ppbv.[24,25,58,59] Atmospheric fog and aqueous aerosol water has been found to have relatively high concentrations of acetic and formic acids.[60] The tropospheric lifetime of formic and acetic acids is two to six days with the primary cause of turnover being wet and dry deposition.[58]

One class of biogenic VOCs that is just now coming to the forefront of emissions research is the sesquiterpenes (C_{15} hydrocarbons).[61] Sesquiterpenes are emitted at relatively high rates from plants,[62] which sequester the compounds for the purpose of defence against herbivores or emit them as signalling compounds to attract herbivore parasitoids.[63] Some of the more commonly emitted sesquiterpenes [e.g. β-caryophyllene (**9**), Figure 14.1] have tropospheric lifetimes of 1–2 min,[64] and are not detected at significant concentrations during atmospheric sampling. Emphasis on observations of sesquiterpene emissions from landscapes has strengthened over the past decade following the discovery that they play a key role in controlling the rate of SOA formation in key airsheds.[65]

14.5 Gas Phase Oxidation of BVOCs

Most gas phase oxidation of alkenes occurs through reaction with •OH radicals. In order to illustrate the oxidation mechanism, the case of isoprene oxidation is considered in Scheme 14.2. The reaction sequence is initiated by addition of •OH to one of the double bonds, and addition is favoured for the terminal carbons. Following addition of •OH, O_2 is added in the presence of a third organizing molecule or surface (denoted as *M*), to form two principal organic peroxy radicals. When atmospheric NO concentrations are greater than about 30 pptv, the peroxy radicals will react with NO (see Reaction 14.16 above), initiating a series of oxidative rearrangements to form the two main products, MVK (**3**) and MACR (**2**) in the approximate ratio 6:4, respectively. These products are each four carbons in length. In both reaction channels, the fifth carbon from the original isoprene molecule is oxidatively cleaved to form a •CH_2OH radical, which rapidly reacts with O_2 to form formaldehyde (CH_2O) plus a hydroperoxy radical (HO_2•).

The fate of BVOCs once they enter the atmosphere is difficult to determine with precision. So, while we can have some relatively high degree of confidence that 'most' of the isoprene molecules begin their photochemical 'journey' by reacting with •OH, our confidence in predicting the fate of these compounds after this initial reaction is low. We are forced to evaluate likely reactions paths with complex models. Our model predictions contain uncertainties due to unknowns about concentration profiles, oxidation efficiencies, interactions between gas phase and aerosol phase chemistry and loss rates to surface deposition. At the global scale it would not be unreasonable to expect errors in the range of $\pm35\%$ in the predicted rate of isoprene oxidation to its ultimate product of CO_2.[66] This large error is superimposed on predictions that only allow for 20–40% of the isoprene carbons to be oxidized completely to CO_2; with most studies converging toward a global average of 30%.[29,67] The value may be higher (near 40%) above continental regions, such as the United States, where pollution enhances the oxidation efficiency.[29] Thus, with numbers like 20–40%, of the isoprene carbon taking part in oxidative photochemistry, and with an overall uncertainty for product formation rates in the range of $\pm$ 35%, one can appreciate that our confidence in predicting global patterns of photochemistry remains somewhat limited.

In addition to reacting with •OH, the double bonds of alkenes can also be oxidized directly by O_3 and NO_3• radicals. The reaction with O_3 is called 'ozonolysis'. The rate coefficients for the alkene-O_3 reaction tend to be low; however, tropospheric O_3 concentrations are high enough that, even with low rate coefficients, oxidation via O_3 can occur at significant rates. Whereas the photochemical production of •OH, and subsequent oxidation of alkenes, occurs predominantly during the day when O_3 photolysis produces O• (^{1}D) radicals, background concentrations of O_3 are typically sufficient to allow ozonolysis to proceed through the night; meaning that the oxidation of those alkenes susceptible to ozonolysis can proceed day (by •OH and O_3) and night (by O_3). Among terpenes, those with internal double bonds are most susceptible to ozonolysis.[68] The exothermic addition of O_3 to the double bonds of alkenes produces a cyclic intermediate capable of decomposing to a carbonyl molecule and a biradical carbonyl oxide, known as a Criegee Intermediate (CI). The CI retains much of the energy of the initial cyclo-adduct and this energy can be channelled into additional reactions. The decomposition of energized CIs often produces •OH as a byproduct,[69] and there is evidence that this source of •OH is significant in further promoting the oxidation of BVOCs above some terpene-emitting forests, especially at night.[70,71] The production of •OH through ozonolysis is likely most relevant to the oxidation of monoterpenes and sesquiterpenes. Isoprene undergoes ozonolysis and •OH can be produced through this reaction;[72] however, the •OH yield from the ozonolysis of isoprene is low compared to that for monoterpenes,[73] and the reaction of isoprene with O_3 is considerably slower than the reaction with •OH (see Table 14.1).[28,74]

In addition to reacting with •OH and O_3, alkenes are also capable of reacting with nitrate radicals (NO_3•).[75] NO_3• radicals undergo rapid photo-dissociation during the day,[76] so that they are most important as atmospheric oxidants at night. Observations at several forested sites have led to the suggestion that oxidation by NO_3• can contribute to the disappearance of isoprene near sunset,[26,77] although other observations suggest that most isoprene oxidation, even at sunset, can be accounted for by reaction with •OH.[78] To a large extent, the significance of NO_3• oxidation as a factor leading to the evening disappearance of isoprene is more academic than practical – given that the same process that enhances NO_3• lifetime late in the day (decrease in solar photon flux), also works to diminish isoprene emission. What may be a more important perspective on the isoprene-NO_3• reaction is its potential to produce organic peroxy radicals (RO_2•) capable of self-reaction to form SOA particles.[79]

14.6 Gas Phase Chemistry of BVOCs in Urban and Suburban Airsheds

In urban and suburban airsheds emitted BVOCs have the potential to enhance tropospheric O_3 production rates. Thus, while the alkene bonds in these compounds have the potential to react with O_3 through ozonolysis, high concentrations of NO_x tend to shift the reaction of organic peroxy radicals (RO_2•) to favour the catalytic cycling of NO_x between forms that support O_3 production; O_3 is destroyed and produced in the same airshed, but the net result is enhanced O_3 production (see Reactions 14.18 to 14.24 above). To review the essential steps:

$$RO_2\bullet + NO \rightarrow RO\bullet + NO_2 \tag{14.25}$$

$$NO_2 + h\nu\,(+O_2) \rightarrow NO + O_3 \tag{14.26}$$

The alkoxy radical (RO•) that is formed is highly reactive and can go on to promote even further photochemistry. The degree to which net O_3 production occurs, rather than net O_3 destruction depends on the concentration of NO_x (concentrations greater than 30 pptv tend to promote net O_3 production), which are principally produced through combustion processes. The regional mixing of air masses from forested suburban forests and urban/suburban population centres can catalyse the photochemistry that results in net O_3 production.

The capacity for BVOCs to catalyse the net production of O_3 is defined by the ozone production efficiency (OPE). The OPE describes the capacity for a mole of NO_x to catalyse the formation of O_3 in a given air mass. Formally, the OPE is defined as the moles of O_3 produced per mole of NO_x oxidized (to other forms of NO_y) [$NO_y = NO + NO_2 + PAN + HNO_3$] and removed from further photochemical activity. (Forms of NO_y such as HNO_3 and PAN are removed from further photochemistry due to their tendencies for deposition to the surface or transport to other regions). In order to appreciate the concept of the OPE, we recognize that a unit of NO_x has two principal reactive paths in an air mass: (*a*) catalytic cycling between NO and NO_2 with concomitant production of O_3, or (*b*) oxidation to other forms of NO_y, and subsequent removal from local photochemical processes due to deposition or transport. The emission of BVOCs potentially increases the OPE by favouring the first path.[80,81] Due to its high reactivity, isoprene is the most active BVOC in promoting

O_3 formation in the presence of high NO_x; monoterpenes and oxygenated BVOCs are less important.[81] The presence of reactive BVOCs affects the OPE in several ways. Of primary importance, as discussed above, is the formation of organic peroxy radicals, capable of oxidizing NO to NO_2. However, the presence of BVOCs also tends to reduce the concentration of •OH which slows the conversion of NO_2 to HNO_3 and enhances the photochemical efficiency of each unit of NO_x in producing O_3. Clearly, the potential for BVOCs to stimulate O_3 production is dependent on a complex set of first- and higher-order relationships among the various forms of NO_y, •OH and the unique photochemical properties of specific BVOCs; while simple relationships can be stated with regard to whether NO_x or BVOCs ultimately limit O_3 production, quantitative metrics, such as the OPE, are strongly influenced by these higher-order processes.

14.7 Gas Phase Chemistry Within and Above Forests

In the vicinity of forests gas phase atmospheric chemistry is driven by fundamental first-order relationships between emitted BVOCs and the primary oxidants that are present; usually •OH, but also O_3 and NO_3•. In addition to being a primary source of emitted BVOCs, forests exert considerable control over the production of •OH, the principal oxidant that reacts with BVOCs. To understand this, we must recognize that the production of •OH requires atmospheric humidity which, in turn, is dependent on ecosystem evapotranspiration.[82] As shown in Scheme 14.1 and Reactions 14.1 to 14.3 (see above), the initiation of •OH formation involves the photo-dissociation of O_3 to produce the O•(^{1}D) radical plus O_2. The O•(^{1}D) radical is short-lived with most being de-energized to the alternative state O•(^{3}P) by collision with N_2 and O_2. In a tropospheric domain with humidity, however, some of the O•(^{1}D) radicals will react with H_2O; for example, in an atmosphere of 50% relative humidity and at 20 °C, approximately 9% of the O•(^{1}D) radicals react with H_2O to form •OH.[83] This is the primary source of •OH, and the reactions leading to its formation occur frequently in the humid air above forests, especially above tropical forests.[84]

Evidence has been presented recently that the chemistry that occurs in the vicinity of forests has the potential to generate even more •OH than can be accounted for by the traditional path involving O_3 photolysis. Recent aircraft measurements above tropical forests have shown enhanced rates of •OH production relative to what is predicted from traditional photochemical models.[85] Conventional wisdom has dictated that in unpolluted ecosystems, such as tropical forests, BVOC emissions (principally isoprene) should reduce tropospheric •OH concentrations due to oxidation and subsequent deposition of reaction products back to the surface.[86] The results of Lelieveld *et al.*,[85] however, challenge the assumption of rapid deposition rates, and suggest that the oxidized products of isoprene-•OH chemistry, principally peroxy radicals, react with HO_2• in a recycling reaction that produces •OH, prior to deposition back to the surface.[87] Observations and modelling at a hardwood forest in Michigan also led to the suggestion that processes other than the O•(^{1}D) reaction with H_2O, generated •OH.[71] Chemistry models simply could not be balanced unless a light-independent missing •OH sink were invoked, and a process could be identified to produce higher than predicted nighttime •OH concentrations. After systematic consideration of all known possibilities, the authors hypothesized that the presence of unobserved terpenes (probably monoterpenes or sesquiterpenes) produces the missing •OH sink and, at the same time, contributes to the nighttime production of •OH through ozonolysis. This hypothesis was consistent with an increased need to invoke the missing •OH sink as temperature increased; presumably because terpene emission rates and associated ozonolysis rates also increased with temperature.

At some forest sites, especially those located in relatively polluted areas, oxidative chemistry can be complex. For example, in an oak-hickory forest in Tennessee, a location that receives significant pollution impacts due to automobile traffic, •OH, O_3 and NO_3• all have major roles as BVOC oxidants (Figure 14.2).[26] Oak and pine trees in this forest emit

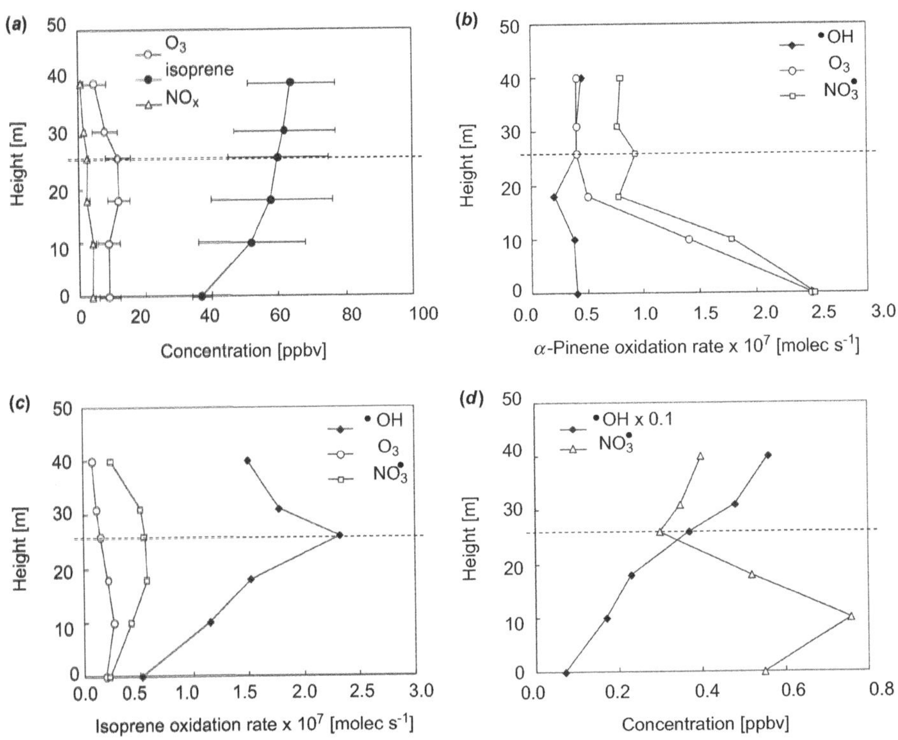

Figure 14.2 (a) Vertical profiles of O_3, isoprene and NO_x concentrations measured in Oak Ridge, Tennessee within and above an oak-hickory forest. (b) Vertical profile of modelled daytime oxidation rates of the monoterpene α-pinene (most likely emitted from pine needles) due to reaction with •OH, O_3 or NO_3•. (c) Vertical profile of modelled daytime oxidation rates of isoprene (most likely emitted from oak leaves) due to reaction with •OH, O_3 or NO_3•. This forest is moderately influenced by NO_x deposition from high volumes of automobile traffic near the forest. (d) Daytime concentration profiles of •OH and NO_3• in Oak Ridge, Tennessee within and above an oak-hickory forest. Note that the •OH concentration has been reduced by 0.1 in order to normalize its scale relative to that for NO_3•. The horizontal dashed line indicates the canopy height. From ref. 26

10
(α-Pinene) (β-Pinene)

Figure 14.3 Structures of α- and β-pinene

isoprene and monoterpenes at high rates. Above the forest canopy, where the daytime •OH production rate is high (due to high rates of O_3 photo-dissociation), most of the emitted isoprene is oxidized by •OH radicals (Figure 14.2c). However, deeper within the canopy, where the photon fluxes that drive •OH production are low, and the lifetime of NO_3• radicals is relatively high, isoprene losses due to oxidation by NO_3• increase and those due to oxidation by •OH decrease (Figure 14.2c); •OH remains the primary oxidant within the canopy, but the ratio of oxidation by NO_3•, relative to •OH increases. Within the canopy air space, NO_3• radical production rates can be high due to high concentrations of anthropogenically produced NO_x and O_3, which can react according to:

$$O_3 + NO_2 \rightarrow NO_3\bullet + O_2 \tag{14.27}$$

and the resulting NO_3• radicals can be 'stored' in a nitrogen pentoxide reservoir:

$$NO_3\bullet + NO_2 \leftrightarrow N_2O_5 \tag{14.28}$$

Thus, a unique combination exists in this canopy of high NO_3• concentrations relative to •OH concentrations (Figure 14.2d). This situation renders the canopy environment particularly conducive to terpene oxidation by NO_3•. Due to higher reaction coefficients of the monoterpenes toward O_3 and NO_3•, compared to isoprene, the oxidation of BVOCs by these alternative reactants is especially important in the case of emitted monoterpenes, such as α-pinene (**10**) (Figures 14.2b, 14.3). In fact, the oxidation of α-pinene within the canopy occurs predominantly by O_3 and NO_3•.

14.8 BVOC Emissions and SOA Formation

The existence of a compound in the gas phase versus liquid phase of the atmosphere depends on temperature and the vapour pressure of the compound. Compounds tend to be in the gas phase if their vapour pressure is $> 10^{-4}$ Pa at ambient temperature;[88] at vapour pressures below that threshold, they are more likely to be semi-volatile or in the liquid phase of aerosol particles. Generally, the progressive oxidation of hydrocarbons renders them more polar and thus less volatile. This process of oxidative hydrocarbon processing feeds the growth of SOA particles and, depending on the polarity of the oxidized products, the hygroscopicity of the particles is also potentially altered. Thus, it is generally through the oxidation products of BVOC chemistry that we find the precursors for SOA formation, not the original emitted compounds. If organic compounds are sufficiently polar to push the hygroscopicity of SOA surfaces beyond certain thresholds, the particles can take up water,[89] dissolved organic acids and other compounds and potentially force those particles to function as cloud condensation nuclei.[90] The formation of SOA from emitted BVOCs can be especially high above terpene-emitting forests.[91]

The formation of SOA particles from the photo-oxidation of monoterpenes has been well studied.[92] Some monoterpenes, such as cyclic dialkenes, can produce aerosol mass yields at 40% efficiency when oxidized by O_3 in controlled smog chambers. However, the conventional wisdom until 2004 was that BVOCs like isoprene, with its terminal double bonds, were insignificant as SOA precursors; instead being oxidized principally through gas phase chemistry. This was a significant bit of conventional wisdom because of the high emission rates of isoprene, compared to monoterpenes. If even a small fraction of the annually emitted isoprene was oxidized to SOA precursors, then the entire global SOA budget would have to be reconsidered. In 2004, Claeys and coworkers published a paper in *Science* in which a careful analytical analysis had been conducted of the organic compounds composing SOA above an Amazonian rain forest.[93] The analysis revealed rather high concentrations of the isoprene oxidation products, 2-methylthreitol (**11**) and 2-methylerythritol (**12**, Scheme 14.3); these compounds belong to the general group known as methyltetrols and occur with low volatility that allows them to easily partition into SOA particles. From knowledge of the photo-oxidation chemistry of isoprene, these workers estimated that emitted isoprene could account for up to 2 Tg $year^{-1}$ of global SOA mass. With the publication of this paper, an active discussion was initiated as to the role of isoprene emissions in SOA dynamics. The exact pathway by which the methyltetrols are formed is still being investigated. In an atmosphere with low NO_x, the peroxy radicals formed from the initial reaction of isoprene with •OH can potentially react among themselves and with other compounds to form the methyltetrols.[93] However, in the presence of high NO_x, which includes periods of biomass burning, peroxy radicals are more likely to react with NO, limiting methyltetrol formation;[94] it was during periods of relatively high NO_x that the original observations of methyltetrols in SOA were made.[93] Thus, it was initially difficult to reconcile the observations of atmospheric methyltetrols during periods when their formation would have been less favoured. Now we recognize that under higher NO_x conditions, a type of acid-catalysed gas phase reaction involving hydrogen peroxide may be more important for the production of the methyltetrols (see Scheme 14.3).[95] The progressive conversion of non-polar isoprene to polar oxidized products, capable of partitioning into aerosol particles is clearly evident as shown in Scheme 14.3. There are other pathways for isoprene oxidation to contribute to SOA formation, but they involve aqueous phase (cloud and fog) reactions, and are less likely to occur than the gas phase pathways.[96]

Recent analyses have also sharpened estimates of the potential contributions of isoprene oxidation to global SOA formation. In one recent study, the channelling of oxidized isoprene to SOA formation was shown to occur with an efficiency (mass of aerosol per unit mass of isoprene) of approximately 3%[94] and given total global isoprene emissions of approximately 500 Tg $year^{-1}$, the annual production of SOA from isoprene on a mass basis would be approximately 15 Tg $year^{-1}$.[97] Lower estimates of the conversion efficiency of isoprene to SOA, however, have been generated from other modelling efforts, including 4.6 Tg $year^{-1}$.[2]

Shortly after Claeys *et al.* published their results in 2004,[93] an intriguing analysis by Goldstein and Galbally was released in which the mass of annual SOA production was estimated to be between 140–910 Tg $year^{-1}$;[98] this is considerably higher than estimates up to that time which had ranged from 10–70 Tg $year^{-1}$. The analysis of Goldstein and Galbally emphasized constraint on mass balance using top-down approaches, whereas previous analyses had emphasized bottom-up accounting of products from observed emission rates

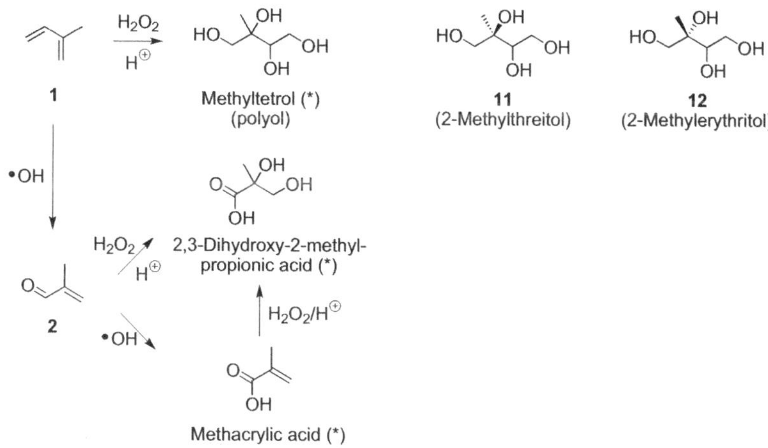

Scheme 14.3 *Oxidation scheme (under acidic conditions) of isoprene (**1**) and its first-generation product, methacrolein (**2**), to form semi-volatile products capable of partitioning into secondary organic aerosol (SOA) particles. Polar compounds that are likely to partition into SOA are noted with asterisks. This scheme is based on refs. 93 and 95a, b*

and modelled chemistry. With the publication of the Goldstein and Galbally analysis, the community had a new challenge: to account for the immense gap between the top-down and bottom-up estimates, and to re-evaluate the potential for unknown compounds and chemical pathways that produce SOA. The results also had important implications for climate modelling; if we increase rates of atmospheric SOA loading tenfold in earth energy budget models, we are forced to reconsider the fundamental sensitivity of surface warming to increases in greenhouse gas concentrations.

Part of the missing aerosol mass may be due to poor understanding of the chemical processing of BVOCs that produces SOA precursors. The growth of SOA particles occurs at the expense of mass produced during the progressive oxidation of BVOCs, and this process can be characterized by a species-dependent aerosol growth curve (Figure 14.4). The kinetics of SOA growth appears to sort out into groups depending on the number of double bonds in the alkene undergoing ozonolysis.[79,99] The oxidation of compounds with a single double bond causes SOA growth that is limited by the rate of the first oxidation step, and particle growth ceases after the initial compound is consumed. The oxidation of compounds with two double bonds, however, can continue even after complete consumption of the initial compound; in these compounds, higher-order products retain the potential for further oxidation. As a result, the growth curves for SOA particles produced from compounds with two double bonds exhibit a vertical phase (increase in SOA mass with no further change in mass of initial compound).

In recent years we have gained a better understanding of which products from the oxidation of terpenes are most active as SOA precursors. Major breakthroughs in this area have occurred through smog chamber studies, in which controlled oxidation can be imposed on the chemistry, and products can be sampled at fairly high temporal frequency.[100] With regard to the oxidation of monoterpenes, common first generation products that feed the growth of SOA particles are: formaldehyde, acetaldehyde, formic acid, acetone, acetic acid

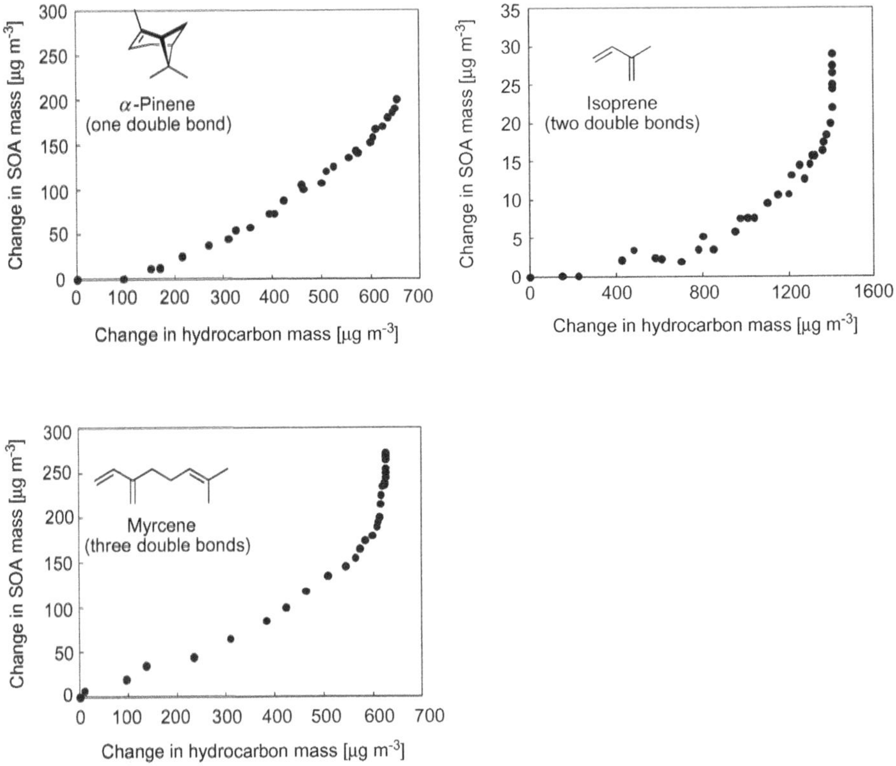

Figure 14.4 Stoichiometric trajectories for the conversion of primary hydrocarbon mass for three different terpenes (with varying numbers of double bonds) into SOA mass. The tendency for the aerosol mass to continue increasing even after all of the initial mass has disappeared is seen as a distinct vertical phase, which is amplified as the number of double bonds in the original molecule increases. Redrawn from parts of a figure in ref. 99. Adapted with permission from [99]. Copyright 2006 American Chemical Society

and nopinone (**13**, Figure 14.5).[101] The oxidative processing of compounds beyond the first generation of products is still highly uncertain for most SOA precursors. This results in a large knowledge gap that could underlie the gap in estimates of SOA formation rates. Heald *et al.*[102] observed concentrations of organic carbon above the north western Pacific Ocean that were 10–100 times higher than those predicted by a state of the art global chemical transport model. Given the transport processes that would have to be invoked to explain these observations, the results are best explained as reflecting inadequacies in the oxidative modelling of long-lived atmospheric BVOCs.

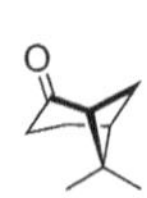

(1*R*)-**13**
(Nopinone)

Figure 14.5 Structure of nopinone (**13**)

14.9 Conclusion

Over the past decade, the topics of biogenic VOC emissions and atmospheric chemistry have merged to form a fully integrated discipline focused on understanding the influences of ecosystem processes on the oxidative power of the atmosphere. The evolution of this emergent discipline has forced biologists studying controls over BVOC emission rates to think more like atmospheric chemists, and vice versa, for chemists studying controls over atmospheric chemistry to think more like biologists. Ultimately, the thinking of both groups is constrained by the need for mass balance in the budgets of ecosystem and atmosphere carbon budgets – ecosystems can only emit BVOCs up to the limit explained by atmospheric oxidation, and atmospheric oxidation must be limited by the mass of organic carbon emitted from ecosystems. The need for full closure of these budgets has forced biologists and chemists to communicate more frequently and in much more quantitative terms. As these communications have progressed, the large gaps in our understanding, and therefore in our BVOC budgets, have been clearly exposed. We know very little about: (*a*) the loss of first- and higher-generation emission products through wet and dry surface deposition; (*b*) the influence of long-term (seasonal to decadal) growth environments on BVOC emission factors; (*c*) the emissions magnitude and reaction kinetics of very short-lived BVOC species, which tend to elude our sensors; and (*d*) the relative roles of urban, suburban and remote airsheds to the processing of BVOCs at regional to global scales. These are some of the topics that will likely be of high priority as we proceed to the next few decades of research in this nascent discipline. Never before has the case been easier to make for the importance of biologists learning chemistry and chemists learning biology. The reality of this necessity doesn't make life easier, but it does make it more interesting.

References

1. R. K. Monson, E. A. Holland, Biospheric trace gas fluxes and their control over tropospheric chemistry, *Annu. Rev. Ecol. Syst.*, **32**, 547–576 (2001).
2. K. Tsigaridis, M. Kankidou, Secondary organic aerosol importance in the future atmosphere, *Atmos. Environ.*, **41**, 4682–4692 (2007).
3. (*a*) S. D. Piccot, J. J. Watson, J. W. Jones, A global inventory of volatile organic compound emissions from anthropogenic sources, *J. Geophys. Res., [Atmos.]*, **97**, 9897–9912 (1992); (*b*) R. Friedrich, A. Obermeier, Anthropogenic emissions of volatile organic compounds, in *Reactive Hydrocarbons in the Atmosphere*, C. N. Hewitt (Ed.), Academic Press, San Diego, 1999, pp. 1–39.
4. A. Guenther, C. N. Hewitt, D. Erickson, R. Fall, C. Geron, T. Graedel, P. Harley, L. Klinger, M. Lerdau, W. McKay, T. Pierce, B. Scholes, R. Steinbrecher, R. Tallamraju, J. Taylor, P. Zimmermann, A global model of natural volatile organic compound emissions, *J. Geophys. Res., [Atmos.]*, **100**, 8873–8892 (1995).
5. A. Guenther, T. Karl, P. Harley, C. Wiedinmyer, P. I. Palmer, C. Geron, Estimates of global terrestrial isoprene emissions using MEGAN (Model of Emissions of Gases and Aerosols from Nature), *Atmos. Chem. Phys.*, **6**, 3181–3210 (2006).
6. W. Winiwarter, H. Haberl, D. Simpson, On the boundary between man-made and natural emissions: problems in defining European ecosystems, *J. Geophys. Res., [Atmos.]*, **104**, 8153–8159 (1999).

7. J. Lathiere, D. A. Hauglustaine, N. De Noblet-Ducoudre, G. Krinner, G. A. Folbert, Past and future changes in biogenic volatile organic compound emissions simulated with a global dynamic vegetation model, *Geophys. Res. Lett.*, **32**, L20818 (2005).
8. *The Emissions Database for Global Atmospheric Research (EDGAR)*, http://www.mnp.nl/edgar/.
9. J. Olivier, A. F. Bouwman, C. W. M. van der Maas, J. J. M. Berdowski, C. Veldt, J. P. J. Bloos, A. J. H. Visschedijk, P. Y. J. Zandveld, J. L. J. Haverlag, Description of EDGAR version 2.0: a set of global emission inventories of greenhouse gas and ozone depleting substance for all anthropogenic and most natural sources on a per country basis on 1 degree × 1 degree grid, RIVM rep. 771060 O02/TNO MEP, Natl. Inst. Public Health Environ., Bilthoven, The Netherlands, 1996.
10. (*a*) A. Srivastava, Source apportionment of ambient VOCS in Mumbai city, *Atmos. Environ.*, **38**, 6829–6843 (2004); (*b*) T. Karl, E. Apel, A. Hodzic, D. D. Riemer, D. R. Blake, C. Wiedinmyer, Emissions of volatile organic compounds inferred from airborne flux measurements over a megacity, *Atmos. Chem. Phys.*, **9**, 271–285 (2009).
11. P. C. Harley, R. K. Monson, M. T. Lerdau, Ecological and evolutionary aspects of isoprene emission from plants, *Oecologia*, **118**, 109–123 (1999).
12. (*a*) C. D. Geron, A. B. Guenther, T. E. Pierce, An improved model for estimating emissions of volatile organic compounds from forests in the eastern United States, *J. Geophys. Res., [Atmos.]*, **99**, 12773–12791 (1994); (*b*) D. Simpson, W. Winiwarter, G. Borjesson, S. Cinderby, A. Ferreiro, A. Guenther, C. N. Hewitt, R. Janson, M. A. K. Khalil, S. Owen, T. E. Pierce, H. Puxbaum, M. Shearer, U. Skiba, R. Steinbrecher, L. Tarrason, M. G. Oquist, Inventorying emissions from nature in Europe, *J. Geophys. Res., [Atmos.]*, **104**, 8113–8152 (1999); (*c*) X. X. Tie, G. H. Li, Z. M. Ying, A. Guenther, M. Madronich, Biogenic emissions of isoprenoids and NO in China and comparison to anthropogenic emissions, *Sci. Total Environ.*, **371**, 238–251 (2006); (*d*) A. K. Baker, A. J. Beyersdorf, L. A. Doezema, A. Katzenstein, S. Meinardi, I. J. Simpson, D. R. Blake, F. S. Rowland, Measurements of nonmethane hydrocarbons in 28 United States cities, *Atmos. Environ.*, **42**, 170–182 (2008).
13. US National Academy of Sciences, *Rethinking the Ozone Problem in Urban and Regional Air Pollution*, National Academy Press, Washington, D. C., 1991.
14. P. Ciccioli, E. Brancaleoni, M. Frattoni, Reactive hydrocarbons in the atmosphere at urban and regional scales, in *Reactive Hydrocarbons in the Atmosphere*, C. N. Hewitt (Ed.), Academic Press, San Diego, 1999, pp. 160–209.
15. D. Helmig, J. Greenberg, A. Guenther, P. Zimmerman, C. Geron, Volatile organic compounds and isoprene oxidation products at a temperate deciduous forest site, *J. Geophys. Res., [Atmos.]*, **103**, 22397–22414 (1999).
16. R. Atkinson, J. Arey, Gas phase tropospheric chemistry of biogenic volatile organic compounds: a review, *Atmos. Environ.*, **37**, S197–S219 (2003).
17. D. J. Jacob, B. D. Field, E. M. Jin, I. Bey, Q. B. Li, J. A. Logan, R. M. Yantosca, H. B. Singh, Atmospheric budget of acetone, *J. Geophys. Res., [Atmos.]*, **107**, 4100 (2002).
18. D. B. Millet, D. J. Jacob, T. G. Custer, J. A. de Gouw, A. H. Goldstein, T. Karl, H. B. Singh, B. C. Sive, R. W. Talbot, C. Warneke, J. Williams, New constraints on terrestrial and oceanic sources of atmospheric methanol, *Atmos. Chem. Phys.*, **8**, 6887–6905 (2008).
19. B. J. Finlayson-Pitts, J. N. Pitts, *Chemistry of the Upper and Lower Atmosphere. Theory, Experiments and Applications*, Academic Press, San Diego, 2000.
20. G. P. Brasseur, J. J. Orlando, G. S. Tyndall, *Atmospheric Chemistry and Global Change*, Oxford University Press, New York, 1999.
21. (*a*) K. Y. Wang, D. E. Shallcross, Modelling terrestrial biogenic isoprene fluxes and their potential impact on global chemical species using a coupled LSM-CTM model, *Atmos. Environ.*, **34**, 2909–2925 (2000); (*b*) C. S. Potter, S. E. Alexander, J. C. Coughlan, S. A. Klooster, Modeling biogenic emissions of isoprene: exploration of model drivers, climate control algorithms, and use of global satellite observations, *Atmos. Environ.*, **35**, 6151–6165, (2001); (*c*) C. Shim, Y. Wang, Y. Choi, P. Palmer, D. Abbot, K. Chance, Constraining global isoprene emissions with Global Ozone Monitoring Experiment

(GOME) formaldehyde column measurements, *J. Geophys. Res., [Atmos.]*, **110**, D24301 (2005).

22. A. Arneth, R. K. Monson, G. Shurgers, U. Niinemets, P. I. Palmer, Why are estimates of global terrestrial isoprene emissions so similar (and why is this not so for monoterpenes)? *Atmos. Chem. Phys.*, **8** 4605–4620 (2008).
23. R. Atkinson, Atmospheric chemistry of VOCs and NO_x, *Atmos. Environ.*, **34**, 2063–2101 (2000).
24. R. S. Martin, H. Westberg, E. Allwine, L. Ashman, J. C. Farmer, B. Lamb, Measurement of isoprene and its atmospheric oxidation products in a central Pennsylvania deciduous forest, *J. Atmos. Chem.*, **13**, 1–32 (1991).
25. J. Kesselmeier, U. Kuhn, S. Rottenberger, T. Biesenthal, A. Wolf, G. Schebeske, M. O. Andrae, P. Ciccioli, E. Brancaleoni, M. Frattoni, S. T. Oliva, M. L. Botelho, C. M. A. Silva, T. M. Tavares, Concentrations and species composition of atmospheric volatile organic compounds (VOCs) as observed during the wet and dry season in Rondonia (Amazonia), *J. Geophys. Res., [Atmos.]*, **107**, 8053 (2002).
26. J. D. Fuentes, D. Wang, D. R. Bowling, M. Potosnak, R. K. Monson, W. S. Goliff, W. R. Stockwell, Biogenic hydrocarbon chemistry within and above a mixed deciduous forest, *J. Atmos. Chem.*, **56**, 165–185 (2007).
27. (*a*) E. C. Tuazon, R. Atkinson, A product study of the gas phase reaction of isoprene with the OH radical in the presence of NO_x, *Int. J. Chem. Kinet.*, **22**, 1221–1236 (1990); (*b*) L. Ruppert, K. H. Becker, A product study of the OH radical-initiated oxidation of isoprene: formation of C_5-unsaturated diols, *Atmos. Environ.*, **34**, 1529–1542 (2000); (*c*) M. Sprengnether, K. L. Demerjian, N. M. Donahue, J. G. Anderson, Product analysis of the OH oxidation of isoprene and 1,3-butadiene in the presence of NO, *J. Geophys. Res., [Atmos.]*, **107**, D000716 (2002).
28. S. E. Paulson, R. C. Flagan, J. H. Seinfeld, Atmospheric photo-oxidation of isoprene part I: the hydroxyl radical and ground state atomic oxygen reactions, *Int. J. Chem. Kinetics*, **24**, 79–101 (1992).
29. A. Miyoshi, S. Hatakeyama, N. Washida, OH radical-initiated photo-oxidation of isoprene: an estimate of global CO production, *J. Geophys. Res., [Atmos.]*, **99**, 18779–18788 (1994).
30. M. Karl, H. P. Dorn, F. Holland, R. Koppmann, D. Poppe, L. Rupp, A. Schaub, A. Wahner, Product study of the reaction of OH radicals with isoprene in the atmosphere simulation chamber SAPHIR, *J. Atmos. Chem.*, **55**, 167–187 (2006).
31. E. S. C. Kwok, R. Atkinson, J. Arey, Observation of hydroxycarbonyls from the OH radical-initiated reaction of isoprene, *Environ. Sci. Technol.*, **29**, 2467–2469 (1995).
32. F. W. Went, Blue hazes in the atmosphere, *Nature*, **187**, 641–643 (1960).
33. H. A. Buttkus, R. J. Bose, D. A. Shearer, Terpenes in the essential oil of sagebrush, *J. Agric. Food Chem.*, **25**, 288–291 (1977).
34. D. M. Pinto, P. Tiiva, P. Miettinen, J. Joutsensaari, H. Kokkola, A. M. Nerg, A. Laaksonen, J. K. Holopainen, The effects of increasing atmospheric ozone on biogenic monoterpene profiles and the formation of secondary aerosols, *Atmos. Environ.*, **41**, 4877–4887 (2007).
35. T. M. Cahill, V. Y. Seaman, M. J. Charles, R. Holzinger, A. H. Goldstein, Secondary organic aerosols formed from oxidation of biogenic volatile organic compounds in the Sierra Nevada mountains of California, *J. Geophys. Res., [Atmos.]*, **111**, D16312 (2006).
36. H. Hakola, V. Tarvainen, T. Laurila, V. Hiltunen, H. Hellen, P. Keronen, Seasonal variation of VOC concentrations above a boreal coniferous forest, *Atmos. Environ.*, **37**, 1623–1634 (2003).
37. (*a*) J. Rinne, T. M. Ruuskanen, A. Reissell, R. Taipale, H. Hakola, M. Kulmala, On-line PTR-MS measurements of atmospheric concentrations of volatile organic compounds in a European boreal forest ecosystem, *Boreal Environ. Res.*, **10**, 425–436 (2005); (*b*) S. Haapanala, J. Rinne, H. Hakola, H. Hellen, L. Laasko, H. Lihavainen, R. Janson, C. O'Dowd, M. Kulmala, Boundary layer concentrations and landscape scale emissions of volatile organic compounds in early spring, *Atmos. Chem. Phys.*, **7**, 1869–1878 (2007).
38. (*a*) U. Kuhn, S. Rottenberger, T. Biesenthal, A. Wolf, G. Schebeske, P. Ciccioli, E. Brancaleoni, M. Frattoni, T. M. Tavares, J. Kesselmeier, Seasonal differences in isoprene and light-dependent

monoterpene emission by Amazonian tree species, *Global Change Biol.*, **10**, 663–682 (2004); (*b*) T. Karl, A. Guenther, R. J. Yokelson, J. Greenberg, M. Potosnak, D. R. Blake, P. Artaxo, The tropical forest and fire emissions experiment: emission, chemistry, and transport of biogenic volatile organic compounds in the lower atmosphere over Amazonia, *J. Geophys. Res., [Atmos.]*, **112**, D18302 (2007); (*c*) Y. F. Wang, S. M. Owen, Q. J. Li, J. Penuelas, Monoterpene emissions from rubber trees (*Hevea brasiliensis*) in a changing landscape and climate: chemical speciation and environmental control, *Global Change Biol.*, **13**, 2270–2282 (2007).

39. R. K. Monson, M. T. Lerdau, T. D. Sharkey, D. S. Schimel, R. Fall, Biological aspects of constructing volatile organic compound emissions inventories, *Atmos. Environ.*, **29**, 2989–3002 (1995).
40. R. Seco, J. Penuelas, I. Filella, Short-chain oxygenated VOCs: emission and uptake by plants and atmospheric sources, sinks, and concentrations, *Atmos. Environ.*, **41**, 2477–2499 (2007).
41. G. W. Schade, A. H. Goldstein, D. W. Gray, M. T. Lerdau, Canopy and leaf level 2-methyl-3-buten-2-ol fluxes from a ponderosa pine plantation, *Atmos. Environ.*, **34**, 3535–3544 (2000).
42. I. E. Galbally, W. Kirstine, The production of methanol by flowering plants and the global cycle of methanol, *J. Atmos. Chem.*, **43**, 195–229 (2002).
43. (*a*) W. V. Kirstine, I. E. Galbally, A simple model for estimating emissions of volatile organic compounds from grass and cut grass in urban airsheds and its application to two Australian cities, *J. Air Waste Manage. Assoc.*, **54**, 1299–1311 (2004); (*b*) B. Davison, A. Brunner, C. Ammann, C. Spirig, M. Joche, A. Neftel, Cut-induced VOC emissions from agricultural grasslands, *Plant Biol.*, **10**, 76–85 (2008).
44. S. N. Matsunaga, A. B. Guenther, M. J. Potosnak, E. C. Apel, Emission of sunscreen salicylic esters from desert vegetation and their contribution to aerosol formation, *Atmos. Chem. Phys.*, **8**, 7367–7371 (2008).
45. H. Singh, Y. Chen, A. Staudt, D. Jacob, D. Blake, B. Heikes, J. Snow, Evidence from the Pacific troposphere for large global sources of oxygenated organic compounds, *Nature*, **410**, 1078–1081 (2001).
46. S. A. McKeen, T. Gierczak, J. B. Burkholder, P. O. Wennberg, T. F. Hanisco, E. R. Keim, R. S. Gao, S. C. Liu, A. R. Ravishankara, D. W. Fahey, The photochemistry of acetone in the upper troposphere: a source of odd-hydrogen radicals, *Geophys. Res. Lett.*, **24**, 3177–3180 (1997).
47. H. B. Singh, L. J. Salas, R. B. Chatfield, E. Czech, A. Fried, J. Walega, M. J. Evans, B. D. Field, D. J. Jacob, D. Blake, B. Heikers, R. Talbot, G. Sachse, J. H. Crawford, M. A. Avery, S. Sandholm, H. Fuelberg, Analysis of the atmospheric distribution, sources, and sinks of oxygenated volatile organic chemicals based on measurements over the Pacific during TRACE-P, *J. Geophys. Res., [Atmos.]*, **109**, D15S07 (2004).
48. D. J. Jacob, B. D. Field, Q. B. Li, D. R. Blake, J. de Gouw, C. Warneke, A. Hansel, A. Wisthaler, H. B. Singh, A. Guenther, Global budget of methanol: constraints from atmospheric observations, *J. Geophys. Res., [Atmos.]*, **110**, D08303 (2005).
49. L. J. Carpenter, A. C. Lewis, J. R. Hopkins, K. A. Read, I. D. Longley, M. W. Gallagher, Uptake of methanol to the North Atlantic ocean surface, *Global Biogeochem. Cycles*, **18**, GB4027 (2004).
50. T. J. Held, F. L. Dryer, A comprehensive mechanism for methanol oxidation, *Int, J. Chem. Kinet.*, **30**, 805–830 (1998).
51. K. I. Scott, M. T. Benjamin, Development of a biogenic volatile organic compounds emission inventory for the SCOS97-NARSTO domain, *Atmos. Environ.*, **S37** S39–S49 (2003).
52. A. Guenther, C. Geron, T. Pierce, B. Lamb, P. Harley, R. Fall, Natural emissions of non-methane volatile organic compounds; carbon monoxide, and oxides of nitrogen from North America, *Atmos. Environ.*, **34**, 2205–2230 (2000).
53. (*a*) Y. Rudich, R. Talukdar, J. B. Burkholder, A. R. Ravishankara, Reaction of methylbutenol with hydroxyl radical – mechanism and atmospheric implications, *J. Phys. Chem.*, **99**, 12188–12194 (1995); (*b*) C. Ferronato, J. J. Orlando, G. S. Tyndall, Rate and mechanism of the reactions of OH and Cl with 2-methyl-3-buten-2-ol, *J. Geophys. Res.*, **103**, 25579–25586 (1998).

54. N. Carrasco, J. F. Doussin, M. O. O'Conner, J. C. Wegner, B. Picquet-Varrault, R. Durand-Jolibois, P. Carlier, Simulation chamber studies of the atmospheric oxidation of 2-methyl-3-buten-2-ol: reaction with hydroxyl radicals and ozone under a variety of conditions, *J. Atmos. Chem.*, **56**, 33–55 (2007).
55. (*a*) H. B. Singh, Reactive nitrogen in the troposphere, *Environ. Sci. Technol.*, **21**, 320–327 (1987); (*b*) G. S. Tyndall, J. J. Orlando, T. J. Wallington, M. D. Hurley, M. Goto, M. Kawasaki, Mechanism of the reaction of OH radicals with acetone and acetaldehyde at 251 and 296 K, *Phys. Chem. Chem. Phys.*, **4**, 2189–2193 (2002).
56. (*a*) M. Staudt, A. Wolf, J. Kesselmeier, Influence of environmental factors on the emissions of gaseous formic and acetic acids from orange (*Citrus sinensis* L.) foliage, *Biogeochemistry*, **48**, 199–216 (2000); (*b*) J. Kesselmeier, Exchange of short-chain oxygenated volatile organic compounds (VOCs) between plants and the atmosphere: a compilation of field and laboratory studies, *J. Atmos. Chem.*, **39**, 219–233 (2001); (*c*) U. Kuhn, S. Rottenberger, T. Biesenthal, C. Ammann, A. Wolf, G. Schebske, S. T. Oliva, T. M. Tavares, J. Kesselmeier, Exchange of short-chain monocarboxylic acids by vegetation at a remote tropical forest site in Amazonia, *J. Geophys. Res., [Atmos.]*, **107**, 8069 (2002).
57. (*a*) S. Madronich, R. B. Chatfield, J. G. Calvert, G. K. Moorgat, B. Veyret, R. Lesclaux, A photochemical origin of acetic acid in the troposphere, *Geophys. Res. Lett.*, **17**, 2361–2364 (1990); (*b*) G. S. Tyndall, R. A. Cox, C. Granier, R. Lesclaux, G. K. Moortgat, M. J. Pilling, A. R. Ravishankara, T. J. Wallington, Atmospheric chemistry of small organic peroxy radicals, *J. Geophys. Res., [Atmos.]*, **106**, 12157–12182 (1997); (*c*) M. Glasius, C. Boel, N. Bruun, L. M. Easa, P. Hornung, H. S. Klausen, K. C. Klitgaard, C. Lindeskov, C. K. Moller, H. Nissen, A. P. F. Petersen, S. Kleefeld, E. Boaretto, T. S. Hansen, J. Heinemeier, C. Lohse, Relative contribution of biogenic and anthropogenic sources to formic and acetic acids in the atmospheric boundary layer, *J. Geophys. Res., [Atmos.]*, **106**, 7415–7426 (2001).
58. E. Sanhueza, L. Figueroa, M. Santana, Atmospheric formic and acetic acids in Venezuela, *Atmos. Environ.*, **30**, 1861–1873 (1996).
59. E. Sanhueza, M. Santana, D. Trapp, C. de Serves, L. Figueroa, R. Romero, A. Rondon, L. Donoso, Field measurement evidence for an atmospheric chemical source of formic and acetic acids in the tropic, *Geophys. Res. Lett.*, **23**, 1045–1048 (1996).
60. (*a*) S. Preunkert, M. Legrand, B. Jourdain, I. Dombrowski-Etchevers, Acidic gases (HCOOH, CH_3COOH, HNO_3, HCl and SO_2) and related aerosol species at a high mountain Alpine site (4360 m elevation) in Europe, *J. Geophys. Res., [Atmos.]*, **112**, D23S12 (2007); (*b*) J. L. Collett, P. Herckes, S. Youngster, T. Lee, Processing of atmospheric organic matter by California radiation fogs, *Atmos. Res.*, **87**, 232–241 (2008).
61. T. R. Duhl, D. Helmig, A. Guenther, Sesquiterpene emissions from vegetation: a review, *Biogeosciences*, **5**, 761–777 (2008).
62. C. Holzke, T. Hoffmann, L. Jaeger, R. Koppmann, W. Zimmer, Diurnal and seasonal variation of monoterpene and sesquiterpene emissions from Scots pine (*Pinus sylvestris* L.), *Atmos. Environ.*, **40**, 3174–3185 (2006).
63. A. X. Cheng, Y. G. Lou, Y. B. Mao, S. Lu, L. J. Wang, X. Y. Chen, Plant terpenoids: biosynthesis and ecological functions, *J. Integ. Plant Biol.*, **49**, 179–186 (2007).
64. Y. H. Shu, R. Atkinson, Atmospheric lifetimes and fates of a series of sesquiterpenes, *J. Geophys. Res.*, **100**, 7275–7281 (1995).
65. T. Sakulyanontvittaya, A. Guenther, D. Helmig, J. Milford, C. Wiedinmyer, Secondary organic aerosol from sesquiterpene and monoterpene emissions in the United States, *Environ. Sci. Technol.*, **42**, 8784–8790 (2008).
66. R. von Kuhlmann, M. G. Lawrence, U. Pöschl, P. J. Crutzen, Sensitivities in global scale modeling of isoprene, *Atmos. Chem. Phys.*, **4**, 1–17 (2004).
67. (*a*) P. Bergamaschi, R. Hein, M. Heimann, P. J. Crutzen, Inverse modeling of the global CO cycle 1. Inversion of CO mixing ratios, *J. Geophys. Res., [Atmos.]*, **105**, 1909–1927 (2000); (*b*) G. G. Pfister, L. K. Emmons, P. G. Hess, J.-F. Lamarque, J. J. Orlando, S. Walters, A. Guenther, P. I. Palmer, P. J. Lawrence, Contribution of isoprene to chemical budgets: a

model tracer study with the NCAR CTM MOZART-4, *J. Geophys. Res., [Atmos.]*, **113**, D05308, (2008).

68. D. Johnson, G. Marston, The gas phase ozonolysis of unsaturated volatile organic compounds in the troposphere, *Chem. Soc. Rev.*, **37**, 699–716 (2008).
69. T. Pfeiffer, O. Forberich, F. J. Comes, The contribution of the ozonolysis of terpenes to tropospheric OH concentrations, *Can. J. Phys.*, **79**, 131–142 (2001).
70. (*a*) P. D. Goldan, W. C. Kuster, F. C. Fehsenfeld, S. A. Montzka, Hydrocarbon measurements in the southeastern United States: the Rural Oxidants in the Southern Environment (ROSE) Program 1990, *J. Geophys Res., [Atmos.]*, **100**, 25945–25963 (1995); (*b*) T. A. Biesenthal, J. W. Bottenheim, P. B. Shepson, S.-M. Li, P. C. Brickell, The chemistry of biogenic hydrocarbons at a rural site in eastern Canada, *J. Geophys Res., [Atmos.]*, **103**, 25487–25498 (1998); (*c*) D. Tan, I. Faloona, J. B. Simpas, W. Brune, P. B. Shepson, T. L. Couch, A. L. Summer, M. A. Carroll, T. Thornberry, E. Apel, D. Riemer, W. Stockwell, HO_x budgets in a deciduous forest: results from the PROPHET summer 1998 campaign, *J. Geophys. Res., [Atmos.]*, **106**, 24407–24427 (2001).
71. P. Di Carlo, W. H. Brune, M. Martinez, H. Harder, R. Lesher, X. Ren, T. Thornberry, M. A. Carroll, V. Young, P. B. Shepson, D. Riemer, E. Apel, C. Campbell, Missing OH reactivity in a forest: evidence for unknown reactive biogenic VOCs, *Science*, **304**, 722–725 (2004).
72. T. E. Kleindienst, M. Lewandowski, J. H. Offenberg, M. Jaoui, E. O. Edney, Ozone-isoprene reaction: re-examination of the formation of secondary organic aerosol, *Geophys. Res. Lett.*, **34**, L01805 (2007).
73. (*a*) P. Neeb, G. K. Moortgat, Formation of OH radicals in the gas phase reaction of propene, isobutene, and isoprene with O_3: yields and mechanistic implications, *J. Phys. Chem. A*, **103**, 9003–9012 (1999); (*b*) S. E. Paulson, M. Y. Chung, A. S. Hasson, OH radical formation from the gas phase reaction of ozone with terminal alkenes and the relationship between structure and mechanism, *J. Phys. Chem. A*, **103**, 8125–8138 (1999); (*c*) D. Zhang, W. F. Lei, R. Y. Zhang, Mechanism of OH formation from ozonolysis of isoprene: kinetics and product yields, *Chem. Phys. Lett.*, **358**, 171–179 (2002).
74. (*a*) R. Atkinson, W. P. L. Carter, Kinetics and mechanisms of the gas phase reactions of ozone with organic compounds under atmospheric conditions, *Chem. Rev.*, **84**, 437–470 (1984); (*b*) S. E. Paulson, R. C. Flagan, J. H. Seinfeld, Atmospheric photo-oxidation of isoprene part II: the ozone-isoprene reaction, *Int. J. Chem. Kinet.*, **24**, 103–125 (1992).
75. (*a*) R. Atkinson, S. M. Aschmann, A. M. Winer, J. N. Pitts, Kinetics of the gas phase reactions of NO_3 radicals with a series of dialkenes, cycloalkenes and monoterpenes at 295 $\pm$ 1 K, *Environ. Sci. Technol.*, **18**, 370–375 (1984); (*b*) T. Ellermann, O. J. Nielsen, H. Skov, Absolute rate constants for the reaction of NO_3 radicals with a series of dienes at 295 K, *Chem. Phys. Lett.*, **200**, 224–229 (1992); (*c*) J. R. Stabel, M. S. Johnson, S. Langer, Rate coefficients for the gas phase reaction of isoprene with NO_3 and NO_2 *Int. J. Chem. Kinet.*, **37**, 57–65 (2004).
76. J. R. Huber, Photochemistry of molecules relevant to the atmosphere: photo-dissociation of nitric acid in the gas phase, *ChemPhysChem*, **5**, 1663–1669 (2004).
77. (*a*) T. Starn, P. B. Shepson, S. B. Bertman, D. D. Riemer, R. Zika, K. Olszyna, Nighttime isoprene chemistry at an urban-impacted forest site, *J. Geophys. Res., [Atmos.]*, **103**, 22437–22447 (1998); (*b*) C. A. Stroud, J. M. Roberts, E. J. Williams, D. Hereid, W. M. Angevine, F. C. Fehsenfeld, A. Wisthaler, A. Hansel, M. Martinez-Harder, H. Harder, W. H. Brune, G. Hoenninger, J. Stutz, A. B. White, Nighttime isoprene trends at an urban forested site during the 1999 Southern Oxidant Study, *J. Geophys. Res., [Atmos.]*, **107**, 4291 (2002); (*c*) M. Steinbacher, J. Dommen, C. Ordonez, S. Reimann, F. C. Gruebler, J. Staehelin, S. Andreani-Aksoyoglu, A. S. H. Prevot, Volatile organic compounds in the Po bason. Part B: biogenic VOCs, *J. Atmos. Chem.*, **51**, 293–315 (2005); (*d*) J. L. Ambrose, H. Mao, H. R. Mayne, J. Stutz, R. Talbot, B. C. Sive, Nighttime nitrate radical chemistry at Appledore Island, Maine during the 2004 International Consortium for Atmospheric Research on Transport and Transformation, *J. Geophys. Res., [Atmos.]*, **112**, D21302 (2007).
78. J. M. Hurst, D. J. Barket, O. Herrera-Gomez, T. L. Couch, P. B. Shepson, I. Faloona, D. Tan, W. Brune, H. Westberg, B. Lamb, T. Biesenthal, V. Young, A. Goldstein, J. W. Munger,

T. Thornberry, M. A. Carroll, Investigation of the nighttime decay of isoprene, *J. Geophys. Res., [Atmos.]*, **106**, 24335–24346 (2001).

79. N. L. Ng, A. J. Kwan, J. D. Surratt, A. W. H. Chan, P. S. Chhabra, A. Sorooshian, H. T. Pye, J. D. Crounse, P. O. Wennberg, R. C. Flagan, J. H. Seinfeld, Secondary organic aerosol (SOA) formation from reaction of isoprene with nitrate radicals (NO_3), *Atmos. Chem. Phys.*, **8**, 4117–4140 (2008).
80. (*a*) D. J. Jacob, L. W. Horowitz, J. W. Munger, B. G. Heikes, R. R. Dickerson, R. S. Artz, W. C. Keene, Seasonal transition from NO_x- to hydrocarbon-limited conditions for ozone production over the eastern United States in September, *J. Geophys. Res., [Atmos.]*, **100**, 9315–9324 (1995); (*b*) A. I. Hirsch, J. W. Munger, D. J. Jacob, L. W. Horowitz, A. H. Goldstein, *J. Geophys. Res., [Atmos.]*, **101**, 12659–12666 (1996); (*c*) P. H. Daum, L. Kleinmann, D. G. Imre, L. J. Nunnermacker, Y. N. Lee, S. R. Springston, L. Newman, J. W. Lloyd, Analysis of the processing of Nashville urban emissions on July 3 and July 18, 1995, *J. Geophys. Res.*, **105**, 9155–9164 (2000); (*d*) F. Solmon, C. Sarrat, D. Serca, P. Tulet, R. Rosset, Isoprene and monoterpenes biogenic emissions in France: modeling and impact during a regional pollution episode, *Atmos. Environ.*, **38**, 3853–3865 (2004); (*e*) M. Bell, H. Ellis, Sensitivity analysis of tropospheric ozone to modified biogenic emissions for the mid-Atlantic region, *Atmos. Environ.*, **38**, 1879–1889 (2004).
81. G. Li, R. Zhang, J. Fan, X. Tie, Impacts of biogenic emissions on photochemical ozone production in Houston, Texas, *J. Geophys. Res., [Atmos.]*, **112**, D10309 (2007).
82. C. A. Nobre, P. J. Sellers, J. Shukla, Amazonian deforestation and regional climate change, *J. Climate*, **4**, 957–988 (1991).
83. J. H. Seinfeld, Global atmospheric chemistry of reactive hydrocarbons, in *Reactive Hydrocarbons in the Atmosphere*, C. N. Hewitt (Ed.), Academic Press, San Diego, 1999, pp. 1–39.
84. J. Lelieveld, W. Peters, F. J. Dentener, M. Krol, Stability of tropospheric hydroxyl chemistry, *J. Geophys. Res., [Atmos.]*, **107**, 4715 (2002).
85. J. Lelieveld, T. M. Butler, J. N. Crowley, T. J. Dillion, H. Fischer, L. Ganzeveld, H. Harder, M. G. Lawrence, M. Martinez, D. Taraborrelli, J. Williams, Atmospheric oxidation capacity sustained by a tropical forest, *Nature*, **452**, 737–740 (2008).
86. (*a*) H. Levy II, Normal atmosphere: large radical and formaldehyde concentrations predicted, *Science*, **173**, 141–143 (1971); (*b*) Y. Wang, D. J. Jacob, J. A. Logan, Global simulation of tropospheric O_3-NO_x-hydrocarbon chemistry, 3. Origin of tropospheric ozone and effects of nonmethane hydrocarbons, *J. Geophys. Res., [Atmos.]*, **103**, 10757–10767 (1998); (*c*) J. Fan, R. Zhang, Atmospheric oxidation mechanism of isoprene, *Environ. Chem.*, **1**, 140–149 (2004).
87. A. S. Hasson, G. S. Tyndall, J. J. Orlando, A product yield study of the reaction of HO_2 radicals with ethyl peroxy ($C_2H_5O_2$), acetyl peroxy ($CH_3C(O)O_2$), and acetonyl peroxy ($CH_3C(O)CH_2O_2$) radicals, *J. Phys. Chem. A*, **108**, 5979–5989 (2004).
88. T. F. Bidleman, Atmospheric processes – wet and dry deposition of organic compounds are controlled by their vapor particle partitioning, *Environ. Sci. Technol.*, **22**, 361–367 (1988).
89. D. A. Hegg, D. S. Covert, H. H. Jonsson, Measurements of size-resolved hygroscopicity in the California coastal zone, *Atmos. Chem. Phys.*, **8**, 7193–7203 (2008).
90. (*a*) S. M. Kreidenweis, M. D. Petters, P. J. DeMott, Single-parameter estimates of aerosol water content, *Environ. Res. Lett.*, **3**, 035002 (2008); (*b*) N. C. Shantz, W. R. Leaitch, L. Phinney, M. Mozurkewich, D. Toom-Sauntry, The effect of organic compounds on the growth rate of cloud droplets in marine and forest settings, *Atmos. Chem. Phys.*, **8**, 5869–5887 (2008).
91. P. Tunved, H.-C. Hansson, V.-M. Kerminen, J. Strom, M. Dal Maso, H. Lihavainen, Y. Viisanen, P. P. Aalto, M. Komppula, M. Kulmala, High natural aerosol loading over boreal forests, *Science*, **312**, 261–263 (2006).
92. T. Hoffmann, J. R. Odum, F. Bowman, D. Collins, D. Klockow, R. C. Flagan, J. H. Seinfeld, Formation of organic aerosols from the oxidation of biogenic hydrocarbons, *J. Atmos. Chem.*, **26**, 189–222 (1997).
93. M. Claeys, B. Graham, G. Vas, W. Wang, R. Vermeylen, V. Pashynska, J. Cafmeyer, P. Guyon, M. O. Andreae, P. Artaxo, W. Maenhaut, Formation of secondary organic aerosols through photo-oxidation of isoprene, *Science*, **303**, 1173–1176 (2004).

94. J. H. Kroll, N. L. Ng, S. M. Murphy, R. C. Flagan, J. H. Seinfeld, Secondary organic aerosol formation from isoprene photo-oxidation, *Environ. Sci. Technol.*, **40**, 1869–1877 (2006).
95. (*a*) M. Claeys, W. Wang, A. C. Ion, I. Kourtchev, A. Gelencser, W. Maenhaut, Formation of secondary organic aerosols from isoprene and its gas phase oxidation products through reaction with hydrogen peroxide, *Atmos. Environ.*, **38**, 4093–4098 (2004); (*b*) E. O. Edney, T. E. Kleindienst, M. Jaoui, M. Lewandowski, J. H. Offenberg, W. Wang, M. Claeys, Formation of 2-methyl tetrols and 2-methylglyceric acid in secondary organic aerosol from laboratory irradiated isoprene/NO_x/SO_2/air mixtures and their detection in ambient PM2.5 samples collected in the eastern United States, *Atmos. Environ.*, **39**, 5281–5289 (2005); (*c*) O. Böge, Y. Miao, A. Plewka, H. Herrmann, Formation of secondary organic particle phase compounds from isoprene gas phase oxidation products: an aerosol chamber and field study, *Atmos. Environ.*, **40**, 2501–2509 (2006).
96. (*a*) H.-J. Lim, A. G. Carlton, B. J. Turpin, Isoprene forms secondary organic aerosol through cloud processing: model simulations, *Environ. Sci. Technol.*, **39**, 4441–4446 (2005); (*b*) A. G. Carlton, B. J. Turpin, H.-J. Lim, K. E. Altieri, S. Seitzinger, Link between isoprene and secondary organic aerosol (SOA): pyruvic acid oxidation yields low volatility organic acids in clouds, *Geophys. Res. Lett.*, **33**, L06822 (2006).
97. (*a*) D. K. Henze, J. H. Seinfeld, Global secondary organic aerosol from isoprene oxidation, *Geophys. Res. Lett.*, **33**, L09812 (2006); (*b*) D. K. Henze, J. H. Seinfeld, N. L. Ng, J. H. Kroll, T.-M. Fu, D. J. Jacob, C. L. Heald, Global modeling of secondary organic aerosol formation from aromatic hydrocarbons: high- versus low-yield pathways, *Atmos. Chem. Phys.*, **8**, 2405–2420 (2008).
98. A. H. Goldstein, I. E. Galbally, Known and unexplored organic constituents in the earth's atmosphere, *Environ. Sci. Technol.*, **41**, 1514–1521 (2007).
99. N. Ng, J. H. Kroll, M. D. Keywood, R. Bahreini, V. Varutbangkul, R. C. Flagan, J. H. Seinfeld, A. Lee, A. H. Goldstein, Contribution of first- versus second generation products to secondary organic aerosols formed in the oxidation of biogenic hydrocarbons, *Environ. Sci. Technol.*, **40**, 2283–2297 (2006).
100. (*a*) V. Varutbangkul, F. J. Brechtel, R. Bahreini, N. L. Ng, M. D. Keywood, J. H. Kroll, R. C. Flagan, J. H. Seinfeld, A. Lee, A. H. Goldstein, Hygroscopicity of secondary organic aerosols formed by oxidation of cycloalkenes, monoterpenes, sesquiterpenes, and related compounds, *Atmos. Chem. Phys.*, **6**, 2367–2388 (2006); (*b*) A. Lee, A. H. Goldstein, J. H. Kroll, N. L. Ng, V. Varutbangkul, R. C. Flagan, J. H. Seinfeld, Gas phase products and secondary aerosol yields from the photo-oxidation of 16 different terpenes, *J. Geophys. Res., [Atmos.]*, **111**, D17305 (2006).
101. A. Lee, A. H. Goldstein, M. D. Keywood, S. Gao, V. Varutbangkul, R. Bahreini, N. L. Ng, R. C. Flagan, J. H. Seinfeld, Gas phase products and secondary aerosol yields from the ozonolysis of ten different terpenes, *J. Geophys. Res.*, **111**, D07302 (2006).
102. C. L. Heald, D. J. Jacob, R. J. Park, L. M. Russell, B. J. Huebert, J. H. Seinfeld, H. Liao, R. J. Weber, A large organic aerosol source in the free troposphere missing from current models, *Geophys. Res. Lett*, **32**, L18809 (2005).

Index

Page numbers in *italics* refer to figures, schemes and tables.

Acetaldehyde, 176, *177*, 243, *243*, 372, 379
Acetate/mevalonate (acetate/MVA) pathway, 13, 15–16, *18*, 36, 37
Acetates, 12, *134*, 135–6, 143–4, *144*
Acetic acid, 372, 379
Acetone, 165, *166*, *369*, 372, 379
Acetoxylactone, *135*, 137
2-Acetyl-1-pyrroline, *238*, 241, *242*
Acetyl-coenzyme A (acetyl-CoA), 12
Acetylmethyl phosphinate (AMPI), 29, *30*, 31, *32*
Acoradiene, *133*
Adenosine triphosphate (ATP)
  isoprenoid biosynthesis, 17, *20*, 22, *23*, 39
    diuron inhibition, 27
  and stability of photosynthetic apparatus, 32, 33
  and sulfur assimilation in plants, 204, *205*
Adoxophyes species (tortrix moths) pheromones, *133*, 135, *140*, 142, 162, *162*
Adrian, Edgar Douglas, First Baron, 254
Aerosol insecticides, 158
Aerosols, secondary organic (SOAs) *see* Secondary organic aerosols (SOAs)
Age and olfactory perception, 267
Alcohols
  as pheromones, 132, *132–4*, 135–6
  release
    enzymatic hydrolysis *see* Hydrolysis: enzymatic
    neighbouring-group assisted hydrolysis *see* Hydrolysis:neighbouring-group-assisted
Aldehydes, *134–5*, 136, 235, *237*, 248
  see also Carbonyl derivatives, release of
Alkanes, 132, *133*, 142, 364, 365, 367
Alkenes, 364
  see also Biogenic VOCs:oxidation, gas phase
Alkenylpyrazines, 239–41, *240*
Alk(en)yl cysteine sulfoxides (CSOs), 206–7, *207*
  biological role, 215–16
  biosynthesis, 207–8
    via cysteine synthases, *208*, 209–10
    via γ-glutamyl peptides, *208*, 208–9
  formation of volatiles from, 210–11, *211*
  release, *211*, 211–12, *212*
Alkyl phenyl ketones, *348*, 348
n-Alkylbenzenes, 259, *260*
Alkylpyrazines, 239–41, *240*
Allelochemicals, 123, *124*, *165*, 165–6, *166*
Alliin, 206, *207*
Alliinase, 210, *211*, 211–12, *212*, 213–14, 215
Allium flavour volatiles, 212–13
  see also Alk(en)yl cysteine sulfoxides (CSOs)
Allomones, *124*, 165
Aluminium phosphide, 155, *155*, 156
Alzheimer's disease, 275
Amadori compounds, 235, *236*, *237*
Amadori, Mario, 235
Amadori rearrangement, *234*, 234–5
Amino acids, 31
  fatty acid-amino acid conjugates (FACs), 102, *103*
  glucosinolate precursors, 216, 217, 218, *218*
Aminoacylase, Zn^{2+}-dependent, 336–7
1-Aminocyclopropane-1-carboxylic acid, 101, *101*
Aminoketose, *234*, 235
AMPI (Acetylmethyl phosphinate), 29, *30*, 31, *32*

The Chemistry and Biology of Volatiles Edited by Andreas Herrmann

Analysis of volatile fractions, 4, 59–61, 86
advances in, 76–80, 83–5
data elaboration, 77–8, 86–7
gas chromatography *see* Gas chromatography (GC)
MS profile, 78
multivariate analysis, 77
princial component analysis (PCA), 77
qualitative analysis, 65–6
quantitative analysis, 66–8, *69*, 70
sample preparation *see* Sample preparation of volatile fractions
separation profiles
Artemisia umbelliformis, *78*
banana-breath analysis, *81*
bergamot, *74*
cyclodextrins, *72*
goldenrod, *98*
hazelnuts, *77*, *82*, *84*, *85*
juniper berries, *53*
linalool, 73
peppermint, *62*
sage, *57–8*
wine, *79*
speed, 61, *62*, 63–5
standards, 68
stationary phases, 60
technique development, 76
in vivo analysis, 78–80
Andrena wilkella bee pheromone, *136*
5α-Androst-16-en-3-one, *3*, *135*
Androstadienone, *272*, 272
Androstenone, 264–5, *265*, 272
Anosmias, 264–5
Anthonomus grandis (cotton boll weevil) pheromone, *133*, 135
Aonidiella species (scales) pheromones, *134*, 160
Aphis gossypii (cotton aphid) pheromone, *133*
Apis mellifera (Honeybee) pheromone, *124*, 125
Aroma compounds, *98*
in *Allium*, 213
defined, 232
GC detection, 232
in roasted coffee, *233*
thermal generation *see* Thermal processing of food
in wine, 6, 37–8
Artemisia tridentata (sagebrush), 36, 112
Artemisia umbelliformis, 77, *78*
Aspidiotus nerii (oleander scale) pheromone, *134*
Astaxanthin, 12, *14*
Atmosphere, VOCs in *see* Biogenic VOCs
ATP (Adenosine triphosphate) *see* Adenosine triphosphate
Atta texana (Leaf cutting ant) pheromone, 124–5
Azadirachta trees, 154

Banana breath analysis, 79–80, *80*, *81*
Benzaldehyde, 272, *272*
Benzoates, *340*, 340–2, *341*, *342*, 345, *345*
2-Benzoylbenzoates, *345*, 345–6
Benzyl salicylate, *294*
Beroza, M., 127
Biodegradation, *7*, 7
Biogenetic isoprene rule, 12
Biogenic VOCs, 363–4, *369*
chemistry, gas phase
in urban/suburban airsheds, 374–5
within/above forests, 375–7, *376*
emission of, vs. anthropogenic VOCs, 364–5
oxidation, 365, *366*, 367–9
oxidation, gas phase
•OH radicals, reaction with, 372–3
NO_3 radicals, reaction with, 374
O_3 radicals, reaction with, 373–4
secondary organic aerosol (SOA) formation, 377–80
types, *370*, 370–2, *371*
see also Floral VOCs; Vegetative VOCs
Biprorulus bibax (spined citrus bug) pheromone, *128*, 140
Bisabolene epoxide, 132, *133*
Bisabolol, 33
α-Bisabolol, *13*
Bisabolol oxide A, *35*, 36
Blattella germanica (German cockroach) pheromone, 125, *135*, 137
Blattellaquinone, *135*, 137
Blindness and olfaction, 276
Bombykol, *3*, 124, 137, 162
isolation and structure determination, *126*, 126–7, 158
structure, *124*, *126*, *128*, *159*
Bombyx mori (Silkworm moth) pheromone *see* Bombykol
Borneol, 33, *34*, *58*, 185, *185*

Bornyl acetate, *58*
Bourgeonal, *261*, 262
Brassicaceae, 206, 210–11, 221–2
  glucosinolates *see* Glucosinolates
exo-Brevicomin, *124*, 124, 139, *140*
Browning, non-enzymatic *see* Maillard reaction
Bruchins, *103*, 104
Bucculatrix thurberiella (cotton leaf perforator)
    pheromone, *134*, 135, 136
2,3-Butanedione, 235, *237*
Butanoic acid, 261, *261*
Butanone, 165, *166*
Butenandt, Adolf, 126
n-Butyl alcohol, 259, *259*
Butyl mercaptan, *264*, 264

C-DOX (C-labelled Deoxyxylulose), 20–1, *21*
δ-Cadinene, *53*
Caeliferins, *103*, 103
Callosobruchusic acid, *140*, 141
Calone, *300*, 300, *301*
Calvin cycle, 23–4, 29, *30*, 30–1
Camphor, 33, *34*, *58*, 370, *371*
Camphorene, *3*, 12, *13*, 34
Canthaxanthin, 12, *14*
Caramelization, *244*, 244–6, *245*
Carane, 70, *70*
Carbamate, *336*, 336–7
Carbon dioxide, 165, *166*, 323
  see also Biogenic VOCs:oxidation
Carbon fixation, 23–4, 26, 29, 31
Carbonates, 338–9, *339*
Carbonyl derivatives, release of
  oxidations, 346–50, *347*
  retro 1,4-additions, *354*, 354–6, *355*
  reversible systems, 350–4, *351*, *352*, *353*
  Schiff bases, hydrolysis of, *350*
Carbonyl sulfide, 223
Carboxylates, 337, *337*
Carene, 33, *34*
β-Carotene, 12, *14*, 32
Carotenoids, 12, 14, 39
  biosynthesis, 15, 16, *17*, *18*, *20*, 27–8
  structures, *14*, 16
Carposia niponensis (peach fruit moth)
    pheromone, *134*, 136
Carvacrol, 33, *34*, *58*, *153*, 153
L-Carvone, 261, *261*
Caryophyllene, 33
α-Caryophyllene, *98*
β-Caryophyllene, *35*, *53*, 113, *113*, *371*
d,l-Caryophyllene, 174–5, *175*, 176
Cashmeran, 188–9, *189*
Cassis (blackcurrant), *182*, 182–4, *183*, *184*
Chamazulene, *35*, 36
Chilo suppressalis (rice stem borer)
    pheromone, *134*, 136
Chiral separation, *70*, 70–5, *72*, *73*, *74*
Chlorophylls, biosynthesis of, 15, *18*,
    20, 28
Chloroplasts
  IPP biosynthesis, 15–16, *17*, 21–2
  metabolism, *30*, 30–1
Chlorpyrifos methyl, 155, *155*, 157
Cholesterol, 12
Chromatography *see* Gas chromatography (GC)
Chrysanthemum, 151, 154
Ciliopathies, 276
Cineol, *17*, 25, 33
1,4-Cineol, 12, *13*
1,8-Cineol (eucalyptol), *3*, 33, 34, *69*
  as insect repellent, 152, 153, *153*
  structure, *34*, *58*
Cinnamic aldehyde, *314*, 314
Citral, 5, 33, 34, 178, 311
  olfactory perception, 260, *260*
  structure, *34*, *179*, *312*
  synthesis, 177–9, *178*, *179*
Citralva, 272, *272*
Citronella, 152–3, 312
Citronellal, 152, 178, 347
  structures, *130*, *131*, *152*, *179*
  vapour pressure and polarity, *3*
Citronellol, 33, 34, 152, 338, 342–3
  structure, *34*, *152*
Citrus bergamia (bergamot), 72, *73*, *74*, 74
Clepsis spectrana (cyclamen tortrix)
    pheromone, *162*, 162
Coffee, 232–3, *233*, 237, *238*
Colletes cunicularius (Vernal solitary bee), 145
α-Copaene, *53*
Corylus avellana (Hazelnuts) *see* Hazelnuts
    (*Corylus avellana*)
Corynebacteria, 336–7
Cosmone, *188*, 188
Coumarin, *262*, 262, 344, *344*
Crocetin, 12, *14*
CSOs *see* Alk(en)yl cysteine sulfoxides (CSOs)
α-Cubebene, *53*
Cucujolide IX, *135*

Culex pipiens fatigans (southern house mosquito) pheromone, *135*, 137
Cycloartenol, 12, *14*, 16
α-Cyclodextrin, *70*, *315*, 315–16
β-Cyclodextrin, *70*, *315*, 315–16
γ-Cyclodextrin, *315*, 315–16
Cyclodextrins, *70*, 70–1, 314–16, *315*
Cyclomusk, 188, *188*
Cydia pomonella (codling moth) pheromone, 132, *133*, *160*
p-Cymene, 33, *34*, *53*, *153*, 153
Cysteine, 204, *205*, 216
  see also Alk(en)yl cysteine sulfoxides (CSOs)
Cysteine synthases, *205*, *208*, 209–10
Cystine lyase, 211

δ-Damascone, *355*, 355–6
Danaus chrysippus (African monarch butterfly) pheromone, *136*, 137
Data elaboration, 50, 77–8, 86–7
Data processing, 77, 80, 83, 87
(2*E*,4*E*)-Decadienal, *250*
(2*E*,4*Z*)-Decadienal, *250*
γ-Decalactone, *3*, *72*
δ-Decalactone, *72*
DEET, *152*, 152, 154
Deltamethrin, *155*
Dendroctonus species (pine beetles) pheromones, 124, *136*, *140*, 145
Deodorants, 336
1-Deoxy-D-xylulose (DOX), *21*, 24–5, *25*
Deoxyosones, 235, *236*, *237*
C-labelled Deoxyxylulose (C-DOX), 20–1, *21*
Deoxyxylulose phosphate (DOXP), *19*, 24–5, *25*, 26
Deoxyxylulose phosphate/methylerythritol phosphate (DOXP/MEP) pathway, 15–16
  compartmentation, *18*
  diterpene biosynthesis, 36
  enzyme steps, 17–19, *20*, 25
  inhibition, 16, *18*
  isoprene biosynthesis, 22, 24–5
  MBO biosynthesis, 28
  monoterpene biosynthesis, 35, 37
  plastidic environment, 16, 31
Detergents, 295, 337, 339
Diallyl disulfide, *3*, *212*, 213, 216
α-Dicarbonyl compounds, 235, *236*, 241–2
β-Dicarbonyl compounds, 235, *237*
Dichlorobenzene, *154*, 154
Dihydromyrcenol, *300*, 300, *301*, 345
Dimethyl disulphide, *211*, 211
Dimethyl sulfide, *96*, 223, *223*, 224, 243, *244*
3,3-Dimethylallyl diphosphate (DMAPP), 2, 16, 25, 26, 28, 31–3
7,11-Dimethylheptadecane, 142
13,23-Dimethylpentatriacontane, 142
Disease and olfaction, 275–6
Disparlure, *3*, *127*, 139, *140*, 145–6, *146*
Diterpenes, 11, 12, *13*, 34–5
  biosynthesis, *17*, *18*, *20*, 36
1,3-Dithiabutane, 223, *223*
Diuron, 27, *27*
(*E*)-7-Dodecenyl acetate, 143, *144*
(*Z*)-7-Dodecenyl acetate, 143, *144*
DOXP reductoisomerase (DXR), 16, 18, 26
DOXP synthase (DXS), 16, 18, 26

Elater ferrugineus (click beetle) kairomone, 142, *143*
Elatophilus species kairomones, *143*, 143
Eldana saccharina (African sugarcane borer) pheromone, 145
Eldanolide, 145, *146*
β-Elemene, *53*
Elephas maximus (Asian elephant) pheromone, 137, 143–4, *144*
Elicitor compounds, 102–4, *103*
Emissions Database for Global Atmospheric Research, 364–5
Emotions and olfaction, 274–5
Enantiomer separation, *70*, 70–5, *72*, *73*, *74*
Enzymes
  of DOXP/MEP pathway, 18–19, *20*, 25
  lytic, 102
    alliinase, 210, *211*, 211–12, *212*, 213–14, 215
    myrosinase, 216, 218–19
  in monoterpene biosynthesis, 35
  see also Hydrolysis:enzymatic; *specific enzymes*
Epithiospecifier proteins (ESPs), 219
Epoxides, 132, *133*
4,5-Epoxy-(2*E*)-decenal, 242, *250*
Essential oils, 33
  defined, 49
  insect repellent properties, 5, *152*, 152–3
  juniper (*Juniper communis* L), *53*

peppermint (*Mentha x piperita L.)*, *62*, *69*
quantitative analysis, 66–8
Esters, *133–5*, 135, 137
Estra-1,3,5(10),16-tetraen-3-ol, *272*, 273
Etaspirene, 182, *183*
2-Ethenyl-3,5-dimethylpyrazine, *238*, 240
Ethyl 2-methylbutyrate, 294, *294*
3-Ethyl-2,5-dimethylpyrazine, *238*, 240
Ethyl butyrate, 311, *312*
Ethyl caprylate, 311, *312*
Ethyl hexanoate, *317*, 317–18
Ethyl mercaptan, *264*, 264
Ethylene, 99, 101, *101*
Eucalyptol (1,8-cineol) *see* 1,8-Cineol (eucalyptol)
Eucalyptus globulus (eucalyptus), *153*, 153
Eugenol, *159*, *161*, 321, *321*
Eysarcoris lewisi pheromone, 129, *130*, *131*, 132

Fabre, Jean Henri, 125–6
Fabric softeners, 337, *337*
Faranal, *124*, 125, 139, *140*
Farnesene, *164*, 164
(*E*)-*β*-Farnesene, *53*, *133*
Farnesol, *13*, 33, 37, *159*
Farnesyl diphosphate, 16, *17*, 36
Fat oxidation *see* Oxidation:lipids
Fatty acid-amino acid conjugates (FACs), 102, *103*
Fatty acids, *30*, 247–8
Ferulic acid, 246–7, *247*
Firsantol, 189, *190*
Flavour compounds, defined, 232
Flavour microencapsulation, 307–8
coating materials, 308, *310*, 311, *311*
fluidized bed coating, 318, *319*, 320
techniques, *309*, *310*, 325
co-crystallization, 322
coacervation, 320–1
cyclodextrin inclusion complexes, 314–16
electrostatic extrusion, 325
extrusion, 312–14, *313*
freeze drying, 321–2
helical inclusion complexes, 316–18, *317*
silica gels, 324–5
spray chilling/cooling, 322
spray drying, 308, 310–12, *311*
supercritical fluids use, 323, *324*
temperature-responsive gels, 325
yeast cells, 325
Flavours, 6, 205–6, 276–7, 307
see also Flavour microencapsulation
Floral VOCs, 95–6, *96*, 112–13
see also Biogenic VOCs; Vegetative VOCs
Food
cooking *see* Thermal processing of food
spoilage, 205–6
Formaldehyde, 235, *237*, 367–8, 379
Formic acid, 372, 379
Fosmidomycin, *19*, 26, *27*, 28
DOXP/MEP pathway inhibition, 16, *18*, 19, *20*
Fourier transfrom infrared spectroscopy (FT-IR), 60
Fragrances, 37–8
duration, 295, 335–6, 339
release, controlled, 333–5, *334*
alcohols *see* Hydrolysis:enzymatic; Hydrolysis:neighbouring-group-assisted
carbonyl derivatives *see* Carbonyl derivatives, release of
Freeman, Walter J., 254–5
Frontalin
Asian elephant pheromone, 137, 143–4, *144*
insect pheromone, *124*, 124–5, 137, *163*, 163
mimics, 145, *146*
vapour pressure and polarity, *3*
Fumigation, insecticidal, 156–8
Furaneol, *238*, 239, *239*
2-Furfurylthiol, *237*, 237, *238*

Gas chromatography (GC), 59–61, *62*
aroma compound detection, 232
enantioselective, *70*, 70–5, *72*, *73*, *74*
fast-GC (F-GC), 61, 63–5
heart-cut GC-GC, 75–6
improvements, 64
isotope ratio mass spectrometry (IRMS), 75
mass spectrometers, compatibility with, 64–5
method translation, 64
multidimensional, 75–6, *77*, 83–5
fingerprint analysis, 83–5, *84*, *85*
group type analysis, *82*, 83
template matching, 85, *85*, *86*
parameters, translatable and nontranslatable, 64
qualitative analysis, 65–6
retention indices, 65

Gas chromatography (GC) (*Continued*)
retention locking time, 66
speed
effects on profile, 62
factors affecting, 61
ultra fast-GC (UFM-GC), 61
Gender and olfaction, 272–3
Genes
DOXP/MEP enzymes, 19
in MHC, 265–6
olfactory receptors, 255, 264–5
Geocoris pallens (big-eyed bug), 101, 106, *109*, 110
Geraniol, 33, 35, 178, *179*
biosynthesis, 2, *4*, *17*
degradation, *7*, 7
flavour/aroma, 6
as HIPV, 5
insect repellent properties, 5
oxidation, *6*, 6–7
as pheromone, 5
release, 7, *7*
smell, 6
structure, *13*
synthesis, *5*, 5–6
vapour pressure and polarity, *3*
in wine, 37
Geranyl diphosphate (GPP), 2, 16, *17*
Geranyl diphosphate (GPP) synthetase, 2
Geranyl methyl sulfide, 137
Geranylgeranyl diphosphate (GGPP), 16, *17*
Geranyllinalool, 34, *35*
Germacrene B, *53*
Germacrene D, *53*
Germacrene-D-4-ol, *53*
Gibberellic acid, *35*, 36
Ginkgolide A, *13*, 36
Glucosidic sulfonic acid, 125, *125*
Glucosinolates, 204, 216, *220*, 220–2
biosynthesis, 216–18, *217*, *218*, *219*
release, 218–20
γ-Glutamyl peptides, *207*, *208*, 208–9
Glutathione, *205*, *208*, 208–9
Glyceraldehyde-3-phosphate (GA-3-P), 16, 17, *19*, 24, *30*
Glycolaldehyde, 235, *237*
Glycosidases, 335–6, *336*
Glycoside conjugates, 7, *7*
Glycosides, 335–6, *336*
Gnatocerus cornutus (broad-horned flour beetle) pheromone, *133*
Grandisol, *133*
Green leaf volatiles, *98*, 98
Guaiacol, 247, *247*
Gyptol, 127, *127*

Habanolide, 180, *180*
Hazelnuts (*Corylus avellana*)
2-D profiles, *82*, *84*, *85*, *86*
analysis of, 83–4
pyrazine structures in, *83*
Headspace sampling *see* Sample preparation of volatile fractions:headspace sampling
Heat protection of leaves, 31
Helvetolide, 188, *188*
Hemerobius stigma kairomone, *143*, 143
Hemiterpenes, 12, *13*, *17*
Hepialus hecta (swift moth) pheromone, *136*, 137
(2*E*)-Heptenal, *250*
Herbivore-induced plant volatiles (HIPVs), 5, 38, 96–7, *166*, 166–7
adaptive function, 97
common compounds, *98*
emission of, influences on, 99–100
future research directions, 113
geraniol, 5
inhibition, 100–1
vs. JA-induce volatiles, 101
and plant fitness, 97
responses to
birds, 109
insects, carnivorous, 108–11, *109*
specificity, 98
separation profile, *98*
tissue damage, response to, 99–102, *100*
whole-community effects, 111
Herbivores
feeding patterns, 101–2
host selection, 107–8
sequestration of plant toxins, 106
stress-related behaviours, 110
VOC elicitor compounds, 102–4, *103*
see also Herbivore-induced plant volatiles (HIPVs); Plant defense; Vegetative VOCs
Hexanal, 80, *80*, *250*
(*Z*)-3-Hexen-1-ol, *98*, *371*, 371
(*E*)-2-Hexenal, 80, *80*, *98*

(*Z*)-3-Hexenal, 371, *371*
(*Z*)-3-Hexenol, *3*, 176, *176*, *177*, 178–9, 192
(*Z*)-3-Hexenyl acetate, 165, 167, 371, *371*
HIPVs *see* Herbivore-induced plant volatiles (HIPVs)
Human immunodeficiency virus (HIV), 275
Human leukocyte antigen (HLA), 265
α-Humulene, *53*
Hunger and olfaction, 273–4
Hydrocarbons as pheromones, 132, *133*
Hydrogen sulfide, 223
Hydrolysis
  enzymatic
    aminoacylases, *336*, 336–7
    disadvantages of, 340
    glycosidases, 335–6, *336*
    lipases, *337*, 337–9, *338*, *339*
  neighbouring-group-assisted
    of 2-carbamoylbenzoates, 341–2, *342*
    of benzoates, *340*, 340
    light-induced, 343–6, *344*, *345*, *346*
    release rates, factors influencing, 341–3, *342*, *343*, *344*
    second-order rate constants, 340–1, *341*
  reversible systems, 349–54
    dynamic mixtures, 351–4, *352*, *353*
    release rates, factors influencing, 352–3, *353*
    Schiff bases, hydrolysis of, *350*, 350–1
2-Hydroxy-2-cyclopenten-1-one derivatives, *245*, 245
3-Hydroxy-4,5-dimethyl-2(5*H*)-furanone, *245*, 245–6
4-Hydroxy-3-methyl-2-(E)-butenyl-diphosphate (HMBPP), 19, *21*
Hydroxyl radical (•OH), 364
  atmospheric lifetime, *369*
  BVOC oxidation, 365, 367, 369
  in forestry airsheds, 375–6, *376*, 377
  in urban/suburban airsheds, 375
Hyphantria cunea (fall webworm moth) pheromone, 132, *133*, 137–8, *138*

Idea leuconoe (danaine butterfly) pheromone, *136*, 137
Inceptins, 102, *103*
Indole, *98*
Insect repellents, 5, 151–4, *152*
Insecticides, 154–8, *155*
Insects *see* Herbivores; Pest control; Pheromones
Invictolide, *135*, *137*
IPP *see* Isopentenyl diphosphate (IPP)
IPP isomerase, 19, *20*, 26
Ips paraconfusus (California five-spined ips) pheromone, *133*, 141
Ipsdienol, *140*, 141, 163, *163*
Ipsenol, *133*
IR3535, 152, *152*
Isoalliin, 206, *207*
Isoamyl acetate, *317*, 317–18
Isobutyl acetate, *80*, 80
Isomenthone, *64*
Isopentenyl diphosphate (IPP), 2, 12
  biosynthesis from
    of diterpenes, 36
    of monoterpenes, 35
    of sesquiterpenes, 36
  biosynthesis of, 15–16, *17*, 21–2
  structure, *13*
  see also Isoprenoids
Isopentyl, 80, *80*
Isoprene, *3*, 11–12, *13*, 31–3, *369*
  biosynthesis, 16, *17*, *20*, *23*, 23–5
    inhibition, 26–7
    regulation, 36
  emission, 22–5, *24*, 370
    global, 365
    inhibition, *27*
    physiological function, 31–3
    regulation, 25–6, *27*
  oxidation, 370, *370*
    and SOA formation, 378–9, *379*, *380*
Isoprenoids, 11–12, *17*, 38, 364
  biosynthesis, 12–14
    prenyl chain formation/elongation, 16
  biosynthesis pathways, 15–16
    compartmentation of, 16, *18*
    crosstalk, *18*, 19–22
  structures, 12–16, *13*
Isopulegol, *64*, *69*, 179, *179*
Isovaleric acid, 261, *261*, 264, 271

Japonilure, *135*, 137, 139, *140*
Jasmonic acid, 99, *101*, 101, *166*, 166
Javanol, 189, *190*
Juniper communis (Juniper), 52, *53*

Kairomones, 123, *124*
pest control, *159*, *160–1*, 163, *165*, 165, *166*
pheromones acting as, 142–3, *143*
Kaurene, *35*, 36
Kenyan ‘push-pull’ intercropping system, 113–14
α-Keto esters, *347*, 347
5-Ketoclomazone, 18, *18*, *19*, *20*, 26, 27–8
Ketones, *134–5*, 136–7, 259, *260*
see also Carbonyl derivatives, release of
Klauscenone, 188, *189*

Lachrymatory factor synthase (LFS), 212, *212*, 212–13, 214–15, *215*
Lactic acid, 165, *166*
Lactones, *135*, 137
Lamiaceae, 153
Lardoglyphus konoi (acarid mite) pheromone, *134*
Lardolure, *134*
Lasius fulginosus ant, pheromone, *135*
Leptinotarsa decemlineata (Colorado potato beetle) pheromone, *135*, 136, 165
Leucoma salicis (satin moth) pheromone, *133*
Leucomalure, 132, *133*
Leucoptera species (leaf miner moths) pheromones, *133*
Light-induced volatiles release, 343–6, *344*, *345*, *346*
of carbonyl derivatives, 346–9, *347*
Lilial, *300*, 300, *301*
Limonene, *3*, 33, 35, *69*, *74*
in *Eucalyptus globulus*, *153*, 153
structure, *34*, *53*
Linalool, 5, 33, 178, *179*
biosynthesis, *17*
chiral recognition, *73*
insect attractant, 165, *165*
insect repellent, 152, *152*
microencapsulation, 317–18
odour perception tests, 261, *261*
in peppermint, *64*, *69*
pheromone mimic, 145, *145*
structure, *13*, *317*
in wine, 37
Linalool oxide, *37*
Linalyl acetate, *72*, 311, *312*
Lineatin, *136*, 137
Linoleic acid, 248, *249*, *250*
Lipase cleavage, 337
of carbonates, 338–9, *339*
of carboxylates, *337*, 337–8
of monosuccinates, *338*
Lipid oxidation *see* Oxidation:lipids
Lipoxygenase 3 (LOX3) gene, 100–1
Lobesia botrana (grape vine moth) pheromone, *134*, 135
Lutzomyia longipalpis (sand fly) pheromone, *133*
Lymantria dispar (gypsy moth) pheromone, *3*, *127*, 139, *140*, 145-6, *146*
Lyonetia clerkella (peach miner moth) pheromone, 132, *133*

Macrolide, 137
Magnesium phosphide, 155
Maillard, Louis Camille, 233
Maillard reaction, 232, 233–4
Amadori rearrangement, *234*, 234–5
compounds formed, *238*
2-acetyl-1-pyrroline, 241, *242*
aldehyde, 235, *237*
alkylpyrazines/alkenylpyrazines, 239–41, *240*
Amadori compounds, *234*, 234–5
2,3-butanedione, 235, *237*
deoxysones, 235, *236*, *237*
formaldehyde, 235, *237*
furaneol, *239*, 239
2-furfurylthiol, 237, *237*
glycolaldehyde, 235, *237*
oxopropanal, 235, *237*
Major histocompatibility complex (MHC), *144*, 265–6
Malodours, human, *336*, 336
Maple lactone, *159*, *261*, 261
Marrubiin, *35*, 36
Mass spectral libary software, 65–6
Mass spectrometry (MS), 60
enantiomer separation, 71–2, *72*, *73*
isotope ratio mass spectrometry (IRMS), 75
mass spectral libary software, 65–6
MS fragmentation, 78, *79*
and multidimensional GC, 76
nosespace (NS) analysis, 78
proton transfer reaction-MS (PTR-MS), 80, *81*
quantitation, detector for, 67
TIC-MS, 78, *79*

Matsucoccus species (pine scales)
pheromones, 141, 142–3, *143*
Megacyllene caryae pheromone, 138, *138*
Melia azaderach tree, 154
Mellein, *135*, 137
Mentha x piperita L. (peppermint), *62*, *64*, *69*
Menthofuran, *64*, *69*
Menthol, *13*, *17*, 33, *64*, *69*, *315*
(–)-Menthol, *179*, 179–80
Menthone, 33, *34*, 35, *64*, *69*
Metabolites, secondary, 6, 96, 105–6, 203–4
Methacrolein, 370, *370*, *379*
Methane, 368–9
Methanethiol, *3*, *208*, 223, 224, *244*
Methanol, *369*, 372
Methiin, 206, *207*, *208*
Methional, 243, *243*
Methionine, *205*, 216–17, 224, 243
2-Methyl-3-furanthiol, 246, *246*
4-Methyl-3-heptanone, *124*, 124
5-Methyl-3-heptanone, 142
4-Methyl-4-sulfanylpentan-2-one, 223, *223*
3-Methyl-α-himachalene, *133*
Methyl bromide, *155*, 156
Methyl cysteine sulfoxide (MCSO), 210–11
CDP-Methyl-D-erythritol (CDP-ME), 19, *20*, *21*
2-C-Methyl-D-erythritol-2,4-cyclo-diphosphate (MEcPP), 19, *21*
Methyl jasmonate (MeJA), *3*, *37*, *101*, 101, 166, *166*
Methyl linoleate, 321, *321*
Methyl phosphine, 157
Methyl salicylate (MeSA), *96*, 96, *98*, 101
Methyl vinyl ketone, *3*, 370, *370*
2-Methylbutanal, 243, *243*
3-Methylbutanal, 243, *243*
Methylbutenol *see* 2-Methylen-3-buten-2-ol (MBO)
2-Methylen-3-buten-2-ol (MBO), 31–3, *371*
biosynthesis, *17*, *20*, *23*, 28, 36
emission, 28–9, 31–3, 372
2-C-Methylerythritol-4-phosphate (MEP), 18, *19*
see also Deoxyxylulose phosphate/ methylerythritol phospate (DOXP/MEP) pathway
(*S*)-9-Methylgermacrene-B, *133*
7-Methylheptadecane, 142
2-Methylpropanal, *243*, 243
Methyltetrols, 378, *379*
Mevalonic acid (MVA), 12, *15*
Mevalonolactone (MVL), 20, *21*
Mevinolin, 13–14, *15*, 16, *18*
Microencapsulation *see* Flavour microencapsulation
Midge magnets, 165–6, *166*
Monomorine I, *136*, 137
Monomorium pharaonis (pharaoh's ant) pheromone, 125, *136*, 137, 139, *140*
Monosuccinates, 338, *338*
Monoterpenes, 12, 32–5, 37–8, *369*
biosynthesis, *17*, *18*, *20*, 35, 36
emission, 35, 370–1
oxidation, 370–1, 378
Moxalone, 188, *189*
α-Multistriatin, *136*, 137
Mus musculus (house mouse) pheromone, *136*, 137, *144*, 144–5
Musca domestica (house fly) pheromone, 132, *133*
Muscalure, 132, *133*
(5Z)-Muscenone, 187, *188*, 188
Muscone, 187, *187*
Musicophilia (Sacks), 303
Musk odorants, *187*, 187–9, *188*, *189*
Mustard oils *see* Glucosinolates
Myrcene, 5, 33, *34*, *53*, *179*, *380*
Myrosinase, 216, 218–19, *220*, 221

NADPH *see* Nicotinamide adenine dinucleotide phosphate (NADPH)
Naphthalene, *154*, 154
Nasutitermes exitosus (termite) pheromone, *133*
National Geographic Smell Survey, 267
Neocembrene, *133*
Neodiprion pinetum (white pine sawfly) pheromone, *134*, 135
Neomenthol, *64*
Neotoxoptera formosana (onion aphid), 215
Nepeta cataria (catnip), 153
Nepetalactone, *152*, 153
Nerol, 5, 33, *34*, 37, *138*, 138
Nerolidol, 33, *35*, *159*
Nezara viridula (green stink bug) pheromone, 132, *133*
Nicotiana attenuata (wild tobacco), 100–1, 106, 107, 112

Nicotinamide adenine dinucleotide phosphate (NADPH)
DOXP/MEP pathway, 17–18, *20*, 22, *23*
diuron inhibition, 27
and stability of photosynthetic apparatus, 32, 33
Nicotine, *107*, 107
Nitrate radical (NO_3•)
atmospheric lifetime, *369*, 369
BVOC oxidation, 374
in forestry airsheds, *376*, 377
Nitric oxide, 367–8
Nitrogen dioxide, 367–8, *369*
(2*E*)-Nonenal, 248, *250*
(3*Z*)-Nonenal, 248, *250*
Nopinone, *380*, 380
Norrish type II photofragmentation, 346–7, *347*

Octanoic acid, 262, *262*
3-Octanol, *64*, *69*
1-Octen-3-ol, *161*, 165, *166*, *250*
(2*E*)-Octenal, *250*
(2*Z*)-Octenal, *250*
Odour deprivation, 266–7
Odour facets, 294
Olean, *140*, 142
Oleanolic acid, 12, *14*
Olfaction, 6, 78, 181–2, 194, 292
see also Olfactory perception, human
Olfactocerebellar pathway, 270
Olfactomotor system, 270
Olfactory bulb, 254, 255, 257, 268
Olfactory perception, human
active sampling, 269–70
anosmias, 264–5
brain exercise, 304
central processing of stimuli, 257
conscious vs. unconscious responses, 258
detection thresholds, 259
factors affecting
age, 267, 276–7
emotions, 274, 274–5
health status, 275–6
hunger/satiety, 273–4
odour names, 274–5, 302
reproductive status, 272–3
top-down influences, 271
imagery, 270–1
importance, 254
mixture perception, 259–62, *260*, *261*
nasal pathways, 262–3
neuron formation, continuous, 268
odor identification tests, 268, 269
odour deprivation, 266–7
odour preferences, MHC-correlated, 265–6
olfactory bulb, rhythms of, 254
olfactory pathway, 255–6, *256*
and Parkinson's disease, 268
rivalry, 258
speed-accuracy trade-off, 260
study of, 253–4
advances, 277
epidemiological studies, 268–9
functional studies, 256–8
functional studies in brain-damaged subjects, 259
history, 254–5
three-dimensional nature, 293
vs. trigeminal odorant identification, 262–3, 265
and working memory, 258
Olfactory receptor neurons (ORNs), 269–70
Olfactory receptors, 254, 255, 264–5
Organophosphates, *155*, 155
Orgyia species pheromones, *128*, *140*, 141
Oryzaephillus surinamensis (sawtoothed grain beetle) pheromone, *135*, 137
Osmoderma eremita (scarab beetle) pheromone, 137, 142, *143*
Oviposition-induced VOC emission, 104
Oxidation
biogenic VOCs, *366*, 367–9
•OH radicals, reaction with, 372–3
NO_3 radicals, reaction with, 374
O_3 radicals, reaction with, 373–4
carbonyl derivatives, release of
non-photochemical pathways, *349*, 349–50
photochemical pathways, 346–9, *347*
lipids
classes, 247–8
fatty acids, order of oxidation, 248
and flavour, 248–50
linoleic acid degradation, 248, *249*, *250*
photo-oxidation protection, 31–2
of terpenes, *6*, 6–7
Oxopropanal, 235, *237*
Ozone, 11–12, 363, *369*
in forestry airsheds, *376*, 377

oxidation path, 367, 368
production/destruction, 374–5
Ozonolysis, 373–4

Parkinson's disease, 268, 271, 275
Patchouli odorants, *184*, *187*
synthesis, 184–6, *185*, *186*, *187*
Patchouli oil, 184, *301*
Patchoulol, *181*, *185*, *186*
Pectinophora gossypiella (Pink bollworm moth) pheromone, *133*, 135, 163–4
Peppermint (*Mentha x piperita L.*), *62*, *64*, *69*
Perfumers, 294–5, 298, 302, 303
Perfumery, 6, 291–2
categorizing volatiles, 296–8, *297*
criteria for volatiles used, 298–9
emotions, creating, 294–5, 302
full hologram effect, 292–5
ingredients, 295
interactions between, 299–300
language of, 296–8
molecule variations, 296
and music, *292*, 303
odour facets, 294
odour names biases, 302
olfactive profiles, *301*
and other sensory arts, *292*
scent endurance, 295, 300
scent layers, 295
scents, interpretations of, 294
volatile characteristics, 298
see also Profragrances/Properfumes
Periplaneta americana pheromone, *128*
Periplanone-B, *128*
Permethrin, 152, *152*
Peroxyacetyl nitrate, 368, 372
Pest control
allelochemicals, 165–6
kairomones, *159*, *160–1*, 163, 165, *165*, *166*
pheromones, 158, *160–1*, 162–5
behavioural modification, 166–7
insect repellents, 151–4
insecticides, 154–8
research, current, 167
α-Phellandrene, 33, *34*
β-Phellandrene, 33, *34*, *53*
Phenethyl alcohol, 261, *261*
Phenylacetaldehyde, *243*, 243
Pheromones, 123–5, *124*, 128, *159*
collection, 127
field bioassay, 129
with kairomonal activities, 142–3, *143*
mammalian, 143–5, *144*
mimics, *145*, 145–7, *146*
multicomponent activity, 137–8, *138*
nonvolatile, 125, *125*
pest control, use in, 158, *160–1*, 162–5
research, history of, 125–7
specificity, 162, *162*
stereochemistry, 139–42, *140*
structural diversity, 132, *133–5*, 135–7, *136*
structure determination and synthesis, 128, *128*
Eysarcoris lewisi, 129, *130*, *131*, 132
volatile, *124*, 125
websites, 147
Phosphine, 155–6, 156–7
Phosphoenol pyruvate (PEP), *30*, 31
Phosphoglyceric acid (PGA), 24, *24*
Photo-oxidation protection, 31–2
Photofragmentation, *346*, 346
nitrobenzyl group cleavage, 349
Norrish type II, 346–8, *347*
photolabile profragrances and properfumes, *348*, 348–9, *349*
Photosynthetic apparatus stability, 31–3
Phragmites australis (Common reed), 31–2
Phyllophaga species (scarab beetles) pheromones, *135*, 137
Phylloquinone K_1, *14*, 15
Phytohormones, 99–101, *101*
Phytol, *13*, 36
Picaridin, 152, *152*
Pinane, 70, *70*
α-Pinene, *3*, 33, *98*, *153*, 153, *380*
structure, *34*, *53*, *377*
β-Pinene, 33, *69*
structure, *34*, *53*, *377*
Pinker, Steven, 303
Pirimiphos methyl, *155*, 155, 156, 157
Pityol, *136*, 137
Pityophthorus pityographus (fir bark beetle) pheromone, *136*, 137
Plamococcus citri (citrus mealybug) pheromone, *134*
Plant defense, 38
induced direct defences, 97–8
metabolism, changes in, 97
physiological trade-offs, 106–7

Plant defense (*Continued*)
VOC emission and seconday metabolite production, 105–6
VOC-mediated, 96–7
agricultural uses, 113–14
beneficial predators, attraction of, 108–11
characteristics of, 109–13
plant-plant interactions, 111–12
within-plant defence signalling, 104–5
wound signalling, 99–102
see also Herbivore-induced plant volatiles (HIPVs); Vegetative VOCs
Plastoquinol-9, 15, *15*, *20*
Plastoquinone-9, 12, *14*, *18*, *20*, 39
Platypus quercivorus (Ambrosia beetle) pheromone, *133*, *136*, 137, 141
Polarity, 2, *3*
Pollinators, specialization of, 96
Polyterpenes, 12, *17*, *18*
Popillia japonica (Japanese beetle)
kairomone, *161*, 165
pheromone, *135*, 137, 139, *140*
Posticlure, *128*
Prenylquinones, *18*
Principal component analysis (PCA), 77, 78, *78*
Profragrances/Properfumes, 333–5, 356–7
alcohols
enzymatic hydrolysis *see* Hydrolysis: enzymatic
neighbouring-group-assisted hydrolysis *see* Hydrolysis:neighbouring-group-assisted
carbonyl derivatives *see* Carbonyl derivatives, release of
photolabile, *348*, 348–9, *349*
Propiin, 206, *207*
Pseudococcus species (mealybugs) pheromones, *134*
Pulegone, 33, *34*, 35
'Push-pull' intercropping system, 113–14
Pyrazines, *82*, *83*
Pyrethrins, *155*, 155
Pyridine, 260, *260*
Pyridoxal-5′-phosphate (PLP), 210, *211*
Pyruvate, 17, *18*, 18, *19*, *20*, 24
isoprene and MBO biosynthesis, 29–30

Quercivorol, *133*
Quercus robur (Pedunculate oak), 26

Retention locking time software, 66
Retro 1,4-additions, *354*, 354–6, *355*
Retro-aldolization, 235, *237*
Rhagoletis cerasi (European cherry fruit fly) pheromone, *125*, 125, 164
Romandolide, *188*, 188
RubisCO, 29

Sabinene, 33, *34*, *53*
Sacks, Oliver, 303
Salicylic acid, 99, *101*
Salvia lavandulifolia Vahl. (Sage), *57–8*
Sample preparation of volatile fractions
distillates, 50
extracts, 50
headspace sampling, 49, 50–1, 51–2
dynamic mode headspace sampling (D-HS), 50, 54, *54*
headspace liquid-phase microxtraction (HS-LPME), 56, 57–9, *58*
headspace-silicone membrane sorptive extractions (HS-SMSE), 55–6, *56*, *57–8*
headspace-solid phase microextration (HS-SPME), *52*, 52, *53*
headspace sorptive extraction (HSSE), *55*, 55–6, *56*, *57–8*
high-concentration capacity headspace sampling (HCC-HS), 51, 51–2, 59
in-tube sorptive extration, 54, *54*
large surface area high concentration capacity, 59
membrane extraction sorbent interface (MESI) method, 59
solid-phase aroma concentrate extraction (SPACE), 56
solid-phase dynamic extraction (SPDE), *54*, 54
sortpive tape extraction (STE) method, 59, *60*
static and trapped headspace (S&T-HS), 56
static mode headspace sampling (S-HS), 50
liquid phase sampling, 51
polydimethylsiloxane (PDMS), 54, 55, 59, *60*
solid-phase aroma concentrate extraction (SPACE), 56
twisters, dual phase, 55, *55*
Sandalwood, 182, 189–92, *190*, *191*, *192*
α-Santalol, *190*, 190
β-Santalol, *181*, *190*, 190–2, *191*, 192, *192*

Santalum album (East Indian sandalwood), 189
Saponines, *17*
Saturnia pyri (Peacock moth)
pheromone, 125–6
Schizophrenia, 275–6
Sclerotium cepivorum fungus, 215–16
Scolytus multistriatus (smaller European elm bark beetle) pheromone, *136*, 137, 164
Secondary organic aerosols (SOAs), 364, *366*, 369, 377, *379*
and isoprene, 378–9, *379*
and monoterpenes, 371, 378, 379–80
and salicylic esters, 371
and sesquiterpenes, 372
Semiochemicals, 123, *124*
see also Kairomones; Pheromones
Serricornin, 139, *140*
Sesquiterpenes, 12, 15, 32–5, 37–8, *369*, 372
biosynthesis, *17*, *18*, 36
Sila-linalool, 145, *145*
α-Sinensal, *177*, 177
Sitophilate, *135*, 137
Sitophilure, *135*
Sitophilus species (weevils) pheromones, *135*, 136, 137
Smell, sense of *see* Olfactory perception, human
SOAs *see* Secondary organic aerosols (SOAs)
Solenopsis invicta (red imported fire ant) pheromone, *135*, 137
Sorption, 51
Spathulenol, *53*
Spodoptera litura (cotton leafworm) pheromone, *134*, 135
Squalene, 16, *17*
Staphylococci, 336
Stegobinone, *136*, 137, 139, *140*, *146*, 147
Stegobium paniceum (drugstore beetle) pheromone, *136*, 137, 139, *140*, 147
Steroids, *17*
Sterols, 15, 16, *17*, *18*
Strecker, Adolf, 241
Strecker aldehydes, 242, *243*
Strecker degradation, 241–3, *244*
Stress-related odours, 273
Sulcatol, *140*, 141
3-Sulfanylhexan-1-ol, *223*, 223
Sulforaphane, 221–2, *222*
Sulfur
assimilation in plants, 204, *205*
flavour and odour compounds, 205–6, *223*, 223–4
Sulfuryl fluoride, *155*
Supella longipalpa (brown-banded cockroach) pheromone, *135*, 137
Supellapyrone, *135*, 137
Sus scrofa (boar) pheromone, *135*, 137
Synomones, *124*, 165
Synthesis of organic compounds
advances, 173
challenges, industrial, *193*
design and synthesis of new odorants
cassis, *182*, 182–4, *183*
musk, *187*, 187–9, *188*, *189*
patchouli, 184–6, *185*, *186*, *187*
sandalwood, 189–92, *190*, *191*, *192*
large-scale
citral, 177–9, *178*, *179*
habanolide, 180, *180*
(Z)-3-hexenol, 176, *176*, *177*
(–)-menthol, *179*, 179–80
methodology, improvements to, 181
strategies for new volatiles, 180–1
organoleptic purity, challenge of, 193
small-scale
D,L-caryophyllene, 174–5, *175*
β-vetivone, *175*, 175–6

Taxol, *13*, 36
Terpenes, *17*, 132, *133*
Terpenoids, 12, 37–8, *98*
biosynthesis, 13–14, 36
structures, 12, *13*, *14*
Terpinen-4-ol, *53*, *153*, 153
α-Terpinene, *53*, 153, *153*
β-Terpinene, *53*
γ-Terpinene, *69*
α-Terpineol, 33, *34*, 37, *58*, *153*, 153
4-Terpineol, *37*
Terpinolene, *53*
Tetraterpenes, 12, *14*, *17*
Thermal processing of food, 231
caramelization, *244*, 244–6, *245*
fat oxidation, 247–50
linoleic acid degradation, 248, *249*, *250*
ferulic acid degradation, 246–7, *247*
Maillard reaction *see* Maillard reaction
Strecker aldehydes, 242–4, *243*, *244*
Strecker degradation, 241–4

Thermal processing of food (*Continued*)
thiamine degradation, 246, *246*
volatiles generated, 231–2, *232*
Thiamin (Vitamin B_1), 246, *246*
Thiobenzaldehyde-*S*-oxide, *223*, 223
α-Thujene, *53*
Thujol, 33, *34*
α-Thujone, *67*
β-Thujone, *67*
Thymol, 33, *34*, 35, *58*
α-Tocopherol, 15, *15*
Tocopherols, *20*
Toxotrypana curvicauda (papaya fruit fly) pheromone, *136*, 137
Transgenic technology, 222
Tribolium species (flour beetles) pheromones, 141, 146
Tribolure, 141, 146
Trigeminal odorant identification, 262–3, 265
Trimedlure, *159*, 165
2,3,5-Trimethylpyrazine, *238*
Triplal, 300, *300*, *301*
Trirhabda virgata (leaf beetle), *99*
Triterpene cycloartenol, 16
Triterpenes, 12, *14*, *17*, *18*
1,2,4-Trithiolane, *223*, 223
Tritrophic interactions, 96, 111, 113–14
Trogoderma granarium (khapra beetle) pheromone, *134*, 136
Trypodendron lineatum (striped ambrosia beetle) pheromone, *136*, 137
Tryptophan, 218, *219*

Ubiquinones, 12, *14*, 15, 16, *18*
Undecanal, *261*, 262

Valeric acid, *261*, 261
Valine methyl ester, *135*, 137
Vanillic acid, 247, *247*
Vanillin
ferulic acid degradation, 247, *247*
microencapsulation, 312, 325
odour/odour threshold, *247*
olfactive profiles, *301*
olfactory perception, *262*, 262
perfumery, use in, 294, 295
structure, *262*, *294*, *312*
vapour pressure and polarity, *3*
Vapour pressures, 2, *3*, 333, 377
Vegetative VOCs, 95
activation, differential, 101
attacker phenotype, 98–9
diversity, 98–9
floral VOCs, similarities to, 112–13
and plant fitness, 97, 111
plant genotype, 98–9
plant interactions
with distant plant parts, 105
with herbivorous insects, 96
with insectivorous insects, 96, 108–11, 167, 215
with other plants, 96, 111–12, 113
tritrophic, 113–14
and plant quality, 107–8
and pollinator specificity, 95–6
response specificity, 101–2
signalling pathway crosstalk, 101
within-plant defence signalling, 104–5
see also Biogenic VOCs; Floral VOCs; Herbivore-induced plant volatiles (HIPVs); Plant defense
Vesperal, *135*, 136
Vesperus xatarti (longhorn beetle) pheromone, *135*, 136
β-Vetivone synthesis, *175*, 175–6
4-Vinylguaiaco, 246–7, *247*
Vitamins, 173, 178, 192, *246*, 246
Volatile fractions, analysis of *see* Analysis of volatile fractions
Volatiles, general description, 1–2
Volicitin, *37*, 38, 102, *103*

Wine, 6, 37–8, 78, *79*, 223

Xanthophylls, *17*, 26, 28, 32, 39

Yeast, 223, 241

Zea mays (Maize), 113–14
Zeaxanthin, 12, *14*, 26, 32
Zingiberene, 33, *35*